All-Russian Conference on Nonlinear Waves (Wave16)

Theory and New Applications

Journal of Physics: Conference Series Volume 722

Novosibirsk, Russia
29 February – 2 March 2016

Editors:

Alexander Chesnokov
Alexander Chupakhin
Valery Liapidevskii
Nikolay Makarenko

ISBN: 978-1-5108-2538-3
ISSN: 1742-6588

Printed from e-media with permission by:

Curran Associates, Inc.
57 Morehouse Lane
Red Hook, NY 12571

Some format issues inherent in the e-media version may also appear in this print version.

Printed by Curran Associates, Inc. (2016)

For permission requests, please contact the Institute of Physics
at the address below.

Institute of Physics
Dirac House, Temple Back
Bristol BS1 6BE UK

Phone: 44 1 17 929 7481
Fax: 44 1 17 920 0979

techtracking@iop.org

Additional copies of this publication are available from:

Curran Associates, Inc.
57 Morehouse Lane
Red Hook, NY 12571 USA
Phone: 845-758-0400
Fax: 845-758-2633
Email: curran@proceedings.com
Web: www.proceedings.com

Table of contents

Volume 722

All-Russian conference on Nonlinear Waves: Theory and New Applications (Wave16)

29 February to 2 March 2016, Novosibirsk, Russia

Accepted papers received: 24 May 2016
Published online: 16 June 2016

Preface

Papers

Journal of Physics: Conference Series **722** (2016) 011001 doi:10.1088/1742-6596/722/1/011001

Preface

The present volume of *Journal of Physics: Conference Series* represents contributions from participants of the Russian conference "Nonlinear waves: theory and new applications" held at the Technopark of Novosibirsk Akademgorodok, Russia, 29 February – 2 March 2016, organised by the Lavrentyev Institute of Hydrodynamics and the Novosibirsk State University.

The conference is longstanding and attracts the attention of mathematicians and physicists. This year the conference was dedicated to the 70th anniversary of the birth of an outstanding scientist in the field of mechanics, Professor Vladimir M. Teshukov (2 March 1946 – 22 April 2008), who was a director of the Lavrentyev Institute of Hydrodynamics from 2004–2008.

Scientific activities of the conference included a discussion of a wide range of problems, mainly related to mathematical theory of nonlinear wave processes in heterogeneous fluids, construction of new models of multiphase fluids and development of analytical and numerical methods for solving problems of mechanics. The conference programme included oral presentations (20 minutes) given by more than 100 participants from 15 cities of Russia (including Moscow, Novosibirsk, Kazan, Vladivostok, Ufa, Yekaterinburg, Krasnoyarsk and Irkutsk), and France (Marseille, Toulouse). A significant part of the talks (about 30%) were presented by young researchers and postgraduate students.

We would like to thank the speakers for their significant contributions to the conference. We also would like to thank the members of the Organizing Committee. We cannot end without expressing our many thanks to our sponsors (Russian Foundation for Basic Research and the technology company Schlumberger) for their financial support.

Novosibirsk, May 2016

Alexander Chesnokov, Alexander Chupakhin,
Valery Liapidevskii, Nikolay Makarenko

Journal of Physics: Conference Series **722** (2016) 011002 doi:10.1088/1742-6596/722/1/011002

Peer review statement

All papers published in this volume of *Journal of Physics: Conference Series* have been peer reviewed through processes administered by the proceedings Editors. Reviews were conducted by expert referees to the professional and scientific standards expected of a proceedings journal published by IOP Publishing.

Conservation laws and symmetries of the shallow water system above rough bottom

A V Aksenov and K P Druzhkov

Lomonosov Moscow State University, Moscow, Russia

E-mail: aksenov.av@gmail.com, Konstantin.Druzhkov@gmail.com

Abstract. The system of one-dimensional shallow water equations above the rough bottom is considered. All its hydrodynamic conservation laws are found, and a group classification is performed. A new conservation law additional to the two basic conservation laws is found. It is shown that the system of shallow water equations can be linearized by a point change of variables only in cases of constant and linear bottom profiles.

1. Introduction

In dimensionless variables, the system of one-dimensional shallow water equations above the rough bottom has the form [1]

$$\begin{aligned} &u_t + uu_x + \eta_x = 0\,, \\ &\eta_t + [(\eta + h)u]_x = 0\,. \end{aligned} \tag{1}$$

Here $y = -h(x)$, $h(x) \geqslant 0$ is the bottom profile, $u = u(x,t)$ is the average speed on the horizontal depth, and $\eta = \eta(x,t)$ is the deviation of the free surface.

It was observed in [2] that the system of shallow water equations with the bottom profile $h(x) = -x$ can be linearized in characteristic variables. Later in [3], it was shown that in this case the original system (1) can be reduced to the following form via a point change of variables,

$$\begin{aligned} &U_T + N_X = 0\,, \\ &N_T - XU_X - U = 0\,, \end{aligned} \tag{2}$$

this form can be seen as formally discarding the nonlinear terms in the system (1). Classes of exact solutions of linear system (2) are given in [4, 5].

In this paper, we find all hydrodynamic conservation laws for all bottom profiles, and solve the problem of group classification.

2. Conservation laws

We will seek hydrodynamic conservation laws of system (1) as a pair of functions $P(x,t,u,\eta)$, $Q(x,t,u,\eta)$ which satisfy the equation

$$D_t(Q) + D_x(P) = 0 \tag{3}$$

on solutions of the system (1). Here

$$D_t = \frac{\partial}{\partial t} + u_t \frac{\partial}{\partial u} + \eta_t \frac{\partial}{\partial \eta}, \qquad D_x = \frac{\partial}{\partial x} + u_x \frac{\partial}{\partial u} + \eta_x \frac{\partial}{\partial \eta}$$

are operators of the total derivative.

Comment 1. Note that both equations of the system (1) have the form (3). The corresponding conservation laws,

$$\begin{aligned} Q_1 &= u, \qquad & P_1 &= \frac{u^2}{2} + \eta, \\ Q_2 &= \eta, \qquad & P_2 &= (\eta + h)u, \end{aligned} \tag{4}$$

have the meaning of the conservation laws of momentum and mass.

Comment 2. From (3), it follows that $Q' = Q + f_x$, $P' = P - f_t$ and $Q'' = CQ$, $P'' = CP$ are also conservation laws for arbitrary function $f = f(x,t)$ and constant C. Our classification is performed modulo this natural equivalence.

Using the expression (3) and the system of equations (1), one can obtain the following overdetermined system for the unknown function P and Q,

$$\begin{aligned} P_x &= uh'Q_\eta - Q_t, \\ P_u &= uQ_u + (\eta + h)Q_\eta, \\ P_\eta &= Q_u + uQ_\eta. \end{aligned} \tag{5}$$

Eliminating P from equations (5) one can get the system for the function Q,

$$\begin{aligned} Q_{xu} &= uh'Q_{\eta\eta} - uQ_{x\eta} - Q_{t\eta}, \\ Q_{tu} &= -u^2h'Q_{\eta\eta} + uh'Q_{u\eta} + \left(u^2 - \eta - h\right) Q_{x\eta} + uQ_{t\eta}, \\ Q_{uu} &= (\eta + h)Q_{\eta\eta}. \end{aligned} \tag{6}$$

The system (6) can be tested for consistency. Here are the results of this test.

2.1. The case of an arbitrary function $h(x)$. In this case, the solution of system (5) has the form

$$\begin{aligned} P &= C_1(\eta + h)\left(u^3 + 2u\eta\right) + C_2\left(\frac{u^2}{2} + \eta\right) + C_3u(\eta + h), \\ Q &= C_1\left[u^2(\eta + h) + \eta^2\right] + C_2u + C_3\eta. \end{aligned} \tag{7}$$

Here, C_1, C_2, C_3 are arbitrary constants. In view of comments 2 and 2, we find a new conservation law from (7),

$$Q_3 = u^2(\eta + h) + \eta^2, \qquad P_3 = (\eta + h)(u^3 + 2u\eta),$$

supplementing the basic conservation laws (4).

2.2. The case of a linear function $h(x) = a_1x + a_2$. In this case, the solution of the system (5) has the form

$$\begin{aligned} P = C_1 \Big(&-a_1xt\beta - x\alpha(\beta + a_2) + \frac{3}{2}a_1t^2\alpha(\beta + a_2) + \frac{1}{2}t\beta^2 + a_2\beta t + \\ &+ \frac{1}{2}{a_1}^2t^3\beta + a_2t\alpha^2 + t\alpha^2\beta\Big) + a_1tQ_1 + P_1, \\ Q = C_1 &\left(a_2t\alpha + t\alpha\beta - x\beta + \frac{1}{2}a_1t^2\beta\right) + Q_1. \end{aligned}$$

Here, $\alpha = u - a_1 t$, $\beta = \eta + a_1 x$; C_1 is an arbitrary constant; $P_1 = P_1(\alpha, \beta)$, $Q_1 = Q_1(\alpha, \beta)$ is an arbitrary solution of the system

$$P_{1\alpha} = \alpha Q_{1\alpha} + (\beta + a_2) Q_{1\beta}\,,$$
$$P_{1\beta} = Q_{1\alpha} + \alpha Q_{1\beta}\,.$$

2.3. The case $h(x) = \dfrac{b_1}{2}x^2 + b_2 x + b_3,\ b_1 > 0$. In this case, the solution of the system (5) has the form

$$\begin{aligned}
P &= C_1(\eta + h(x))\left(u^3 + 2u\eta\right) + C_2\left(\frac{u^2}{2} + \eta\right) + C_3 u(\eta + h(x)) + \\
&+ C_4 e^{-\sqrt{b_1}t}\left[\left(\sqrt{b_1}u^2 + uh'(x)\right)(\eta + h(x)) + \frac{\sqrt{b_1}}{2}\eta(\eta + 2h(x))\right] + \\
&+ C_5 e^{\sqrt{b_1}t}\left[\left(-\sqrt{b_1}u^2 + uh'(x)\right)(\eta + h(x)) - \frac{\sqrt{b_1}}{2}\eta(\eta + 2h(x))\right], \\
Q &= C_1\left(u^2(\eta + h(x)) + \eta^2\right) + C_2 u + C_3\eta + \\
&+ C_4 e^{-\sqrt{b_1}t}\left(\sqrt{b_1}u(\eta + h(x)) + \eta h'(x)\right) + \\
&+ C_5 e^{\sqrt{b_1}t}\left(-\sqrt{b_1}u(\eta + h(x)) + \eta h'(x)\right).
\end{aligned} \tag{8}$$

Here, C_1, C_2, C_3, C_4, C_5 are arbitrary constants. From the solution (8) we find that in this case there are two additional conservation laws,

$$Q_4 = e^{-\sqrt{b_1}t}\left(\sqrt{b_1}u(\eta + h(x)) + \eta h'(x)\right),$$
$$P_4 = e^{-\sqrt{b_1}t}\left[\left(\sqrt{b_1}u^2 + uh'(x)\right)(\eta + h(x)) + \frac{\sqrt{b_1}}{2}\eta(\eta + 2h(x))\right],$$

and

$$Q_5 = e^{\sqrt{b_1}t}\left(-\sqrt{b_1}u(\eta + h(x)) + \eta h'(x)\right),$$
$$P_5 = e^{\sqrt{b_1}t}\left[\left(-\sqrt{b_1}u^2 + uh'(x)\right)(\eta + h(x)) - \frac{\sqrt{b_1}}{2}\eta(\eta + 2h(x))\right].$$

2.4. The case $h(x) = \dfrac{b_1}{2}x^2 + b_2 x + b_3,\ b_1 < 0$. In this case, the solution of the system (5) has the form

$$
\begin{aligned}
P = {} & C_1(\eta + h(x))\left(u^3 + 2u\eta\right) + C_2\left(\frac{u^2}{2} + \eta\right) + C_3 u(\eta + h(x)) + \\
& + C_4\Big[\sin(\sqrt{-b_1}t)\Big(\sqrt{-b_1}u^2(\eta + h(x)) + \frac{\sqrt{-b_1}}{2}\eta(\eta + 2h(x))\Big) + \\
& + \cos(\sqrt{-b_1}t)\cdot uh'(x)(\eta + h(x))\Big] + C_5\Big[-\cos(\sqrt{-b_1}t)\left(\sqrt{-b_1}u^2(\eta + h(x)) + \right.\\
& \left. + \frac{\sqrt{-b_1}}{2}\eta(\eta + 2h(x))\right) + \sin(\sqrt{-b_1}t)\cdot uh'(x)(\eta + h(x))\Big], \qquad (9)\\
Q = {} & C_1\left(u^2(\eta + h(x)) + \eta^2\right) + C_2 u + C_3\eta + \\
& + C_4\left(\sin(\sqrt{-b_1}t)\cdot\sqrt{-b_1}u(\eta + h(x)) + \cos(\sqrt{-b_1}t)\cdot\eta h'(x)\right) + \\
& + C_5\left(-\cos(\sqrt{-b_1}t)\cdot\sqrt{-b_1}u(\eta + h(x)) + \sin(\sqrt{-b_1}t)\cdot\eta h'(x)\right).
\end{aligned}
$$

From the solution (9) we find that in this case there are two additional conservation laws,

$$
\begin{aligned}
Q_4 = {} & \sin(\sqrt{-b_1}t)\cdot\sqrt{-b_1}u(\eta + h(x)) + \cos(\sqrt{-b_1}t)\cdot\eta h'(x), \\
P_4 = {} & \sin(\sqrt{-b_1}t)\Big(\sqrt{-b_1}u^2(\eta + h(x)) + \frac{\sqrt{-b_1}}{2}\eta(\eta + 2h(x))\Big) + \\
& + \cos(\sqrt{-b_1}t)\cdot uh'(x)(\eta + h(x)),
\end{aligned}
$$

and

$$
\begin{aligned}
Q_5 = {} & -\cos(\sqrt{-b_1}t)\cdot\sqrt{-b_1}u(\eta + h(x)) + \sin(\sqrt{-b_1}t)\cdot\eta h'(x), \\
P_5 = {} & -\cos(\sqrt{-b_1}t)\left(\sqrt{-b_1}u^2(\eta + h(x)) + \frac{\sqrt{-b_1}}{2}\eta(\eta + 2h(x))\right) + \\
& + \sin(\sqrt{-b_1}t)\cdot uh'(x)(\eta + h(x)).
\end{aligned}
$$

3. Group classification

The task of the group classification is to find Lie symmetries admitted by the considered system of equations depending on the unknown function entering the system [6]. We seek symmetry operators of the system (1) as

$$
X = \xi^1(x,t,u,\eta)\frac{\partial}{\partial x} + \xi^2(x,t,u,\eta)\frac{\partial}{\partial t} + \eta^1(x,t,u,\eta)\frac{\partial}{\partial u} + \eta^2(x,t,u,\eta)\frac{\partial}{\partial \eta}.
$$

Applying the criterion of invariance [6], we obtain the following overdetermined linear homogeneous system of determining equations,

$$
\begin{aligned}
& \xi^1_\eta - u\xi^2_\eta + \xi^2_u = 0, \\
& \xi^1_u - u\xi^2_u + (\eta + h)\,\xi^2_\eta = 0, \\
& \eta^1 - u\xi^1_x + u^2\xi^2_x - (\eta + h)\,\eta^1_\eta - \xi^1_t + u\xi^2_t + \eta^2_u + (\eta + h)\,\xi^2_x = 0, \\
& h'\xi^1 + \eta^2 + (\eta + h)\left(\eta^1_u - \xi^1_x + 2u\xi^2_x - \eta^2_\eta + \xi^2_t - 2uh'\xi^2_\eta\right) = 0, \\
& 2u\xi^2_x - \eta^1_u + \xi^2_t + \eta^2_\eta - \xi^1_x = 0, \\
& 2(\eta + h)\eta^1_\eta - 2\eta^2_u + uh'\left(\xi^1_\eta - \xi^2_u - u\xi^2_\eta\right) = 0, \\
& u\eta^1_x + \eta^1_t + \eta^2_x + uh'\left(\xi^2_x - \eta^1_\eta\right) = 0, \\
& uh''\xi^1 + h'\eta^1 + (\eta + h)\eta^1_x + u\eta^2_x + \eta^2_t + uh'\left(u\xi^2_x - \eta^2_\eta + \xi^2_t - uh'\xi^2_\eta\right) = 0.
\end{aligned} \qquad (10)
$$

The system (10) can be tested for consistency. Here are the results of this test.

3.1. $h(x)$ – arbitrary function.

$$X_1 = \frac{\partial}{\partial t}.$$

This operator will be present for all bottom profiles.

3.2. $h(x) = n_2 e^{n_1 x} + n_3,\ n_1 \neq 0,\ n_2 \neq 0$. In this case, we obtain the additional symmetry operator

$$X_2 = \frac{\partial}{\partial x} - \frac{n_1}{2} t \frac{\partial}{\partial t} + \frac{n_1}{2} u \frac{\partial}{\partial u} + n_1(\eta + n_3)\frac{\partial}{\partial \eta}.$$

3.3. $h(x) = m_2 \ln|x + m_1| + m_3,\ m_2 \neq 0$. In this case, we obtain the additional symmetry operator

$$X_2 = (x + m_1)\frac{\partial}{\partial x} + t\frac{\partial}{\partial t} - m_2 \frac{\partial}{\partial \eta}.$$

3.4. $h(x) = c_3(x + c_1)^{c_2} + c_4,\ c_2 \neq 0,\ c_2 \neq 1,\ c_2 \neq 2,\ c_3 \neq 0$. In this case, we obtain the additional symmetry operator

$$X_2 = 2(x + c_1)\frac{\partial}{\partial x} - (c_2 - 2)t\frac{\partial}{\partial t} + c_2 u \frac{\partial}{\partial u} + 2c_2(\eta + c_4)\frac{\partial}{\partial \eta}.$$

3.5. $h(x) = b_1 x^2/2 + b_2 x + b_3,\ b_1 > 0$. In this case, we obtain the additional symmetry operators

$$X_2 = \left(x + \frac{b_2}{b_1}\right)\frac{\partial}{\partial x} + u\frac{\partial}{\partial u} + \left(2\eta - \frac{{b_2}^2}{b_1} + 2b_3\right)\frac{\partial}{\partial \eta},$$
$$X_3 = e^{\sqrt{b_1}t}\left(\frac{\partial}{\partial x} + \sqrt{b_1}\frac{\partial}{\partial u} - (b_1 x + b_2)\frac{\partial}{\partial \eta}\right),$$
$$X_4 = e^{-\sqrt{b_1}t}\left(\frac{\partial}{\partial x} - \sqrt{b_1}\frac{\partial}{\partial u} - (b_1 x + b_2)\frac{\partial}{\partial \eta}\right).$$

3.6. $h(x) = b_1 x^2/2 + b_2 x + b_3,\ b_1 < 0$. In this case, we obtain the additional symmetry operators

$$X_2 = \left(x + \frac{b_2}{b_1}\right)\frac{\partial}{\partial x} + u\frac{\partial}{\partial u} + \left(2\eta - \frac{{b_2}^2}{b_1} + 2b_3\right)\frac{\partial}{\partial \eta},$$
$$X_3 = \cos(\sqrt{-b_1}t)\frac{\partial}{\partial x} - \sqrt{-b_1}\sin(\sqrt{-b_1}t)\frac{\partial}{\partial u} - (b_1 x + b_2)\cos(\sqrt{-b_1}t)\frac{\partial}{\partial \eta},$$
$$X_4 = \sin(\sqrt{-b_1}t)\frac{\partial}{\partial x} + \sqrt{-b_1}\cos(\sqrt{-b_1}t)\frac{\partial}{\partial u} - (b_1 x + b_2)\sin(\sqrt{-b_1}t)\frac{\partial}{\partial \eta}.$$

3.7. $\boldsymbol{h(x) = a_2}$ **.** In this case, the basis of symmetry operators can be written as

$$Y_1 = \left(\frac{3}{2}t(\eta + a_2) - \frac{3}{4}tu^2\right)\frac{\partial}{\partial x} + \left(\frac{1}{2}x - \frac{3}{2}tu\right)\frac{\partial}{\partial t} + \\ + \left(\frac{1}{4}u^2 + \eta + a_2\right)\frac{\partial}{\partial u} + u(\eta + a_2)\frac{\partial}{\partial \eta},$$

$$Y_2 = \frac{1}{2}x\frac{\partial}{\partial x} + \frac{1}{2}u\frac{\partial}{\partial u} + (\eta + a_2)\frac{\partial}{\partial \eta},$$

$$Y_3 = t\frac{\partial}{\partial x} + \frac{\partial}{\partial u},$$

$$Y_4 = x\frac{\partial}{\partial x} + t\frac{\partial}{\partial t},$$

$$Y_\infty = (u\ w_1(u,\eta) + w_2(u,\eta))\frac{\partial}{\partial x} + w_1(u,\eta)\frac{\partial}{\partial t},$$

where $w_1(u,\eta)$, $w_2(u,\eta)$ are arbitrary solutions of the system

$$w_{1u} + w_{2\eta} = 0\,, \\ w_{2u} + w_1 + (\eta + a_2)w_{1\eta} = 0\,.$$

Note that the symmetry operator X_1 is contained in the operator Y_∞ if one puts $w_1 = 1$, $w_2 = -u$.

3.8. $\boldsymbol{h(x) = a_1x + a_2,\ a_1 \neq 0}$ **.** In this case, the basis of the symmetry operators can be written as

$$Y_1 = \left(-4a_1xt - a_2t - {a_1}^2t^3 + \frac{3}{2}tu^2 - 3t\eta\right)\frac{\partial}{\partial x} + \left(3tu - x - \frac{5}{2}a_1t^2\right)\frac{\partial}{\partial t} + \\ + \left(4a_1tu - 3a_1x - 3{a_1}^2t^2 - \frac{1}{2}u^2 - 2\eta\right)\frac{\partial}{\partial u} + \\ + \left(6{a_1}^2xt + 3a_1a_2t + {a_1}^3t^3 - 2a_1xu - \frac{3}{2}a_1tu^2 + 5a_1t\eta - 2a_2u - 2u\eta\right)\frac{\partial}{\partial \eta},$$

$$Y_2 = \left(x + \frac{1}{2}a_1t^2\right)\frac{\partial}{\partial x} + t\frac{\partial}{\partial t} + a_1t\frac{\partial}{\partial u} - \left(a_1x + \frac{1}{2}a_1^2t^2\right)\frac{\partial}{\partial \eta},$$

$$Y_3 = 2x\frac{\partial}{\partial x} + t\frac{\partial}{\partial t} + u\frac{\partial}{\partial u} + 2(\eta + a_2)\frac{\partial}{\partial \eta},$$

$$Y_4 = -t\frac{\partial}{\partial x} - \frac{\partial}{\partial u} + a_1t\frac{\partial}{\partial \eta},$$

$$Y_\infty = (u\ w_1(u - a_1t, \eta + a_1x) + w_2(u - a_1t, \eta + a_1x))\frac{\partial}{\partial x} + w_1\frac{\partial}{\partial t} + \\ + a_1\ w_1\frac{\partial}{\partial u} + (a_1u\ w_1 + a_1\ w_2)\frac{\partial}{\partial \eta},$$

where $w_1(u - a_1t, \eta + a_1x)$, $w_2(u - a_1t, \eta + a_1x)$ are arbitrary solutions of the system

$$w_{1\alpha} + w_{2\beta} = 0\,, \\ w_{2\alpha} + w_1 + (\beta + a_2)w_{1\beta} = 0\,.$$

Here, the situation is the same as in the search for conservation laws of the system of equations with a linear bottom profile,

$$\alpha = u - a_1t\,, \\ \beta = \eta + a_1x\,.$$

Journal of Physics: Conference Series **722** (2016) 012001 doi:10.1088/1742-6596/722/1/012001

Comment 3. In paper [5], a point change of variables that links the systems with constant and linear bottom profiles was found. Thus, the Lie algebras of symmetry operators of these systems are isomorphic.

On the basis of group classification results, we can conclude that the system (1) can be linearized by a point change of variables only in the case of constant and linear bottom profiles (otherwise, the Lie symmetry algebra of the system is finite).

4. Conclusion

The main results of the paper are the new basic conservation law supplementing the laws of conservation of momentum and mass, as well as the solution to the problem of group classification. The results can be used for the construction of new exact solutions, as well as for numerical modeling of shallow water motion over an uneven bottom.

Acknowledgments

This work was supported by RFBR grants No. 15-01-00361 and No. 15-01-04677.

References

[1] Stoker J J 1948 *Comm. Pure Appl. Math.* **1** 1–87
[2] Carrier G F and Greenspan H P 1958 *J. Fluid Mech.* **4** 97–109
[3] Pelinovsky E N and Mazova R Kh 1992 *Natural Hazards* **6** 227–49
[4] Dobrokhotov S Yu and Tirozzi B 2010 *Russian Mathematical Surveys* **65** 177–9
[5] Chirkunov Yu A, Dobrokhotov S Yu, Medvedev S B and Minenkov D S 2014 *Theoretical and Mathematical Physics* **178** 278–98
[6] Ovsiannikov L V 1982 *Group Analysis of Differential Equations* (Academic Press)

Analysis of the 3D acoustic cloaking problems using optimization method

G V Alekseev[1,2] **and Yu E Spivak**[2]

[1] Institute of Applied Mathematics FEB RAS, 690041, Vladivostok, Russia
[2] Far Eastern Federal University, 690922, Vladivostok, Russia

E-mail: alekseev@iam.dvo.ru

Abstract. Control problems for the 3D model of acoustic scattering which describes scattering acoustic waves by a permeable obstacle with the form of a spherical layer are considered. These problems arise while developing the design technologies of acoustic cloaking devices using the wave flow method. The solvability of direct and control problems for the acoustic scattering model under study is proved. The sufficient conditions which provide local uniqueness and stability of optimal solutions are established.

1. Introduction

Recent significant research has focused on design of invisibility cloaking devices for material bodies. Beginning with pioneering papers [1, 2, 3], a large number of publications was devoted to developing different methods of solving the cloaking problems. The transformation optics (TO) method proposed in [2] is the most popular cloaking method lying on the base of the so-called passive cloaking strategy. The first applications of the TO method were connected with solving electromagnetic cloaking problems. Then after proof of the equivalence of the 2D Maxwell equations to the 2D equations of acoustics of anisotropic medium, this theory was expanded to acoustic cloaking at first in 2D case [4] and then in general 3D case [5, 6]. We also mention the papers [7, 8, 9] devoted to the proof of existence of acoustic cloaking shells and to the studies of properties of solutions of the acoustic cloaking problem. In [10], the invisibility problem in X-ray tomography was studied.

It should be emphasized that the technical realization of solutions obtained in cited references is connected with substantial difficulties. One of approaches of overcoming these difficulties consists of replacing the exact cloaking problem by the approximate cloaking problem for which solutions admit relatively simple technical realization (see, e.g., [11, 12, 13, 14] and references therein). Another approach is based on using the optimization method of solving inverse problems. This method was applied in papers [15, 16] devoted to numerical analysis of 2D cloaking problems and in [17, 18, 19] when studying impedance cloaking problems.

Optimization method is also applied in this paper while theoretical analysis of acoustic cloaking problems in which the cloaking effect is achieved due to the choice of variable parameters of inhomogeneous isotropic liquid medium filling the acoustic cloaking shell. We begin with formulation of the direct scattering problem. Let Ω be the bounded domain in R^3 which is topologically equivalent to the spherical layer with the interior domain Ω_i, the external domain Ω_e^∞ and, in the general case, with the curvilinear boundary Γ consisting of two components:

Journal of Physics: Conference Series **722** (2016) 012002 doi:10.1088/1742-6596/722/1/012002

internal Γ_i and external Γ_e (see Figure 1). We assume that the domains Ω_i, Ω and Ω_e^∞ are filled with liquid medium, and let ρ and c be functions given in Ω which have the sense of the density of the medium and the speed of sound in the domain Ω, ρ_0=const and c_0=const are the constant density of the medium and the speed of sound in Ω_i and Ω_e^∞, p, p_i, and p_e are the sound pressure in the domains Ω, Ω_i and Ω_e^∞ respectively, ω is the angular frequency. We introduce a constant wave number $k = \omega/c_0$. For brevity, we refer to triple (Ω, ρ, η) as an acoustic shell.

Let us assume that the field p^{inc} arises in the external part Ω_e^∞ of the region Ω. The incidence of this field to the shell results in the appearance of the refracted field p in the domain Ω, the internal field p_i in Ω_i, and the field p_s scattered by this shell in the domain Ω_e^∞. The problem of determining the fields p_i in Ω_i, p in Ω and p_s in Ω_e^∞ is reduced to finding the fields p_i in Ω_i, p in Ω and $p_e = p^{inc} + p_s$ in Ω_e^∞ from the following relations:

$$\Delta p_i + k^2 p_i = 0 \text{ in } \Omega_i, \ \ \Delta p_e + k^2 p_e = 0 \text{ in } \Omega_e^\infty, \ \ \rho \operatorname{div}(\frac{1}{\rho}\nabla p) + k^2 \eta p = 0 \text{ in } \Omega, \tag{1}$$

$$p = p_i, \ \ \frac{1}{\rho}\frac{\partial p}{\partial n} = \frac{1}{\rho_0}\frac{\partial p_i}{\partial n} \text{ on } \Gamma_i, \ \ p = p_e, \ \ \frac{1}{\rho}\frac{\partial p}{\partial n} = \frac{1}{\rho_0}\frac{\partial p_e}{\partial n} \text{ on } \Gamma_e, \tag{2}$$

$$\frac{\partial p^s(\mathbf{x})}{\partial r} - ikp^s(\mathbf{x}) = o(r^{-1}) \ \text{ as } \ r = |\mathbf{x}| \to \infty. \tag{3}$$

Here, in particular, $\eta = c_0^2/c^2$ is a refraction index and (3) has the meaning of the Sommerfeld radiation condition in R^3.

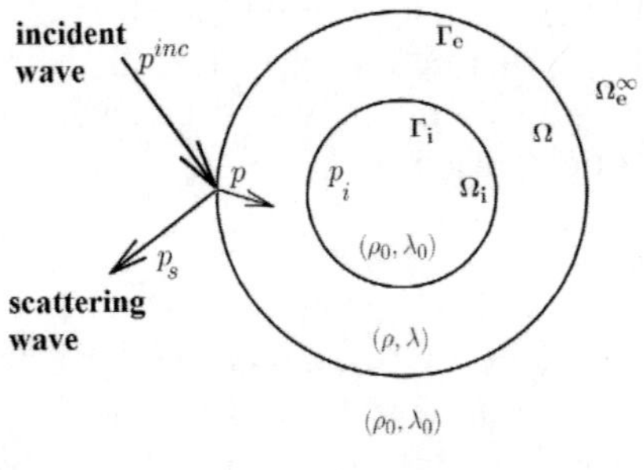

Figure 1. Geometry of cloaking shell.

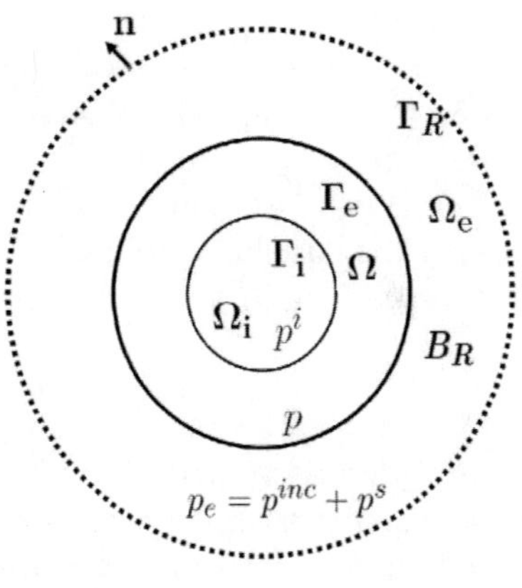

Figure 2. Geometry of bounded problem.

In this paper the inverse problem for the model (1)–(3) arising while developing the design technologies of cloaking devices using the wave flow method is considered. This problem in exact formulation consists of finding variable parameters ρ and η of the medium filling the domain Ω from the condition of perfect cloaking any object placed in the internal domain Ω_i. By optimization method, this problem is reduced to the problem of minimization of the following cost functional:

$$I_0(P) = \|P - p^d\|_Q^2 \equiv \int_Q |P - p^d|^2 dx. \tag{4}$$

Here, P is the function equalled to p_i in Ω_i, p in Ω and $p^{inc} + p_s$ in Ω_e^∞, function $p^d \in L^2(Q)$ describes the field measured in some subset $Q \subset \Omega_i \cup \Omega_e^\infty$. As controls, we take the pair $\lambda = 1/\rho$, $\mu = \lambda\eta$ uniquely connected with the pair (ρ, η) of physical parameters at $\rho > 0, \eta > 0$.

The main goal of this paper is to perform theoretical analysis of control problems with tracking-type functionals of type (4) for the model (1)–(3). Keeping this goal in the mind, we first reduce unbounded problem (1)–(3) to an equivalent problem considered in bounded domain, and it is shown that under appropriate assumptions the weak solution of the latter problem exists, it is unique and depends continuously on the datum (incident field p^{inc}). Then, we prove the existence of a solution of our control problem and deduce the optimality system. Based on its' analysis, we prove the uniqueness results for concrete control problems and derive the stability estimates of optimal solutions. It is to note that this paper contains detailed proofs of some theorems formulated in [20].

2. Study of the direct problem

Let us introduce certain functional spaces which will be used while studying the direct and control problems for model (1)–(3). Let B_R be the ball of radius R with the boundary Γ_R containing Ω, $\Omega_e \equiv \Omega_e^\infty \cap B_R$ (see Figure 2). We use the spaces $H^1(\Omega)$, $H^1(\Omega_e)$, $H^1(B_R)$, $L^2(\Omega)$, $L^2(Q)$, $L^2(\Gamma_r)$, $H^{1/2}(\Gamma_R)$, $H^{-1/2}(\Gamma_R)$, $L^\infty(\Omega)$ and $H^s(\Omega)$ with the norms $\|\cdot\|_{1,\Omega}$, $\|\cdot\|_{1,\Omega_e}$, $\|\cdot\|_{1,B_R}$, $\|\cdot\|_\Omega$, $\|\cdot\|_Q$, $\|\cdot\|_{\Gamma_r}$, $\|\cdot\|_{1/2,\Gamma_R}$, $\|\cdot\|_{-1/2,\Gamma_R}$, $\|\cdot\|_{L^\infty(\Omega)}$ and $\|\cdot\|_{s,\Omega}$ respectively. We put $L^\infty_{\eta_0}(\Omega) = \{\eta \in L^\infty(\Omega) : \eta(x) \geq \eta_0 \text{ in } \Omega\}$, $\eta_0 = \text{const} > 0$. We also need the space $H^1(\Delta, \Omega_e) = \{v \in H^1(\Omega_e) : \Delta v \in L^2(\Omega_e)\}$ equipped with the norm $\|v\|^2_{H^1(\Delta,\Omega_e)} = \|v\|^2_{1,\Omega_e} + \|\Delta v\|^2_{\Omega_e}$ and the space $\mathcal{P}^{inc} = \{v \in H^1(\Omega_e) : \Delta v + k^2 v = 0 \text{ in } \Omega_e\} \subset H^1(\Delta, \Omega_e)$ equipped with the norm $\|\cdot\|_{1,\Omega_e}$. The latter will serve for describing the restrictions of incident waves to the domain Ω_e. The space $X = H^1(B_R)$ equipped with the Hilbert norm $\|\cdot\|_X = \|\cdot\|_{B_R}$ is defined and denoted by X^* the space antidual of X.

As in [17, 18], we introduce the Dirichlet-to-Neumann operator $T \in \mathcal{L}\ (H^{1/2}(\Gamma_R), H^{-1/2}(\Gamma_R))$, which maps an arbitrary function $h \in H^{1/2}(\Gamma_R)$ to the function $\partial\tilde{p}/\partial n \in H^{-1/2}(\Gamma_R)$. Here, $\tilde{p}$ is the solution of the Dirichlet external problem $\Delta\tilde{p} + k^2\tilde{p} = 0$ in $\Omega_e^\infty \setminus B_R$, $\tilde{p}|_{\Gamma_R} = g$ satisfying the radiation condition (3). It is known that problem (1)–(3) considered in unbounded domain R^3 is equivalent in a certain sense to boundary value problem (1), (2), considered in the ball B_R under the following additional condition for p_s on Γ_R:

$$\partial p_s/\partial n = T p_s \text{ on } \Gamma_R. \tag{5}$$

We begin with derivation of the weak formulation of problem (1)–(3). To this end, we multiply the first two equations in (1) considered in $\Omega_i \cup \Omega_e$ by the function $\overline{\Phi}/\rho_e$ where $\Phi \in X$ is the test function, and $\overline{\Phi}$ denotes the complex conjugate function to Φ, then we multiply the last equation in (1) considered in Ω by $\overline{\Phi}/\rho$, integrate over $\Omega_i \cup \Omega_e$ or Ω, apply Green's formulas in Ω_i, Ω_e and Ω, and combine the obtained identities. Using the relation $p = p^{inc} + p_s$ in Ω_e and the boundary conditions in (2), (5) we arrive to the following identity for triple $P = (p_i, p, p_e) \in X$:

$$a_u(P, \Phi) \equiv a_0(P, \Phi) + a^\mu_\lambda(P, \Phi) = \langle f^{inc}, \Phi\rangle \ \ \forall \Phi \in X. \tag{6}$$

Here and below, u denotes the pair (λ, μ) where $\lambda = \rho^{-1}$, $\mu = \lambda\eta$, $a_u : X \times X \to C$ and $f^{inc} : X \to C$ are the sesquilinear and antilinear forms defined by

$$a^\mu_\lambda(P, \Phi) = a(\lambda, \mu; P, \Phi) = (\lambda\nabla P, \nabla\Phi) - k^2(\mu P, \Phi) \equiv \int_\Omega (\lambda\nabla P \cdot \nabla\overline{\Phi} - k^2\mu P\overline{\Phi})dx, \tag{7}$$

$$a_0(P, \Phi) = \rho_0^{-1}\int_{\Omega_i\cup\Omega_e} (\nabla P \cdot \nabla\overline{\Phi} - k^2 P\overline{\Phi})dx - \rho_0^{-1}\int_{\Gamma_R} (TP)\overline{\Phi}d\sigma, \tag{8}$$

$$\langle f^{inc}, \Phi\rangle = -\rho_0^{-1}\int_{\Gamma_R} (Tp^{inc})\overline{\Phi}d\sigma + \rho_0^{-1}\int_{\Gamma_R} \frac{\partial p^{inc}}{\partial n}\overline{\Phi}d\sigma, \ \ \Phi \in X. \tag{9}$$

The integral $\int_{\Gamma_R}$ in (8) and (9) denotes the duality pairing between the spaces $H^{-1/2}(\Gamma_R)$ and $H^{1/2}(\Gamma_R)$. Identity (6) represents the weak formulation of problem (1), (2), (5) and its solution $P \in X$ is called as a weak solution of problem (1)–(3).

We remind that by trace theorem for any function $\Phi \in X$ there exists the trace $\Phi|_{\Gamma_R} \in H^{1/2}(\Gamma_R)$ while for any function $p^{inc} \in \mathcal{P}^{inc}$ there exists the normal trace $\partial p^{inc}/\partial n|_{\Gamma_R} \in H^{-1/2}(\Gamma_R)$ and the following estimates hold:

$$\|\Phi\|_{1/2,\Gamma_R} \leq C_R\|\Phi\|_X \ \forall\Phi \in X, \ \|\partial p^{inc}/\partial n\|_{-1/2,\Gamma_R} \leq C'_R\|p^{inc}\|_{1,\Omega_e} \ \forall p^{inc} \in \mathcal{P}^{inc}. \tag{10}$$

Here, C_R and C'_R are constants which depend on Ω, R and may be on k but are independent of $\Phi \in X$ and $p^{inc} \in \mathcal{P}^{inc}$. Using (10) and estimating

$$\|\lambda\|_{L^\infty(\Omega)} \leq C_s\|\lambda\|_{s,\Omega} \ \ \forall\lambda \in H^s(\Omega), \ s > 3/2 \tag{11}$$

with the constant C_s depending on Ω which follows from the continuous and compact embedding $H^s(\Omega) \subset L^\infty(\Omega)$ at $s > 3/2$, one can derive the following estimates:

$$|(\lambda\nabla P, \nabla\Phi)| \leq \|\lambda\|_{L^\infty(\Omega)}\|P\|_X\|\Phi\|_X \leq C_s\|\lambda\|_{s,\Omega}\|P\|_X\|\Phi\|_X, \tag{12}$$

$$|(\mu P, \Phi)| \leq \|\mu\|_{L^\infty(\Omega)}\|P\|_X\|\Phi\|_X \leq C_r\|\mu\|_{r,\Omega}\|P\|_X\|\Phi\|_X, \tag{13}$$

$$|a_0(P, \Phi)| \leq C_1\|P\|_X\|\Phi\|_X, \ \ |\langle f^{inc}, \Phi\rangle| \leq C_1\|p^{inc}\|_{1,\Omega_e}\|\Phi\|_X. \tag{14}$$

Here, C_1 is a constant which depends on Ω, R, ρ_0 and k.

We note that the sesquilinear form a_u introduced in (6), (7), (8) defines the pair of mutually adjoint operators $A_u : X \to X^*$ and $A^* : X \to X^*$ by

$$\langle A_u P, \Phi\rangle = a_u(P, \Phi) = \langle A_u^*\Phi, P\rangle \ \forall P, \Phi \in X \tag{15}$$

and variational problem (6) is equivalent to the operator equation $A_u P = f^{inc}$. Using (10)–(14) and properties of the operator T (see, e.g., [17]), one can prove that A_u is a Fredholm operator, and besides it is injective if $\lambda \in H^s_{\lambda_0}(\Omega)$, $\mu \in H^r_{\mu_0}(\Omega)$ where $s > 5/2, r > 3/2, \lambda_0 = \text{const} > 0$, $\mu_0 = \text{const} > 0$. We denote by $A_u^{-1} : X^* \to X$ the inverse operator to A_u and sets $C_u = \|A_u^{-1}\|$. It is clear that for any $f^{inc} \in X^*$ problem (6) has a unique solution $P_u = A_u^{-1}(f^{inc})$ which satisfies the estimate $\|P_u\|_X \leq C_u\|f^{inc}\|_{X^*}$ where constant C_u depends on the pair $u = (\lambda, \mu)$. Furthermore, if λ and μ belong to bounded sets $K_1 \subset H^s_{\lambda_0}(\Omega)$ and $K_2 \subset H^r_{\mu_0}(\Omega)$ respectively then one can prove arguing as in [17] that the following estimate holds for the solution P_u of problem (6):

$$\|P_u\|_X \leq C_2\|p^{inc}\|_{1,\Omega_e} \ \forall u = (\lambda, \mu) \in K_1 \times K_2. \tag{16}$$

Here, the constant C_2 is independent of u. Let us formulate this result as a theorem.

Theorem 1. *Let $\Gamma \in C^{0,1}$, $K_1 \subset H^s_{\lambda_0}(\Omega)$ and $K_2 \subset H^s_{\mu_0}(\Omega)$ be the nonempty bounded sets where $s > 5/2$, $r > 3/2$, $\lambda_0 = const > 0$, $\mu_0 = const > 0$ and let $u = (\lambda, \mu) \in K_1 \times K_2$. Then for an arbitrary incident field $p^{inc} \in \mathcal{P}^{inc}$ problem (6) has a unique solution $P_u \in X$ which satisfies the estimate (16) where constant C_2 depends on Ω, R, ρ_0, k and sets K_1 and K_2 but is independent of $u = (\lambda, \mu)$.*

3. Study of control problem. The optimality system

In this section, a precise formulation of our control problem for acoustic scattering model under study is given, the existence of optimal solutions is proved and the optimality system which describes necessary conditions of extremum for the control problem is derived. We remind that the control problem for scattering model (1)–(3) consists of minimization of certain cost functional depending on the state (acoustic field $P = (p_i, p, p_e) \in X$) and controls $\lambda = \rho^{-1} \in H^s_{\lambda_0}(\Omega)$ and $\mu = \lambda\eta \in H^r_{\mu_0}(\Omega)$ satisfying the state equation. The latter has the form of the weak formulation (6) of direct scattering problem (1)–(3). We shall assume that controls λ and μ are changed over certain sets K_1 and K_2. More precisely, the following conditions are assumed to hold:

(j) $K_1 \subset H^s_{\lambda_0}(\Omega)$ and $K_2 \subset H^r_{\mu_0}(\Omega)$ are nonempty convex closed sets where $s > 5/2, r > 3/2$, $\lambda_0 = \text{const} > 0$, $\mu_0 = \text{const} > 0$; $\alpha_0 = \text{const} > 0$.

Let $K = K_1 \times K_2$, $u = (\lambda, \mu)$. Define operator $G : X \times K \times \mathcal{P}^{inc} \to X^*$ by $\langle G(P, u, p^{inc}), \Phi\rangle = a_u(P, \Phi) - \langle f^{inc}, \Phi\rangle$ for any Φ and rewrite weak formulation (6) of problem (1)–(3) as operator equation $G(P, u, p^{inc}) = 0$. The following constrained minimization problem is considered:

$$J(P, u) \equiv \frac{\alpha_0}{2} I(P) + \frac{\alpha_1}{2}\|\lambda\|^2_{s,\Omega} + \frac{\alpha_2}{2}\|\mu\|^2_{r,\Omega} \to \inf,\ (P, u) \in X \times K,\ \ G(P, u, p^{inc}) = 0. \quad (17)$$

Here, $I : X \to \mathbb{R}$ is a cost functional. Denote by $Z_{ad} = Z_{ad}(p^{inc}) = \{(P, u) \in X \times K : G(P, u, p^{inc}) = 0, J(P, u) < \infty)\}$ the set of admissible pairs for problem (17).

Theorem 2. *Let $I : X \to R$ be a weakly lower semicontinuous functional under assumptions (i) and (j), $p^{inc} \in \mathcal{P}^{inc}$ and Z_{ad} be nonempty set. Let $\alpha_1 \geq 0$, $\alpha_2 \geq 0$ and K_1, K_2 be bounded sets or $\alpha_1 > 0$, $\alpha_2 > 0$ and functional $I(P)$ be bounded below. Then control problem (17) has at least one solution $(P, u) \in X \times K$.*

Proof. Let $\{(P_m, u_m)\}$ where $u_m = (\lambda_m, \eta_m)$, $m = 1, 2, ...$ be a minimizing sequence for which

$$a_0(P_m, \Phi) + a(\lambda_m, \mu_m; P_m, \Phi) = \langle f^{inc}, \Phi\rangle\ \forall\Phi \in X, \quad (18)$$

$$\lim_{m\to\infty} J(P_m, u_m) = \inf_{(P,u)\in Z_{ad}} J(P, u) \equiv J^*. \quad (19)$$

From (19), condition (j) and Theorem 1, the following estimates follow: $\|\lambda_m\|_{s,\Omega} \leq c_1$, $\|\mu_m\|_{r,\Omega} \leq c_2$, $\|P_m\|_X \leq c_3$. Here, c_1, c_2, c_3 are constants which are independent of m. From these estimates, it follows that there exist weak limits $\hat\lambda \in K_1 \subset H^s_{\lambda_0}(\Omega)$, $\hat\mu \in K_2 \subset H^r_{\mu_0}(\Omega)$, $\hat P \in X$ of some subsequences of sequences $\{\lambda_m\}$, $\{\mu_m\}$, $\{P_m\}$. Using these facts, compactness of embedding $H^t(\Omega) \subset L^\infty(\Omega)$ at $t > 3/2$ we conclude (passing if necessary to subsequences) that $P_m \to \hat P \in X$ weakly in X, $\lambda_m \to \hat\lambda \in K_1$, $\mu_m \to \hat\mu \in K_2$ strongly in $L^\infty(\Omega)$. Let us show that triple $(\hat P, \hat\lambda, \hat\mu)$ satisfies the following identity:

$$a_0(\hat P, \Phi) + a(\hat\lambda, \hat\mu; \hat P, \Phi) = a_0(\hat P, \Phi) + \int_\Omega (\hat\lambda\nabla\hat P \cdot \nabla\overline{\Phi} - k^2\hat\mu\hat P\overline{\Phi})dx = 0\ \ \forall\Phi \in X. \quad (20)$$

To this end, we pass to limit in (18) as $m \to \infty$. It is clear that linear term $a_0(P_m, \Phi)$ passes to the term $a_0(\hat P, \Phi)$ as $m \to \infty$ while for the difference $a(\lambda_m, \mu_m; P_m, \Phi) - a(\hat\lambda, \hat\mu; \hat P, \Phi)$ there is

$$a(\lambda_m, \mu_m; P_m, \Phi) - a(\hat\lambda, \hat\mu; \hat P, \Phi) =$$

$$= ((\lambda_m - \hat\lambda)\nabla P_m, \nabla\Phi) + (\hat\lambda(\nabla P_m - \nabla\hat P), \nabla\Phi) - k^2[((\mu_m - \hat\mu)P_m, \Phi) + (\hat\mu(P_m - \hat P), \Phi)]. \quad (21)$$

Using (12), (13), we derive that as $m \to \infty$

$$|((\lambda_m - \hat\lambda)\nabla\tilde P_m, \nabla\Phi)| \leq \|\lambda_m - \hat\lambda\|_{L^\infty(\Omega)}\|P_m\|_X\|\Phi\|_X \to 0,\ |(\hat\lambda(\nabla P_m - \nabla\hat P), \nabla\Phi)| \to 0\ \ \forall\Phi \in X,$$

$$|((\mu_m - \hat\mu)P_m, \Phi)| \leq \|\mu_m - \hat\mu\|_{L^\infty(\Omega)}\|P_m\|_X\|\Phi\|_X \to 0,\ |(\hat\mu(P_m - \hat P), \Phi)| \to 0\ \ \forall\Phi \in X. \quad (22)$$

Thus, passing to the limit in (18) as $m \to \infty$ and using (22), we obtain (20). This means that $G(\hat P, \hat u, p^{inc}) = 0$ where $\hat u = (\hat\lambda, \hat\mu)$. Since J is weakly lower semicontinuous on $X \times K$, we obtain that $J(\hat P, \hat u) = J^*$ which proves the theorem. ■

It should be noted that the assertion of Theorem 2 is valid for functional I_0 defined by (4) since it is weakly lower semicontinuous and nonnegative.

The next step in the study of control problem (17) is to establish sufficient conditions on the input data in problem (17) under which its solution is unique and stable. For this purpose, make use the approach developed in [17, 18] for the 2D impedance cloaking problems is used. It is based on the derivation and analysis of the optimality system describing the first-order necessary conditions of an extremum in the problem (17). However, since problem (17) is stated for complex-valued functions, it has to be preliminarily decomplexified. As a result, it is reduced to equivalent control problem considered in the class of real valued functions. Then based on [21], the optimality system can be derived for the latter "real" control problem. Finally, this "real" optimality system is transformed to "complex" optimality system corresponding to the initial problem (17) . Using this scheme and arguing as in [17, 18] the following theorem can be proved.

Theorem 3. *Let the pair $(\hat{P}, \hat{u}) \equiv (\hat{P}, \hat{\lambda}, \hat{\mu}) \in X \times K$ be the solution of problem (17) under conditions (i), (j) and let the functional $I(P)$ be continuously differentiable with respect to state in the point $\hat{P}$. Then there is a unique Lagrange multiplier $R \in X$ which is the solution of the Euler-Lagrange equation*

$$a_0(\Psi, R) + a(\hat{\lambda}, \hat{\mu}; \Psi, R) = -(\alpha_0/2)\overline{\langle I'_P(\hat{P}), \Psi\rangle} \ \forall \Psi \in X \tag{23}$$

and the following variational inequalities are fulfilled:

$$\alpha_1(\hat{\lambda}, \lambda - \hat{\lambda})_{s,\Omega} + \mathrm{Re}((\lambda - \hat{\lambda})\nabla\hat{P}, \nabla R) \geq 0 \ \forall \lambda \in K_1, \tag{24}$$

$$\alpha_2(\hat{\mu}, \mu - \hat{\mu})_{r,\Omega} - \mathrm{Re}k^2((\mu - \hat{\mu})\hat{P}, R) \geq 0 \ \forall \mu \in K_2. \tag{25}$$

Simple analysis using the definition (15) of operator A_u^* adjoint of A_u shows that the identity (23) is equivalent to the operator equation

$$A_{\hat{u}}^* R = (-\alpha_0/2) I'_P(\hat{P}). \tag{26}$$

Taking into account (26), we shall refer below to (23) as an adjoint problem. We also note that in the particular case when $I(P) = I_0(P)$ (23) takes the form

$$a_0(\Psi, R) + a(\hat{\lambda}, \hat{\eta}; \Psi, R) = -\alpha_0(\Psi, \hat{P} - p^d)_Q \ \forall \Psi \in X.$$

Adjoint problem (23) together with variational inequalities (24), (25) and direct problem (6) form the optimality system for problem (17). The optimality system plays an important role in studying the properties of solutions of the control problem. On its basis, efficient numerical algorithms of solving problem (17) can be developed (please refer to [20] about some of them for the case of layered cloaking shell). In addition, on the base of analysis of the optimality system one can establish the sufficient conditions on the initial data providing the uniqueness and stability of solutions of particular extremal problems.

We restrict ourselves by the formulation of respective result for the control problem

$$J(P, \lambda, \mu) = \frac{\alpha_0}{2}\|P - p^d\|_Q^2 + \frac{\alpha_1}{2}\|\lambda\|_{s,\Omega}^2 + \frac{\alpha_2}{2}\|\mu\|_{r,\Omega}^2 \to \inf,$$

$$G(P, \lambda, \mu, p^{inc}) = 0, \ (P, \lambda, \mu) \in X \times K_1 \times K_2 \tag{27}$$

corresponding to the functional $I_0(P)$ in (4). Denote by (P_1, λ_1, μ_1) an arbitrary solution of problem (27) corresponding to given functions $p^d = p_1^d$ and $p^{inc} = p_1^{inc}$. By (P_2, λ_2, μ_2), the solution of the same problem corresponding to the perturbed functions $\tilde{p}^d = p_2^d$ and $\tilde{p}^{inc} = p_2^{inc}$ is denoted. We put

$$M_P = C_2 \sup_{p^{inc} \in K^{inc}} \|p^{inc}\|_{1,\Omega_e},\ M_P^0 = M_P + \max(\|p_1^d\|_Q, \|p_2^d\|_Q), a = 2C_2 M_P^0, b = 2C_2 M_P^0 M_P^{-1},$$

$$\varphi(\|p^{inc}\|_{1,\Omega_e}) = (a\|p^{inc}\|_{1,\Omega_e} + b\|p^{inc}\|_{1,\Omega_e}^2)^{1/2}$$

and assume that the following conditions are satisfied:

$$\alpha_1(1-\varepsilon) > 5\alpha_0 V C_2^2 C_s^2 M_P M_P^0,\ \alpha_2(1-\varepsilon) > 5\alpha_0 C_2^2 C_r^2 M_P M_P^0, \tag{28}$$

where $\varepsilon \in (0,1)$ is an arbitrary number. The following theorem can be proved.

Theorem 4. *Let, in addition to conditions (i), (j), K_1, K_2 and $\mathcal{K}^{inc} \subset \mathcal{P}^{inc}$ be bounded sets and let the triple (P_l, λ_l, μ_l) be the solution of problem (27) corresponding to given functions $p_l^d \in L^2(Q)$ and $p_l^{inc} \in \mathcal{K}^{inc}$, $l = 1,2$. Assume that conditions (28) are satisfied. Then the following estimates hold true:*

$$\|P_1 - P_2\|_Q \le \|p_1^d - p_2^d\|_Q + \varphi(\|p_1^{inc} - p_2^{inc}\|_{1,\Omega_e}), \tag{29}$$

$$\|\lambda_1 - \lambda_2\|_{s,\Omega} \le \sqrt{\alpha_0/\varepsilon\alpha_1}\Delta,\ \|\mu_1 - \mu_2\|_{r,\Omega} \le \sqrt{\alpha_0/\varepsilon\alpha_2}\Delta, \tag{30}$$

$$\|P_1 - P_2\|_X \le C_2[M_P(C_s\sqrt{\alpha_0/\varepsilon\alpha_1} + k^2 C_r\sqrt{\alpha_0/\varepsilon\alpha_2})\Delta] + C_2\|p_1^{inc} - p_2^{inc}\|_{1,\Omega_e}. \tag{31}$$

Here Δ is defined by

$$\Delta = (1/2)\|p_1^d - p_2^d\|_Q + \varphi(\|p_1^{inc} - p_2^{inc}\|_{1,\Omega_e}). \tag{32}$$

We note that the estimates (30), (31) have the meaning of stability estimates of optimal solutions for control problem (27). These estimates are valid only if the parameters α_1 and α_2 entering into (27) are positive and satisfy conditions (28). This means that terms $(\alpha_1/2)\|\lambda\|_{s,\Omega}^2$ and $(\alpha_2/2)\|\eta\|_{r,\Omega}^2$ in the expression for the minimized functional in (27) have a regularizing effect on control problem under study.

4. Conclusion

In this paper, the control problems for the 3D model of acoustic scattering by permeable inhomogeneous isotropic obstacle having the form of the spherical layer are studied. These problems arise while developing the design technologies of acoustic cloaking devices using the optimization method. The solvability of the general control problem has been proved, the optimality system which describes the necessary conditions of extremum derived and sufficient conditions to the data to provide uniqueness and stability of optimal solutions established. The separate paper of the authors will be devoted to develop numerical algorithms of solving cloaking problems on the base of the optimality system and to analyse results of numerical experiments.

Acknowledgments

This work was financially supported by the Russian Foundation for Basic Research (project No. 16-01-00365-a) and Program of Fundamental Research of Far Eastern Branch of the Russian Academy of Sciences "Far East" (project No. 15-1-4-036).

References

[1] Dolin L S 1961 *Izvestiya Vuzov. Radiophizika* **4** 964–7
[2] Pendry J B, Shurig D and Smith D R 2006 *Science* **312** 1780–2
[3] Leonhardt U 2006 *Science* **312** 1777–80
[4] Cummer S A and Schurig D 2007 *New J. of Phys.* **9** 45
[5] Cummer S A, et al 2008 *Phys. Rev. Lett.* **100** 024301
[6] Chen H and Chan C T 2007 *Appl. Phys. Lett.* **91** 183518
[7] Romanov V G 2010 *Doklady Mathematics* **81** 238–240
[8] Alekseev G V and Romanov V G 2012 *J. Appl. Industr. Math.* **6** 1–5
[9] Romanov V G and Chirkunov Yu A 2013 *Doklady Mathematics* **87** 73–5
[10] Anikonov D S, Nazarov V G and Prokhorov I V 1997 *Dokl. Math.* **56** 955–8
[11] Greenleaf A, Kurylev Y, Lassas M and Uhlmann G 2008 *New J. of Phys.* **10** 115024
[12] Kohn R, Onofrei D, Vogelius M and Weinstein M 2010 *Comm. Pure Appl. Math.* **63** 973–1016
[13] Liu H 2013 *Journal of Differential Equations* **254** 1230–46
[14] Nguyen H-M 2013 *SIAM J. Math. Anal.* **45** 3034–49
[15] Popa B-I and Cummer S A 2009 *Phys. Rev.* A **79** 023806
[16] Yu Z Z, Feng Y J, Xu X F, Zhao J M and Jiang T 2011 *J. Phys. D: Appl. Phys.* **44** 185102
[17] Alekseev G V 2013 *Comp. Math. And Math. Phys.* **53** 1853–69
[18] Alekseev G V 2014 *Appl. Anal.* **93** 254 – 68
[19] Alekseev G V 2016 *Differential Equations* **52**
[20] Alekseev G V and Levin V A 2014 *Doklady Physics* **59** 89–93
[21] Ioffe A D and Tikhomirov V M 1979 *Theory of extremal problems* (Amsterdam: North-Holland) p 480

Modelling of hydraulic fracture propagation in inhomogeneous poroelastic medium

A N Baykin[1] **and S V Golovin**[1,2]

[1] Novosibirsk State University, Novosibirsk, Russia
[2] Lavrentyev Institute of Hydrodynamics SB RAS, Novosibirsk, Russia

E-mail: alexey.baykin@gmail.com, svgolovin@gmail.com

Abstract. In the paper a model for description of a hydraulic fracture propagation in inhomogeneous poroelastic medium is proposed. Among advantages of the presented numerical algorithm, there are incorporation of the near-tip analysis into the general computational scheme, account for the rock failure criterion on the base of the cohesive zone model, possibility for analysis of fracture propagation in inhomogeneous reservoirs. The numerical convergence of the algorithm is verified and the agreement of our numerical results with known solutions is established. The influence of the inhomogeneity of the reservoir permeability to the fracture time evolution is also demonstrated.

1. Introduction

Hydraulic fracturing is an important part of modern technologies for intensification of hydrocarbon production. The propagation of the hydraulic fracture is stimulated by the pumping of viscous fluid which creates pressure on fracture's walls high enough to overcome the rock closure stresses and cause the rock failure. Process of the hydraulic fracture growth is governed by several factors: flow of viscous fluid in a narrow fracture's gap, elastic reaction of the fracture's walls, filtration of fluid from the fracture to the reservoir, rock failure and advance of the fracture tip. Description of the hydraulic fracture dynamics is a complicated problem that is rarely solved in its most general formulation. Recent progress in the mathematical modelling of hydraulic fracture dynamics is described in review papers [1, 2].

In the present work the poroelastic model proposed in [3] is extended to the case of propagating fractures. The model allows determination of the porous pressure and the rock deformation coupled with the fracture disclosure and the pressure of the fracturing fluid. The material of the formation is observed as an inhomogeneous permeable medium governed by the Biot poroelasticity equations [4, 5]. The advantages of this approach in comparison with the classical approaches based on the KGD and PKN models (see [1, 2]) are the correct account for the interaction of the pore fluid with the fracturing fluid, finiteness of fluid pressure in the fracture's tip, readiness for modelling of reservoirs with inhomogeneous physical properties under non-uniform closure stresses.

The numerical solution of the problem was carried out by the finite element method with the use of a modification of the algorithm proposed in [3]. The advantages of this algorithm is lack of necessity for explicit track of the fracture's tip and corresponding remeshing, which is typical for contact problems of this type. The numerical convergence of the algorithm and possibility for computation of problems of the scale required for the engineering practice with a satisfactory accuracy is demonstrated. Also we present some results demonstrating the dependence of the fracture geometry and propagation behavior on the inhomogeneity of reservoir permeability.

2. Mathematical formulation of the problem

In this paper we make use of the model proposed in [3]. The mathematical problem is formulated in the following way. Let us consider a vertical planar fracture of fixed height H, propagating along the x axis (see Figure 1). Oz axis is directed upwards relative to the Oxy-plane. The fracture is opening along y direction due to the pressure generated by the fluid flow inside the fracture. Following [3] we suppose that fracture aperture is constant along z-coordinate and vertical deformations are negligible. This implies, that we can limit ourselves to the plane strain approximation observing only the central cross-section $z = 0$ of the fracture.

The poroelastic medium is characterized by its porosity ϕ and permeability $k_r(\mathbf{x})$, with the solid phase displacement $\mathbf{u}(t, \mathbf{x})$, and the pore pressure $p(t, \mathbf{x})$. Pores are saturated by a single-phase Newtonian fluid with the effective viscosity η_r. The linear Darcy's law for the fluid velocity $\mathbf{q} = -(k_r/\eta_r)\nabla p$ is employed. It is supposed that the fluid filtrating from the fracture to the reservoir has the same viscosity as the pore fluid. However, the fluid injected into the fracture has different viscosity η_f. It corresponds to the normal situation in hydraulic fracturing when the fracturing fluid is a high-viscous gel and only its low-viscous base fluid is filtrated into the reservoir.

For generality, the reservoir is initially subjected to a prestress with the stress tensor $\boldsymbol{\tau}_0(x, y)$. Hereafter, only planar fractures propagating along Ox axis are observed, therefore, the tensor $\boldsymbol{\tau}_0$ satisfies the symmetry conditions relative to this axis.

The governing equation of the quasi-static poroelasticity model is the following:

$$\begin{gathered} \operatorname{div} \boldsymbol{\tau} = 0, \quad \boldsymbol{\tau} = \boldsymbol{\tau}_0 + \lambda \operatorname{div} \mathbf{u}\, \mathrm{I} + 2\mu\, \mathcal{E}(\mathbf{u}) - \alpha p\, \mathrm{I}, \\ S_\varepsilon \frac{\partial p}{\partial t} = \operatorname{div}\Big(\frac{k_r}{\eta_r}\nabla p - \alpha \frac{\partial \mathbf{u}}{\partial t}\Big). \end{gathered} \tag{1}$$

Here $\mathcal{E}(\mathbf{u})$ is the Cauchy's strain tensor $2\mathcal{E}(\mathbf{u})_{ij} = \partial u_i/\partial x_j + \partial u_j/\partial x_i$ $(i, j = 1, 2)$, α is the Biot coefficient, $\lambda(\mathbf{x})$ and $\mu(\mathbf{x})$ are elasticity moduli, I is the identity tensor. The storativity S_ε reflects the dependence of the Lagrangian porosity ϕ on $\epsilon = \operatorname{tr}\mathcal{E}$ and p as in [6]:

$$\frac{\partial \phi}{\partial t} = \alpha \frac{\partial \epsilon}{\partial t} + S_\varepsilon \frac{\partial p}{\partial t}, \quad S_\varepsilon = \frac{(\phi_0 - \alpha)(1 - \alpha)}{K}, \tag{2}$$

where $K = \lambda + \dfrac{2\mu}{3}$ is the bulk modulus, ϕ_0 is the initial porosity. Due to the plane strain approximation, the solid phase displacement vector $\mathbf{u} = (u_1, u_2) = (u, v)$ is two-dimensional, all vector operations are also taken in 2D space of independent variables $(x_1, x_2) = (x, y)$.

Symmetry of the problem with respect to Ox-axis makes it possible solving equations (1) in domain $\Omega = \{(x, y) : |x| \le R, 0 \le y \le R\}$ as shown in Figure 2. Over the outer boundary $\Gamma_R : \ |\mathbf{x}|_\infty = R$, the confining far-field stress $\boldsymbol{\sigma}_\infty$ is applied and the constant pore pressure $p = p_\infty$ is prescribed:

$$\Gamma_R : \quad p = p_\infty, \quad \boldsymbol{\tau}\langle\mathbf{n}\rangle = \boldsymbol{\sigma}_\infty, \quad (\boldsymbol{\tau}\langle\mathbf{n}\rangle)_i = \tau_{ij} n_j. \tag{3}$$

Henceforth, $\mathbf{n}$ and $\mathbf{s}$ denote the outer normal and tangential unit vectors to the boundary of the domain Ω; the summation over the repeating index is implied. We restrict ourselves to the case with $\boldsymbol{\sigma}_\infty = -\sigma_\infty \mathbf{e}_2$, where σ_∞ is a scalar function. Moreover, it is assumed that the presstress $\boldsymbol{\tau}_0$ satisfies the same boundary condition: $\boldsymbol{\tau}_0\langle\mathbf{n}\rangle|_{\Gamma_R} = \boldsymbol{\sigma}_\infty$.

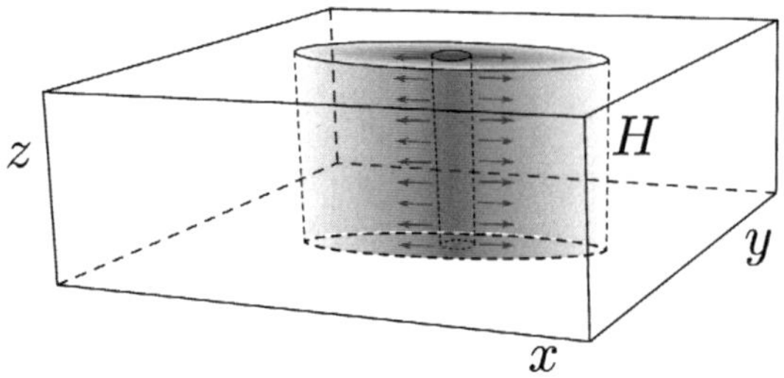

Figure 1. Planar vertical hydraulic fracture in a poroelastic medium

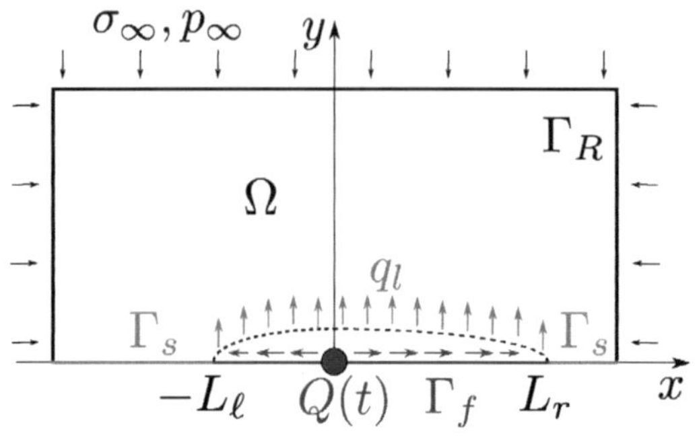

Figure 2. The horizontal cross-section of the fracture by plane $z = 0$

The line $y = 0$ is divided into the part $\Gamma_f = \{-L_\ell(t) \leqslant x \leqslant L_r(t),\ y = 0\}$ occupied by the fracture, and the remaining part $\Gamma_s = \{-R < x < -L_\ell(t),\ y = 0\} \bigcup \{L_r(t) < x < R,\ y = 0\}$. Outside the fracture on Γ_s the symmetry conditions (see [3]) are satisfied:

$$\Gamma_s: \quad \frac{\partial u}{\partial y} = 0, \quad v = 0, \quad \frac{\partial p}{\partial y} = 0. \tag{4}$$

With $p_f(t, x)$ standing for the fluid pressure inside the fracture, the force balance over the fractures walls yields

$$\Gamma_f: \quad p = p_f, \quad \mathbf{n} \cdot \boldsymbol{\tau}\langle \mathbf{n} \rangle = -p_f + \sigma_{coh}, \quad \mathbf{s} \cdot \boldsymbol{\tau}\langle \mathbf{n} \rangle = 0. \tag{5}$$

Here the tangential stress due to the fluid friction on the fractures walls in comparison with the normal stress is neglected.

In order to account for the rock failure during fracturing, the cohesive zone model is adopted as a fracture propagation criterion. Such an approach was initially proposed by Barenblatt [7] and Dugdale [8], where they postulated the existence of cohesive forces σ_{coh} (see Figure 3), which act in the zone of micro-cracking and plastic deformations in the vicinity of the fracture tips and prevent opening of the fracture. On the computational side, presence of cohesive forces removes the stress singularity at the fracture tip inherent to the Linear Elastic Fracture Mechanics (LEFM) by making the fracture aperture smoothly vanishing towards the tip.

We use the following traction/separation bi-linear law [9] to reflect the dependence of σ_{coh} on the half of the fracture aperture w as shown in Figure 4:

$$\sigma_{coh}(w) = \begin{cases} \sigma_c \dfrac{w}{w_m}, & 0 \leqslant w \leqslant w_m, \\ \sigma_c \Big(\dfrac{w_c - w}{w_c - w_m}\Big), & w_m \leqslant w \leqslant w_c, \\ 0, & w \geqslant w_c. \end{cases} \tag{6}$$

The cohesive forces reach their maximum value near the fracture tip. The region of the softening behaviour is limited by w_c. It is calculated from the considerations that the release of energy during the creation of the new fracture surface is equal to the work of the cohesive forces on the fracture opening. Hence,

$$G_c = \sigma_c w_c, \tag{7}$$

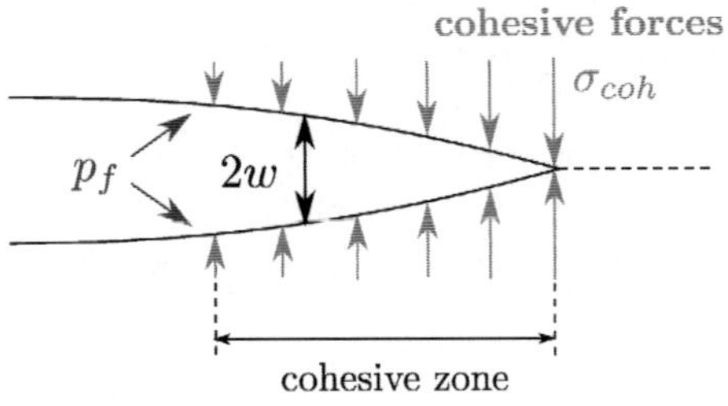

Figure 3. Cohesive zone near the fracture tip

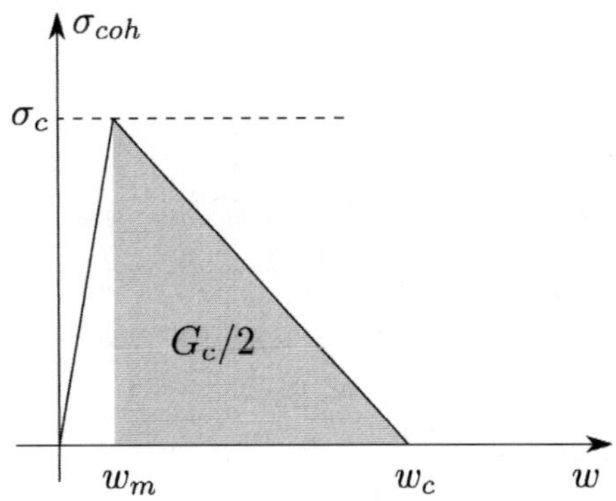

Figure 4. Bi-linear traction/separation law

where G_c is the energy release rate from the Griffiths's theory of brittle fractures [10], σ_c is the critical cohesive stress value. The elastic region of cohesive forces is small, $w_m = 2.5 \times 10^{-4} w_c$. It is required to regularize the cohesive energy near $w = 0$ [11].

If the cohesive zone is small relative to the fracture length, the connection with the fracture toughness K_{Ic} from LEFM is given by the Irvin's formula [12]:

$$K_{Ic} = \sqrt{G_c \frac{E}{1-\nu^2}}, \tag{8}$$

where E is the Young's modulus and ν is the Poisson's ratio.

The fluid flow in the fracture is governed by the mass conservation law complemented with the Poiseuille formula:

$$\frac{\partial w}{\partial t} + \frac{\partial (wq)}{\partial x} = -q_l, \quad w \equiv v|_{y=0}, \quad q = -\frac{(2w)^2}{12\eta_f}\frac{\partial p_f}{\partial x}. \tag{9}$$

Here q is the fluid velocity in x-direction. No fluid lag is assumed at the fracture tip.

The leak-off velocity q_l is given by the Darcy law in the reservoir as

$$q_l = -\left.\frac{k_r}{\eta_r}\frac{\partial p}{\partial y}\right|_{y=0}. \tag{10}$$

The resulting equation governing the flow inside the fracture reads

$$\frac{\partial w}{\partial t} = \frac{\partial}{\partial x}\left(\frac{w^3}{3\eta_f}\frac{\partial p_f}{\partial x}\right) + \left.\frac{k_r}{\eta_r}\frac{\partial p}{\partial y}\right|_{y=0}. \tag{11}$$

The flow rate (per unit height) injected into the fracture upper half-plane is calculated as

$$Q(t) = \frac{Q_v(t)}{2H} = -\left.\frac{w^3}{3\eta_f}\frac{\partial p_f}{\partial x}\right|_{x=0+} + \left.\frac{w^3}{3\eta_f}\frac{\partial p_f}{\partial x}\right|_{x=0-}, \tag{12}$$

where the division by 2 shows that the total flow rate is equally distributed between the symmetric fracture parts, and $Q_v(t)$ denotes the volumetric flow rate injected into the wellbore.

Equation (11) is often referred to as the lubrication theory equation [1]. Note that, due to the right-hand side of (10), equation (11) represents a boundary condition for equations of the main model (1). The leak-off rate q_l arises here naturally in the course of the problems solution, which differentiates the model favourably from the usual artificial approximations like the Carters formula or other similar expressions [13].

In order to close the model it is supplemented with the initial data at some moment t^0:

$$\mathbf{u}|_{t=t^0} = \mathbf{u}^0(x,y), \quad p|_{t=t^0} = p^0(x,y), \quad L_i|_{t=t^0} = L_i^0, \ i = \ell, r. \tag{13}$$

For computational reasons, it is convenient to homogenise the conditions over the outer boundary Γ_R. It can be done by considering stresses inside the reservoir relative to the prestress state $\boldsymbol{\tau}_0$, and taking p_∞ as a reference pressure. Similar to [3] the following new sought functions are introduced:

$$\tilde{\mathbf{u}} = \mathbf{u} - \varkappa \mathbf{x}, \quad \varkappa = \frac{\alpha p_\infty}{2(\lambda+\mu)}, \quad \tilde{\boldsymbol{\tau}} = \boldsymbol{\tau} - \boldsymbol{\tau}_0, \quad \tilde{p} = p - p_\infty. \tag{14}$$

Substituting (14) into equations (1) and taking into account boundary conditions (3), (4), (5), (11), (12), we obtain the following problem

$$\Omega: \quad \operatorname{div} \tilde{\boldsymbol{\tau}} = -\operatorname{div} \boldsymbol{\tau}_0, \quad \tilde{\boldsymbol{\tau}} = \lambda \operatorname{div} \tilde{\mathbf{u}}\, \mathrm{I} + 2\mu\, \mathcal{E}(\tilde{\mathbf{u}}) - \alpha \tilde{p}\, \mathrm{I}, \tag{15}$$

$$\Omega: \quad S_\varepsilon \frac{\partial \tilde{p}}{\partial t} = \operatorname{div}\left(\frac{k_r}{\eta_r}\nabla \tilde{p} - \alpha \frac{\partial \tilde{\mathbf{u}}}{\partial t}\right), \tag{16}$$

$$\Gamma_R: \quad \tilde{p} = 0, \quad \tilde{\boldsymbol{\tau}}\langle \mathbf{n}\rangle = 0, \tag{17}$$

$$\Gamma_s: \quad \tilde{u_y} = 0, \quad \tilde{v} = 0, \quad \tilde{p_y} = 0, \tag{18}$$

$$\Gamma_f: \quad \mathbf{n}\cdot \tilde{\boldsymbol{\tau}}\langle \mathbf{n}\rangle = -(\tilde{p} + p_\infty) - \mathbf{n}\cdot \boldsymbol{\tau}_0\langle \mathbf{n}\rangle + \sigma_{coh}, \quad \mathbf{s}\cdot \tilde{\boldsymbol{\tau}}\langle \mathbf{n}\rangle = 0, \tag{19}$$

$$\Gamma_f: \quad \frac{\partial \tilde{v}}{\partial t} = \frac{\partial}{\partial x}\left(\frac{\tilde{v}^3}{3\eta_f}\frac{\partial \tilde{p}}{\partial x}\right) + \frac{k_r}{\eta_r}\frac{\partial \tilde{p}}{\partial y}; \quad -\frac{\tilde{v}^3}{3\eta_f}\frac{\partial \tilde{p}}{\partial x}\bigg|_{y=0,x=0+} + \frac{\tilde{v}^3}{3\eta_f}\frac{\partial \tilde{p}}{\partial x}\bigg|_{y=0,x=0-} = Q(t). \tag{20}$$

In what follows, we work with the new sought functions skipping the tilde for simplicity of notations.

3. Numerical algorithm

In this Section the numerical method to solve the problem stated in Section 2 is provided. To accomplish this, firstly, we write the variational formulation of the problem. Following [3] let us choose a smooth vector-function $\boldsymbol{\psi} = \big(\psi_1(x,y), \psi_2(x,y)\big)$ and a smooth scalar function $\varphi(x,y)$ such that

$$\psi_2|_{\Gamma_s} = 0, \quad \varphi|_{\Gamma_R} = 0. \tag{21}$$

Then, we multiply equations (15) and (16) by $\boldsymbol{\psi}$ and φ, respectively, and integrate over Ω. Taking into account the boundary conditions (17)–(20), after integration we obtain

$$\int\limits_\Omega \big(\lambda \operatorname{div}(\mathbf{u}) - \alpha\, p\big) \operatorname{div}(\boldsymbol{\psi}) + 2\mu\, \mathcal{E}(\mathbf{u}) : \mathcal{E}(\boldsymbol{\psi})\, dxdy -$$

$$- \int\limits_{\Gamma_f} (p + p_\infty + \mathbf{n}\cdot \boldsymbol{\tau}_0\langle \mathbf{n}\rangle + \sigma_{coh})\psi_2\, dx = 0, \tag{22}$$

$$\int\limits_{\Omega} S_\varepsilon \frac{\partial p}{\partial t}\varphi\, dxdy + \int\limits_{\Omega} \frac{k_r}{\eta_r}\nabla p \cdot \nabla\varphi\, dxdy + \int\limits_{\Omega} \alpha\, \frac{\partial}{\partial t}\Big(\mathrm{div}\,\mathbf{u}\Big)\varphi\, dxdy +$$

$$+ \int\limits_{\Gamma_f} \frac{\partial v}{\partial t}\varphi\, dx + \int\limits_{\Gamma_f} \frac{v^3}{3\eta_f} p_x \varphi_x\, dx - Q(t)\varphi(0,0) = 0. \quad (23)$$

This formulation is not convenient on the computational use because of the necessity to track the fracture tips and change the size of Γ_f at every time step. Therefore, we modify Γ_f by regarding it as a potential fracture, which has closed part with $v = 0$ and the opened part with $v > 0$. However, under such interpretation the absence of the interpenetration of the opposite fracture walls in the computations cannot be guaranteed. In order to avoid this problem we impose the restriction

$$\Gamma_f: \quad v \geqslant 0, \quad (24)$$

by adding the penalty term

$$-\frac{1}{\delta}\int\limits_{\Gamma_f} \chi_{[v<0]}\, v\, \psi_2\, dx, \quad (25)$$

into the variational formulation. Here $\delta \ll 1$ is a small number, $\chi_{[v<0]}$ is the indicator function of set $\{\mathbf{x} : v(\mathbf{x}) < 0\}$.

The problem (23), (23) with (25) is solved using the finite element method via the open-source FEM package FreeFEM++ [14]. For the spatial discretization we use the piecewise-linear P1-elements over the triangulated computational domain Ω. The time derivatives are approximated with the first order of accuracy: $\dfrac{\partial f}{\partial t} \approx \dfrac{f^{n+1} - f^n}{\Delta t}$, where f denotes either of functions u, v or p; Δt is a time step. The upper index designates the number of the time step: $f^n = f(t^n, \mathbf{x})$, $t^n = t^0 + n\Delta t$. The computation starts with the intact state $\mathbf{u}^0 = 0$ and $p^0 = 0$ as the initial data. The non-linearity is resolved by the Newton-Raphson method.

4. Verification of the numerical algorithm

4.1. Convergence Test

In order to verify the algorithm we choose physical parameters typical for hydraulic fracturing problem and check the numerical convergence. For the verification the reservoir is assumed to be homogeneous. The reservoir is loaded by a uniform pre-stress caused by a far-field stress σ_∞:

$$\boldsymbol{\tau}_0\langle\mathbf{n}\rangle = -\sigma_\infty \mathbf{n}.$$

The following physical input parameters are used in simulations: domain size $R = 105$ m, maximal right tip position $L_r^{max} = 40$ m, maximal left tip position $L_r^{max} = 40$ m, Young's modulus $E = 17$ GPa, Poisson's ratio $\nu = 0.2$, energy release rate $G_c = 120$ Pa·m, critical cohesive stress $\sigma_c = 1.25$ MPa, initial porosity $\phi = 0.2$, reservoir permeability $k_r = 10^{-14}$ m^2, Biot coefficient $\alpha = 0.75$, storage coefficient $S_\varepsilon = 1.46 \times 10^{-11}$ Pa^{-1}, far-field stress $\sigma_\infty = 10$ MPa, reservoir pressure $p_\infty = 0$ MPa, reservoir fluid viscosity $\eta_r = 10^{-3}$ Pa·s, fracturing fluid viscosity $\eta_f = 10^{-1}$ Pa·s, injection rate per unit height $2Q = 10^{-3}$ m^2/s. Given Young's modulus E and Poisson's ratio ν, elasticity moduli λ and μ are calculated via the known formulas. The expression for the storage coefficient S_ε is given by formula (2).

Several simulations were conducted on the sequence of refining meshes. For the convergence test we compute the maximal relative difference in L_2-norm between solutions on two successive meshes:

$$\varepsilon_{max}(h) = \max\Big(\frac{||p_h - p_{h/2}||_{L_2}}{||p_h||_{L_2}}, \frac{||u_h - u_{h/2}||_{L_2}}{||u_h||_{L_2}}, \frac{||v_h - v_{h/2}||_{L_2}}{||v_h||_{L_2}}\Big) \times 100\% \quad (26)$$

The result of computations is demonstrated in Figure 5, where $h = \sqrt{S_{max}}$ is a mesh refining parameter and S_{max} is the maximal dimensionless area of all triangles in the corresponding mesh. One can see that at $h \approx 0.13$ the relative difference between solutions is less than 2 %. Therefore, this mesh was chosen for the further simulations.

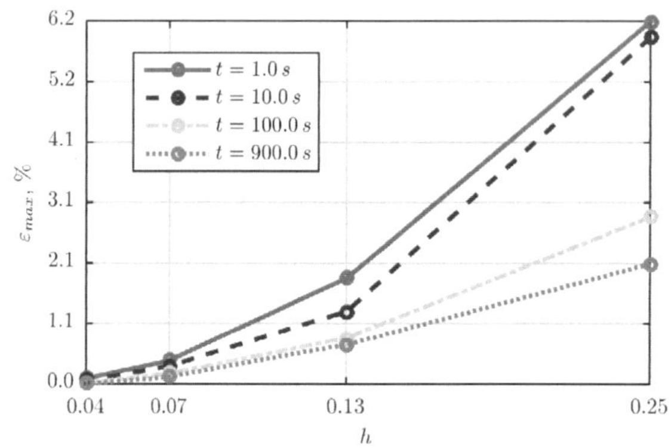

Figure 5. Relative difference in L_2-norm between solutions on two successive meshes for different time moments

4.2. Comparison with existing models

According to [15], [16], in the case of KGD-model the fracture propagation is governed by two competing energy dissipation mechanisms (viscous dissipation in fluid and creation of the new fracture surface) and two fluid storage mechanisms (in the fracture and in the reservoir). Therefore, there are four asymptotic regimes: storage-toughness dominated, leak-off-toughness dominated, storage-viscosity dominated and leak-off-viscosity dominated ones. All these regimes are demonstrated in paper [11] also in comparison with the analytical solution for KGD model. As a part of the model verification, we compare our results with the corresponding calculations of paper [11] in the case of the storage-toughness dominated regime.

The common input parameters used in simulations are the same as in Section 4.1, except for the domain size $R = 45$ m and maximal right and left tip positions $L_r^{max} = L_l^{max} = 15$ m. Also, we use equal viscosities for the reservoir and the fracturing fluids $\eta_r = \eta_f = 10^{-4}$ Pa·s to minimize the viscosity dissipation. Far-field stress is taken as $\sigma_\infty = 3.7$ MPa. Two simulations were carried out for permeabilities $k_r = 10^{-15}$ m^2 and $k_r = 10^{-16}$ m^2 during 14 s and 20 s, respectively, to ensure that the fracture propagates in the storage-toughness dominated regime. According to [11], for this set of parameters the results are in a good agreement with the early-time near-K solution [16] for the KGD model.

Good agreement of the fracture half-length in our model (solid lines 1, 2) and in [11] (marker lines 5, 6) is demonstrated in Figure 6. It confirms the consistency between the resulting fracture geometries and ones presented in [11]. A good match of the boundary of the cohesive zone (solid lines 3, 4 and marker lines 7, 8) indicates compliance in the implementation of the cohesive zone model.

5. Reservoir permeability contrast

In this Section, the model capabilities in the application to the hydraulic fracturing problem in the heterogeneous reservoir are demonstrated. The interesting case occurs when the reservoir

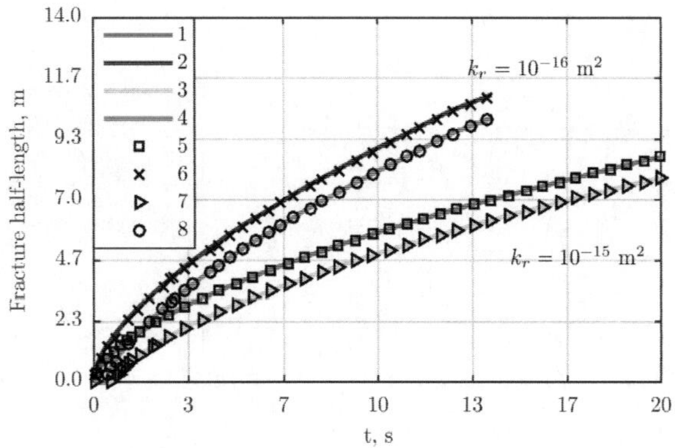

Figure 6. Fracture half-length in the is storage-toughness dominated regime: for $k_r = 10^{-15}$ m^2 (1) in present work, (5) in [11]; for $k_r = 10^{-16}$ m^2 (2) in present work, (6) in [11]. Position of the right boundary of the cohesive zone: for $k_r = 10^{-15}$ m^2 (3) in present work, (7) in [11]; for $k_r = 10^{-16}$ m^2 (4) in present work, (8) in [11].

has two layers (see Figure 7) with different permeabilities:

$$k_r(x) = \begin{cases} 10^{-14}\ \text{m}^2, & x \leqslant x^*, \\ 10^{-16}\ \text{m}^2, & x > x^*. \end{cases} \tag{27}$$

Here, $x = x^*$ denotes the border between layers, pictured as the brown line in Figure 7. Thus,

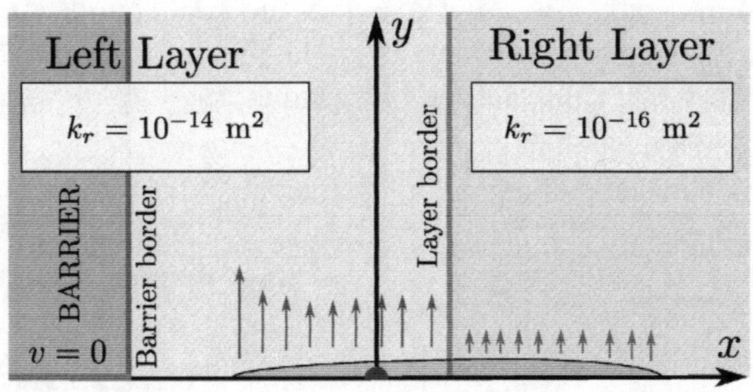

Figure 7. Reservoir with high-permeable (left) and low-permeable (right) layers. The barrier for fracture propagation is located in the high-permeable layer.

the leak-off from the fracture into the reservoir is higher in the left layer. The remaining simulation parameters are the same as in Section 4.1, except for the computational domain size parameters: $R = 150$ m, $L_\ell^{max} = 40$ m, $L_r^{max} = 73$ m. We propose the existence of a barrier located at $x = -L_\ell^{max}$ that prevents propagation of the fracture behind the barrier border.

The goal is to study the fracture dynamics depending on the location of the low-permeable layer relative to the injection point. Firstly, we place the layer border close to the wellbore at $x^* = 3.5$ m. Figure 8 (b) shows that the fracture reaches the layers border at $t \approx 35$ s and then proceeds its propagation further to the right. The leak-off in the low-permeable layer is lower; hence, the fracture becomes wider. The source of fluid at the wellbore near the border between the two layers supplies enough fluid to both sides of the fracture; therefore, the fracture propagates faster into the low-permeable layer where loss of fluid into the reservoir is lower (see Figure 8 (a)).

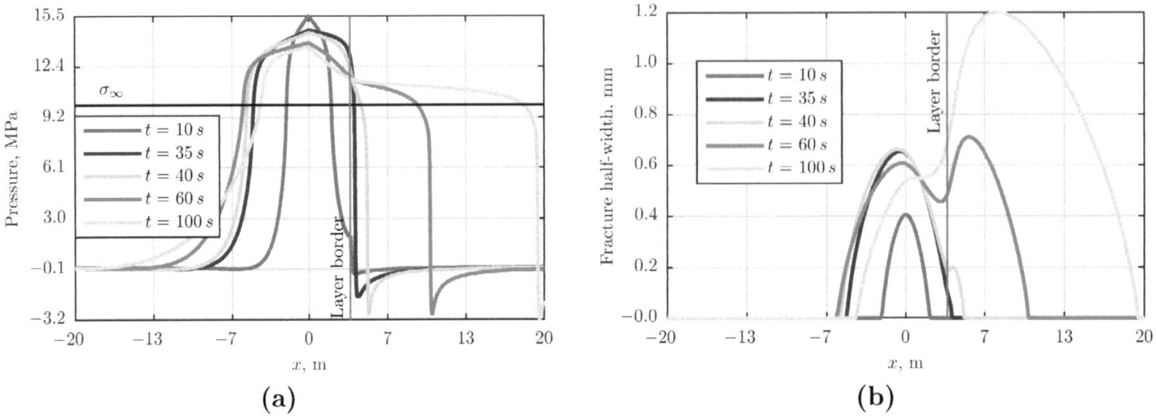

Figure 8. Pressure (a) and fracture half-width (b) along the fracture at different time for reservoir with two zones with high permeability contrast. The zones border (vertical brown line) is located at $x = 3.5$ m

The different situation occurs if we shift the border between the two layers with different permeabilities to $x^* = 10$ m (see Figure 9). The fracture propagates in both directions until the right tip enters the layer border. This occurs at $t \approx 200$ s and after that the fracture propagates to the left until the left fracture tip reaches the barrier at $t \approx 800$ s. For $200 \leq t \leq 800$ the pressure at the injection point stays almost at the constant level and, due to the lower permeability of the right layer, it becomes more difficult to increase the reservoir pressure for the further fracture growth to the right behind layer border. Thus, more fluid flows to the left half-wing of the fracture and only the left tip is moving. As the left tip reaches the barrier, it stops. As a result, the pressure starts growing over the whole fracture. When it overcomes some limit near the right layer border, the fracture breaks into the low-permeable layer and continues its propagation to the right direction.

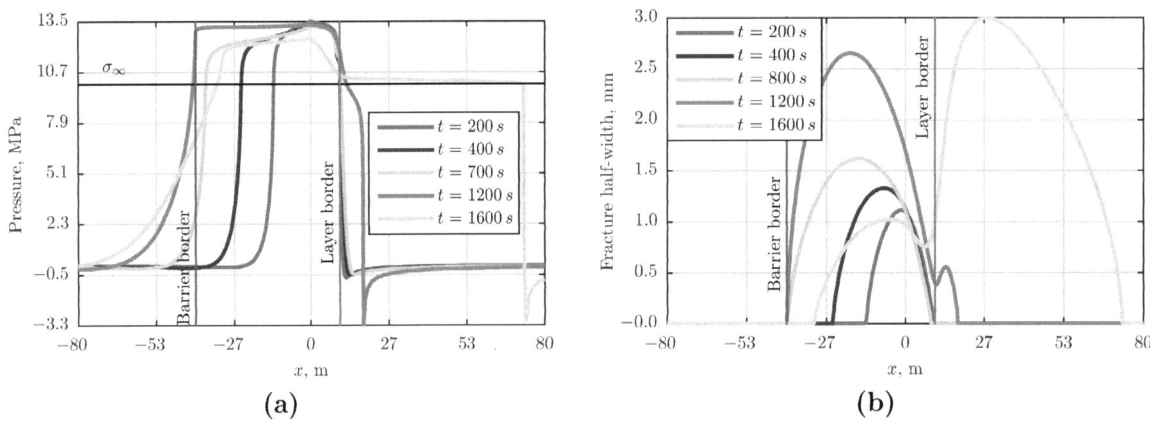

Figure 9. Pressure (a) and fracture half-width (b) along the fracture at different time for reservoir with two zones with high permeability contrast. The border between the permeability zones (vertical brown line) is located at $x = 10$ m

6. Conclusion

A numerical model of a hydraulic fracture propagating in a heterogeneous poroelastic medium is presented in the paper. The proposed approach accounts the influence of the pore fluid and the elastic rock deformation to the growth of the fracture. The fluid exchange between the reservoir and the fracture is naturally incorporated into the model. This makes it possible to state the complete problem in the single variational formulation with the mass conservation equation in the fracture playing the role of a natural nonlinear boundary condition. The cohesive zone model is used as a fracture propagation criterion.

The convergence of the numerical algorithm is checked by the comparison of the difference between solutions on the successively refining grids. Also, the simulation results are compared against the existing hydraulic fracture models, showing good agreement with numerical and analytical solutions.

The model capabilities are demonstrated on the case of reservoir with two layers with a permeability contrast. It is shown that fracture dynamics depends on the position of the border between the layers relative to the injection point. The fracture breaks into the low-permeable layer if the injection point is located near border between the layers. Otherwise, the fracture propagates through the high-permeable layer. The described non-stationary dynamics of the fracture grows demonstrates an importance of taking into account variability in physical properties of the reservoir for the correct modelling of the fracture development.

Acknowledgments

This work was supported by the RFBR grant (Project No. 16-01-00610).

References

[1] Adachi J, Siebrits E, Peirce A and Desroches J 2007 *Int. J. Rock Mech. and Min. Sci.* **44**(5) 739–57
[2] Detournay E 2016 *Ann. Rev. Fluid Mech* **48** 311–39
[3] Shelukhin V V, Baikov V A, Golovin S V, Davletbaev A Y and Starovoitov V N 2014 *International Journal of Solids and Structures* **51**(11) 2116–22
[4] Biot M A 1955 *Journal of Applied Physics* **26**(2) 182–5
[5] Biot M A 1956 *Journal of the Acoustical Society of America* **28**(2) 168–91
[6] Coussy O 2004 *Poromechanics* (John Wiley & Sons)
[7] Barenblatt G I 1962 *Adv Appl Mech* **7** 55–129
[8] Dugdale D 1962 *J Mech Phys* **8**(2) 100–4
[9] Geubelle P H and Baylor J S 1998 *Compos Part B* **29** 589–602
[10] Griffith A A 1921 *Philos T Roy Soc A* **221** 163–98
[11] Carrier B and Granet S 2012 *Engng Fract Mech* **79** 312–28
[12] Irwin G 1957 *J Appl Mech* **24** 361–4
[13] Economides M J and Nolte K G 2000 *Reservoir Stimulation* 3rd ed (John Wiley&Sons, LTD)
[14] Hecht F 2012 *J. Numer. Math.* **20** 251–65 ISSN 1570-2820
[15] Adachi J and Detournay E 2008 *Engng Fract Mech* **75**(16) 4666–94
[16] Bunger A, Detournay E and Garagash D 2005 *Int J Fract* **134**(2) 175–90

Torsion of circular rods at anisotropic creep

I A Banshchikova[1], D M Petrov [1,2] and I Yu Tsvelodub [1]

[1] Lavrentyev Institute of Hydrodynamics SB RAS, Novosibirsk, Russian Federation
[2] Novosibirsk State Technical University, Novosibirsk, Russian Federation

E-mail: binna@ngs.ru

Abstract. The torsion problems for aluminum alloy rods with circular cross section are solved. The rods are cut out from a transversely-isotropic plate with reduced resistance to creep strain in the direction of 45° to the normal direction of plate. Approximate estimates and finite element analysis show that the warping of the cross-section occurs when the rods are cut out in the longitudinal direction. In this case the value of torsion angle exceeds more than three times the value of torsion angle calculated in solving the isotropic problem. The analytical solution and finite element analysis for the rods cut out in the normal direction of plate show that the warping of the cross-section is missing, but the value of torsion angle s greater by an order of the value of torsion angle calculated in solving of the isotropic problem.

1. Introduction

Most modern structural materials have the properties of anisotropy at creep. The anisotropy can arise due to inhomogeneity of the material, which is typical for composites, or after such technological processes as rolling or pressing. Numerical and experimental verification of anisotropic models under complex stress state requires significant time and material expenses. A detailed study of the anisotropic continuum at creep published in recent years can be found in [1, 2, 3, 4]. Mostly, sheet materials have the properties of anisotropy in different directions (along, across, normal to the plate), different resistance in tension and compression, hardening and softening at creep. These features create certain difficulties in solving problems of the shaping parts and their further successful exploitation predictions.

For the alloy V95pcT2 (almost pure alloy V95 treated in mode T2) at $T = 180°$C (plate thickness 50 mm) based on experiments in tension and compression a considerable anisotropy and different resistance in tension and compression at creep [5] has been shown. The direction at the angle of 45° to the normal direction of plate was found as the weakest. The results of experiments on dependence of creep at constant stress σ on tension in the longitudinal (along), the transverse and normal directions of plate, compression in the longitudinal direction and tensile samples cut out at 45° to normal direction of plate were presented.

Power dependence for the creep strain rate $\eta = \dot{\varepsilon}^c = B\sigma^n$ approximates the experimental data of the steady state phase with coefficients: $B_\circ = 6.3 \cdot 10^{-31}$ MPa^{-n}s^{-1} for the longitudinal, transverse and normal directions of plate in tension; $B_\Delta = 3.9 \cdot 10^{-30}$ MPa^{-n}s^{-1} for the direction of 45° to the normal direction of plate in tension; $B_\bullet = 2 \cdot 10^{-31}$ MPa^{-n}s^{-1} in the longitudinal direction of plates in compression. The creep exponent $n = 10$ is the same for all directions and is independent of the sign of the applied load.

The intensity of the process for the longitudinal direction is three times higher for tension then compression. Comparison of experimental results for tensile samples cut out at 45° to the normal direction of plate with the results of experiments in tension for longitudinal, transverse and normal directions of plate shows almost a six-time increase in the intensity of the creep process.

Problems of torsion of arbitrary cross section rods under creep conditions were solved in [6, 7] using the methods of finite element discretization and finite difference approximations for isotropic alloys. The problem of restrained torsion of a rod of rectangular cross-section and its comparison with experimental data is dicussed in [8]. Various approximate methods including the method of combining of the elastic and plastic solutions for isotropic creep medium is considered in [9]. A detailed analysis of torsion of the anisotropic elastic rods is described in [10]. The ideal plasticity ratio of anisotropic medium in torsion is investigated in [11]. Due to the physical nonlinearity the solution of the problem of torsion in anisotropic creep is possible only approximately even for circular cross-section rod.

2. Theory

At an arbitrary stress state the process of creep can be described in a general view

$$\eta_{ij} = \partial\Phi/\partial\sigma_{ij}, \;\; \Phi = T^{n+1}/(n+1),$$

where η_{ij} are the components of creep strain rates tensor, σ_{ij} are the components of stresses tensor, Φ is the scalar potential function of stresses tensor, T is the positively defined quadratic form of stresses, n is a constant of a material [12]. For orthotropic incompressible material in the axes of the coordinates coinciding with the main axes of anisotropy, this quadratic form looks like form used by Hill for the description of anisotropic plasticity and includes six coefficients A_{ij} determined experimentally

$$T(\sigma_{ij}) = (A_{11}(\sigma_{22}-\sigma_{33})^2 + A_{22}(\sigma_{33}-\sigma_{11})^2 + A_{33}(\sigma_{11}-\sigma_{22})^2 + 2A_{12}\sigma_{12}^2 + 2A_{23}\sigma_{23}^2 + 2A_{31}\sigma_{31}^2)^{0.5}.$$

The coefficients of the quadratic form A_{ij} are defined as follows

$$2A_{11} = B_2^m + B_3^m - B_1^m, \quad 2A_{12} = 4B_{12}^m - A_{11}^m - A_{22}^m.$$

The other coefficients and components of creep strain rates are obtained by a cyclic permutation of the indices. Here B_1, B_2, B_3 are the constant one-dimensional creep in three main directions; B_{12}, B_{23}, B_{31} are similar constants in three directions along the axes in the coordinate system obtained by rotating the original coordinate system by 45°; $m = 2/(n+1)$. Then the creep strain rates are

$$\eta_{11} = T^{n-1}\Big((A_{22}+A_{33})\sigma_{11} - A_{33}\sigma_{22} - A_{22}\sigma_{33}\Big), \quad \eta_{12} = 2T^{n-1}A_{12}\sigma_{12}.$$

The dependence $G = T^n$ exists similarly to the isotropic creep, where G is the quadratic form of creep strain rates and coefficients of anisotropy.

In case of the unconstrained torsion of arbitrary cross-section rod the relationships between the creep strain rates and stresses are [9]:

$$\eta_{23} = \theta(W_{,2} + x_1) = 2T^{n-1}A_{23}\sigma_{23}, \quad \eta_{13} = \theta(W_{,1} - x_2) = 2T^{n-1}A_{31}\sigma_{31}, \tag{1}$$

where $T = (2A_{23}\sigma_{23}^2 + 2A_{31}\sigma_{31}^2)^{0.5}$; $W(x_1, x_2)$ is the function corresponding to the displacement of cross-section points (warping) in the direction x_3 along rod in torsion; θ is the rate of torsion angle per unit length. Stresses are expressed through the components of the creep strain rates

$$\sigma_{23} = \frac{\eta_{23}}{2A_{23}}G^{(1-n)/n}, \quad \sigma_{13} = \frac{\eta_{13}}{2A_{31}}G^{(1-n)/n}, \tag{2}$$

where $G = (\eta_{13}^2/(2A_{31}) + \eta_{23}^2/(2A_{23}))^{0.5}$.

The equation of equilibrium for a rod with free ends is

$$\partial\sigma_{31}/\partial x_1 + \partial\sigma_{23}/\partial x_2 = 0,$$

with the boundary condition on the contour of cross section $\sigma_{31}n_1 + \sigma_{23}n_2 = 0$.

From conditions of compatibility of creep strain rates (1) follows that

$$\frac{\partial}{\partial x_2}\left(2A_{31}T^{n-1}\sigma_{31}\right) - \frac{\partial}{\partial x_1}\left(2A_{23}T^{n-1}\sigma_{23}\right) = -2\theta. \tag{3}$$

The solution of torsion problem can be obtained by converting permitting ratios to the differential equation with respect to function of warping $W(x_1, x_2)$ [6, 7], or with respect to function of stresses $F(x_1, x_2)$, such that $\sigma_{13} = \partial F/\partial x_2, \quad \sigma_{23} = -\partial F/\partial x_1$ [9].

Torque is

$$M = \int\int(\sigma_{23}x_1 - \sigma_{31}x_2)dx_1dx_2 = 2\int\int F dx_1dx_2.$$

Warping of the cross section is absent in torsion of a solid circular isotropic rod. Since $2A_{31} = 2A_{23} = 3B^m$ and in all directions $B = B_\circ$, the solution of (3) with the boundary condition $F(a) = 0$ on the steady-state stage of creep can be written as [9]

$$F(r) = \frac{3+1/n}{1+1/n}\frac{M}{2\pi a^2}\left(1 - \left(\frac{r}{a}\right)^{1+1/n}\right), \quad \theta = (\sqrt{3})^{n+1}\frac{B}{a}\left(\frac{3+1/n}{2\pi a^3}M\right)^n, \tag{4}$$

where a is the radius of the rod cross section, $r = (x_1^2 + x_2^2)^{1/2}$.

For rod which is cut out in the normal direction to plate (axis x_2 coincides with the normal direction of plate) the equation of equilibrium is converted to the form $\partial\sigma_{12}/\partial x_1 + \partial\sigma_{23}/\partial x_3 = 0$. In this case the warping is absent for transversely-isotropic alloy with the reduced resistance to creep strain in the direction 45° to the normal direction of plate, i.e. when

$$B_1 = B_2 = B_3 = B_{31} = B_\circ, \quad B_{12} = B_{23} = B_\Delta. \tag{5}$$

After solving the equation similar to (3) the rate of torsion angle can be written as

$$\theta = (2A_{12})^{(n+1)/2}\frac{1}{a}\left(\frac{3+1/n}{2\pi a^3}M\right)^n. \tag{6}$$

For a rod which is cut out in the longitudinal direction x_3 from transversely-isotropic alloy with constants (5) we can assume that the function of warping approximately has the form [10] $W(x_1, x_2) = Cx_1x_2$, where C is a constant that can be found through a variational principle [9]. In this case

$$\theta = \left(\frac{M}{R}\right)^n, \quad R = 4\int_0^a\int_0^{f(x_2)}\left(\frac{x_1^2(C+1)}{2A_{23}} - \frac{x_2^2(C-1)}{2A_{31}}\right)\left(\frac{x_1^2(C+1)^2}{2A_{23}} + \frac{x_2^2(C-1)^2}{2A_{31}}\right)^k dx_1dx_2, \tag{7}$$

where $f(x_2) = (a^2 - x_2^2)^{1/2}, \quad k = (1-n)/(2n)$. If $C = 0$, then (7) gives a lower estimate of the torsion angle rate.

For numerical simulations of isotropic and anisotropic material the Solid45 element of the ANSYS was tested. The hill option of TB command activates the properties of the material anisotropy in plasticity and creep. The coefficients R_{xy}, R_{yz} are connected with coefficients (5) by the relations $R_{xy} = R_{yz} = \left(3/(4(B_\Delta/B_\circ)^m - 1)\right)^{1/2}$, $R_{xx} = R_{yy} = R_{zz} = R_{xz} = 1$. The problem is solved in a geometrically linear statement.

3. Results and discussion

The calculations are performed for material V95pcT2 at a temperature of $T = 180°$ C based on the material properties only in tension with parameters: the radius of the rod $a = 0.02$ m, the length $L = 0.1$ m, torque $M = 3500$ N·m, the Poisson's ratio $\nu = 0.4$, Young's modulus $E = 55000$ MPa, the time of torsion $t = 600$ s.

Table 1 includes a full angle of torsion (radians) calculated analytically ($\varphi = Lt\theta$) and with finite element method (angle estimated by the displacement) without taking into account the elastic angle at $t = 0$ for the rods in the assumption of isotropic and transversely-isotropic alloy.

Table 1. Full angle of torsion.

type of anisotropy/ direction of rod cut	φ / formula	φ Solid45
isotropic	0.18 / (4)	0.20
transversely-isotropic/ longitudinal direction	0.51 / (7)	0.75
transversely-isotropic/ normal direction	1.77 / (6)	1.98

Figure 1 *a* and *b* show the isolines of the stresses σ_{13} and σ_{23} (MPa) at the torsion of isotropic rod.

Figure 2 *a* and *b* show the isolines of the same stresses at the torsion of rod cut out in the longitudinal direction x_3 (coincide with Z) made of an alloy with the reduced resistance to creep strain in the direction 45° to the normal direction of plate x_2 (coincide with Y). Numerical simulations demonstrated that the cross-section warping (displacement in direction Z, m) occurs during the torsion of circular rods (Figure 2 *c*). In experiments, this is visually confirmed by the appearance of spiral lines on the samples [13]. Similar warping occurs at torsion of anisotropic elastic circular rods [10].

Figure 3 *a* and *b* show the diagrams of contour stresses σ_{13} and σ_{23} calculated by an approximate method (2), (7). In this case, maximum stresses (2) are $\sigma_{13} = 239.9$ MPa and $\sigma_{23} = 190.6$ MPa, the torsion angle $\varphi = 0.51$. The work [14] contains a description of the test element Beam189 in ANSYS. The resulting torsion angle φ calculated by the finite element method close to the angle derived from (7) at $C = 0$ equals 0.5, i.e. the warping is absent.

The values and contours of the shear stresses of rod cut out in a normal direction of plate made of transversely-isotropic alloy almost coincide with those demonstrated in Figure 1 *a, b*. In this case, the warping of the cross section is absent, and the lines-notches in torsion remain parallel cross sections on the surface of the samples, as in case of isotropic material [13].

Note that the torsion angle obtained using Solid45 is more than 10% greater than the angle calculated analytically. It is related apparently to peculiarities of calculations of shear deformations in the physically nonlinear statement. The elastic torsion angle $\varphi_0 = 7.09 \cdot 10^{-2}$ at the initial time of loading varies for less than 0.01%.

4. Conclusions

The analytical calculations and the finite element method show a significant effect of reduced resistance to creep strain in shear direction: the torsion angle may differ by an order of magnitude

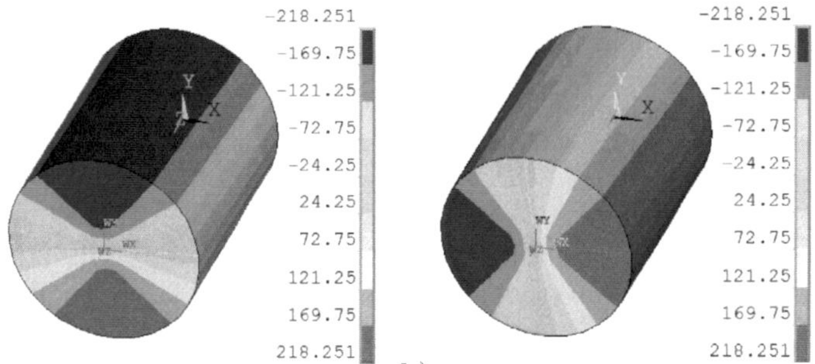

a) b)

Figure 1. Isolines of the stresses σ_{13} – *a*) and σ_{23} – *b*) at the torsion of the isotropic rod

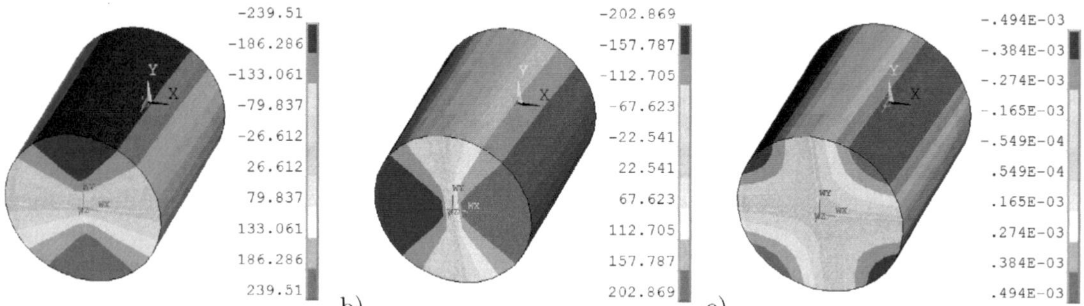

a) b) c)

Figure 2. Isolines of the stresses σ_{13} – a), σ_{23} – b) and displacement (warping) – c) at the torsion of the rod cut out in the longitudinal direction x_3 of the transversely-isotropic plate

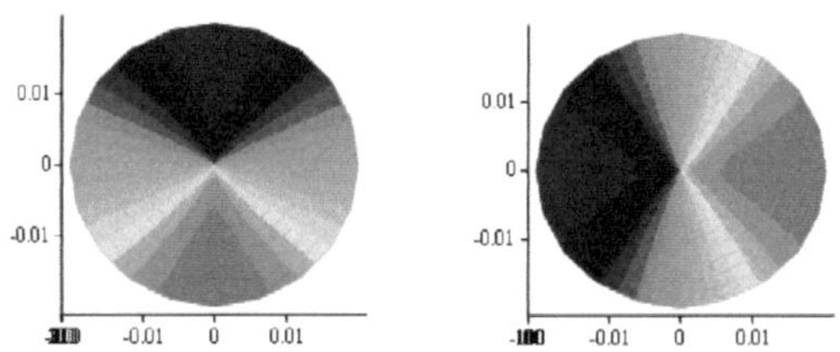

a) b)

Figure 3. Isolines of the stresses σ_{13} –a) σ_{23} – b) calculated by approximate method (2),(7).

compared with the angle calculated in the isotropic formulation. Thus, the reduced resistance to creep strain in the transverse shear direction may cause reduced stiffness and strength properties of structures that need to be considered in their design and operation.

Acknowledgments
This work was supported by the Russian Foundation for Basic Research (project codes 15-01-07631-a and 16-08-00713-a).

References
[1] Altenbach H 2003 *International Applied Mechanics* **39** 631
[2] Betten J 2008 *Creep Mechanics* (Berlin Heidelberg: Springer-Verlag) p 367
[3] Zolochevsky A, Sklepus S, Kozmin Yu, Kozmin A, Zolochevsky D and Betten J 2004 *Forschung im Ingenieurwesen* **68** 182
[4] Tsvelodub I Yu 1991 *Stability postulate and its applications in the creep theory of metallic materials* (Novosibirsk: Inst. of Hydrodynamics) p 201

[5] Gorev B V and Masanov I 2009 *Tekhnologiya Mashinostroeniya* **7** 13
[6] Banshchikova I A, Gorev B V and Sukhorukov I V 2002 *Appl. Mech. Tech. Phys.* **43** 448
[7] Banshchikova I A and Sukhorukov I V 2003 *Computational Technologies - Regional Bulletin Of The Est* **8** 104
[8] Banshchikova I A, Sukhorukov I V and Biao Yuan 2004 *Computational Technologies - The KazNU Journal, Mathematics, Mechanics and Computer Science Edition* **9** 248
[9] Kachanov L M 1960 *The Theory of Creep* (Moscow: Fizmatgiz) p 455
[10] Lekhnitskii S 1971 *Torsion of anisotropic and inhomogeneous rods* [in Russian] (Moscow: Nauka) p 240
[11] Mironov B G and Derevyannykh E A 2012 *I. Yakovlev Chuvash State Pedagogical University Bulletin* **4** 108
[12] Sosnin O 1965 *Appl. Mech. Tech. Phys.* **6**(6) 67–70
[13] Larichkin A Yu 2013 *Fizicheskoe i chislennoe modelirovanie deformirovaniya materialov s uchetom bolshih deformacij [Physical and numerical modeling of material deformation taking into account large deformations]: dissertation* (Novosibirsk: Lavrentyev Institute of Hydrodynamics).
[14] Banshchikova I A, Tsvelodub I Yu and Petrov D M 2015 *Physics and mathematics, Kazan. Gos. Univ. Uchen. Zap. Ser. Fiz.-Mat. Nauki* **157**(3) 34

Using non-linear analogue of Nyquist diagrams for analysis of the equation describing the hemodynamics in blood vessels near pathologies

A A Cherevko[1,2], E E Bord[2], A K Khe[1,2], V A Panarin[3], K J Orlov[3] and A P Chupakhin[1,2]

[1] Lavrentyev Institute of Hydrodynamics of the Siberian Branch of the Russian Academy of Sciences, Novosibirsk, Russia
[2] Novosibirsk State University, Novosibirsk, Russia
[3] Meshalkin Research Institute of Blood Circulation Pathology, Novosibirsk, Russia

E-mail: cherevko@mail.ru, bord.elizaveta@yandex.ru

Abstract. This article considers method of describing the behaviour of hemodynamic parameters near vascular pathologies. We study the influence of arterial aneurysms and arteriovenous malformations on the vascular system. The proposed method involves using generalized model of Van der Pol–Duffing to find out the characteristic behaviour of blood flow parameters. These parameters are blood velocity and pressure in the vessel. The velocity and pressure are obtained during the neurosurgery measurements. It is noted that substituting velocity on the right side of the equation gives good pressure approximation. Thus, the model reproduces clinical data well enough. In regard to the right side of the equation, it means external impact on the system. The harmonic functions with various frequencies and amplitudes are substituted on the right side of the equation to investigate its properties. Besides, variation of the right side parameters provides additional information about pressure. Non-linear analogue of Nyquist diagrams is used to find out how the properties of solution depend on the parameter values. We have analysed 60 cases with aneurysms and 14 cases with arteriovenous malformations. It is shown that the diagrams are divided into classes. Also, the classes are replaced by another one in the definite order with increasing of the right side amplitude.

1. Introduction

The cerebral blood vessel network is a complex system with blood circulating. The walls of a blood vessel are viscoelastic and the outer part of the vessel consists of muscles which react reflexively to the pulse wave. Besides, the vessel is surrounded by the brain substance. Sometimes different pathological defects, for example aneurysms [2] and arteriovenous malformations (AVM), [1] may occur in the cerebral blood vessel network. The AVM is a pathological plexus of the veins and arteries. This leads to violation of blood circulation. Moreover, AVM leads to the direct emission of blood from arterial pool to vein pool. Another pathology is the arterial aneurysm. It is the bulging of artery wall as a result of its extension. The material of the aneurysm wall is different from the material of the health vessel. Intra-vascular surgery is used for the treatment of patients with these pathologies. Embolization which means selective sealing up of blood vessels is applied in case of AVM or intra-vascular setting of the stent in case of aneurysm.

2. Materials and methods

During the neurosurgical operations, the hemodynamic parameters are monitored. ComboMap device with ComboWire sensor ($d \approx 0.3\,\mathrm{mm}$, $L > 1\,\mathrm{m}$) is used for this purpose. It is unique equipment for complex measurement inside of small vessels. The sensor measures simultaneously the pressure and velocity in the blood vessel. During surgery, series of measurements are made in different areas of the vasculature. The measurements are made both in arterial and venous pool. Data collection frequency is about 200 Hz. The obtained data are written into the file with ADC device [3, 4].

The obtained data namely pressure and velocity are cleaned by using software noise wavelet filters. This is done by the expending signals in Gabor wavelet of frequency 6. Then the highest harmonics of wavelet expanding are removed. The signals are recovered from the remaining coefficients of wavelet decomposition.

The authors use PV-diagrams to present the obtained data. The practice of investigation has shown the high efficiency of this representation. PV-diagram is a parametric curve in the plane. The velocity is marked on the horizontal axis, the pressure is marked on the vertical axis, the time is the parameter along the curve. Also, we use QU-diagrams built on normalized dimensionless values of pressure and velocity. Important to note that the forms of diagrams constructed according to data obtained near aneurysms and AVM qualitatively differ (see Figure 1 and Figure 2).

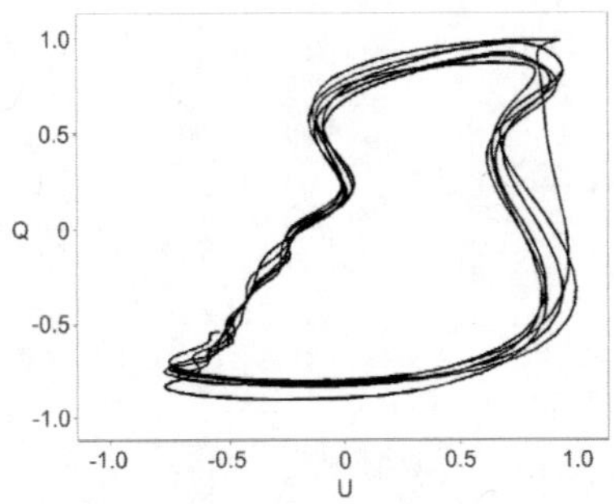

Figure 1. QU-diagram in case of aneurysm.

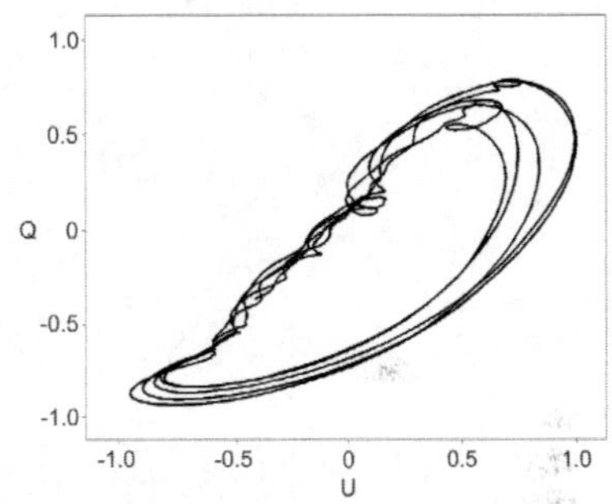

Figure 2. QU-diagram in case of AVM.

Diagrams in the venous and arterial pools differ too. The image point moves the diagram anticlockwise in arterial part, and clockwise in venous part with time.

To identify the characteristic behaviour of hemodynamic parameters in the surroundings of vascular pathologies the model of generalized Van der Pol–Duffing equation [6] was suggested.

$$\varepsilon q''(t) + P_2(q)q' + P_3(q) = ku(t). \qquad (1)$$

The normalized velocity u is a given value measured during the operation. The normalized pressure q is the solution of the equation. In the equation (1) $P_2(q) = a_1 + a_2q + a_3q^2$, $P_3(q) = b_1q + b_2q^2 + b_3q^3$. The coefficients are constructed based on the experimental data [5]. Coefficients b_i respond for the elastic properties of the vessels, a_i respond for the viscous friction, ε corresponds to the relaxation oscillation character. The various clinical cases are characterized by different coefficients of the equation. These coefficients are founded by the methods of inverse problem theory.

The model was confirmed experimentally with large amount of clinical data (about 300). The pressure values obtained in the experiment coincide well with the pressure values predicted by the model. Figure 3 shows typical QU-diagrams constructed on experimental data (black) and ones obtained by solving the equation (red).

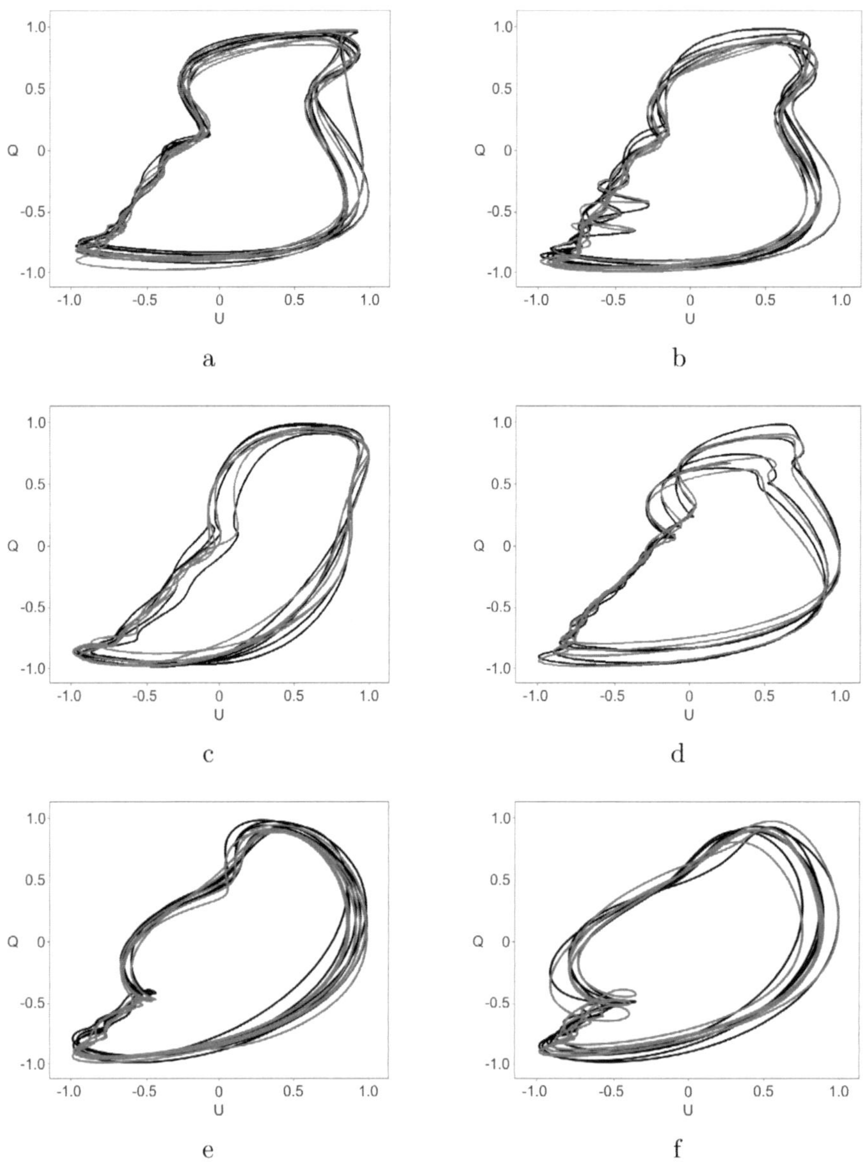

Figure 3. QU-diagrams. a–d: typical for aneurysms, e–f: typical for AVM. Black – experimental data, red — data obtained by solving the equation (1).

3. Results

From the medical point of view it seems important to know the state of the patient vascular bed before operation and monitor changes in this state at time of surgery. In this regard it is necessary not only to know the current response of blood vessels to pulse wave, but also be able to predict their response to the changes in blood flow. For ethical reasons if the effects on patients are not justified medically, they are considered to be intolerable. Thus, it is impossible to conduct research on real patient aiming to find out the reaction of bloodstream to the changes in blood flow. That is why it seems to be natural to investigate the model of vascular system of that patient.

To accomplish this we analyse the solutions of generalized Van der Pol–Duffing equation built for certain patients. The analysis is carried out by replacing the right side by a harmonic function, which parameters are the amplitude and frequency. We relate the solution behaviour with these two parameters. The analysis of the solution behaviour is important to predict how the system will react to outer force.

The generalized Van der Pol–Duffing equation is non-linear. So the solutions are not harmonic functions and have fairly complex behaviour even in case of harmonic right side. The analysis of main solution characteristics is necessary to find out the key influence patterns of right side amplitude and frequency on the solution. Such characteristics are primarily the solution amplitude and the solution phase change relative to the phase of the right side.

Due to the large amount of built solutions, direct behaviour analysis of the amplitude and phase change is difficult. Therefore, the authors suggested and used the non-linear generalization of the Nyquist diagrams (ND). In linear case ND is the amplitude-phase-frequency characteristic of the system. ND is depicted as a curve in polar coordinates and built for the fixed value of the right side amplitude. The right side frequency is the parameter along the curve. Polar radius is the amplitude and angle is the phase change of the appropriate solution.

But since in the non-linear case the oscillations are not harmonic, the modules of the solution maximum and minimum do not coincide. In this regard two curves in the polar coordinates are suggested to be used as a non-linear analogue of ND. The first curve is for module of the solution maximum and the second is for minimum. On figures in the paper the curve of maximum is red and the curve of minimum is blue. The number on plot corresponds to the right side frequency value. Figure 4 shows typical ND.

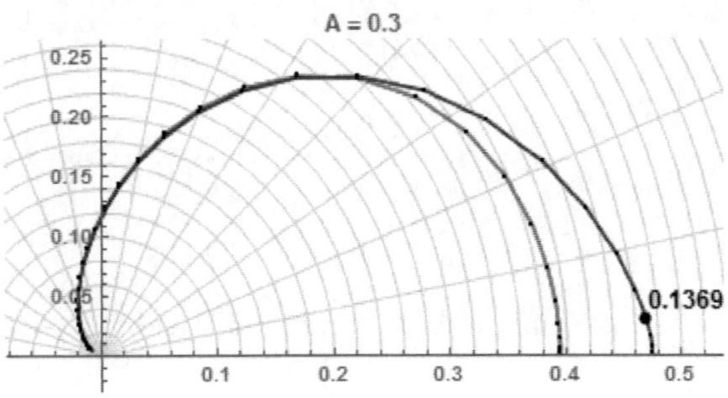

Figure 4. Example of Nyquist Diagram.

Table 1. The number of measurements for different patients.

Gu	Ko	Pi	Po	Si	Ch
15	15	6	10	14	14

For the construction of the equation (1) the clinical data of 6 patients were used. It was aneurysms in 5 cases and AVM in 1 case. The measurements were made at various sensor positions in the surroundings of the pathologies for each patient. Table 1 shows the number of measurements for each patient.

Generalized Van der Pol–Duffing equation characterized by its own set of coefficients was obtained for all measurements [3]. Further, we constructed ND for all of the coefficient sets using 4 fixed right side amplitude values. When constructing ND the following set of amplitudes was used: 0.3, 0.5, 0.75, 1. This choice of amplitude is due to the fact that dimensionless values in the equation vary from 0 to 1. In addition, the right side frequency varies from 0.01 Hz to 50 Hz. As a result we have analysed about 300 ND.

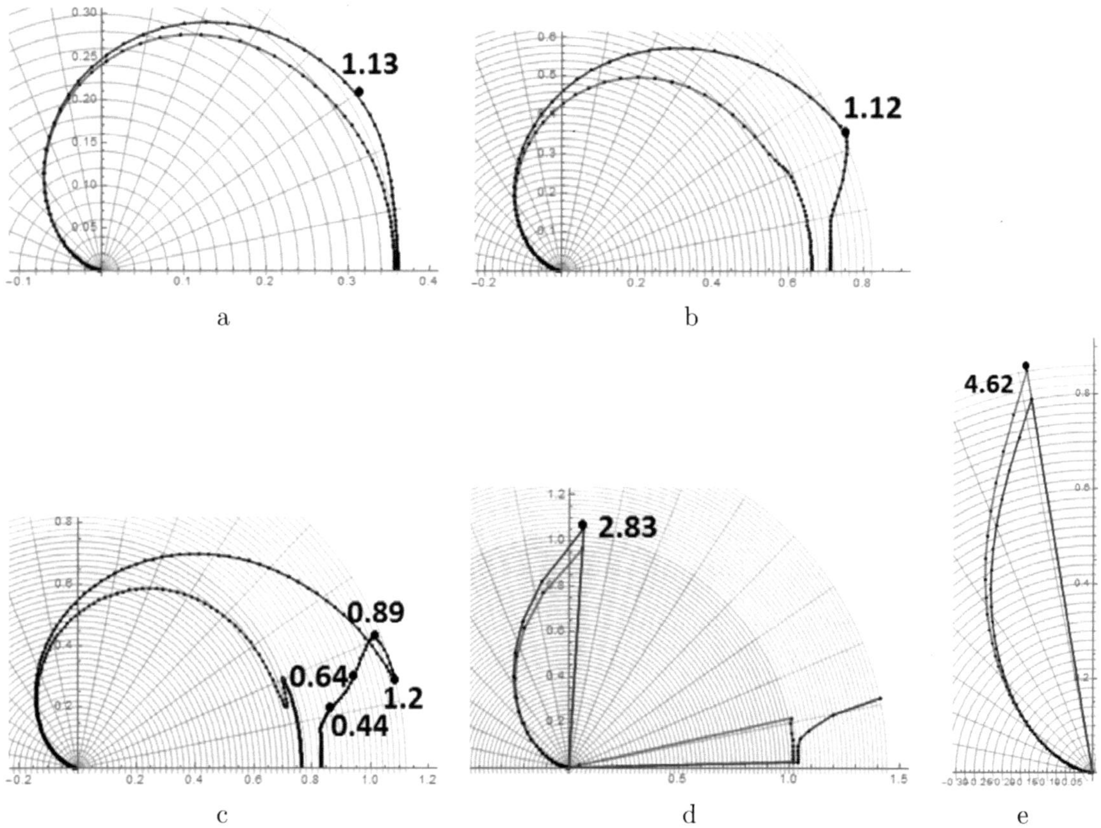

Figure 5. Classes of ND. a – normal case, b – divergent case, c – loop case, d – gap case, e – break case.

4. Discussion

While analysing the obtained ND we revealed that diagrams may be classified. For the time being it is possible to identify the following classes: normal case, divergent case, loop case, gap case, break case. These classes are often found between different patients. Typical ND for these classes are shown on Figure 5. It was observed that diagrams move from class to class in the following order with the increasing of the amplitude.

$$normal\,case \longrightarrow divergent\,case \longrightarrow loop\,case \longrightarrow gap\,case \longrightarrow break\,case \qquad (2)$$

If the right side frequency value changes continuously, the graph of the relevant solution transforms in a continuous manner too.

Next, we consider the solution behaviour corresponding to different classes of ND (Figure 6).

In normal case ND resembles the diagram for linear system. Besides, the solution of the equation (1) with a harmonic right side is close to a harmonic function. In the picture the blue curve denotes the right side of the equation (1), the yellow curve denotes corresponding solution. The plot of solution is built for the fixed frequency value in the point of the greatest distance between the curves of ND.

In the divergent case the difference between solution deviation up and down from the horizontal axis is increasing. As the amplitude grows, the solution differs more from a harmonic function.

In the loop case the typical situation is the occurrence of subharmonics in the decision. These subharmonics appear in the surroundings of the loops on ND.

In the gap case subharmonics are also observed. Moreover, the oscillation breakdown takes place in the gap area.

The break case is characterized by a long transient regime of oscillation establishment.

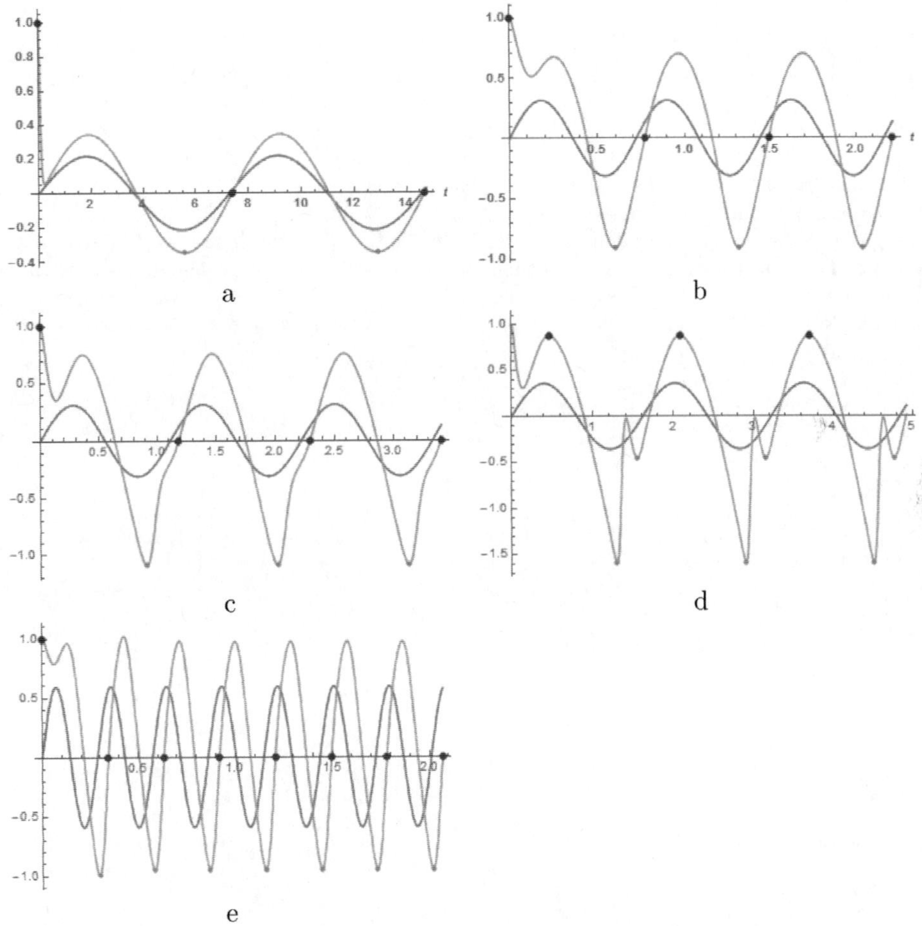

Figure 6. Typical solutions of equation (1) in different cases: a – normal case, b – divergent case, c – loop case, d – gap case, e – break case.

ND runs the chain (2) entirely not for the all sensor positions. There are cases when the ND remains within the first two classes for all the amplitude values from the set.

As the authors suppose, complicated non-linear oscillations may indicate a pathological state of vascular system in the surroundings of the sensor position. Such oscillations are mostly peculiar to the last three classes in the chain (2).

5. Conclusion

The study of generalized Van der Pol–Duffing equations built on the clinical data of real patients is presented in this paper. It is shown that ND constructed on the solutions of these equations are divided into several classes. The diagrams move from one class to another in definite order with the increasing of the right side amplitude. The dependence of the characteristic solution behaviour on the form of ND is found out.

Acknowledgments

The study was completed thanks to the support of Russian Foundation for Basic Research (project No. 14-01-00036).

References

[1] Klimov A B, Ryabuhin V E, Orlov K Yu and Krilov V V 2012 *Surgery of the Cerebral Aneurysms* **3** 231–68
[2] Krylov V V (ed) 2011 *Surgery of the Cerebral Aneurysms* vol 1 (Moscow: T. A. Alexeeva)
[3] Chupakhin A P, Cherevko A A, Khe A K, Telegina N J, Krivoshapkin A L, Orlov K J, Panarin V A and Baranov V I 2012 *Circulation Pathology and Cardiac Surgery* **16** No **44** 27–31
[4] Khe A K, Cherevko A A, Chupakhin A P, Krivoshapkin A L and Orlov K J 2016 *Journal of Physics: Conference Series* Endovascular blood flow measurement system
[5] Cherevko A A, Mikhaylova A V, Chupakhin A P, Ufimtseva I V, Krivoshapkin A L and Orlov K J, 2016 *Journal of Physics: Conference Series* Relaxation oscillation model of hemodynamic parameters in the cerebral vessels
[6] Butenin N V, Nejmark Yu I and Fufaev N L (ed) 1987 *Introduction in Nonlinear Oscillation Theory* (Moscow: Nauka)

Development and analysis of computational algorithm of the Maxwell's equations in flat domains

M A Boronina[1] and V A Vshivkov[2,3]

[1] Institute of Computational Mathematics and Mathematical Geophysics SB RAS, Novosibirsk, Russia
[2] Novosibirsk State University, Novosibirsk, Russia
[3] Lavrentyev Institute of Hydrodynamics SB RAS, Novosibirsk, Russia

E-mail: boronina@ssd.sscc.ru

Abstract. We present a new scheme for the Maxwell's equations computations in three-dimensional domains, where size in one direction is much smaller than the other sizes. The scheme is based on the Langdon-Lasinski scheme, which is standard for numerical experiments in plasma physics. Our study is devoted to analysis of correct wave propagation due to the effects of using a finite-difference approximation. To show the main dependencies we present numerical results in one-dimensional case. The results demonstrate, that the new scheme maintains the wave amplitude, the propagation speed and allows using of bigger time step in comparison with the Langdon-Lasinski scheme.

1. Introduction

The scheme of A. Langdon and B. Lasinski [1] is widely used in numerical experiments in plasma physics. The advantages of the scheme are the second order of accuracy in space and time and simplicity of its realization. However, the stability condition defines the maximal ratio of the time step to the spatial step. In cases of flat domains (where one domain size is much smaller than the other sizes), the time step depends on the minimal spatial step and leads to a bigger number of the time steps with the spatial grid refinement. The computations with the unnecessary small time step become longer and ineffective. For example, in problem of ultrarelativistic beam dynamics in supercolliders, the ratio of the beam sizes may achieve $1 : 200 : 60000$ [2] and the domain sizes depend on the beam sizes. A three-dimensional parallel algorithm with mixed decomposition and load-balancing makes it possible to perform numerical experiments with the beam sizes ratio $1 : 50 : 500$ [3].

A new scheme for the Maxwell's equations is presented. The scheme is based on the Langdon-Lasinski scheme. The weakened stability condition allows using the bigger time steps and decreasing of the computation time. The new scheme must be validated in terms of correct wave propagation: the numerical errors may change the solution significantly, in the same time it is impossible to distinguish the numerical errors from the physical effects [4]. The scheme effects for the simplest solutions must be studied to predict the scheme influence in real numerical experiments.

The implicit scheme was analyzed and compared with the Langdon-Lasinski scheme. It was shown that the new scheme maintains the wave amplitude, the propagation speed and allows using the bigger time step in comparison with the Langdon-Lasinski scheme. The implicit scheme may be applied in a variety of problems with the Maxwell's equations solutions in flat domains.

2. The scheme development and analysis

Let one consider the Maxwell's equations in dimensionless units:

$$rot\vec{E} = -\frac{\partial \vec{B}}{\partial t}$$
$$rot\vec{B} = \frac{\partial \vec{E}}{\partial t} + \vec{j}$$
$$div\vec{E} = \rho$$
$$div\vec{B} = 0$$

The Langdon-Lasinski scheme uses grids, which are shifted in half-step in time and space:

$$\frac{H^{m+1/2} - H^{m-1/2}}{\tau} = -rot_h E^m \tag{1}$$

$$\frac{E^{m+1} - E^m}{\tau} = j^{m+1/2} + rot_h H^{m+1/2}, \tag{2}$$

where

$$rot_h H = \begin{vmatrix} \frac{Hz_{i,k,l-1/2} - Hz_{i,k-1,l-1/2}}{h_y} - \frac{Hy_{i,k-1/2,l} - Hy_{i,k-1/2,l-1}}{h_z} \\ \frac{Hx_{i-1/2,k,l} - Hx_{i-1/2,k,l-1}}{h_z} - \frac{Hz_{i,k,l-1/2} - Hz_{i-1,k,l-1/2}}{h_x} \\ \frac{Hy_{i,k-1/2,l} - Hy_{i-1,k-1/2,l}}{h_x} - \frac{Hx_{i-1/2,k,l} - Hx_{i-1/2,k-1,l}}{h_y} \end{vmatrix},$$

$$rot_h E = \begin{vmatrix} \frac{Ez_{i-1/2,k+1/2,l} - Ez_{i-1/2,k-1/2,l}}{h_y} - \frac{Ey_{i-1/2,k,l+1/2} - Ey_{i-1/2,k,l-1/2}}{h_z} \\ \frac{Ex_{i,k-1/2,l+1/2} - Ex_{i,k-1/2,l-1/2}}{h_z} - \frac{Ez_{i+1/2,k-1/2,l} - Ez_{i-1/2,k-1/2,l}}{h_x} \\ \frac{Ey_{i+1/2,k,l-1/2} - Ey_{i-1/2,k,l-1/2}}{h_x} - \frac{Ex_{i,k+1/2,l-1/2} - Ex_{i,k-1/2,l-1/2}}{h_y} \end{vmatrix}.$$

The scheme may be reduced to the following:

$$\frac{y_i^{m+1} - 2y_i^m + y_i^{m-1}}{\tau^2} = \frac{y_{i+1}^m - 2y_i^m + y_{i-1}^m}{h^2}, \tag{3}$$

where y is any of the electric or magnetic field components.

The stability condition for scheme (1-2) $\tau/h1 \leq 1$ provides the amplitude conservation.

Let one consider scheme [5], which is implicit in the direction of the smallest domain size:

$$\frac{y_i^{m+1} - 2y_i^m + y_i^{m-1}}{\tau^2} = \delta\Delta_{xx}y_i^{m+1} + (1 - 2\delta)\Delta_{xx}y_i^m + \delta\Delta_{xx}y_i^{m-1} \tag{4}$$

where

$$\Delta_{xx}y_i^m = \frac{y_{i+1}^m - 2y_i^m + y_{i-1}^m}{h^2},$$

and $0 \le \delta \le 1/2$. For $\delta = 0$ the original Langdon-Lasinski scheme is obtained. For $\delta \ge 1/4$, the scheme is absolutely stable and maintains the wave amplitude.

Considering the solution as $y_l^m = A\exp(-i(\omega tm - klh))$, one obtains the dispersion relations for the Langdon-Lasinski scheme:

$$\sin^2\frac{\omega\tau}{2} = \frac{\tau^2}{h^2}\sin^2\frac{kh}{2},$$

and for the second scheme:

$$\sin^2\frac{\omega\tau}{2} = \frac{\tau^2}{h^2}\frac{\sin^2\frac{kh}{2}}{1 + 4\delta\frac{\tau^2}{h^2}\sin^2\frac{kh}{2}}.$$

Having expressed ω through k and having used Taylor series for τ and h we obtain the relations for the wave speed respectively:

$$u_1 = \frac{\omega}{k} = 1 + k^2\left(\frac{\tau^2 - h^2}{24}\right) + O(h^3, \tau^3, h\tau^2, h^2\tau) \tag{5}$$

$$u_2 = \frac{\omega}{k} = 1 + k^2\left(\frac{\tau^2 - h^2}{24} + \frac{\delta\tau^2}{2}\right) + O(h^3, \tau^3, h\tau^2, h^2\tau) \tag{6}$$

The formula (5) for the Langdon-Lasinski scheme demonstrates the fact, that the numerical wave speed is always smaller than 1 in dimensionless units (which corresponds to the light speed in dimensional units), because $\tau/h \le 1$. The opposite conclusion is held for the second scheme (6) when $\tau > h/\sqrt{1+12\delta} \ge h/2$ the wave has the speed bigger then 1, which is not a physical effect. For both schemes the error arises with increasing of the wavenumber k. Thus, the high-frequency modes of numerical solutions must be described with a proper spatial grid. However, the difference between the two scheme errors is not big, and the new scheme is very promising.

The same idea can be used to obtain the corresponding schemes in a three-dimensional case. Let x be the direction of the smallest domain size. Then the Langdon-Lasinski scheme turns into

$$\frac{y^{m+1} - 2y^m + y^{m-1}}{\tau^2} = \Delta_{xx}y^m + \delta\Delta_{xx}y^{m-1} + \Delta_{yy}y^m + \Delta_{zz}y^m \tag{7}$$

and the second scheme turns into the following:

$$\frac{y^{m+1} - 2y^m + y^{m-1}}{\tau^2} = \delta\Delta_{xx}y^{m+1} + (1-2\delta)\Delta_{xx}y^m + \delta\Delta_{xx}y^{m-1} + \Delta_{yy}y^m + \Delta_{zz}y^m \tag{8}$$

The stability condition of the explicit scheme (7) is $\tau^2/h_x^2 + \tau^2/h_y^2 + \tau^2/h_z^2 \le 1$, while the stability condition of the second scheme (8), which contains the implicitness, does not depend on h_x for $\delta \ge 1/4$: $\tau^2/h_y^2 + \tau^2/h_z^2 \le 1$. The symmetrical scheme ($\delta = 1/4$) provides the second order of accuracy in time, and both schemes (7,8) have the second order of approximation in space and time.

To create a new scheme for the Maxwell's equations, one applies the changes of the magnetic field in the original scheme (1,2) using $\delta = 1/4$ [6]:

$$\begin{aligned}
&\frac{Hx^{m+1/2}-Hx^{m-1/2}}{\tau}=\Delta_z Ey^m-\Delta_y Ez^m\\
&\frac{Hy^{m+1/2}-2Hy^{m-1/2}+Hy^{m-3/2}}{\tau^2}=\Delta_{yy}Hy^{m-1/2}+\Delta_{zz}Hy^{m-1/2}+\\
&+\frac{1}{4}\Delta_{xx}Hy^{m+1/2}+\frac{1}{2}\Delta_{xx}Hy^{m-1/2}+\frac{1}{4}\Delta_{xx}Hy^{m-3/2}\\
&\frac{Hz^{m+1/2}-2Hz^{m-1/2}+Hz^{m-3/2}}{\tau^2}=\Delta_{yy}Hz^{m-1/2}+\Delta_{zz}Hz^{m-1/2}+\\
&+\frac{1}{4}\Delta_{xx}Hz^{m+1/2}+\frac{1}{2}\Delta_{xx}Hz^{m-1/2}+\frac{1}{4}\Delta_{xx}Hz^{m-3/2}\\
&\frac{Ex^{m+1}-Ex^m}{\tau}=\Delta_y Hz^{m+1/2}-\Delta_z Hy^{m+1/2}-jx^{m+1/2}\\
&\frac{Ey^{m+1}-Ey^m}{\tau}=\Delta_z Hx^{m+1/2}-\Delta_x Hz^{m+1/2}-jy^{m+1/2}\\
&\frac{Ez^{m+1}-Ez^m}{\tau}=\Delta_x Hy^{m+1/2}-\Delta_y Hx^{m+1/2}-jz^{m+1/2}
\end{aligned} \tag{9}$$

The Gauss laws in this case for the dimensionless units are following:

$$\frac{1}{4}\Delta_x Hx^{m+1/2}+\frac{1}{2}Hx^{m-1/2}+\frac{1}{4}Hx^{m-3/2}+\Delta_y Hy^{m-1/2}+\Delta_z Hz^{m-1/2}=0$$

for $divH=0$ and

$$\Delta_x Ex^m+\Delta_y Ey^m+\Delta_z Ez^m=\rho^m$$

for $divE=\rho$.

Another scheme, based on the changes of electric the field computation

$$\begin{aligned}
&\frac{Hx^{m+1/2}-Hx^{m-1/2}}{\tau}=\Delta_z Ey^m-\Delta_y Ez^m\\
&\frac{Hy^{m+1/2}-Hy^{m-1/2}}{\tau}=\Delta_x Ez^m-\Delta_z Ex^m\\
&\frac{Hz^{m+1/2}-Hz^{m-1/2}}{\tau}=\Delta_y Ex^m-\Delta_x Ey^m\\
&\frac{Ex^{m+1}-Ex^m}{\tau}=\Delta_y Hz^{m+1/2}-\Delta_z Hy^{m+1/2}-jx^{m+1/2}\\
&\frac{Ey^{m+1}-2Ey^m+Ey^{m-1}}{\tau^2}=\frac{1}{4}\Delta_{xx}Ey^{m+1}+\frac{1}{2}\Delta_{xx}Ey^m+\frac{1}{4}\Delta_{xx}Ey^{m-1}+\\
&+\Delta_{yy}Ey^m+\Delta_{zz}Ey^m-\Delta_y\rho^m-\frac{jy^{m+1/2}-jy^{m-1/2}}{\tau}\\
&\frac{Ez^{m+1}-2Ez^m+Ez^{m-1}}{\tau^2}=\frac{1}{4}\Delta_{xx}Ez^{m+1}+\frac{1}{2}\Delta_{xx}Ez^m+\frac{1}{4}\Delta_{xx}Ez^{m-1}+\\
&+\Delta_{yy}Ez^m+\Delta_{zz}Ez^m-\Delta_z\rho^m-\frac{jz^{m+1/2}-jz^{m-1/2}}{\tau}
\end{aligned} \tag{10}$$

with corresponding Gauss laws

$$\Delta_x Hx^{m+1/2} + \Delta_y Hy^{m+1/2} + \Delta_z Hz^{m+1/2} = 0$$

for $divH = 0$ and

$$\frac{1}{4}\Delta_x Ex^{m+1} + \frac{1}{2}Ex^m + \frac{1}{4}Ex^{m-1} + \Delta_y Ey^m + \Delta_z Ez^m = \rho^m$$

for $divE = \rho$ is advantageous for the cases with boundary conditions for electric field. For example, cycle of the original beam dynamics algorithm [7] consists of the following computational steps:

1. Magnetic field for time moment m: $\frac{H^m - H^{m-1/2}}{\tau/2} = -rot_h E^m$.
2. Particle motion.
3. Magnetic field for time moment $m + 1/2$: $\frac{H^{m+1/2} - H^{m-1/2}}{\tau} = -rot_h E^m$.
4. Boundary conditions for the electric field.
5. The electric field in the whole domain.

The two new equations in finite differences on the step 5 may be sold with tridiagonal matrix algorithm. This method requires O(n) times more operations in comparison with the Langdon-Lasinski scheme. Besides, the three-dimensional array of the currents $j^{m-1/2}$ on the previous times step must be stored. The Langdon-Lasinski doesn't require the density computation, but the array is needed to calculate the boundary conditions for the electric field [8]. In the case of ultrarelativistic beam dynamics code, the main group processors compute the density arrays and send the corresponding contributions into the electric field. The interprocessor communications increase significantly the total computation time. It is effective the electric field grid to have 2 additional nodes more in each direction, then the magnetic field can be computed as the electric field derivative, and there is no need to set the boundary conditions for the magnetic field. The new scheme will allow performing the numerical experiments with bigger time step and thus with more seldom computations of the boundary conditions in comparison with the original scheme. The application of the implicit scheme makes possible significant decreasing of the calculation time, especially in cases of computationally difficult time steps.

3. Numerical results

The basic study requires the consideration of the one-dimensional case. The solution of flat wave $E_z(x,t) = sin(kx - \omega t)$ in domain $[0 : L], L = 0.01$ without charges was studied. The boundary conditions are considered as periodic ones. The cyclic tridiagonal matrix algorithm was used to apply the implicit scheme. The solution in finite differences was taken as $Ez_i^m = sin\left(\frac{2\pi}{L}\left(k(i-1.5)h - \omega m\tau\right)\right)$. In the dimensionless units $c = 1$. The numerical wave speed was calculated with second order of accuracy as the speed of the solution root point. One considers the error $\epsilon_{num} = \omega/k - 1$ for the numerical solution, $\epsilon_{theor_1} = k^2(h^2 - \tau^2)/24$ for the original Langdon-Lasinski scheme (1-2,5) and $\epsilon_{theor_2} = k^2(h^2 + 2\tau^2)/24$ (10,6) for the new scheme with $\delta = 1/4$.

Figure 1 demonstrates the errors ϵ_{num_1} with color lines and ϵ_{th_1} with thin horizontal lines for k=10 and different steps with $\tau/h = 0.4$. The red line corresponds to $h = 2 \cdot 10^{-4}$, the blue line to $h = 10^{-4}$, the black line to $h = 5 \cdot 10^{-5}$. The same notations are used for figure 2, where the errors ϵ_{num_2} and ϵ_{th_2} are presented. For $k = 1$ with the same steps the errors of both schemes do not exceed 0.001, for $k = 10$ and $\tau/h = 0.4$ the errors are higher. The numerical and theoretical errors are in good coincidence and decrease with the step decreasing. The differences between the two schemes are low.

When $k = 100$, the errors for both schemes become extremely high. The minimal spatial resolution for description of the wave is 400 nodes (that is $h = 2.5 \cdot 10^{-5}$). Figure 3 and figure

Journal of Physics: Conference Series **722** (2016) 012006 doi:10.1088/1742-6596/722/1/012006

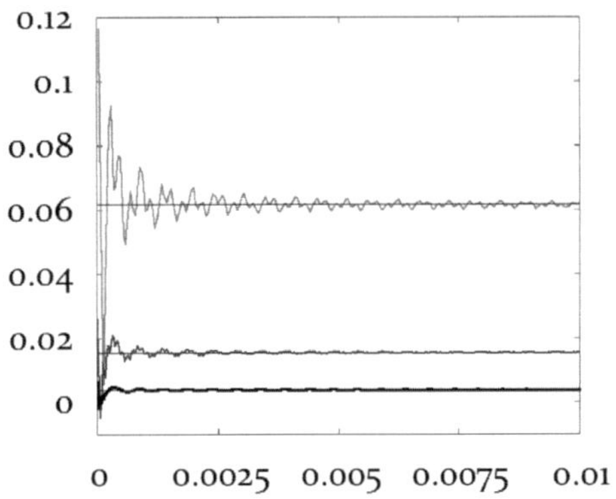

Figure 1. The errors ϵ_{num_1} and ϵ_{theor_1} for different step values, k=10.

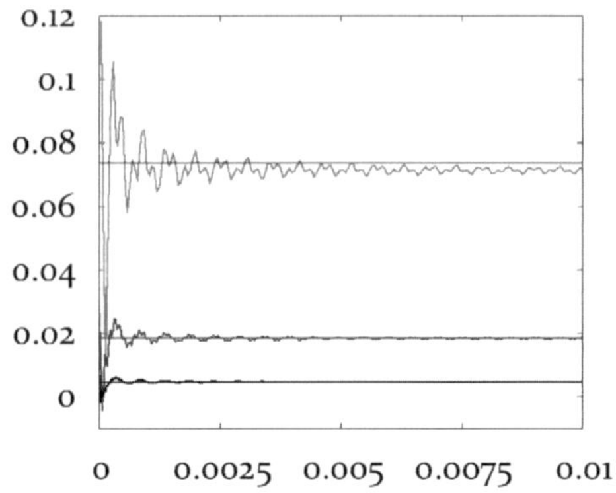

Figure 2. The errors ϵ_{num_2} and ϵ_{theor_2} for different step values, k=10.

4 demonstrate the numerical and theoretical error behaviour for $k = 100$ and $\tau/h = 0.4$ for the explicit and implicit schemes correspondingly. The red lines denote $h = 5 \cdot 10^{-5}$, the blue lines denote $h = 1.25 \cdot 10^{-5}$, the black lines denote $h = 6.25 \cdot 10^{-6}$. The second order of accuracy leads to the error decrease with the spatial step refinement. The differences between the two schemes are also low.

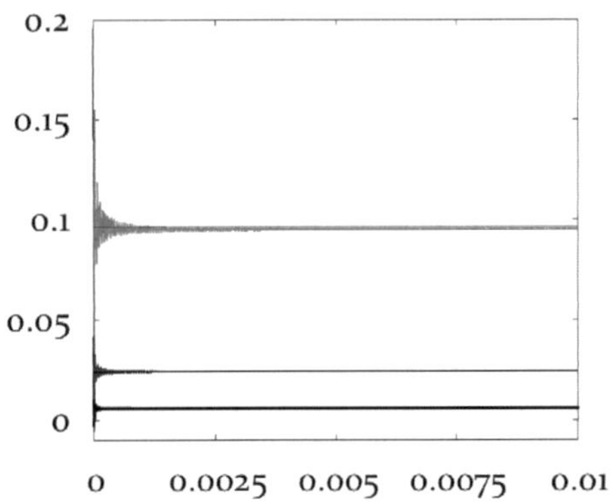

Figure 3. The errors ϵ_{num_1} and ϵ_{theor_1} for different step values, k=100.

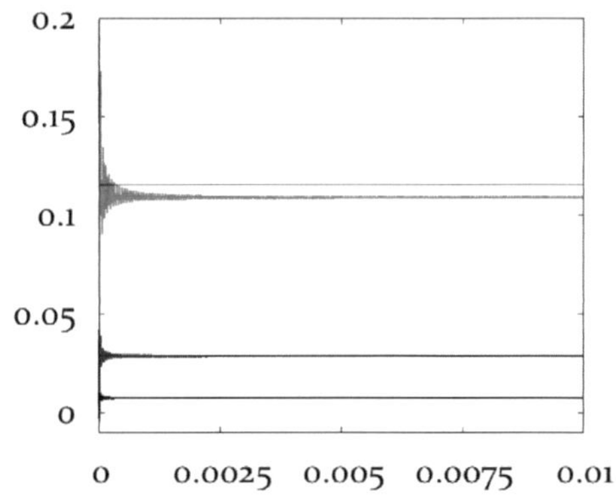

Figure 4. The errors ϵ_{num_2} and ϵ_{theor_2} for different step values, k=100.

Figure 5 demonstrates the analytical solution (black line), the numerical solution with the original scheme (blue line), and the solution with the new scheme (red line) after 6400 time steps with $k = 100, h = 6.25 \cdot 10^{-6}, \tau/h = 0.4$. One can see that the numerical solution accumulates the errors, and although the amplitude is precise enough, the numerical results are far away from the analytical solution. However, the numerical solutions obtained with the two schemes coincide the better, the smaller is the spatial step.

The advantage of the implicit scheme is a possibility to perform calculations with $\tau/h > 1$. This fact is demonstrated on figure 6, where the ϵ_{num_2} with $k = 10$ for $h = 2.5 \cdot 10^{-5}, \tau/h = 2$ (red line) $\tau/h = 4$ (black line) are shown.

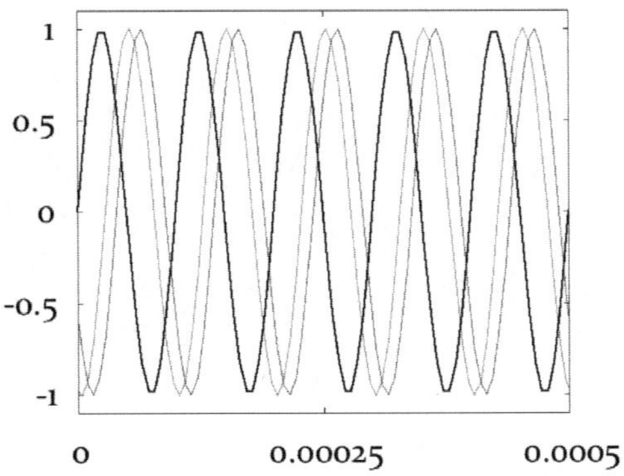

Figure 5. The numerical solutions with considered schemes and the analytical solution, k=100.

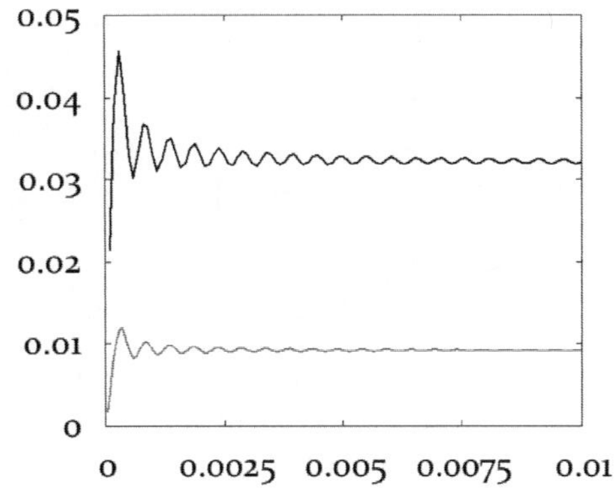

Figure 6. The errors ϵ_{num_2} for different Courant numbers, k=10.

4. Conclusion

The new scheme for the Maxwell's equations computations in three-dimensional flat domains is presented. The scheme is based on the Langdon-Lasinski scheme. The theoretical and numerical analyses of the new scheme demonstrate the maintenance of the wave amplitude and the propagation speed with the second order of accuracy. The error increases with the increase of the wave number for both schemes, and the spatial step in numerical experiments must be taken appropriately small to describe correctly the high-frequency modes. The difference between the numerical errors of the two schemes is low, but the main advantage of the new scheme is a possibility to use a bigger time step in comparison with the Langdon-Lasinski scheme, which is standard for numerical experiments in plasma physics.

Acknowledgments

The theoretical part of the work was done under Russian Science Foundation grant No. 14-11-00485 "High performance methods and technologies of modelling the electrophysical processes and devices". The numerical experiments were performed with support of Russian Foundation for Basic Research grants No. 16-31-00301 "Development of efficient parallel algorithms for numerical simulation of ultrarelativistic beams of charged particles", No. 16-01-00209 "Computer simulation unstable modes of interaction relativistic electron bunches with plasma in the CNF facilities", No. 14-01-00392 "Development of the effective parallel algorithms for modeling the processes in the relativistic plasma physics", and No. 14-07-00241 "Template implementation of Particle-In-Cell method for solution of plasma physics problems with hybrid supercomputers" .

References

[1] Langdon A and Lasinski B 1976 *Electromagnetic and relativistic plasma simulation models Methods in Computational Physics* vol 16 (New York: Academic Press) p 327
[2] Adolphsen C et al. 2013 *The International Linear Collider Technical Design Report - Volume 3.II: Accelerator Baseline Design*
[3] Boronina M and Vshivkov V 2014 *Proc. of IPAC14* (Dresden) pp 439–41
[4] Boronina M and Vshivkov V 2015 *J. of Plasma Phys.* **81** 06 495810605
[5] Samarskiy A and Gulin A 1989 *Numerical methods* (Moscow: Nauka)
[6] Boronina M and Vshivkov V 2016 *J. Phys.: Conf. Series* **681** 012032
[7] Boronina M and Korneev V 2013 *Bull. Nov. Comp. Center, Num. Anal.* **16** 15–22

[8] Boronina M and Vshivkov V 2012 Three-dimensional simulation of ultrarelativistic charged beams dynamics: study of initial and boundary conditions *Matem. Mod.* **24**(2) 67–83

Journal of Physics: Conference Series **722** (2016) 012007 doi:10.1088/1742-6596/722/1/012007

On solvability of inverse coefficient problems for nonlinear convection–diffusion–reaction equation

R V Brizitskii[1,2] and Zh Yu Saritskaya[2]

[1] Institute of Applied Mathematics FEB RAS, Vladivostok, Russia
[2] Far Eastern Federal University, Vladivostok, Russia

E-mail: `mlnwizard@mail.ru`, `zhsar@icloud.com`

Abstract. An inverse coefficient problem is considered for a stationary nonlinear convection–diffusion–reaction equation, in which reaction coefficient has a rather common dependence on substance concentration and on spacial variable. The solvability of the considered nonlinear boundary value problem is proved in a general case. The existence of solutions of the inverse problem is proved for the reaction coefficients, which are defined by the product of two functions. The first function depends on a spatial variable, the second one depends nonlinearly on the solution of the boundary value problem. The mentioned inverse problem consists in reconstructing the first function with the help of additional information provided by the solution of the boundary value problem.

1. Introduction. Solvability of the boundary value problem

Inverse problems research for linear and nonlinear heat-and-mass transfer models has been an urgent issue during a long period of time. One of the main problems is an identification problem of unknown densities of boundary and distributed sources or of models differential equations coefficients or boundary conditions with the help of additional information about systems conditions, which is described by the model. The studying of such inverse problems can be reduced to the studying of corresponding extremal problems using the optimization method [1]. A number of papers is dedicated to the description of this method for heat-and-mass transfer and other hydrodynamic models [2]–[10].

In this paper an inverse problem is studied in the case when the recovering reaction coefficient depends nonlinearly on both the boundary value problems solution and the spatial variable. The last one generalizes the statements of inverse coefficient problems from [5]–[9], where coefficients are searched in case of spatial variable dependence, and also of identification problem studied in [11, 12], where reaction coefficients depending only on solution were examined and at the same time the right part of the equation was recovered.

In a bounded domain $\Omega \subset \mathbf{R}^3$ with the boundary Γ the following boundary value problem is considered

$$-\lambda\Delta\varphi + \mathbf{u}\cdot\nabla\varphi + \tilde{k}(\varphi,\mathbf{x})\varphi = f \text{ in } \Omega, \quad \varphi = 0 \text{ on } \Gamma. \tag{1}$$

Here function φ means polluting substance concentration, $\mathbf{u}$ is a given vector of velocity, f is a volume density of external sources of substance, λ is a constant diffusion coefficient, function $\tilde{k}(\varphi,\mathbf{x})$, where $\mathbf{x}\in\Omega$, is a reaction coefficient. This problem (1) will be called problem 1 below.

In this paper firstly a global solvability and local uniqueness of the problem 1 solution is proved, when the reaction coefficient depends on either substance concentration or spatial variables and functions belonging to a rather wide class. Then an identification problem is formulated for the reaction coefficients in the form of $\tilde{k} = \beta(\mathbf{x})k(\varphi)$, which consists of function $\beta(\mathbf{x})$ recovered using the measured substance concentration in a subdomain $Q \subset \Omega$. At that, the conditions on functions $\beta(\mathbf{x})$ and $k(\varphi)$ are also rather general. It should be mentioned that the multiplicative structure of the reaction coefficient gives an opportunity to take into account a rather wide-spread dependence of a reaction velocity on a concentration of a reacting substance from the one side, and from the other one allows us to take into consideration the effect of different types of chemical reaction behavior in the considered domain.

While studying problem 1 and optimal control problems, Sobolev spaces will be used: $H^s(D)$, $s \in \mathbf{R}$ $L^r(D)$, $1 \le r \le \infty$, where D is either a domain Ω or its boundary Γ. Inner products in $L^2(\Omega)$ and $H^1(\Omega)$ are denoted by $(\cdot,\cdot)$ $(\cdot,\cdot)_1$, inner products in $L^2(\Gamma)$ is denoted by $(\cdot,\cdot)_\Gamma$, norm in $L^2(\Omega)$ is denoted by $\|\cdot\|_\Omega$, norm or semi-norm in $H^1(\Omega)$ is denoted by $\|\cdot\|_{1,\Omega}$ or $|\cdot|_{1,\Omega}$. Also let $\mathbf{Z} = \{\mathbf{v} \in \mathbf{L}^4(\Omega) : \operatorname{div} \mathbf{v} = 0 \text{ in } \Omega\}$.

It will be assumed that the following conditions hold:

(i) Ω is a bounded domain in $\mathbf{R}^3$ with boundary $\Gamma \in C^{0,1}$;

(ii) $\mathbf{u} \in \mathbf{Z}$, $f \in L^2(\Omega)$;

In the papers [11, 12] the reaction coefficient $k(\varphi)$ was considered in case, when it was not a function of φ in ordinary sense, as in [13], but was denoted as an operator $k(\varphi) : H^1(\Omega) \to L^p_+(\Omega)$, where $p \ge 3/2$, such as for any $w_1, w_2 \in B_r = \{w \in H^1(\Omega) : \|w\|_{1,\Omega} \le r\}$ an estimate takes place:

$$\|k(w_1) - k(w_2)\|_{L^p(\Omega)} \le L\|w_1 - w_2\|_{L^4(\Omega)}, \qquad (2)$$

were L is a constant, depending on r, but not depending on w_1, w_2.

Let us mention that power functions $k(w) = w^2$ and $k(w) = w^2|w|$ also satisfy the inequality (2). Indeed, for the function $k(w) = w^2|w|$ from [14, 15] the estimate is true:

$$\|k(w_1) - k(w_2)\|^{3/2}_{L^{3/2}(\Omega)} = \int_\Omega (w_1^2|w_1| - w_2^2|w_2|)^{3/2} d\Omega \le \int_\Omega |w_1 - w_2|^{3/2}(w_1^2 + |w_1 w_2| + w_2^2)^{3/2} d\Omega \le$$

$$\le \|w_1 + w_2\|^3_{L^6(\Omega)}\|w_1 - w_2\|^{3/2}_{L^3(\Omega)} \le (2C_6 r)^3\|w_1 - w_2\|^{3/2}_{L^3(\Omega)} \le (2C_6 r)^3\|w_1 - w_2\|^{3/2}_{L^4(\Omega)},$$

which means that this function satisfy the inequality (2). For $k(w) = w^2$ the reasoning is the same.

Further let us make an example of the operator $k(\varphi)$, satisfying (2) and not being an ordinary function of φ: $k(\varphi) = \varphi^2$ in a subdomain $Q \subset \Omega$ and $k(\varphi) = k_0$ in $\Omega \setminus \overline{Q}$, where $k_0 \in L^{3/2}_+(\Omega)$.

In the present paper the reaction coefficient dependence on either problem 1 solution or the spatial variable $\mathbf{x} \in \Omega$ is emphasized. Let us suppose that it is true for the reaction coefficient $\tilde{k}(w, \mathbf{x})$:

(iii) for any $w_1, w_2 \in B_r = \{w \in H^1(\Omega) : \|w\|_{1,\Omega} \le r\}$ the estimate takes place:

$$\|k(w_1, \mathbf{x}) - k(w_2, \mathbf{x})\|_{L^p(\Omega)} \le L\|w_1 - w_2\|_{L^4(\Omega)}, \quad \mathbf{x} \in \Omega,$$

where L is a constant, which depends on r, but does not depend on w_1, w_2.

Let us note that in the condition (iii) the constant L can depend not only on r. For example, the constant L also depends on $\|\beta_1(\mathbf{x})\|_{H^2(\Omega)}$ for the reaction coefficient $\tilde{k}(\varphi, \mathbf{x}) = \beta_1(\mathbf{x})k(\varphi) + \beta_2(\mathbf{x})$, where $\beta_1(\mathbf{x}) \in H^2_+(\Omega)$, $\beta_2(\mathbf{x}) \in L^2_+(\Omega)$, and the operator $k(\varphi)$ was denoted above.

Let $\mathcal{D}(\Omega)$ be the space of infinitely differentiable, finite functions in Ω, $H^1_0(\Omega)$ be the closure of $\mathcal{D}(\Omega)$ in $H^1(\Omega)$, $H^{-1}(\Omega) = H^1_0(\Omega)^*$. The Poincare-Friedrichs inequality holds:

$|\varphi|^2_{1,\Omega} \geq \delta\|\varphi\|^2_{1,\Omega}$ $\forall\varphi \in H^1_0(\Omega)$, where $\delta = \text{const} > 0$, $L^p_+(\Omega) = \{k \in L^p(\Omega) : k \geq 0\}$, $p \geq 3/2$, $H^s_+(\Omega) = \{h \in H^s(\Omega) : h \geq 0\}$, $s \geq 0$.

Let us also remind that on the strength of Sobolev embedding theorem the space $H^1(\Omega)$ embeds in the space $L^s(\Omega)$ continuously at $s \leq 6$ and compactly at $s < 6$, and the estimate takes place with some constant C_s, depending on s and Ω

$$\|\varphi\|_{L^s(\Omega)} \leq C_s\|\varphi\|_{1,\Omega} \ \ \forall\varphi \in H^1(\Omega). \tag{3}$$

The following Green's theorem will be used (see [10, p. 128] for more details about it):

$$(\Delta\varphi, \eta) = -(\nabla\varphi, \nabla\eta) + \langle\partial\varphi/\partial n, \eta\rangle_\Gamma \ \ \forall\varphi \in H^1(\Delta, \Omega),\ \eta \in H^1(\Omega), \tag{4}$$

$$(\mathbf{u}\cdot\nabla\varphi, \eta) = -(\mathbf{u}\cdot\nabla\eta, \varphi) \ \ \forall\mathbf{u} \in \mathbf{Z},\ \varphi, \eta \in H^1_0(\Omega). \tag{5}$$

Here and below $\langle\cdot,\cdot\rangle_\Gamma$ means the relation of duality between $H^{1/2}(\Gamma)$ and $H^{-1/2}(\Gamma)$.

For any function $k_0 \in L^p_+(\Omega)$, $p > 1$ Holder's inequality for three functions holds:

$$|(k_0\varphi, \eta)| \equiv |\int_\Omega k_0\,\varphi\,\eta d\mathbf{x}| \leq \|k_0\|_{L^p(\Omega)}\|\varphi\|_{L^q(\Omega)}\|\eta\|_{L^r(\Omega)},\ \ \frac{1}{p}+\frac{1}{q}+\frac{1}{r} = 1,\ q > 1,\ r > 1. \tag{6}$$

Let us denote bilinear forms $a_\mathbf{u}$ and a_0: $H^1_0(\Omega) \times H^1_0(\Omega) \to \mathbf{R}$ by formulae

$$a_\mathbf{u}(\varphi, \eta) = (\mathbf{u}\cdot\nabla\varphi, \eta) = \int_\Omega(\mathbf{u}\cdot\nabla\varphi)\eta d\mathbf{x},\ a_0(\varphi, \eta) = \lambda(\nabla\varphi, \nabla\eta) + a_\mathbf{u}(\varphi, \eta) + (k_0\varphi, \eta). \tag{7}$$

Using (3), (5), (6), it is not difficult to check that when conditions (i), (ii) and $k_0 \in L^p_+(\Omega), p \geq 3/2$ hold, then the forms, introduced in (7), are continuous and

$$|a_\mathbf{u}(\varphi, \eta)| \leq |(\mathbf{u}\cdot\nabla\varphi, \eta)| \leq \gamma_1\|\mathbf{u}\|_{\mathbf{L}^4(\Omega)}\|\varphi\|_{1,\Omega}\|\eta\|_{1,\Omega} \ \ \forall(\varphi, \eta) \in H^1_0(\Omega)^2, \tag{8}$$

$$|(k_0\varphi, \eta)| \leq \gamma_p\|k_0\|_{L^p(\Omega}\|\varphi\|_{1,\Omega}\|\eta\|_{1,\Omega} \ \ \forall(\varphi, \eta) \in H^1_0(\Omega)^2, \tag{9}$$

$$a_\mathbf{u}(\varphi, \varphi) = 0,\ a_0(\varphi, \varphi) = \lambda(\nabla\varphi, \nabla\varphi) + (k_0\varphi, \varphi) \geq \lambda_*\|\varphi\|^2_{1,\Omega} \ \ \forall\varphi \in H^1_0(\Omega),\ \lambda_* \equiv \delta\lambda. \tag{10}$$

Here γ_1 or γ_p are some constants, which depend on Ω or on Ω and p correspondingly. From (8)–(10) it follows that for any function $k_0 \in L^p_+(\Omega), p \geq 3/2$ the form $a_0(\cdot,\cdot)$ is continuous and coercitive with the constant λ_* on $H^1_0(\Omega)$. In its turn, this estimate results from (9) and (iii):

$$|((k(\varphi_1, \mathbf{x}) - k(\varphi_2, \mathbf{x}))\varphi, \eta)| \leq \gamma_p L\|\varphi_1 - \varphi_2\|_{L^4(\Omega)}\|\varphi\|_{1,\Omega}\|\eta\|_{1,\Omega} \ \forall\varphi_1, \varphi_2 \in B_r,\ \varphi, \eta \in H^1_0(\Omega). \tag{11}$$

Let us multiply the equation in (1) by $h \in H^1_0(\Omega)$ and integrate over Ω, applying the formula (4). Then this will be obtained:

$$\lambda(\nabla\varphi, \nabla h) + (k(\varphi, \mathbf{x})\varphi, h) + (\mathbf{u}\cdot\nabla\varphi, h) = (f, h) \ \ \forall h \in H^1_0(\Omega). \tag{12}$$

The function $\varphi \in H^1_0(\Omega)$, which satisfies (12), will be called a weak solution of problem 1.

The proof of problem (12) solvability will be conducted with the help of Schauder fixed point theorem according to the scheme, which is described in [10] for nonlinear hydrodynamic models. For this purpose let us define a mapping $G: H^1_0(\Omega) \to H^1_0(\Omega)$, which sticks with formula $G(w) = \varphi$ for $w \in H^1_0(\Omega)$. Here function $\varphi \in H^1_0(\Omega)$ is the solution of the linear problem

$$a_w(\varphi, h) = \lambda(\nabla\varphi, \nabla h) + (k(w, \mathbf{x})\varphi, h) + (\mathbf{u}\cdot\nabla\varphi, h) = (f, h) \ \forall h \in H^1_0(\Omega). \tag{13}$$

Since $k(w, \mathbf{x}) \in L^p_+(\Omega)$, $p \geq 3/2$ for any $w \in H^1(\Omega)$ on the strength of conditions (iii), then according to (10) the form $a_w : H^1_0(\Omega) \times H^1_0(\Omega) \to \mathbf{R}$, introduced in (13), is continuous and coercitive with the constant $\lambda_* = \delta\lambda$, and the problem (13) has a unique solution $\varphi \in H_0^1(\Omega)$, for which the estimate holds

$$\|\varphi\|_{1,\Omega} \le M_\varphi \equiv C_* \|f\|_\Omega. \tag{14}$$

Let us inject in the space $H_0^1(\Omega)$ the sphere $B_r = \{w \in H_0^1(\Omega) : \|w\|_{1,\Omega} \le r\}$, where $r = M_\varphi$. From the construction of the sphere B_r and from (14) follows that the operator G, which was defined above, is a mapping of B_r into itself. Let us prove that G is continuous and compact on B_r. Let $\{w_n\}_{n=1}^\infty$ be a random sequence from B_r. On the strength of reflexivity of the space $H_0^1(\Omega)$ and of embedding compactness $H_0^1(\Omega) \subset L^4(\Omega)$ there is a subsequence of the subsequence $\{w_n\}_{n=1}^\infty$, which can be again denoted by $\{w_n\}_{n=1}^\infty$, and there is also a function $w \in B_r$ such as $w_n \to w$ weakly in $H^1(\Omega)$, $w_n \to w$ strongly in $L^4(\Omega)$ at $n \to \infty$. Let $\varphi_n = G(w_n)$, $\varphi = G(w)$. These equalities mean that $\varphi \in H_0^1(\Omega)$ is the solution of problem (13) and $\varphi_n \in H_0^1(\Omega)$ is the solution of problem

$$\lambda(\nabla\varphi_n, \nabla h) + (k(w_n, \mathbf{x})\varphi_n, h) + (\mathbf{u} \cdot \nabla\varphi_n, h) = (f, h) \ \ \forall h \in H_0^1(\Omega), \tag{15}$$

which is obtained from (13) by substituting w for w_n, φ for φ_n. Let us demonstrate that $\varphi_n \to \varphi$ strongly in $H^1(\Omega)$ at $n \to \infty$. For that purpose let us subtract (13) from (15). Taking into consideration that $(k(w_n, \mathbf{x})\varphi_n, h) - (k(w, \mathbf{x})\varphi, h) = (k(w, \mathbf{x})(\varphi_n - \varphi), h) + ((k(w_n, \mathbf{x}) - k(w, \mathbf{x}))\varphi_n, h)$, the following will be got

$$\lambda(\nabla(\varphi_n - \varphi), \nabla h) + (k(w, \mathbf{x})(\varphi_n - \varphi), h) + (\mathbf{u} \cdot \nabla(\varphi_n - \varphi), h) =$$

$$= -((k(w_n, \mathbf{x}) - k(w, \mathbf{x}))\varphi_n, h) \ \ \forall h \in H_0^1(\Omega). \tag{16}$$

Using the estimate (11) at $\varphi_1 = w_n$, $\varphi_2 = w$, $\varphi = \varphi_n$ and the estimate $\|\varphi_n\|_{1,\Omega} \le M_\varphi$ for $n = 1, 2...$, resulting from (14), it can be obtained that

$$|((k(w_n, \mathbf{x}) - k(w, \mathbf{x}))\varphi_n, h)| \le \gamma_p L M_\varphi \|w_n - w\|_{L^4(\Omega)} \|h\|_{1,\Omega} \to 0 \ \ n \to \infty \ \forall h \in H_0^1(\Omega). \tag{17}$$

Then from (16) on the strength of (17) and (10) follows that $\|\varphi_n - \varphi\|_{1,\Omega} \to 0$ at $n \to \infty$. This implies continuity and compactness of the operator G. In such case it follows from the Schauder fixed point theorem that the operator G has a fixed point $\varphi = G(\varphi) \in H_0^1(\Omega)$. That is the point φ, which is the searched solution of problem (12).

Let us determine the sufficient conditions for the uniqueness of problem (12) solution. Let $\varphi_1, \varphi_2 \in H_0^1(\Omega)$ be two solutions of problem (12), for which the estimate (14) takes place. Obviously, the difference $\varphi = \varphi_1 - \varphi_2$ satisfies the relation

$$\lambda(\nabla\varphi, \nabla h) + (k(\varphi_2, \mathbf{x})\varphi, h) + (\mathbf{u} \cdot \nabla\varphi, h) = -((k(\varphi_1, \mathbf{x}) - k(\varphi_2, \mathbf{x}))\varphi_1, h) \ \ \forall h \in H_0^1(\Omega). \tag{18}$$

Reasoning like above, when the estimate (17) was obtained, it can be shown that

$$|((k(\varphi_1, \mathbf{x}) - k(\varphi_2, \mathbf{x}))\varphi_1, h)| \le \gamma_p C_4 L M_\varphi \|\varphi_1 - \varphi_2\|_{1,\Omega} \|h\|_{1,\Omega}. \tag{19}$$

Here C_4 is a constant, which is included in (3) at $s = 4$. As on the strength of (iii) $k(\varphi_2, \mathbf{x}) \in L_+^p(\Omega)$, then it follows from (10) subject to (19) and from the equality $M_\varphi = C_* \|f\|_\Omega$ that $\|\varphi\|_{1,\Omega} \le C_*^2 \gamma_p C_4 L \|f\|_\Omega \|\varphi\|_{1,\Omega}$. Let the condition hold

$$\gamma_p C_4 L \|f\|_\Omega < \lambda_*^2 = C_*^{-2}. \tag{20}$$

Then from the previous inequality it follows that $\|\varphi\|_{1,\Omega} = 0$, and, accordingly, $\varphi_1 = \varphi_2$. Let us formulate the received results as the following theorem.

Theorem 1. *Let conditions (i)–(iii) hold. Then there is a weak solution $\varphi \in H_0^1(\Omega)$ of the problem 1 and the estimate (14) is met. If, besides, the condition (20) is satisfied, then the problem 1 weak solution is unique.*

While studying extremal problems, it will be assumed that the function $\tilde{k}(\varphi, \mathbf{x})$ satisfies the condition

(iv) $\tilde{k}(\varphi, \mathbf{x}) = \beta(\mathbf{x})k(\varphi)$, where $\beta(\mathbf{x}) \in H_+^1(\Omega)$, while the function $k(\varphi) \in L_+^2(\Omega)$ and on any sphere $B_r = \{\varphi \in H^1(\Omega) : \|\varphi\|_{1,\Omega} \le r\}$ the inequality holds

$$\|k(\varphi_1) - k(\varphi_2)\|_\Omega \le L\|\varphi_1 - \varphi_2\|_{L^4(\Omega)} \ \forall \varphi_1, \varphi_2 \in B_r. \tag{21}$$

Here L is a constant, depending on r, but not depending on $\varphi_1, \varphi_2 \in B_r$.

Conditions (iv) describe a particular type of function $\tilde{k}(\varphi, \mathbf{x})$, satisfying the conditions (iii). Indeed,

$$\|\beta(\mathbf{x})(k(\varphi_1) - k(\varphi_2))\|_{L^{3/2}(\Omega)} \le \|\beta(\mathbf{x})\|_{L^6(\Omega)}\|k(\varphi_1) - k(\varphi_2)\|_\Omega \le$$
$$\le L\|\beta(\mathbf{x})\|_{1,\Omega}\|\varphi_1 - \varphi_2\|_{L^4(\Omega)}.$$

It is clear that the function $\tilde{k}(\varphi, \mathbf{x}) = \beta(\mathbf{x})\varphi^2$ satisfies the condition (iv), because for any φ^2 the estimate (21) is true. It is interesting to mention that at $\tilde{k}(\varphi, \mathbf{x}) = \beta(\mathbf{x})\varphi^2$ there is a nonlocal (i.e. without smallness condition (20)) uniqueness of problem 1 solution. Actually, let $\varphi_1, \varphi_2 \in H_0^1(\Omega)$ be two solutions of the problem 1 at $\tilde{k}(\varphi, \mathbf{x}) = \beta(\mathbf{x})\varphi^2$. Then their difference $\varphi = \varphi_1 - \varphi_2$ satisfies the relation

$$\lambda(\nabla\varphi, \nabla h) + (\beta(\mathbf{x})(\varphi_1^2 + \varphi_1\varphi_2 + \varphi_2^2), \varphi h) + (\mathbf{u} \cdot \nabla\varphi, h) = 0 \ \forall h \in H_0^1(\Omega).$$

As $\varphi_1^2 + \varphi_1\varphi_2 + \varphi_2^2 \ge 0$ a.e. in Ω and $\beta(\mathbf{x}) \in H_+^1(\Omega)$, then it follows form (10) that $\varphi = 0$.

Let us note another possible kind of dependence $k(\varphi) = \varphi^2|\varphi|$, considered in [14, 15]. However, to deal with such coefficient, stricter conditions for the function $\beta(\mathbf{x})$ should be used, for example, $\beta(\mathbf{x}) \in H^2(\Omega)$.

2. Statement of the optimal control problem and its solvability

There are a lot of situations in practice, when some parameters of problem 1 are unknown and it is required to determine them together with the solution φ according to some information about the solution. On the capacity of the mentioned information about solution, the values $\varphi^d(\mathbf{x})$ of the concentration φ are often taken, which can be measured in points of some set $Q \subset \Omega$.

The case will be studied below, when the function $\beta(\mathbf{x})$ is unknown, should be searched together with the solution φ and is multiplicatively included in reaction coefficient $\tilde{k}(\varphi, \mathbf{x}) = \beta(\mathbf{x})k(\varphi)$, will be studied. To solve this inverse problem the optimization method will be used, according to which the considered problem is reduced to solving the control problem. In compliance with this method the whole set of initial data of problem 1 is divided into two groups: the group of fixed data, where functions $\mathbf{u}$, f and $k(\varphi)$ will be included, and the group of controls, which consists of function $\beta(\mathbf{x})$, assuming that it can be changed in some set K.

Let us introduce an operator $F : H_0^1(\Omega) \times K \to H^{-1}(\Omega)$ by formula

$$\langle F(\varphi, \beta), h\rangle = \lambda(\nabla\varphi, \nabla h) + (\beta(\mathbf{x})\, k(\varphi)\varphi, h) + (\mathbf{u} \cdot \nabla\varphi, h) - (f, h).$$

Then (12) can be rewritten in the following form:

$$F(\varphi, \beta) = 0. \tag{22}$$

Treating (22) as a conditional restriction on the state $\varphi \in H_0^1(\Omega)$ and on the control $\beta \in K$, the problem of conditional minimization can be formulated as follows:

$$J(\varphi, \beta) \equiv \frac{\mu_0}{2}I(\varphi) + \frac{\mu_1}{2}\|\beta\|_{1,\Omega}^2 \to \inf,\ F(\varphi, \beta) = 0,\ (\varphi, \beta) \in H_0^1(\Omega) \times K. \tag{23}$$

Here $I : H_0^1(\Omega) \to \mathbf{R}$ is a weakly semicontinuous below functional.

The set of possible pairs for problem (23) is denoted by $Z_{ad} = \{(\varphi, \beta) \in H_0^1(\Omega) \times K : F(\varphi, \beta) = 0, J(\varphi, \beta) < \infty\}$ and let us suppose that these conditions hold

(j) $K \subset H_+^1(\Omega)$ is a nonempty convex closed set;

(jj) $\mu_0 > 0$, $\mu_1 \geq 0$ and K is a bounded set or $\mu_0 > 0$, $\mu_1 > 0$ and functional I is bounded below.

The following cost functionals can be used in the capacity of the possible ones [10]:

$$I_1(\varphi) = \|\varphi - \varphi^d\|_Q^2 = \int_\Omega |\varphi - \varphi^d|^2 d\mathbf{x}, \quad I_2(\varphi) = \|\varphi - \varphi^d\|_{1,Q}^2. \tag{24}$$

Here $\varphi^d \in L^2(Q)$ (or $\varphi^d \in H^1(Q)$) is a given function in some subdomain $Q \subset \Omega$.

Theorem 2. *Let conditions (i), (ii), (iv) and (j), (jj) hold, $k(\varphi) \in L^3(\Omega)$, $I : H_0^1(\Omega) \to \mathbf{R}$ is a weakly semicontinuous below functional, and Z_{ad} is nonempty set. Then there is at least one solution $(\varphi, \beta) \in H_0^1(\Omega) \times K$ of optimal control problem (23).*

Proof. Let $(\varphi_m, \beta_m) \in Z_{ad}$ be a minimizing sequence for the functional J, for which the following is true:

$$\lim_{m\to\infty} J(\varphi_m, \beta_m) = \inf_{(\varphi,\beta)\in Z_{ad}} J(\varphi, \beta) \equiv J^*.$$

That and the conditions of theorem for the functional J from (23) imply the estimate $\|\beta_m\|_{1,\Omega} \leq c_1$. From theorem 1 follows directly that $\|\varphi_m\|_{1,\Omega} \leq c_2$, where constants c_1, c_2 do not depend on m.

Then the weak limits $\varphi^* \in H_0^1(\Omega)$ and $\beta^* \in K$ of some subsequences of the sequences $\{\varphi_m\}$ and $\{\beta_m\}$ exist. Corresponding sequences will be also denoted by $\{\varphi_m\}$ and $\{\beta_m\}$. With this in mind it can be considered that

$$\varphi_m \to \varphi^* \in H_0^1(\Omega) \text{ weakly in } H^1(\Omega) \text{ and strongly in } L^s(\Omega),\ s < 6, \tag{25}$$

$$\beta_m \to \beta^* \in K \text{ weakly in } H^1(\Omega) \text{ and strongly in } L^s(\Omega),\ s < 6. \tag{26}$$

Let us show that $\mathcal{F}(\varphi^*, \beta^*) = 0$, i.e.

$$\lambda(\nabla\varphi^*, \nabla h) + (\beta^* k(\varphi^*)\varphi^*, h) + (\mathbf{u}\cdot\nabla\varphi^*, h) = (f, h) \ \ \forall h \in H_0^1(\Omega). \tag{27}$$

And it should be taken into account that the pair (φ_m, β_m) satisfies the relations

$$\lambda(\nabla\varphi_m, \nabla h) + (\beta_m k(\varphi_m)\varphi_m, h) + (\mathbf{u}\cdot\nabla\varphi_m, h) = (f, h) \ \ \forall h \in H_0^1(\Omega),\ m = 1, 2, \ldots. \tag{28}$$

Let us pass to the limit in (28) at $m \to \infty$. All linear summands in (28) turn into corresponding ones in (27).

From condition (iv) follows that $k(\varphi_m) \to k(\varphi)$ strongly in $L^2(\Omega)$. Taking into account (26), it is not difficult to show that

$$\beta_m k(\varphi_m) \to \beta k(\varphi) \text{ strongly in } L^{3/2}(\Omega).$$

Really,

$$\int_\Omega (\beta_m k(\varphi_m) - \beta k(\varphi))^{3/2} d\Omega \leq \int_\Omega \tilde{C} k(\varphi)^{3/2}(\beta_m - \beta)^{3/2} d\Omega + \int_\Omega \tilde{C}\beta_m^{3/2}(k(\varphi_m) - k(\varphi))^{3/2} d\Omega \leq$$

$$\leq \tilde{C}\|k(\varphi)\|_{L^3(\Omega)}^{1/2}\|\beta_m - \beta\|_{L^3(\Omega)}^{1/2} + \tilde{C}\|\beta_m\|_{L^6(\Omega)}^{1/4}\|k(\varphi_m) - k(\varphi)\|_{L^2(\Omega)}^{3/4} \to 0 \text{ at } m \to \infty.$$

On the strength of (25) $\varphi_m h \to \varphi h$ weakly in $L^3(\Omega)$ at $m \to \infty$ for all $h \in H_0^1(\Omega)$. This allows to pass to the limit in the nonlinear summand in (28).

As the functional J is weakly semicontinuous below on $H_0^1(\Omega) \times H^1(\Omega)$, then from aforesaid follows that

$$J^* = \lim_{m\to\infty} J(\varphi_m, \beta_m) = \underline{\lim}_{m\to\infty} J(\varphi_m, \beta_m) \geq J(\varphi^*, \beta^*) \geq J^*. \ \blacksquare$$

In more general case, when $k(\varphi) \in L_+^2(\Omega)$, the theorem 2 should be proved further as in [12].

Acknowledgments
This work was supported by the Russian Sciences Foundation (project no. 14-11-00079).

References
[1] Tikhonov A N and Arsenin V Ya 1986 *Methods for solving ill–posed problems* (Moscow: Nauka)
[2] Ito K and Kunish K 1997 *Inverse Problems* **13** 995–1013
[3] Agoshkov, V I, Minuk F P, Rusakov A S and Zalesny V B 2005 *Russ. J. Numer. Anal. Math. Modelling* **20** 19–43
[4] Alekseev G V and Adomavichus E A 2001 *J. Inverse Ill-Posed Probl.* **9** 435–68
[5] Alekseev G V 2002 *Comp. Math. Math. Phys.* **42** 363–76
[6] Alekseev G V 2007 *Comp. Math. Math. Phys* **47** 1007–28
[7] Alekseev G V and Tereshko D A 2011 *Comp. Math. Math. Phys.* **51** 1539–57
[8] Alekseev G V, Soboleva O V and Tereshko D A 2008 *J. Appl. Mech. Tech. Phys.* **49** 537–47
[9] Alekseev G V, Vakhitov I S and Soboleva O V 2012 *Comp. Math. Math. Phys.* **52** 1635–49
[10] Alekseev G V 2010 *Optimization in stationary problems of heat and mass transfer and magnetohydrodynamics* (Moscow: Nauchnyi Mir)
[11] Brizitskii R V and Saritskaya Zh Yu 2015 *Siberian Electronic Mathematical Reports* **12** 447–56
[12] Alekseev G V, Brizitskii R V and Saritskaya Zh Yu 2016 *Sib. J. Industrial Math.* **19** 3–16
[13] Fursikov A V 2000 *Optimal control of distributed systems. Theory and aplications* (Providence: American Mathematical Society)
[14] Kovtanyuk A E, Chebotarev A Yu, Botkin N D and Hoffman K-H 2015 *Commun Nonlinear Sci. Number Simulat.* **20** 776–84
[15] Kovtanyuk A E and Chebotarev A Yu 2014 *Diff. Eq.* **50** 1592–99

Determination of hydraulic fracture parameters using a non-stationary fluid injection

A V Valov [2] and S V Golovin[1,2]

[1] Lavrentyev Institute of Hydrodynamics SB RAS, Novosibirsk, Russia
[2] Novosibirsk State University, Novosibirsk, Russia

E-mail: tehnik.819@gmail.com, svgolovin@gmail.com

Abstract. In this paper, one provides a theoretical justification of the possibility of hydraulic fracture parameters determination by using a non-stationary fluid injection. It is assumed that the fluid is pumped into the fractured well with the time-periodic flow rate. It is shown that there is a phase shift between waves of fluid pressure and velocity. For the modelling purposes, the length and width of the fracture are assumed to be fixed. In the case of infinite fracture, one constructs an exact solution that ensures analytical determination of the phase shift in terms of the physical parameters of the problem. In the numerical calculation, the phase shift between pressure and velocity waves is found for a finite fracture. It is shown that the value of the phase shift depends on the physical parameters and on the fracture geometry. This makes it possible to determine parameters of hydraulic fracture, in particular its length, by the experimental measurement of the time shift and comparison with the numerical solution.

1. Introduction

One of the important components of hydraulic fracturing procedure is the control of the fracture consisting of the determination of its basic geometrical parameters: size and position. The control is usually performed during its production and based on the analysis of data of the microseismic monitoring [1]. However, this method is not precise and gives only an estimation of the fracture positioning. Thus, any complementary studies for the control of hydraulic fracture are highly relevant.

In this paper, one provides a theoretical justification of the possibility to use a non-stationary fluid injection to determine the parameters of hydraulic fracture. It is assumed that the fluid is injected into the fractured well with a time-periodic flow rate. It is shown that a phase shift between the waves of fluid pressure and velocity exists in this case. The phase shift depends on the formation parameters as well as the fracture length. Thus, the simultaneous measurement of pressure and flow rate at the borehole provides additional information about the formation and the fracture properties.

Mathematical modelling is based on the Biot's poroelastic medium model [2]. Boundary conditions on the fracture are adopted from paper [3]. Both the length and the aperture of the fracture are assumed to be fixed. In the case of an infinite fracture, one constructs an exact solution that makes it possible to determine the phase shift explicitly as a function of the problem parameters. For the fracture of a finite length, a numerical solution of the problem is provided. It is shown that the value of the phase shift depends on the physical parameters and on the geometry of the fracture in all cases. This observation indicates a principal possibility for the determination of physical parameters of the formation and of the fracture by the experimental measurement of the phase shift and comparison with the numerically predicted phase shift.

2. A mathematical model

One considers a planar vertical hydraulic fracture of fixed height $2H$, width $2w$, and length $2L$ extending along the x-axis in the poroelastic formation (Figure1). The fluid is pumped into the fracture through the wellbore that is located along the vertical axis at $x = y = 0$.

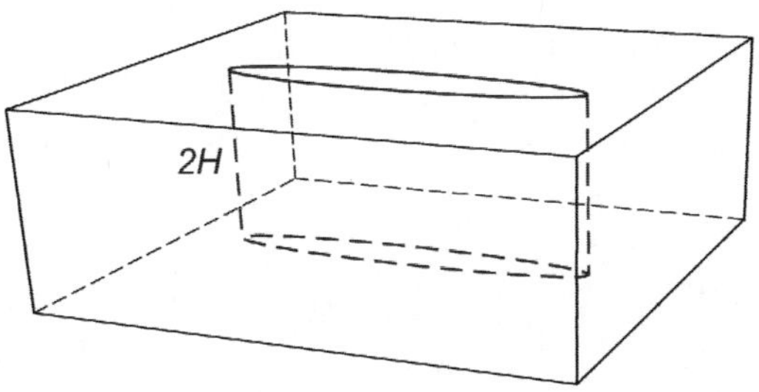

Figure 1. Fracture geometry.

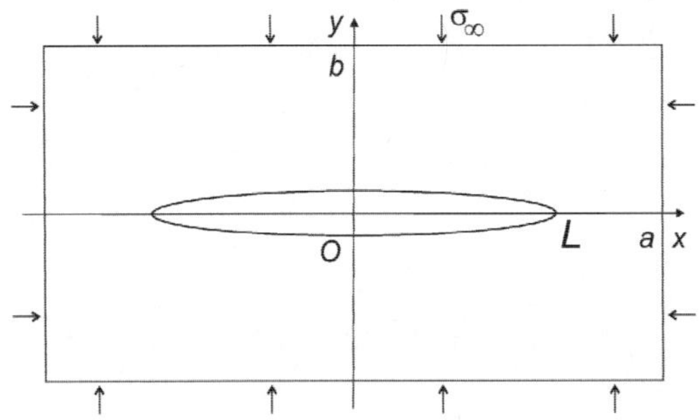

Figure 2. Fracture's cross-section by the plane $z = 0$.

It is assumed that pores are saturated by a single-phase Newtonian fluid with the efficient viscosity η and efficient density ρ. Also, physical properties of the injected fluid and the formation fluid are assumed to be identical.

Due to the statement of the problem, one uses the plane strain approximation, which implies that there is no deformation along the vertical axis, and all cross-sections by the planes $z = H_1, |H_1| \leq H$ are identical. Hence, the 2D problem in the plane $z = 0$ is solved (Figure 2). The displacement vector $\mathbf{u}$ is two-dimensional: $\mathbf{u} = (u_1, u_2) = (u, v)$. Fluid pressure p and displacement $\mathbf{u}$ depend on time t and coordinates $(x_1, x_2) = (x, y)$. The fracture is located along the line $y = 0$ and occupies the segment $-L \leq x \leq L$.

Symmetry with respect to planes $x = 0$ and $y = 0$ makes it possible to solve the problem only in the first quarter of Oxy plane. Consider the domain $\Omega = \{(x, y) : 0 \leq x \leq a, 0 \leq y \leq b\}$. Outside the fracture, functions $\mathbf{u}$ and p satisfy the system

$$\Omega : \quad \operatorname{div} \tau = 0, \tag{1}$$

$$S_\varepsilon \frac{\partial p}{\partial t} = \operatorname{div}\left(\frac{k_r}{\eta}\nabla p - \alpha\frac{\partial \mathbf{u}}{\partial t}\right), \tag{2}$$

where τ is the stress tensor, k_r is the formation permeability, η is the fluid viscosity, S_ε is the fluid yielding capacity coefficient, and α is the Biot coefficient. According to the Biot theory [2], the tensor τ is defined as follows

$$\tau = \lambda\epsilon(\mathbf{u}) \cdot I + 2\mu\mathcal{E}(\mathbf{u}) - \alpha p \cdot I, \quad \epsilon(\mathbf{u}) = \operatorname{tr}\mathcal{E}(\mathbf{u}) = \mathcal{E}(\mathbf{u})_{ii} = \operatorname{div}\mathbf{u}, \tag{3}$$

where $I_{ij} = \delta_{ij}$, $\mathcal{E}(\mathbf{u})$ is the strain tensor related to the field $\mathbf{u}$: $2\mathcal{E}(\mathbf{u})_{ij} = \frac{\partial u_i}{\partial x_j} + \frac{\partial u_j}{\partial x_i}$, λ and μ are the elasticity moduli. Given the Young modulus E and the Poisson ratio ν for the fluid-saturated rock, one can compute the elasticity moduli as

$$\lambda = \frac{\nu E}{(1+\nu)(1-2\nu)}, \quad \mu = \frac{E}{2(1+\nu)}.$$

Equation (1) in view of (3) takes the form

$$(\lambda+\mu)\nabla\operatorname{div}\mathbf{u}+\mu\Delta\mathbf{u}-\alpha\nabla p=0. \tag{4}$$

Over the outer border $\Gamma_e=\partial\Omega\cap(\{x=a\}\cup\{y=b\})$, a load σ_∞ is applied, and a pore pressure p_∞ is prescribed:

$$\Gamma_e: \quad p=p_\infty, \quad \tau\langle\mathbf{n}\rangle=-\sigma_\infty\mathbf{n}, \tag{5}$$

where $\mathbf{n}$ is the outward normal unit vector to Γ_e.

Over the lines $x=0$ and $y=0$, the symmetry conditions are required

$$\Gamma_l=\{x=0,0<y<b\}: \quad u=0, \quad \frac{\partial v}{\partial x}=0, \quad \frac{\partial p}{\partial x}=0, \tag{6}$$

$$\Gamma_c=\{y=0,0<x<L\}: \quad v=0, \tag{7}$$

$$\Gamma_s=\{y=0,L<x<a\}: \quad \frac{\partial u}{\partial y}=0, \quad v=0, \quad \frac{\partial p}{\partial y}=0. \tag{8}$$

Within the fracture, the fluid mass conservation law takes the form

$$\frac{\partial w}{\partial t}+\frac{\partial(wV)}{\partial x}=-q. \tag{9}$$

Here, the fluid velocity $V(t,x)$ in x-direction is obtained by the Darcy's law

$$V(t,x)=-\frac{k_c}{\eta}\frac{\partial p}{\partial x}, \tag{10}$$

where k_c is the fracture permeability. Exchange of fluid between the fracture and the surrounding medium is governed by the leakoff rate q that is also given by the Darcy's law for the formation:

$$q=-\frac{k_r}{\eta}\frac{\partial p}{\partial y}. \tag{11}$$

Thus, equations (9), (10), (11) and the requirement that the fracture has a constant width $w=$ const yield

$$\Gamma_c: \quad -\frac{\partial}{\partial x}\left(\frac{wk_c}{\eta}\frac{\partial p}{\partial x}\right)=\frac{k_r}{\eta}\frac{\partial p}{\partial y}. \tag{12}$$

With $Q(t)$ being the volumetric flow injection rate at the borehole per unit fracture height (m^2/s), there are the following injection condition

$$\Gamma_c: \quad -\frac{wk_c}{\eta}\frac{\partial p}{\partial x}\bigg|_{y=0+,x=0+}=\frac{1}{4}Q(t). \tag{13}$$

To complete the formulation, the initial data are specified in the following form:

$$t=0: \quad \mathbf{u}=\mathbf{u}_0(x,y), \quad p=p_0(x,y).$$

The mathematical model for a non-fixed fracture of a variable aperture is presented in [3].

3. Exact solution for the infinite fracture

Let one assume that the fracture has infinite length $L = \infty$. In this case, the domain is $\Omega = \mathbb{R}^2$, and the fracture position is described as $\Gamma_c = \{y = 0\}$.

The problem (4), (2), (12) is considered. Let one seek a solution in the form of a wave packet

$$\mathbf{u} = \boldsymbol{\varphi} e^{i\omega t - (\mathbf{m}\cdot\mathbf{x})} \tag{14}$$

$$p = p_0 e^{i\omega t - (\mathbf{m}\cdot\mathbf{x})}. \tag{15}$$

Here, $i = \sqrt{-1}$ is the imaginary unit, $\boldsymbol{\varphi} = \begin{pmatrix}\varphi_1\\ \varphi_2\end{pmatrix}, \mathbf{m} = \begin{pmatrix}m_1\\ m_2\end{pmatrix}$ are constant complex vectors, ω is a real constant. By definition, $(\mathbf{a}\cdot\mathbf{b}) = a_1b_1 + a_2b_2$ for all complex vectors $\mathbf{a} = \begin{pmatrix}a_1\\ a_2\end{pmatrix}, \mathbf{b} = \begin{pmatrix}b_1\\ b_2\end{pmatrix}$. Substituting (14) and (15) to the system of equations (4), (2), (12) the system of algebraic equations with four complex unknowns φ_1, φ_2, m_1, m_2 is obtained:

$$(\lambda + \mu)(\boldsymbol{\varphi}\cdot\mathbf{m})\mathbf{m} + \mu(\mathbf{m}\cdot\mathbf{m})\boldsymbol{\varphi} + \alpha p_0\mathbf{m} = 0, \tag{16}$$

$$i\omega p_0 = \frac{k_r}{S_{\mathcal{E}}\eta}(\mathbf{m}\cdot\mathbf{m})p_0 + \frac{\alpha}{S_{\mathcal{E}}} i\omega(\boldsymbol{\varphi}\cdot\mathbf{m}), \tag{17}$$

$$m_1^2 = \sigma m_2, \tag{18}$$

where $\sigma = k_r/(wk)$. One can show that the unique solution of (16)–(18) has the following form

$$\mathbf{m} = \begin{pmatrix}\sigma\sqrt{-\frac{\cos\theta}{\cos 2\theta}}e^{i\frac{\theta}{2}}\\ -\sigma\frac{\cos\theta}{\cos 2\theta}e^{i\theta}\end{pmatrix}, \quad \boldsymbol{\varphi} = \frac{i\alpha p_0}{A(\lambda + 2\mu)}\mathbf{m}, \tag{19}$$

where

$$\theta = \frac{1}{2}\left(\pi - \arcsin\left(\sqrt{\frac{\sigma^4}{16A^2} + 1} - \frac{\sigma^2}{4A}\right)\right), \quad A = \omega\frac{S_{\mathcal{E}}\eta}{k_r}\left(\frac{\alpha^2}{S_{\mathcal{E}}(\lambda + 2\mu)} + 1\right).$$

Since $\mathbf{m}$ is a complex vector, it can be represented in the form $\mathbf{m} = \mathbf{m}_1 + i\mathbf{m}_2$, where $\mathbf{m}_1$ and $\mathbf{m}_2$ are real and imaginary parts of $\mathbf{m}$, respectively. Both the real and the imaginary parts of (4), (2), (12) give the solution of the initial problem:

$$\begin{aligned}\operatorname{Re}\mathbf{u} &= C\big(\mathbf{m}_2\cos(\omega t - \mathbf{m}_2\cdot\mathbf{x}) + \mathbf{m}_1\sin(\omega t - \mathbf{m}_2\cdot\mathbf{x})\big)e^{-\mathbf{m}_1\cdot\mathbf{x}},\\ \operatorname{Re}p &= p_0\cos(\omega t - \mathbf{m}_2\cdot\mathbf{x})e^{-\mathbf{m}_1\cdot\mathbf{x}},\end{aligned} \tag{20}$$

$$\begin{aligned}\operatorname{Im}\mathbf{u} &= -C\big(\mathbf{m}_1\cos(\omega t - \mathbf{m}_2\cdot\mathbf{x}) - \mathbf{m}_2\sin(\omega t - \mathbf{m}_2\cdot\mathbf{x})\big)e^{-\mathbf{m}_1\cdot\mathbf{x}},\\ \operatorname{Im}p &= p_0\sin(\omega t - \mathbf{m}_2\cdot\mathbf{x})e^{-\mathbf{m}_1\cdot\mathbf{x}},\end{aligned} \tag{21}$$

where $C = -\frac{\alpha p_0}{A(\lambda+2\mu)}$,

$$\mathbf{m}_1 = \begin{pmatrix}\sigma\sqrt{-\frac{\cos\theta}{\cos 2\theta}}\cos\frac{\theta}{2}\\ -\sigma\frac{\cos\theta}{\cos 2\theta}\cos\theta\end{pmatrix}, \mathbf{m}_2 = \begin{pmatrix}\sigma\sqrt{-\frac{\cos\theta}{\cos 2\theta}}\sin\frac{\theta}{2}\\ -\sigma\frac{\cos\theta}{\cos 2\theta}\sin\theta\end{pmatrix}$$

are the real vectors; ω and p_0 are free parameters.

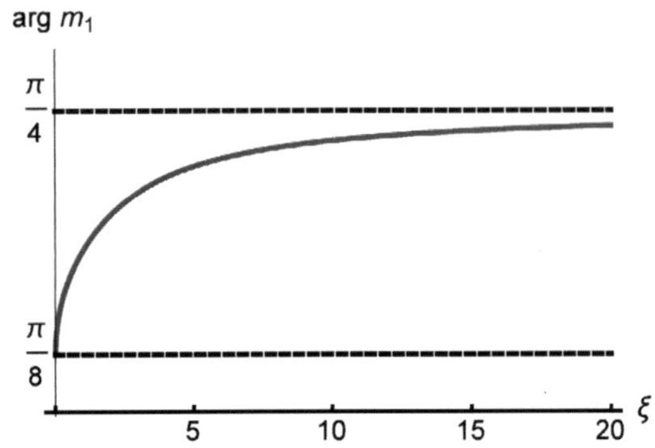

Figure 3. Graph of $\arg m_1(\xi)$

4. The phase shift

An analysis of exact solutions (20), (21) shows that there is a phase shift between fluid pressure and velocity waves. Indeed, according to the Darcy's law, the fluid velocity is proportional to the pressure gradient:

$$\mathbf{V} \sim -\nabla p = \mathbf{m} p_0 e^{i\omega t - (\mathbf{m}\cdot\mathbf{x})}.$$

The fluid velocity within the fracture can be represented as

$$V_1 \sim m_1 p_0 e^{i\omega t - m_1 x_1} = |m_1| p_0 e^{i(\omega t + \arg m_1) - m_1 x_1}.$$

Hence, the number $\arg m_1/\omega$ determines the phase shift between velocity and pressure waves. In view of $\arg m_2 = \theta$ by introduction of the dimensionless complex $\xi = \sigma^2/A$, one can find $\arg m_1$ as a function of ξ:

$$\arg m_1 = \frac{1}{2}\theta = \frac{1}{4}\left(\pi - \arcsin\left(\sqrt{\frac{\sigma^4}{16A^2} + 1} - \frac{\sigma^2}{4A}\right)\right) =$$
$$\frac{1}{4}\left(\pi - \arcsin\left(\sqrt{\frac{\xi^2}{16} + 1} - \frac{\xi}{4}\right)\right) = \arg m_1(\xi).$$

This dependence is presented in Figure 3. Thus, in the case of an infinite fracture, the phase shift is fully determined by the value of ξ.

Note that the constructed exact solution (20), (21) has a disadvantage, because it does not describe the fluid injection at the downhole at $x = y = 0$ (the condition (13) is not satisfied). Therefore, the obtained result regarding the phase shift can be observed as a motivation for the further analysis of the correct statement of the problem for the finite-length fracture.

5. Numerical solution

In this section, one constructs a numerical solution for the problem with a finite-length fracture as formulated in Section 2. The problem is solved by the finite-element method. To this end, one writes a variational formulation of the model developed above. First, one homogenises the boundary conditions over the outer boundary Γ_e. By denoting,

$$\tilde{\mathbf{u}} = \mathbf{u} - \varkappa\mathbf{x}, \quad \tilde{p} = p - p_\infty, \quad \varkappa = \frac{\alpha p_\infty - \sigma_\infty}{2(\lambda + \mu)},$$

it is found that

$$\mathcal{E}(\mathbf{u}) = \mathcal{E}(\tilde{\mathbf{u}}) + \varkappa \cdot I, \quad \tau = \tilde{\tau} - \sigma_\infty \cdot I, \quad \tilde{\tau} = \lambda\epsilon(\tilde{\mathbf{u}}) \cdot I + 2\mu\mathcal{E}(\tilde{\mathbf{u}}) - \alpha\tilde{p} \cdot I.$$

Functions $\tilde{\mathbf{u}}$ and $\tilde{p}$ solve the boundary-value problem

$$\Omega: \quad \operatorname{div}\tilde{\tau} = 0, \quad S_\varepsilon \frac{\partial \tilde{p}}{\partial t} = \operatorname{div}\left(\frac{k_r}{\eta}\nabla\tilde{p} - \alpha\frac{\partial \tilde{\mathbf{u}}}{\partial t}\right), \tag{22}$$

$$\Gamma_e: \quad \tilde{p} = 0, \quad \tilde{\tau}\langle\mathbf{n}\rangle = 0,$$

$$\Gamma_l: \quad \frac{\partial \tilde{v}}{\partial x} = 0, \quad \tilde{u} = 0, \quad \frac{\partial \tilde{p}}{\partial x} = 0,$$

$$\Gamma_s: \quad \frac{\partial \tilde{u}}{\partial y} = 0, \quad \tilde{v} = 0, \quad \frac{\partial \tilde{p}}{\partial y} = 0,$$

$$\Gamma_c: \quad \tilde{v} = 0, \quad -\frac{\partial}{\partial x}\left(\frac{wk_c}{\eta}\frac{\partial \tilde{p}}{\partial x}\right) = \frac{k_r}{\eta}\frac{\partial \tilde{p}}{\partial y}, \tag{23}$$

$$\Gamma_c: \quad -\frac{wk_c}{\eta}\frac{\partial \tilde{p}}{\partial x}\bigg|_{y=0,x=0+} = \frac{1}{4}Q(t). \tag{24}$$

For simplicity, the initial data are chosen as follows:

$$t = 0: \quad \tilde{\mathbf{u}} = 0, \quad \tilde{p} = 0. \tag{25}$$

Let one denote by $\psi(x, y)$ and $\boldsymbol{\varphi}(x, y)$ any smooth scalar and vector test functions such that

$$\psi|_{\Gamma_e} = 0, \quad \boldsymbol{\varphi} \cdot \mathbf{n}|_{\Gamma_s} = 0. \tag{26}$$

Keeping in mind the boundary conditions over Γ_e, Γ_l, Γ_s, Γ_c, one multiplies the first equation in (22) and the second equation in (22) by $\boldsymbol{\varphi}$ and ψ, respectively, and integrates over the domain Ω. In this way by using equation (23) and (24), one arrives at the equalities

$$\int_\Omega \lambda \operatorname{div}\tilde{\mathbf{u}} \operatorname{div}\boldsymbol{\varphi} + 2\mu\mathcal{E}(\tilde{\mathbf{u}}) : \mathcal{E}(\boldsymbol{\varphi}) - \alpha\tilde{p} \operatorname{div}\boldsymbol{\varphi}\, dxdy = 0,$$

$$\int_\Omega \Big[\Big(S_\varepsilon\frac{\partial \tilde{p}}{\partial t} + \alpha \operatorname{div}\frac{\partial \tilde{\mathbf{u}}}{\partial t}\Big)\psi + \frac{k_r}{\eta}\nabla\tilde{p}\cdot\nabla\psi\Big]\, dxdy = \int_{\Gamma_c} \frac{wk_c}{\eta}\frac{\partial \psi}{\partial x}\frac{\partial \tilde{p}}{\partial x}\, dx + \psi(0,0)Q(t)/2,$$

Next, the dimensionless variables are introduced

$$\mathbf{x}' = \frac{\mathbf{x}}{L}, \quad t' = \frac{t}{t_*}, \quad \mathbf{u}' = \frac{\tilde{\mathbf{u}}}{w}, \quad p' = \frac{\tilde{p}}{p_*}, \quad Q'_a = \frac{Q}{Q_*},$$

where t_*, p_*, and Q_* are reference values; $L' = 1$ is the dimensionless length of fracture. In the new variables,

$$\Omega' = \{(x', y') : 0 \leqslant x' \leqslant \frac{a}{L}, 0 \leqslant y' \leqslant \frac{b}{L}\},$$
$$\Gamma'_c = \{(x', y') : y' = 0, 0 \leqslant x' \leqslant 1\}, \quad \Gamma'_s = \{y' = 0, 1 \leqslant x' \leqslant \frac{a}{L}\},$$
$$\Gamma'_e = \partial\Omega' \cap (\{x' = \frac{a}{L}\} \cup \{y' = \frac{b}{L}\}), \quad \Gamma'_l = \{x' = 0, 0 \leqslant y' \leqslant \frac{b}{L}\}.$$

Omitting primes for simplicity, one arrives at the following dimensionless variational formulation

$$\int_\Omega a_1 \operatorname{div} \mathbf{u} \operatorname{div} \boldsymbol{\varphi} + a_2 \mathcal{E}(\mathbf{u}) : \mathcal{E}(\boldsymbol{\varphi}) - \alpha p \operatorname{div} \boldsymbol{\varphi} \, dx dy = 0, \tag{27}$$

$$\int_\Omega \Big[\Big(b_1 \frac{\partial p}{\partial t} + \alpha \operatorname{div} \frac{\partial \mathbf{u}}{\partial t} \Big) \psi + b_2 \nabla p \cdot \nabla \psi \Big] \, dx dy = \int_{\Gamma_c} b_3 \frac{\partial \psi}{\partial x} \frac{\partial p}{\partial x} \, dx + b_4 \psi(0,0) Q. \tag{28}$$

The test functions satisfy the conditions (26), and the dimensionless numbers are defined as follows:

$$a_1 = \frac{\lambda w}{L p_*}, \quad a_2 = \frac{2\mu w}{L p_*}, \quad b_1 = \frac{S_\varepsilon p_* L}{w}, \quad b_2 = \frac{t_* p_* k_r}{\eta w L}, \quad b_3 = \frac{p_* t_* k}{\eta L^2}, \quad b_4 = \frac{Q_* t_*}{2 w L H},$$

where H is the half of the fracture height.

The calculations are performed with the help of the freely accessible finite element PDE soler FreeFEM++ [4]. The time derivatives are approximated by finite differences: $\partial f / \partial t \approx (f^{n+1} - f^n)/\Delta t$ where f is one of functions p or $\mathbf{u}$, and Δt is the time step. The upper index denotes the time instant: $f^n = f(t_n, \mathbf{x})$, $t_n = n \Delta t$.

Multiplication of Eq. (28) by the time step Δt and subtraction from Eq. (27) provides a symmetrical formulation with respect to the unknown function $(\mathbf{u}, p)$ and the test functions $(\boldsymbol{\varphi}, \psi)$. This assures the symmetry of the stiffness matrix, which is important for the correctness of the numerical algorithm.

To perform computations, one takes 30 mesh vertices over the outer boundary Γ_e, 80 over Γ_l, 40 over Γ_f and 17 over Γ_s. The total number of mesh vertices is 1743. Dimensionless time step is equal to 0.0005 which is equivalent to 1.8 s. Since we start from zero initial data (25), at the beginning of calculations the process non-periodic. However, after some initial period of time (about 2 s), the system comes to a periodic state where the pressure waves repeat at every pulsation of the input data.

6. Numerical simulation

One performs calculations using the following data: $L = 150$ m, $w = 0.5$ cm, $H = 21$ m, $\alpha = 0.7$, $k_r = 10$ mD, $k_c = 10^4$ D, 0.3 cp, $E = 15$ GPa, $\nu = 0.18$, $S_\varepsilon^{-1} = 1.754$ GPa, $Q_* = 100$ m^3/day, $a = 200$ m, $b = 200$ m, $t_* = 1$ h, $p_* = 100$ MPa.

In calculations, one uses a periodic function of the flow injection rate at the downhole $Q(t) = Q_* \sin(\omega t)$. By using the numerically calculated pressure p, the flow velocity at the downhole is computed via the Darcy's law (10). Now, one plots the vp-diagram of the process by drawing a curve $\{ (V(t), p(t)),\ t = [t_1, t_2] \}$ in the plane Ovp. Here, t_1 is considered to be great enough for the system to come to the periodic state.

In order to find the dependence of the phase shift on the problem's parameters, calculations for $\omega t_* = 3$, 5, 10 and $L = 10^3$, $2.5 \cdot 10^3$, $1.5 \cdot 10^4$ are done. The resulting vp-diagrams are shown in Figure 4. One can see the elliptical shape of the diagrams, which implies the existence of the phase shift between pressure and velocity waves. In case of zero time shift the diagrams would be flat. The dependence of the time shift on the physical parameters of the problem is also noticed for variation of w, η, k_r, k_c, S_ε.

Explicit numerical calculation of the time shift based on the calculations with a denser set of variation of parameters L and ω makes it possible to represent the graphs of the dependence of the time shift on the parameters as shown in Figure 5.

One can see the strongly pronounced dependence of the phase shift and the shape of vp-diagram to length of the fracture L and frequency of fluid injection ω. These dependences allow solving the parametric optimization problem: determining formation parameters on the base of the experimental determination of the phase shift between pressure and velocity waves.

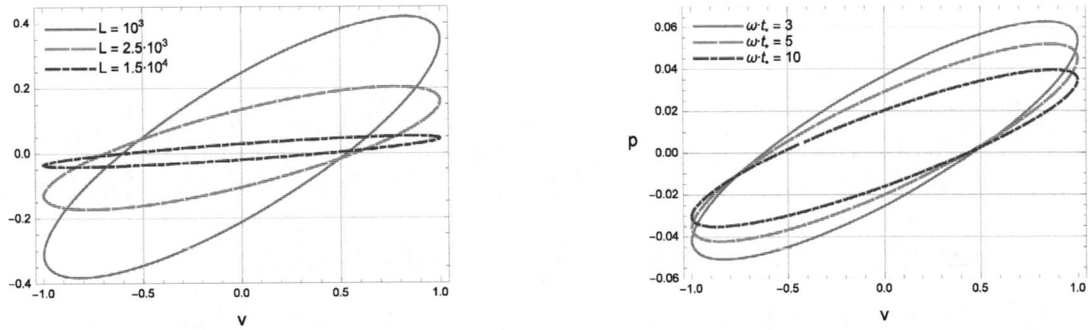

L ω

Figure 4. vp-diagrams for different length of fracture L and frequency ω.

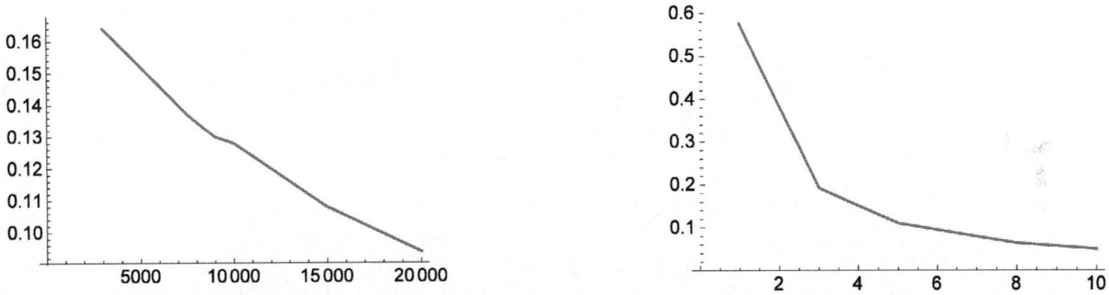

L ω

Figure 5. Phase shift as a function of length of the fracture L and frequency of fluid injection ω.

7. Conclusion

Suppose, we can organize a periodic injection/suction of fluid to the fracture at the downhole according to some periodic law. Doing this experiment with different frequencies $\omega_1, \ldots, \omega_k$ we measure simultaneously pressure and fluid velocity at the downhole. Then, knowing the pressure and velocity as functions of time, we can calculate the values of the phase shifts between pressure and velocity waves for different ω. Using the calculations based on the presented model and methods of mathematical optimization we can determine a parameter of interest, for example length of fracture L with a reasonable accuracy.

Acknowledgments

This work was supported by the Russian Science Foundation (Project No. 15-11-20013).

References

[1] Rothert E and Shapiro S 2003 *Geophysics* **68** 685–9
[2] Biot M 1956 *J. Acoust. Soc. Amer.* **28** 168–78
[3] Shelukhin V, Baikov V, Golovin S, Davletbaev A and Starovoitov V 2014 *Int. J. Sol. Struct.* **51** 2116–22
[4] Hecht F 2012 *J. Numer. Math.* **20** 251–65

Modeling process of embolization arteriovenous malformation on the basis of two-phase filtration model

A A Cherevko[1,2], T S Gologush[2], V V Ostapenko[1,2], I A Petrenko[3] and A P Chupakhin[1,2]

[1] Lavrentyev Institute of Hydrodynamics of the Siberian Branch of the Russian Academy of Sciences, Novosibirsk 630090, Russia
[2] Novosibirsk State University, Novosibirsk 630090, Russia
[3] Vladimir State University, Vladimir 60000, Russia

E-mail: cherevko@mail.ru, tatiana_06.08@mail.ru

Abstract. Arteriovenous malformation is a chaotic disordered interlacement of very small diameter vessels, performing reset of blood from the artery into the vein. In this regard it can be adequately modeled using porous medium. In this model process of embolization described as penetration of non-adhesive substance ONYX into the porous medium, filled with blood, both of these fluids are not mixed with each other. In one-dimensional approximation such processes are well described by Buckley–Leverett equation. In this paper Buckley–Leverett equation is solved numerically by using a new modification of Cabaret scheme. The results of numerical modeling process of embolization of AVM are shown.

1. Introduction

Arteriovenous malformation (AVM) is a pathological connection between arteries and veins. This anomaly leads to a direct bypass (reset) of blood from an arterial pool in the system of superficial and deep veins. The typical size of AVM amounts to centimeters. The most frequent cases, when AVM consists of vessels of very small diameter (0.01mm-0.9mm) and they almost completely fill the volume of AVM [1]. Cerebral AVM is a serious medical problem. More than half of the patients during life get haemorrhage, 10–30% of them die after the first haemorrhage, and from 10 to 20% of survivors are disabled. There are various treatments for AVM: microsurgical, endovascular and radiosurgical. Earlier embolization (endovascular method) was considered as an auxiliary method and used before removing or irradiating the AVM. The situation has changed with the advance of non-adhesive substance ONYX [2] and currently the endovascular methods are more popular. However, the danger of intraoperative rupture during embolization of AVM still exists. So development of modeling methods of embolization is an urgent problem [3, 4].

2. Modeling

2.1. Physical modeling

The typical AVM consists of large amount of strongly intertwined vessels of small diameter and so it can be considered as the porous medium with sufficient accuracy for modeling. In this case the movement of blood and non-adhesive substance ONYX through the AVM can be modeled as a two-phase filtration process, where the displaceable phase is blood and displacing phase is ONYX.

2.2. Mathematical modeling

In order to simplify the problem, we make the following assumptions. We assume that the AVM is a cylinder with length L, and that blood and non-adhesive substance are incompressible fluids. Within the one-dimensional approximation we consider all values as averaged over the cross section of the cylinder. Suppose that capillary forces can be neglected, and hence, the pressure in the phases is the same. Let the porosity of AVM is constant and equal to m. Denote by $S(x,t)$ saturation of blood. Let $k_i = k_i(S),\ i = b, e$, is the phase permeability for the blood and ONYX. Also, let $\mu_i,\ i = b, e$, is the viscosities of the blood and ONYX, which are constant. Because the characteristic time of embolization amounts to several tens of minutes, the change in values during the cardiac cycle is not taken into account. Also denote by $V = V(t)$ total flow volume for two phases, which is independent of x due to the assumption of impermeability of the lateral surface of the AVM. Then the equation for the saturation of blood inside the AVM during embolization can be written in the form of Buckley–Leverett [5, 6]:

$$m\frac{\partial S}{\partial t} + V(t)\frac{\partial\, b(S)}{\partial x} = 0, \tag{1}$$

where the function of the Buckley–Leverett

$$b(S) = \frac{k_b(S)/\mu_b}{k_b(S)/\mu_b + k_e(S)/\mu_e},$$

defining the allocation of the phase flows. It is determined on the interval $[0,1]$. It is assumed that it satisfies the following conditions on this segment:

$$b(S) \in \mathbf{C}^1[0,1],\ \ b(0) = 0,\ b(1) = 1,\ \ b'(0) = b'(1) = 0,\ \ b'(S) > 0\ \ \forall S \in (0,1).$$

By replacing $\theta = \frac{1}{m}\int_0^t V(\xi)d\xi$ the equation (1) can be simplified and reduced to the form:

$$\frac{\partial S}{\partial \theta} + \frac{\partial\, b(S)}{\partial x} = 0,$$

which will be further major for us.

2.3. Initial-boundary problem of embolization

At the initial moment AVM does not contain ONYX, that is $S(0,x) \equiv 1$. ONYX is injected into the AVM through the artery. We assume that the arterial part of AVM has the coordinate $x = 0$ and denote the saturation of blood on the arterial end by $g(\theta)$. According to medical indications, ONYX should never reach the venous end of AVM (high risk of rupture), that is always $S(\theta, L) = 1$.

Thus, the initial-boundary value problem has the form:

$$\frac{\partial S}{\partial \theta} + \frac{\partial\, b(S)}{\partial x} = 0, \tag{2}$$

$$S(0,x) = S_0(x), \quad x \in [0,L] \tag{3}$$

$$S(\theta,0) = g(\theta), \quad \theta \in (0,T]. \tag{4}$$

Moreover, due to the peculiarities of the process of embolization $S_0(x) \equiv 1$, and the moment of time T is determined from the condition that ONYX has reached the venous end of AVM.

3. Finite-difference scheme

For numerical solution of problem (2)–(4) we used the explicit second order CABARET scheme [7]. The main advantages of this scheme owe to the following property: in linear case this scheme is reversible in time and exact for two different Courant numbers $r = 0.5$ and $r = 1$. For this reason, the scheme possesses unique dissipative and dispersive properties [8]. In view of the special flux correction and certain constrains during approximation of initial data, the CABARET scheme is monotonic and strongly monotonic [9] at Courant numbers $r \leq 0.5$. Therefore, the value $r < 0.5$ is used in the computations described below. Moreover, the CABARET scheme has a compact stencil bounded by the size of one spatial-temporal cell of the difference grid.

The problem (2)–(4) is approximated by the two-time-level CABARET scheme [8] on a uniform rectangular grid

$$\{(x_j, \theta_n): \ x_j = jh, \ j = \overline{0, J}; \quad \theta_n = n\tau, \ n = \overline{0, N}\ \}, \tag{5}$$

where $h = L/J$ and $\tau = T/N$ are constant mesh step in space and in time. The time step is determined according to the stability condition by the formula

$$\tau = \frac{zh}{\max\limits_{[0,1]} b'(S)} ,$$

where $z = 0.4$ is the safety factor. This scheme involves the flux variables $u_j^n = S(x_j, \theta_n)$ and the conservative variables $U_{j+1/2}^n = S(x_{j+1/2}, \theta_n)$, which are defined at integer x_j and half-integer $x_{j+1/2} = x_j + h/2$ spatial grid nodes, respectively.

Let u_j^n, $j = \overline{0, J}$ and $U_{j+1/2}^n$, $j = \overline{0, J-1}$ is the known numerical solution of problem (2)–(4) at the n-th time level θ_n; at $n = 0$, this is a grid approximation of the initial function $S_0(x)$. Based on the modified CABARET scheme, the numerical solution u_j^{n+1}, $U_{j+1/2}^{n+1}$ at the $(n+1)$-th time level θ_{n+1} is found in five stages. At the first stage, the difference equations

$$\frac{U_{j+1/2}^{n+1/2} - U_{j+1/2}^n}{\tau/2} + \frac{b_{j+1}^n - b_j^n}{h} = 0, \quad j = \overline{0, J-1},$$

where $b_j^n = b(u_j^n)$, are used to calculate the conservative variables $U_{j+1/2}^{n+1/2} = S(x_{j+1/2}, \theta_{n+1/2})$ at the half-integer time level $\theta_{n+1/2} = \theta_n + \tau/2$. At the second stage, the extrapolation

$$\tilde{u}_{j+1}^{n+1} = 2U_{j+1/2}^{n+1/2} - u_j^n, \quad j = \overline{0, J-1},$$

is employed to find the preliminary flux variables $\tilde{u}_j^{n+1}$, which are corrected as follows

$$\bar{u}_{j+1}^{n+1} = F\left(\tilde{u}_{j+1}^{n+1}, m_{j+1/2}^n, M_{j+1/2}^n\right), \quad j = \overline{0, J-1},$$

$$F(u, m, M) = \begin{cases} u, & m \leq u \leq M, \\ m, & u < m, \\ M, & u > M, \end{cases}$$

$$m_{j+1/2}^n = \min\left(u_j^n, U_{j+1/2}^n, u_{j+1}^n\right), \quad M_{j+1/2}^n = \max\left(u_j^n, U_{j+1/2}^n, u_{j+1}^n\right).$$

At the third stage, we use formulas

$$\bar{b}_j^{n+1/2} = b\left(\bar{u}_j^{n+1/2}\right), \quad \bar{u}_j^{n+1/2} = \frac{u_j^n + \bar{u}_j^{n+1}}{2}, \quad j = \overline{1, J},$$

for calculation of preliminary values $\bar{b}_j^{n+1/2}$ of numerical fluxes, which are corrected as follows

$$b_j^{n+1/2} = F\left(\bar{b}_j^{n+1/2}, \bar{m}_{j-1/2}^n, \bar{M}_{j-1/2}^n\right), \quad j = \overline{1, J},$$

$$\bar{m}_{j-1/2}^n = \min\left(b_{j-1/2}^n, \max(b_j^n, \varphi_{j-1/2}^n)\right), \ \bar{M}_{j-1/2}^n = \max\left(b_{j-1/2}^n, \min(b_j^n, \varphi_{j-1/2}^n)\right),$$

$$b_j^n = b(u_j^n), \quad b_{j-1/2}^n = b(U_{j-1/2}^n), \quad \varphi_{j-1/2}^n = \frac{U_{j-1/2}^n - u_{j-1}^n}{r} + b_{j-1}^n, \quad r = \frac{\tau}{h}.$$

The left boundary numerical flux $b_0^{n+1/2}$ is defined from boundary condition (5) under the formula

$$b_0^{n+1/2} = b(g(\theta_{n+1/2})).$$

At the fourth stage, the difference equations

$$\frac{U_{j+1/2}^{n+1} - U_{j+1/2}^n}{\tau} + \frac{b_{j+1}^{n+1/2} - b_j^{n+1/2}}{h} = 0, \quad j = \overline{0, J-1},$$

are used to calculate the conservative variables $U_{j+1/2}^{n+1} = S(x_{j+1/2}, \theta_{n+1})$ at the $(n+1)$-th time level $\theta_{n+1} = \theta_n + \tau$. At the fifth stage, we restore flux variables

$$\hat{u}_j^{n+1} = 2b^{-1}\left(b_j^{n+1/2}\right) - u_j^n, \quad j = \overline{1, J-1},$$

where b^{-1} is the function inverse to b, which then is corrected as follows

$$u_j^{n+1} = F\left(\hat{u}_j^{n+1}, m_j^{n+1}, M_j^{n+1}\right), j = \overline{1, J-1},$$

$$m_j^{n+1} = \min\left(U_{j-1/2}^{n+1}, U_{j+1/2}^{n+1}\right), \quad M_j^{n+1} = \max\left(U_{j-1/2}^{n+1}, U_{j+1/2}^{n+1}\right).$$

The boundary flux variables at $j = 0, J$ are set by formulas

$$u_0^{n+1} = b(g(\theta_{n+1})), \quad u_J^{n+1} = \bar{u}_J^{n+1},$$

first of them follows from boundary condition (5). As a result we have the finite-difference solution on the $(n+1)$-th time level.

By analogy to paper [10], it is possible to show that the offered modification of the CABARET scheme is monotonous [11] that provides its high resolution at shock capturing calculation.

4. Model calculations

In order to test the suitability of the method for the calculation tasks of embolization, the essential moments of embolization process were modeled in our numerical experiments.

4.1. Model function of Buckley–Leverett and exact solutions

In the calculations Buckley–Leverett function was used and it was defined by analytical :

$$b(S) = \sin^2\left[\frac{\pi S}{2}\right].$$

The plot of this function is shown in Figure 1. Prescribed in such way, the Buckley–Leverett function allows to construct exact solutions of initial-boundary value problem in some cases. It

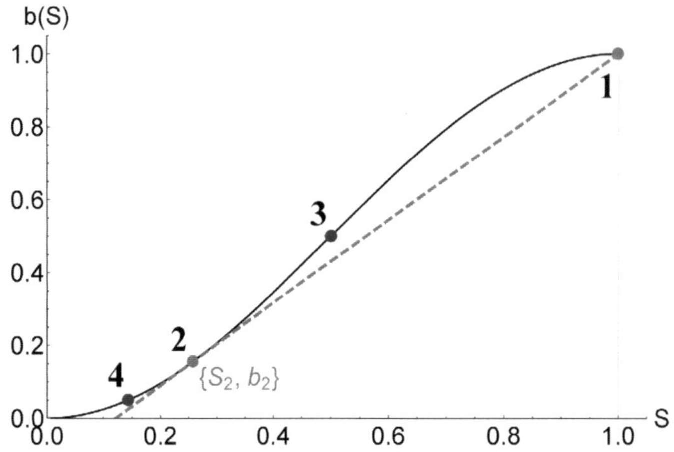

Figure 1. The model form of the Buckley–Leverett function.

must be noted that since the function of Buckley–Leverett is not convex, and the problem is hyperbolic, then compression shock waves as well rarefaction shock waves may be formed in the solution.

Let us consider the behaviour of the solution at the start of embolization. For simplicity, assume that at moment of time $\theta = 0$, ONYX is introduced and, due to it, blood saturation immediately reduced to a value $g_0 < 1$. And this concentration of ONYX at the input is being supported throughout the all time of modeling. Because the sound speed $b'(S)$ is equal to zero, where $S(x, 0) = 1$, no continuous movement is possible, thus information from $x = 0$ can not be spread inside the domain. The only option is the flow with rarefaction shock wave.

There are two qualitatively different cases. Point 2 with coordinates (S_2, b_2) plays an important role on the plot of the function $b(S)$. It is defined as the point of tangency of the curve $b(S)$ and straight line, which passes through the point 1 with coordinates $(1, 1)$. The point 1 corresponds to the initial state before the shock wave.

If condition behind the shock wave is the point 3 on Figure 1, which lies above the point 2, stable shock wave $1 - 3$ is formed, in accordance with the O. A. Oleinik criterion of shock wave stability [12]. The solution consists of two constant values, connected by a running gap. Two families of characteristics come on the shock wave, one is from x-axis and the other is from θ-axis.

This solution is given by the following equation:

$$S = \begin{cases} 1, & x \geq \theta \left(b(S_0) - b(g_0)\right) / \left(S_0 - g_0\right) \\ g_0, & x < \theta \left(b(S_0) - b(g_0)\right) / \left(S_0 - g_0\right) \end{cases}$$

If condition behind the shock wave is the point 4, which lies below the point 2, the shock wave $1 - 4$ will be unstable in accordance with the O. A. Oleinik criterion. Such wave at the time $\theta > 0$ decomposes into shock wave $1 - 2$ with lower intensity and self-similar solution abutting to it (centred rarefaction wave).

The analytical expression for this case has the form:

$$S = \begin{cases} 1, & x \geq b'(S_2)\theta \\ (b')^{-1}(x/\theta), & b'(g_0)\theta < x < b'(S_2)\theta \\ g_0, & x \leq b'(g_0)\theta \end{cases}$$

4.2. Results of numerical modeling

Calculations were carried out on a rectangular grid with constant steps in time and space. The number of partitions in space is equal to 100 points, the number of partition points in time takes the value 300 to 400 in different cases. Courant number, which is equal to $z = 0.4$, was chosen for calculations.

The results are shown in the figures of two types:
1) Plots of conservative quantities U_j^n — where calculated values U_j^n are depicted as circles, and for some cases the line corresponding to the exact solution is drawn.
2) Plots on a plane (x, θ), on which contour lines U_j^n (corresponding to characteristics) and shock waves are shown.

4.2.1. Constant boundary condition $S \equiv g_0 > S_2$. Without loss of generality, we assume $g_0 = 0.5$. The calculation results are shown in Figure 2. It is seen that the calculation reproduces strong discontinuity of solution well and no numerical oscillations near this gap are observed.

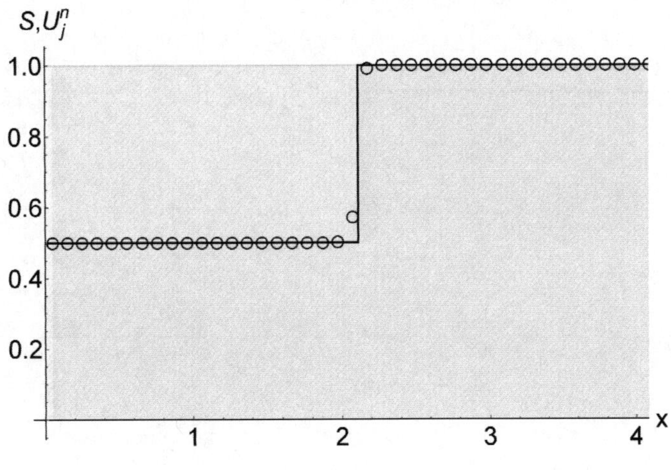

Figure 2. Numeric (circles) and exact (line) solutions for the case $g_0 = 0.5$ at time $\theta = 2.5$.

4.2.2. Constant boundary condition $S \equiv g_0 < S_2$. Condition $g_0 = 0.1$ is selected for calculation. The calculation results are shown in Figure 3. The calculation sufficiently reproduces the discontinuity decay, both in the shock region and the area of centred rarefaction wave.

4.2.3. The collision of two shock waves. Formation of shock waves is possible not only at the first moment of time, but also later, even in the case of continuous boundary conditions. As an example we consider the following boundary condition:

$$g_0 = \begin{cases} 0, & \theta \leq 2 \\ (\theta - 2)/8, & 2 < \theta < 10 \\ 1, & \theta \geq 10. \end{cases}$$

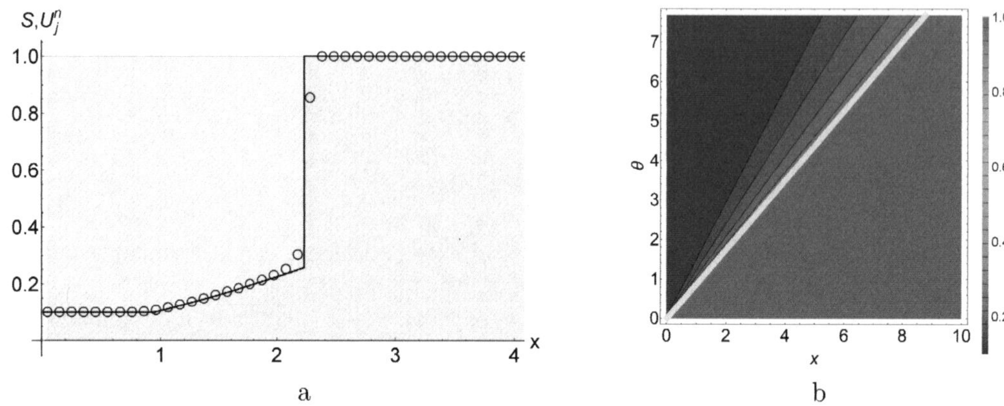

Figure 3. The case $g_0 = 0.1$. **a** – Numeric (circles) and exact (line) solutions at time $\theta = 2.5$. **b** – Result of simulation in (x, θ) plane with exact solution shock wave line (cyan).

In this case at the moment of time $\theta = 2$ a compression shock wave is formed. It catches up a rarefaction shock wave (formed at $\theta = 0$) and merges with it. As result a shock wave of lower intensity is formed. The exact solution is rather complicated, thus we consider only a numeric solution. Corresponding calculation is shown in Figures 4, 5a.

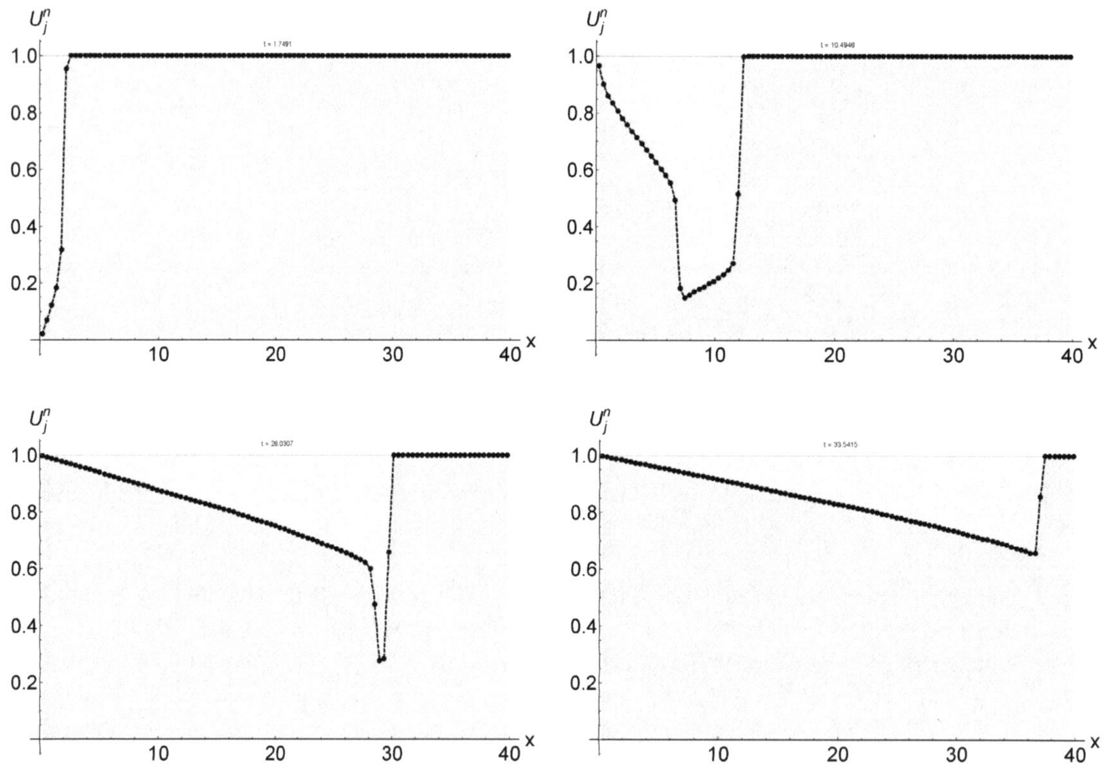

Figure 4. The collision of two shock waves. Successive stages of the process development. Black points – numerical result, dashed line – its linear interpolation.

Journal of Physics: Conference Series **722** (2016) 012009 doi:10.1088/1742-6596/722/1/012009

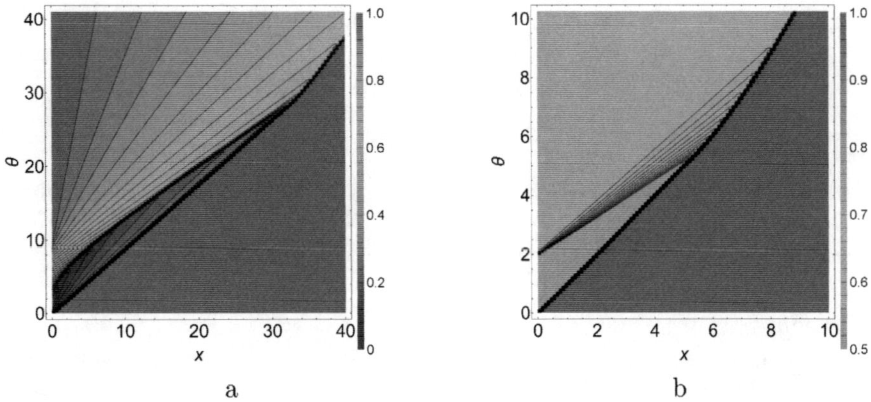

Figure 5. Result of simulation in (x, θ) plane. **a** – the collision of two shock waves, **b** – complicated flow with one shock wave. Shock wave curves roughness is not result of computation, but result of pre-plot interpolation.

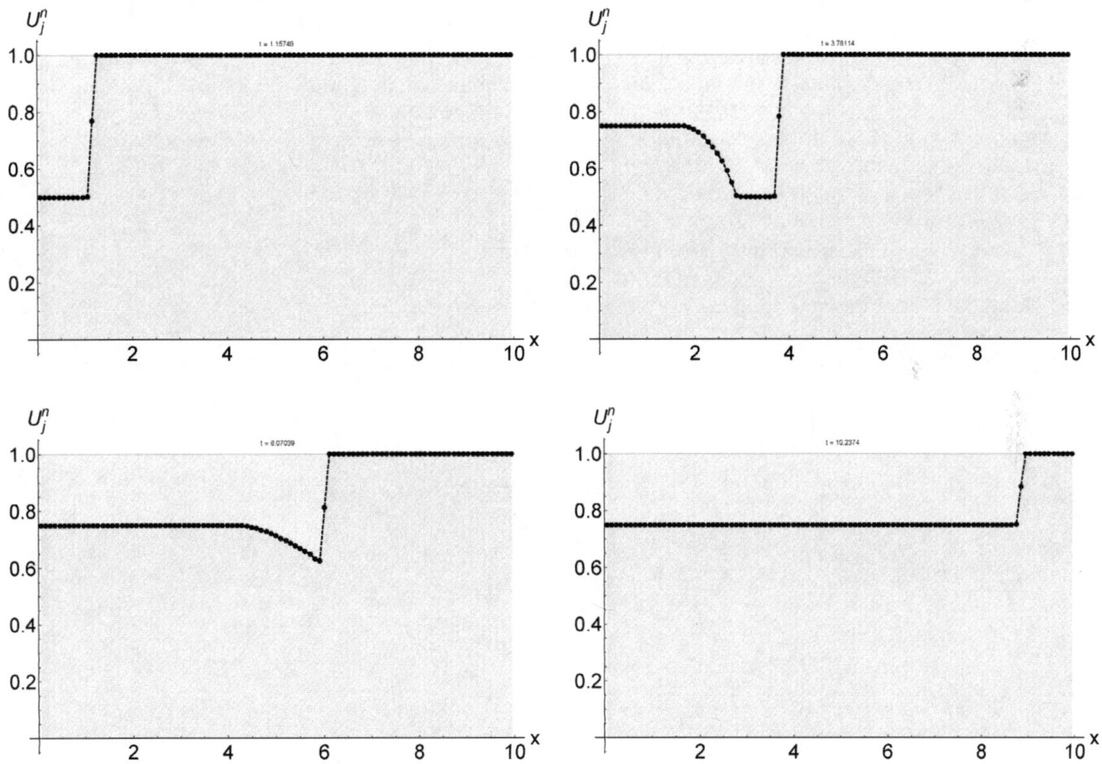

Figure 6. Complicated flow with one shock wave. Successive stages of the process development. Black points – numerical result, dashed line – its linear interpolation.

4.2.4. Complicated flow with one shock wave. Even discontinuous boundary conditions not always lead to the appearance of discontinuity in the solution. We choose function $g(\theta)$ as:

$$g_0 = \begin{cases} 0.5, & \theta < 2 \\ 0.75, & \theta \geq 2 \end{cases}$$

The shock wave formed by the discontinuity in the boundary conditions is unstable, so it breaks up into a centred compression wave when $\theta > 0$. Only numerical simulation is carried out in this case. The results corresponding to the calculation are shown in Figures 5b, 6.

5. Conclusions

The monotonous by Godunov version of Cabaret scheme, proposed by the authors, well reproduces the essential features of discontinuous flows, arising in the embolization problems. It can be used for further study on the process of AVM embolization.

Acknowledgments

The study was completed thanks to the support of Russian Foundation for Basic Research (project No. 14-01-00036).

References

[1] Klimov A B, Ryabuhin V E, Orlov K Y and Krilov V V 2012 *Surgery of the Cerebral Aneurysms* **3** 231–68
[2] Krivoshapkin A L et al 2013 *Bulletin SB RAMS* **33** 65–73
[3] Chupahin A P, Cherevko A A, Khe A K, Telegina N Y, Krivoshapkin A L, Orlov K Y, Panarin V A and Baranov V I 2012 *Patologija krovoobrashenija i kardiohirurgija (in russian)* **16** 27–31
[4] Panarin V A, Orlov K Y, Krivoshapkin A L, Chupahin A P, Cherevko A A, Khe A K, Telegina N Y and Baranov V I 2012 *Patologija krovoobrashenija i kardiohirurgija (in russian)* **16** 39–43
[5] Buckley S E and Leverett M C 1942 *Soc. Pertrol. Eng. J.* **146** 107–16
[6] Basniev K S, Kochina N I and Maximov M V 1993 *Undeground Hydromechanics* (Moscow: Nedra)
[7] Goloviznin V M 2005 *Dokl. Math.* **72** 619–23
[8] Karabasov S A and Goloviznin V M 2009 *J. Comp. Phys.* **228** 7426–51
[9] Ostapenko V V 2012 *Comput. Math. Math. Phys.* **52** 447–60
[10] Zyuzina N A and Ostapenko V V 2016 *Dokl. Math.* **93** 69
[11] Godunov S K 1959 *Mat. Sb. (in Russian)* **47** 271–306
[12] Oleinik O A 1959 *Uspehi Mat. Nauk (in Russian)* **14** 159–164

Gaseous bubble oscillations in anisotropic non-Newtonian fluids under influence of high-frequency acoustic field

R N Golykh

Chair of Measurement and Automation Methods and Tools, Biysk Technological Institute of FSBEI "Altai State Technical University named after I.I. Polzunov", Biysk city, Russia

E-mail: romangl90@gmail.com

Abstract. Progress of technology and medicine dictates the ever-increasing requirements (heat resistance, corrosion resistance, strength properties, impregnating ability, etc.) for non-Newtonian fluids and materials produced on their basis (epoxy resin, coating materials, liquid crystals, etc.). Materials with improved properties obtaining is possible by modification of their physicochemical structure. One of the most promising approaches to the restructuring of non-Newtonian fluids is cavitation generated by high-frequency acoustic vibrations. The efficiency of cavitation in non-Newtonian fluid is determined by dynamics of gaseous bubble. Today, bubble dynamics in isotropic non-Newtonian fluids, in which cavitation bubble shape remains spherical, is most full investigated, because the problem reduces to ordinary differential equation for spherical bubble radius. However, gaseous bubble in anisotropic fluids which are most wide kind of non-Newtonian fluids (due to orientation of macromolecules) deviates from spherical shape due to viscosity dependence on shear rate direction. Therefore, the paper presents the mathematical model of gaseous bubble dynamics in anisotropic non-Newtonian fluids. The model is based on general equations for anisotropic non-Newtonian fluid flow. The equations are solved by asymptotic decomposition of fluid flow parameters. It allowed evaluating bubble size and shape evolution depending on rheological properties of liquid and acoustic field characteristics.

1. Introduction

In recent decades, non-Newtonian fluids are more and more widely used in various areas of human activity. Examples are epoxy resins which are the basis for the preparation of polymeric composite materials. It is well known that medical implants based on such materials are rarely rejected by the body (in comparison with metal-based implants) and tissue irritation on the contact surface with such implants is minimized. Composite materials in aircraft structures make it possible to combine ease of the aircraft with its resistance to extreme mechanical loads as opposed to metal. Along with polymers, other examples of non-Newtonian fluids used in modern technology are coating materials, liquid crystals, etc. Today, progress of technology and medicine dictates the ever-increasing requirements (heat resistance, corrosion resistance, strength properties, impregnating ability, etc.) for non-Newtonian fluids and materials produced on their basis. Obviously, materials with improved properties are possible to obtain by modification of its physicochemical structure.

A number of researchers found that one of the most promising approaches to the restructuring non-Newtonian fluids is cavitation generated by high-frequency acoustic vibrations [1]. It allows to concentrate the low density energy (primary influence) in high density energy in the form of shock waves with the amplitude of pressure up to 100 MPa at bubble collapse and increasing the local temperature to 5000 K in cavitation nuclei. For example, this leads to the dispersion of the filler particles mixed in fluid into smaller ones [2], to reduce the viscosity of the liquid base due to breakdown of molecular bonds, etc. [3] Furthermore, associated effects that are stationary swirls acoustic flows and Eckart flows provide additional medium mixing that ensures uniform treatment.

The degree of appearence of each of the presented physical effects is primarily determined by the dynamics of the cavitation bubble or by the change of its size and shape under an external field influence (in particular, under high-frequency acoustic or ultrasonic field). Thus, the task of investigating the dynamics of a gas bubble in a non-Newtonian fluid is important.

2. Mathematical problem statement

Due to difficulties of experimental observation of bubble dynamics in non-Newtonian fluids caused by high speed of process, the most appropriate study method is mathematical modeling. The main mathematical difficulties of investigation of bubble dynamics in non-Newtonian fluid are caused by viscosity depending on absolute value and direction of shear rate. Today for the particular case of isotropic non-Newtonian fluids that have viscosity depending only on shear rate Euclidean norm and not depending on shear rate direction, some authors [4, 5, 6, 7] obtained ordinary differential equations for the spherical bubble radius as a function of time. However, for an anisotropic non-Newtonian fluid, whose viscosity depends not only on the module, but also on the shear direction, the problem cannot be reduced to the same ordinary differential equation, because gaseous bubble in anisotropic fluids deviates from spherical shape (Figure 1).

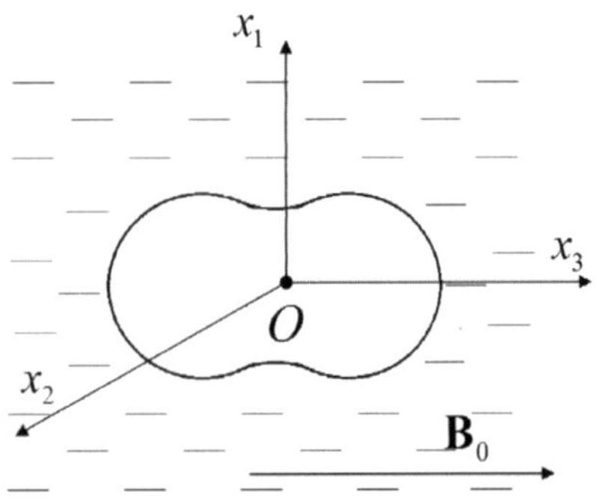

Figure 1. Schematic diagram of deformed bubble in anisotropic fluid.

Therefore, to solve this problem, a consideration of the general equations of non-Newtonian anisotropic medium flow (1–3) was made based on rheological model presented by V.S. Volkov [8]. These equations together with the speed and pressure of the fluid include the unit vector orientation $\mathbf{B}$ of the macromolecules, which actually determines the viscosity dependence on shear rate direction. Due to small bubble sizes compared with acoustic wave length, fluid can be assumed incompressible.

$$\operatorname{div} \mathbf{v} = 0; \tag{1}$$

$$\rho\left(\frac{\partial \mathbf{v}}{\partial t} + (\mathbf{v}, \nabla)\,\mathbf{v}\right) = -\nabla p + \operatorname{div} \mathbf{T}(\mathbf{D}, \mathbf{B}, t)\,. \tag{2}$$

$$\frac{dB_i}{dt} = \lambda\left(\sum_{s=1}^{3} D_{is}B_s - \sum_{m,\ n=1}^{3} D_{mn}B_iB_mB_n\right); \tag{3}$$

where $\mathbf{v}$ is the velocity of fluid, m/s; p is the pressure in fluid; $\mathbf{T}$ is the tensor of viscosity stresses in fluid generally depending on deformation rate tensor $\mathbf{D}$ (s^{-1}), vector of macromolecules orientation $\mathbf{B}$ (non-dimensional) and time t (s), λ is the non-dimensional "tumbling"parameter [8].

It is assumed, that at the initial time the vector of macromolecules orientation is uniform in area of fluid and equals to $\mathbf{B}_0$ (see Figure 1), which is collinear to x_3 axis and has unit norm $|\mathbf{B}_0| = 1$. It is easy to check, that unit norm of $\mathbf{B}$ always remains

$$\frac{d|\mathbf{B}|^2}{dt} = 2\left(\mathbf{B},\ \frac{d\mathbf{B}}{dt}\right) = \lambda\left(\sum_{i,\ s=1}^{3} D_{is}B_sB_i - \sum_{i=1}^{3} B_i^2 \sum_{m,\ n=1}^{3} D_{mn}B_mB_n\right) = \tag{4}$$
$$= \lambda\left(\sum_{i,\ s=1}^{3} D_{is}B_sB_i - \sum_{m,\ n=1}^{3} D_{mn}B_mB_n\right) = 0;$$

These equations are supported by boundary conditions on bubble wall

$$\mathbf{v} = \frac{d\mathbf{r}}{dt}; \tag{5}$$

$$\mathbf{Tn} = (p - p_B + 2\sigma K)\,\mathbf{n}; \tag{6}$$

and infinity

$$\mathbf{v} = 0; \tag{7}$$

$$p = p_{st} + p_A sin\,(\omega t)\,; \tag{8}$$

where $\mathbf{r}$ is the radius vector of bubble wall point, m; $\mathbf{n}$ is the normal vector to bubble wall; K is the curvature of bubble wall, m^{-1}; σ is the surface tension of liquid, N/m; p_{st} is the static pressure, Pa; p_A is the amplitude of acoustic pressure, Pa; p_B is the gaseous pressure in bubble interior, Pa; ω is the angular frequency of acoustic oscillations, s^{-1}.

Tensor of viscosity stresses in anisotropic non-Newtonian fluid is defined as follows:

$$\sum_{k=1}^{3}\sum_{l=1}^{3}\tau_{ijkl}\frac{dT_{kl}}{dt} + T_{ij} = \sum_{k=1}^{3}\sum_{l=1}^{3}\eta_{ijkl}\,(\mathbf{D})\left(\frac{\partial v_k}{\partial x_l} + \frac{\partial v_l}{\partial x_k}\right) \tag{9}$$

The 4th range tensors of viscosity η_{ijkl} and relaxation time τ_{ijkl} included in equation (9) satisfy the conditions of symmetry:

$$\eta_{ijkl} = \eta_{jikl} = \eta_{ijlk} = \eta_{klij}; \qquad \tau_{ijkl} = \tau_{jikl} = \tau_{ijlk} = \tau_{klij}. \tag{10}$$

Depending on subclass of anisotropic non-Newtonian fluid, the viscosity and relaxation time tensors yet satisfy some conditions:

- viscoelastic anisotropic fluids ($\eta_{ijkl}\,(\mathbf{D}) = const$);
- nonlinear viscous anisotropic fluids ($\tau_{ijkl} = 0$).

Finite element solution of these systems of equations consumes many computing time, especially when it comes to finding the evolution of the position of the free boundary. Therefore, the author has proposed a method of solution based on asymptotical decompositions of flow parameters into a series with powers of Reynolds and Eulerian numbers.

The method is described in the next section.

3. Method for solution

For solution of anisotropic non-Newtonian fluid flow equations, firstly the viscosity and relaxation time tensors are decomposed into spectrum (11) that defines viscosity depending on macromolecules orientation vector [8]

$$\eta_{ijkl} = \sum_{\alpha=1}^{3} \eta_\alpha (\mathbf{D})\, a_{ijkl}^{(\alpha)}; \qquad \tau_{ijkl} = \sum_{\alpha=1}^{3} \tau_\alpha a_{ijkl}^{(\alpha)} \tag{11}$$

$$a_{ijkl}^{(1)} = \frac{\delta_{ik}\delta_{je} + \delta_{ie}\delta_{jk}}{2} - \frac{\delta_{ij}\delta_{ke}}{3} - a_{ijkl}^{(2)} - a_{ijkl}^{(3)}; \qquad a_{ijkl}^{(2)} = \frac{3}{2}\left(N_{ij} - \frac{1}{3}\delta_{ij}\right)\left(N_{kl} - \frac{1}{3}\delta_{kl}\right)$$

$$a_{ijkl}^{(3)} = \frac{1}{2}\left(\delta_{ik}^{\perp} N_{je} + N_{ik}\delta_{je}^{\perp} + \delta_{ie}^{\perp} N_{jk} + N_{ie}\delta_{jk}^{\perp}\right)$$

where $N_{ij} = B_i B_j$ is the structural tensor, determined by vector of macromolecules orientation $\mathbf{B}$; $\delta_{ij}^{\perp} = \delta_{ij} - N_{ij}$ is the transverse Kronecker symbol which makes a projection on the direction orthogonal to the unit vector $\mathbf{B}$.

In expression for $a_{ijkl}^{(1)}$, the term $\delta_{ij}\delta_{ke}/3$ can be excluded due to fluid incompressibility. The spectral decomposition follows that viscosity of fluid is determined by four parameters (we assume that fluid has uniaxial anisotropy [8]) $\eta_\perp = \eta_1 = \eta_2$, $\eta_\parallel = \eta_3$, $\tau_\perp = \tau_1 = \tau_2$, $\tau_\parallel = \tau_3$.

Using the viscosity tensor representation, we define Reynolds and Eulerian number as

$$\mathrm{Re} = \rho\omega R_0^2 \sqrt{\frac{2}{\eta_\perp^2 + \eta_\parallel^2}} \tag{12}$$

$$\mathrm{Eu} = \frac{p_A}{(\rho\omega^2 R_0^2)} \tag{13}$$

where R_0 is an equilibrium radius of bubble, m.

According to the offered method for calculation of bubble shape, the parameters of the flow are decomposed into a series of power of Reynolds and Eulerian numbers. Herewith, the tensor of viscous stresses is represented as

$$\sum_{k=1}^{3}\sum_{l=1}^{3} \tau_{ijkl}\frac{dT_{kl}}{dt} + T_{ij} = \frac{1}{\mathrm{Re}} \sum_{k=1}^{3}\sum_{l=1}^{3} E_{ijkl}(\mathbf{D})\left(\frac{\partial v_k}{\partial x_l} + \frac{\partial v_l}{\partial x_k}\right). \tag{14}$$

And the decompositions into the series are

$$\mathbf{v} = \sum_{n=0}^{\infty} \frac{\mathbf{v}_n}{\mathrm{Re}^n}; \qquad \mathbf{v}_n = \sum_{m=1}^{\infty} \mathrm{Eu}^m \mathbf{v}_{nm}; \qquad \mathrm{div}\, \mathbf{v}_{nm} = 0; \tag{15}$$

$$p = \sum_{n=0}^{\infty} \frac{p_n}{\mathrm{Re}^n}; \qquad p_n = \sum_{m=1}^{\infty} \mathrm{Eu}^m p_{nm}; \tag{16}$$

$$\mathbf{B} = \mathbf{B}_0 + \sum_{n=0}^{\infty} \frac{\mathbf{B}^{(n)}}{\mathrm{Re}^n}; \qquad \mathbf{B}^{(n)} = \sum_{m=1}^{\infty} \mathrm{Eu}^m \left(\lambda_{nm}\mathbf{n}_\parallel + \mu_{nm}\mathbf{n}_\perp\right); \tag{17}$$

$$\mathbf{T} = \sum_{n=0}^{\infty} \frac{\mathbf{T}^{(n)}}{\mathrm{Re}^n}; \qquad \mathbf{T}^{(n)} = \sum_{m=1}^{\infty} \mathrm{Eu}^m \mathbf{T}^{(nm)}; \tag{18}$$

where $\mathbf{n}_{\parallel} = \mathbf{B}_0$, $\mathbf{n}_{\perp} = (x_1\mathbf{e}_1 + x_2\mathbf{e}_2) / (x_1^2 + x_2^2)$, λ_{nm} and μ_{nm} are the coefficients satisfying norm conditions

$$\forall N \in \mathrm{N} \cup \{0\}, \forall M \in \mathrm{N} \qquad \left(\sum_{n=0}^{N}\sum_{m=1}^{M} \lambda_{nm}\right)^2 + \left(\sum_{n=0}^{N}\sum_{m=1}^{M} \mu_{nm}\right)^2 = 1 \tag{19}$$

The convergence of the series is guaranteed at $\mathrm{Re} \geq 3$ and $\mathrm{Eu} \geq 1$, that is at high frequency acoustic field.

At $\mathrm{Re} \to \infty$ the equations (1, 2) reduce to the equations of flow of ideal isotropic fluid

$$\mathrm{div}\ \mathbf{v}_0 = 0; \tag{20}$$

$$\rho\left(\frac{\partial \mathbf{v}_0}{\partial t} + (\mathbf{v}_0, \nabla)\,\mathbf{v}_0\right) = -\nabla p_0 \tag{21}$$

with spherical symmetric solution

$$\mathbf{v}_0 = R_0^2 \frac{\partial R_0}{\partial t}\frac{\mathbf{r}}{r^3}; \tag{22}$$

$$p_0 = p_{st} + p_A sin\,(\omega t) + R_0 \frac{p_B - 2\sigma K - p_{st} - p_A sin\,(\omega t)}{|\mathbf{r}|}; \tag{23}$$

Using the power series, the fluid flow around the bubble is calculated by following algorithm.

(i) Choose N and M which are maximum of Re and Eu powers in series respectively.

(ii) To calculate the p_0, $\mathbf{v}_0$ in spherical symmetry representation (22, 23) and $\mathbf{B}^{(0)}$ through $\mathbf{v}_0$ by equation (3).

For (n, m)

(iii) To solve Poisson equation

$$\Delta p_{nm} = \mathrm{div}\ \mathrm{div}\ \mathbf{T}^{(nm)} - \rho \mathrm{div}\left(\sum_{k=0}^{n}\sum_{l=1}^{m-1} (\mathbf{v}_{kl}, \nabla)\,\mathbf{v}_{(n-k)(m-l)}\right); \tag{24}$$

(iv) To calculate viscosity tensor

$$\mathbf{T}^{(nm)} = \mathbf{T}^{(nm)}[\mathbf{v}_0,\ \mathbf{v}_1,\ \ldots,\ \mathbf{v}_{n-1}, \mathbf{v}_{n0},\ \mathbf{v}_{n1},\ \ldots,\ \mathbf{v}_{n(m-1)}] \tag{25}$$

(v) To calculate velocity $\mathbf{v}_{nm}$ through $\mathbf{v}_0,\ \mathbf{v}_1,\ \ldots,\ \mathbf{v}_{n-1}, \mathbf{v}_{n0},\ \mathbf{v}_{n1},\ \ldots,\ \mathbf{v}_{n(m-1)}$ by equation (2) and the bubble wall point radius vector $\mathbf{r}$ by boundary condition (5);

(vi) Go to step iii for $(n + 1, m)$, when $n < N$, for $(0, m + 1)$, when $n = N$ and $m < M$, and stop the algorithm when $n = N$ and $m = M$.

The method allows us obtaining the bubble shapes at different acoustic field characteristics and fluid rheological properties. Obtained results for two subclasses of fluids (nonlinear viscous anisotropic and viscoelastic anisotropic) are presented in next section.

4. Obtained results

4.1. Nonlinear viscous anisotropic fluids case

As previously mentioned, the viscosity tensor in nonlinear viscous anisotropic fluids is presented as

$$T_{ij} = \sum_{k=1}^{3}\sum_{l=1}^{3} \eta_{ijkl}\,(\mathbf{D})\left(\frac{\partial v_k}{\partial x_l} + \frac{\partial v_l}{\partial x_k}\right) \tag{26}$$

and fluid viscosity (η_{ijkl}) is determined by two parameters $\eta_{\parallel}$ and $\eta_{\perp}$. Obtained shapes of gaseous bubble at different frequencies of acoustic field and relations between viscosities $(\eta_{\parallel}/\eta_{\perp})$ in different directions is presented in following Figure 2.

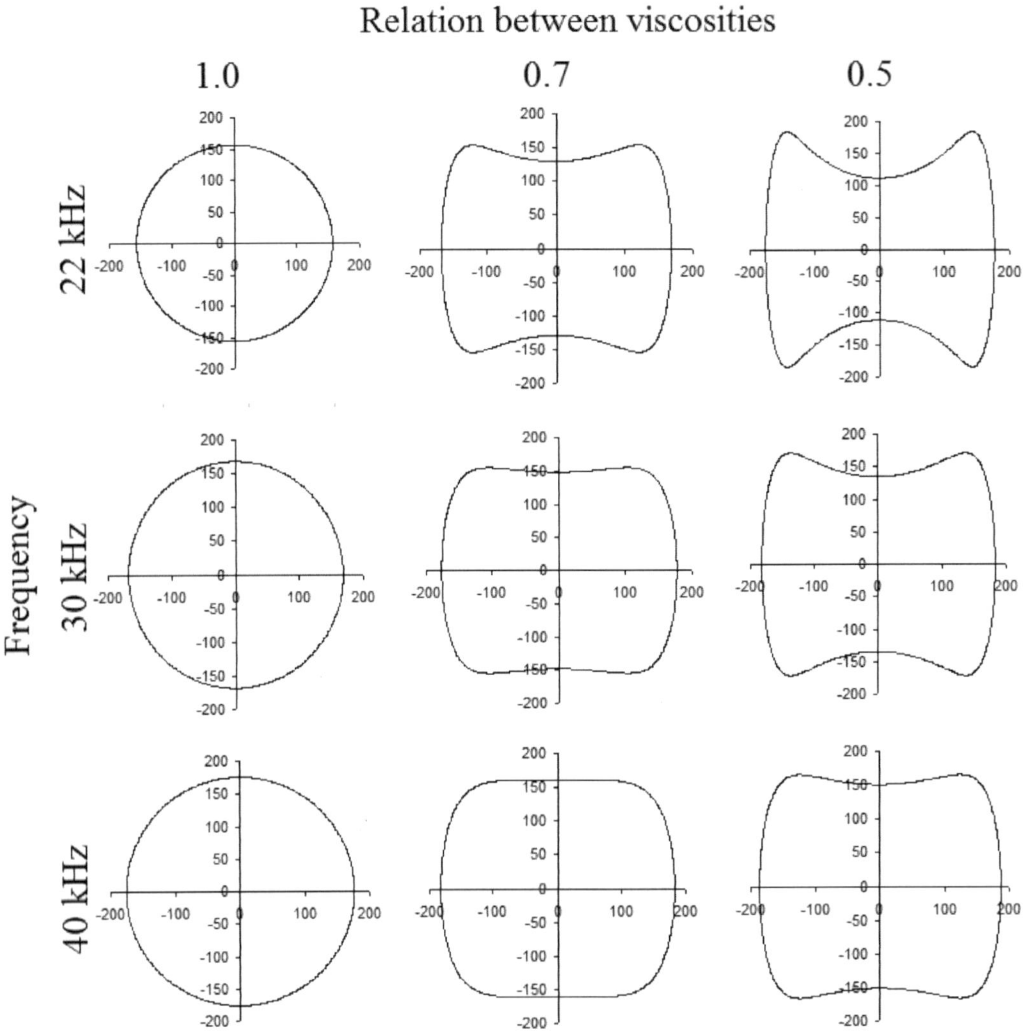

Figure 2. Bubble shapes in nonlinear viscous anisotropic fluids (unit division in axes is 1 μm; horizontal axis is x_3; vertical axis is x_1).

As follows from the figures, decreasing of the ratio between viscosities in the longitudinal and transverse directions leads to increase of deformation of the cavitation bubble. Thus, gaseous bubble has the largest dimensions in the longitudinal direction, since the fluid at shear in a given direction has the smallest viscosity. However, increasing of the acoustic field frequency causes the cavitation bubble shape is close to spherical. This happens due to the Reynolds number increasing and the viscosity influence weakening.

4.2. Viscoelastic anisotropic fluids case

As previously mentioned, the viscosity tensor in viscoelastic anisotropic fluids is presented as

$$\sum_{k=1}^{3}\sum_{l=1}^{3}\tau_{ijkl}\frac{dT_{kl}}{dt}+T_{ij}=\sum_{k=1}^{3}\sum_{l=1}^{3}\eta_{ijkl}\left(\frac{\partial v_k}{\partial x_l}+\frac{\partial v_l}{\partial x_k}\right) \tag{27}$$

and fluid rheological properties are determined by four parameters $\eta_{\|}$, $\eta_{\perp}$, $\tau_{\|}$ and $\tau_{\perp}$.

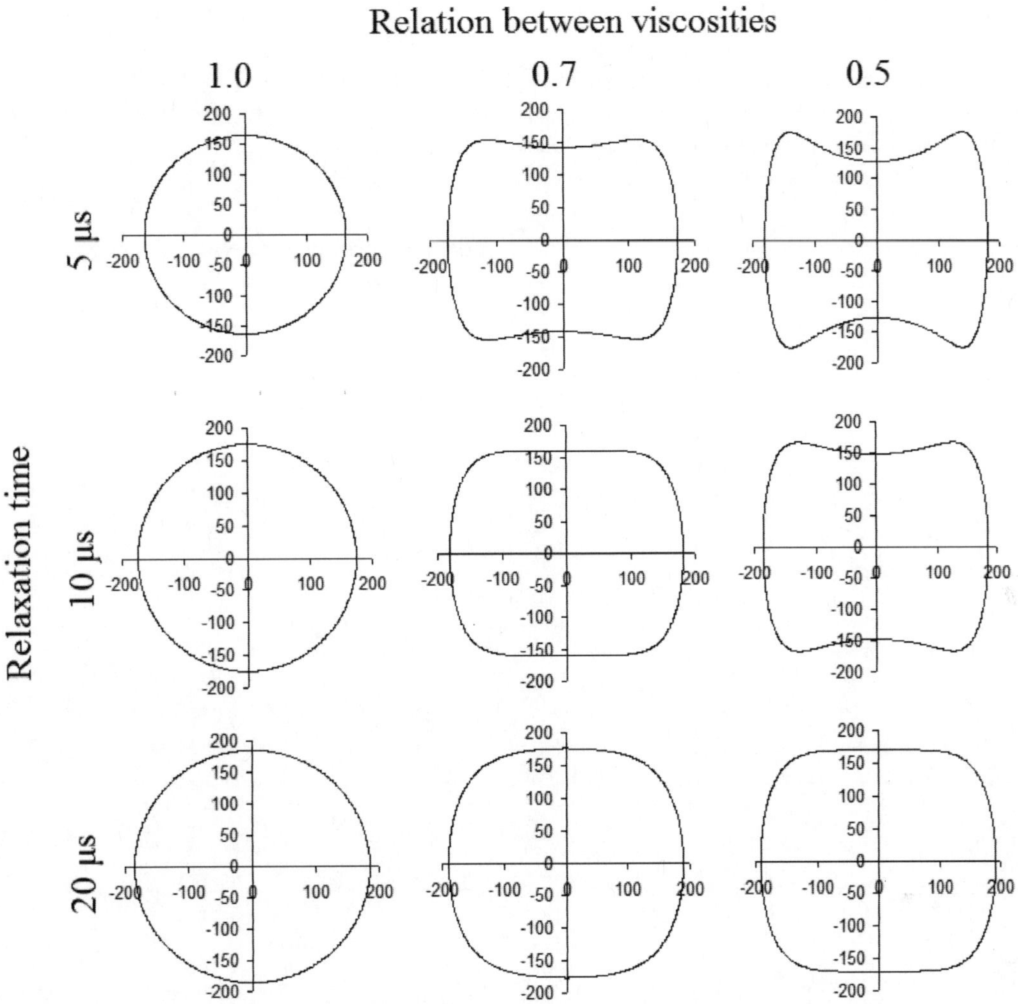

Figure 3. Bubble shapes in viscoelastion anisotropic fluids (unit division in axes is 1 μm; horizontal axis is x_3; vertical axis is x_1).

Obtained shapes of gaseous bubble at different intensities, relaxation times ($\tau_\perp$) and relations between viscosities ($\eta_\parallel/\eta_\perp$) in different directions is presented in following Figure 3.

As follows from the figures, decreasing of the ratio between viscosities in the longitudinal and transverse directions leads to an increase of deformation of the cavitation bubble, that is similar to nonlinear viscous anisotropic fluid case. However, increasing of the relaxation time causes the cavitation bubble shape is close to spherical. This is due to decreasing of viscosity tensor components absolute values, because in this case viscosity stresses is inertial, and its values is $\sim \frac{\eta}{\omega\tau}$.

5. Conclusion

Thus, the model of gaseous bubble oscillations in anisotropic non-Newtonian fluid was developed. To calculate bubble shape, the method based on asymptotical decomposition into Reynolds and Eulerian numbers power series of flow parameters was developed. This allowed evaluating the bubble size and shape achieved during oscillation and depending on rheological properties of anisotropic non-Newtonian fluids and acoustic field characteristics. It can be used for further investigations aimed at determination of charateristics of shock wave generated during cavitation bubble collapse. Shock wave characteristics will allow to analyse cavitation influence on molecular structure of fluid and evaluate acoustic influence modes providing required properties of the fluids and materials produced on their basis.

Acknowledgments
The reported study was financially supported by Russian Foundation for Basic Research (RFBR). Project No. 16-08-01298 a.

References

[1] Khmelev V N, Golykh R N, Shalunov A V, Bazhin V E and Nesterov V A 2015 *16th Int. Conf. of Young Specialists on Micro/Nanotechnologies and Electron Devices EDM Conf. Proc.* (IEEE Russia) pp 208–12
[2] Tolochko N K 2008 *Advanced materials* [In Russian] **2** 5–9
[3] Huang Y D, Liu L, Qiu J H and Shao L 2002 *Composit. Sci. Techn.* **62** 2153–9
[4] Golykh R N, Shalunova A V, Nesterov V A and Ilchenko E V 2014 *Americal Journal of Engineering Research* **12** 124–33
[5] Khmelev V N, Golykh R N, Shalunov A V, Khmelev S S and Karzakova K A 2014 *15th Int. Conf. of Young Specialists on Micro/Nanotechnologies and Electron Devices EDM Conf. Proc.* (IEEE Russia) pp 203–7
[6] Brujan E A and Williams P R 2005 *Rheology reviews* pp 147–72
[7] Tanasawa I and Yang W-J 1970 *Journal of Applied Physics* **41**(11) 124–33
[8] Volkov V S 2010 *Polymer Science Series* A **52**(11) 1105–13

Stratified flows and internal waves in the Central West Atlantic

K S Grigorenko[1], N I Makarenko[2,3], E G Morozov[4], R Yu Tarakanov[4] and D I Frey[4]

[1] Institute of Arid Zones RAS, 344006 Rostov-on-Don, Russia
[2] Lavrentyev Institute of Hydrodynamics SB RAS, 630090 Novosibirsk, Russia
[3] Novosibirsk State University, 630090 Novosibirsk, Russia
[4] Shirshov Institute of Oceanology, 117997 Moscow, Russia

E-mail: makarenko@hydro.nsc.ru

Abstract. In this paper, we study stratified flows and internal waves in the fracture zones of the Mid Atlantic Ridge. The results of measurements carried out in the 39th and 40th cruises of RV *Akademik Sergey Vavilov* in the autumn of 2014 and 2015 are presented. Hydrophysical properties of the near-bottom flows are studied experimentally on the basis of CTD- and LADCP profiling. Theoretical analysis involves mathematical formulation of stratified fluid flow which uses CTD-data obtained from field observation in the Vema Fracture Zone region. Spectral properties and kinematic characteristics of internal waves are calculated by finite element method.

1. Introduction

Deep basins of the Atlantic Ocean are filled with the water of an Antarctic origin which can spread to the mid-latitudes of the Northern Hemisphere [1, 2]. This water is usually called the Antarctic Bottom Water (AABW) which means [3] the water mass limited from above by an isotherm of potential temperature 2.0°C. Abyssal water exchange between basins occurs through the fractures of underwater ridges [2]. The bottom topography data show that the inflow of AABW into the channels of the fractures is most likely localized in a few passages. They include the equatorial Romanche and Chain fracture zones in the Mid-Atlantic Ridge and the Vema Fracture Zone in the Central West Atlantic. These fractures provide the transportation of cold abyssal waters from the West Atlantic to the equatorial region of the East Atlantic.

The fractures and underwater slopes present an active zone in which a sharp increase in the intensity of the turbulent mixing occurs [4]. Internal waves play significant role here due to the wave breaking which intensifies the mixing. Most interesting wave structures arise in powerful cataract flows near the sills of the fractures. These orographic obstacles disturb descending gravity currents by forced lee waves, attached hydraulic jumps, mixing layers etc. All these effects were observed by the authors in the Romanche and Chain fracture zones during the cruises of the RV *Akademik Ioffe* in the equatorial Atlantic [5]. Strong mixing in the sill region often leads to the splitting of inclined continuously stratified flow into the jets with different water densities [6]. In addition, shear flows over the obstacles can form critical layers which separate the flow of bottom cold water from opposite overflow. Such critical layers can reflect or absorb internal waves generated by the topography, so the upward propagation of these perturbations is blocked [7, 8].

In this paper we analyze the results of our measurements carried out recently in the abyssal channels of the Mid-Atlantic Ridge located in the Central West Atlantic. We investigate kinematic properties of short-period internal waves on the basis of the thermohaline characteristics of waters in the Vema Fracture Zone [9]. In this case, the mathematical formulation reduces to the solution of homogeneous eigenvalue problem, which is solved numerically by finite element method as a problem of the modal analysis.

2. Near-bottom stratified flows in the southern part of the Northern Mid-Atlantic Ridge

In the autumn of 2014 and 2015, the expeditions onboard the RV *Akademik Sergey Vavilov* carried out measurements of current velocities and thermohaline properties of bottom water in several quasi-zonal fractures in the southern part of the Northern Mid Atlantic Ridge (see Figure 1). These fractures connect deep basins of the West and East Atlantic, they include the Vema Fracture Zone (FZ) (10°50′ N) and a group of sub-equatorial fractures: Doldrums (8°15′ N), Vernadsky (7°40′ N), Bogdanov (07°09′ N) and Nameless fracture at 7°30′ N. The estimates of bottom water ($\theta < 2.0$°C) transport through this group of minor fractures based on our measurements are approximately 0.40 Sv (1 Sv = $10^6\,m^3/s$), which is close to 33% of the transport estimate through the Vema FZ (1.20 Sv) obtained in the same expeditions. The coldest bottom water temperatures among the investigated fractures were recorded also in the Vema FZ.

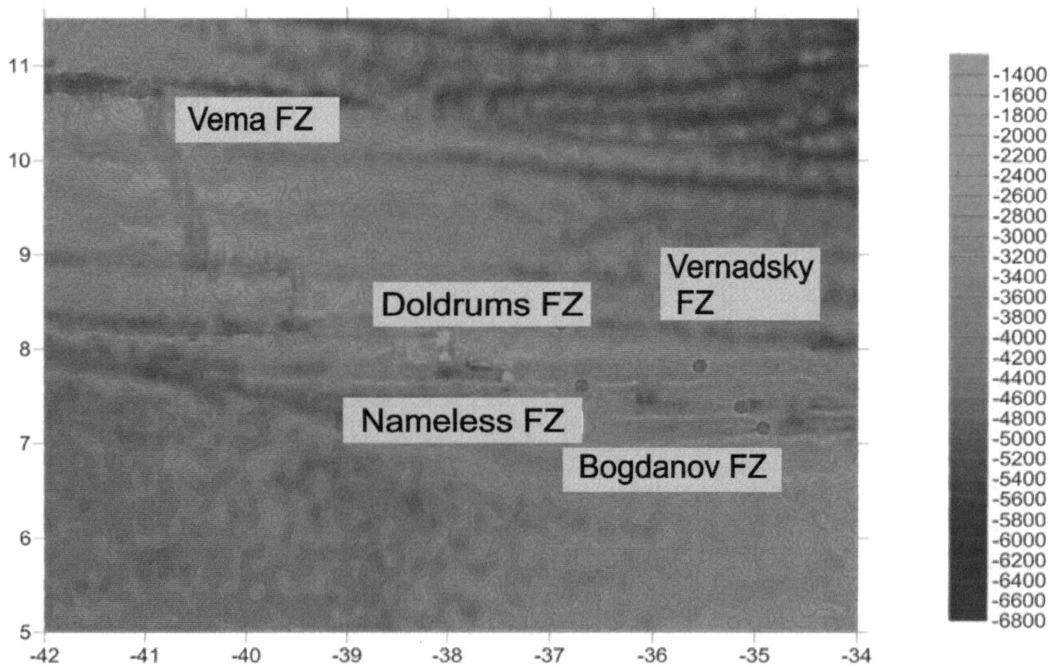

Figure 1. Fractures in the southern part of the North Atlantic Ridge

Hydrological measurements were carried out using the SBE-19plus temperature, salinity, and pressure profiler and acoustic current profiler LADCP (RDI Workhorse Sentinel, 300 kHz). The profiling stations were carried out almost to the bottom (3-5 m to the bottom) from a ship that maintained its position at a station point with an accuracy to 100 m. The echo sounder survey was initially carried out along the meridional section and then on the ship's route from one station to another. A chart of the bottom topography in the region of the Vema Fracture Zone is shown in Figure 2.

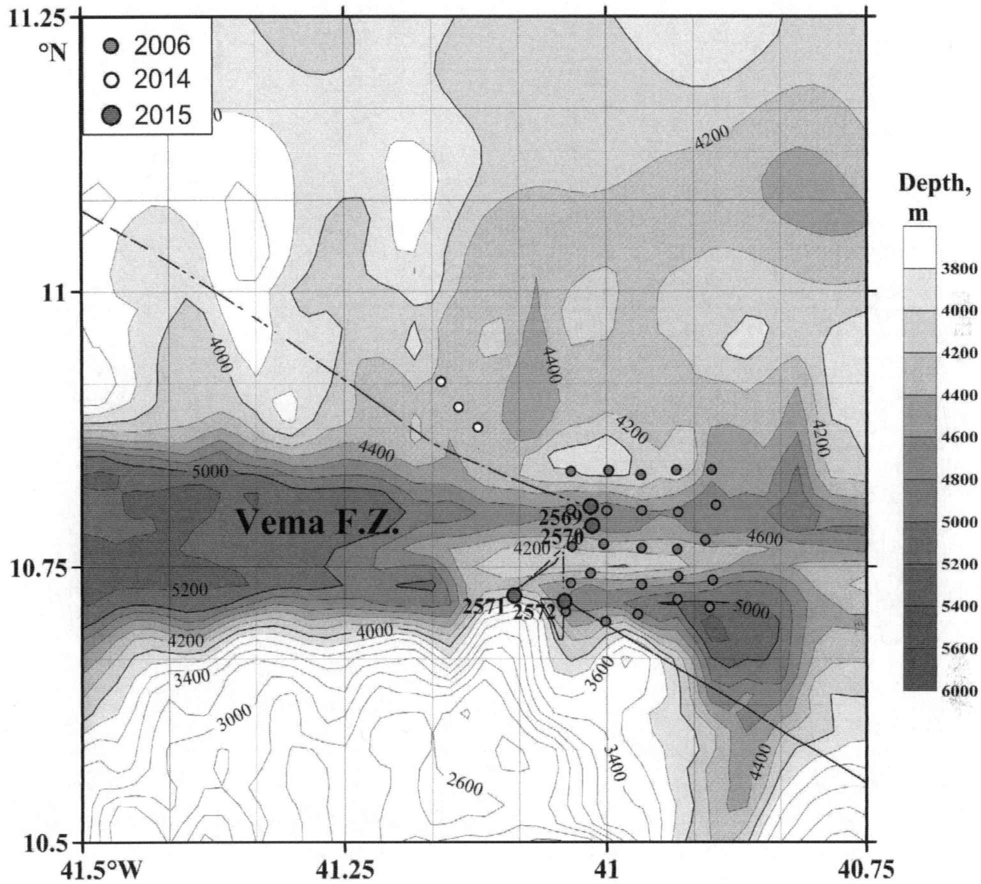

Figure 2. Bottom topography in the region of the Vema Fracture Zone

The pattern of abyssal currents is governed by the complex bottom topography in this region. Measurements in 2014 showed [9] that the bottom water flow in the Vema FZ directed to the East Atlantic splits into three streams. The northernmost flow is the widest. It flows through the channel which crosses the northern wall of the Vema FZ. The flow in this channel is directed to the northeast. Its width is approximately 10 km along the 3950 m isobath, and the depth of the channel reaches 4350 m (in all fractures we estimated their width along the isobaths close to the depth of the 2.0°C isotherm). Our measurements showed that the minimum potential temperature at the bottom here was 1.38°C, and the velocities of the bottom current in the northeastern direction reached 18 cm/s. An estimate of the AABW transport through this channel based on LADCP measurements is 0.08 Sv.

The main channel of the Vema FZ has latitudinal extension, and this channel is divided by a zonal submarine ridge into the northern and southern parallel sub-channels. The southern channel of the fracture appeared shallower than the northern one. The width of the southern channel along the 3780 m isobath in the region of measurements is close to 4 km and the depth is up to 4500 m. The minimum measured temperature at the bottom was 1.47°. The velocities of the easterly current were up to 30 cm/s. The AABW transport was estimated at 0.26 Sv.

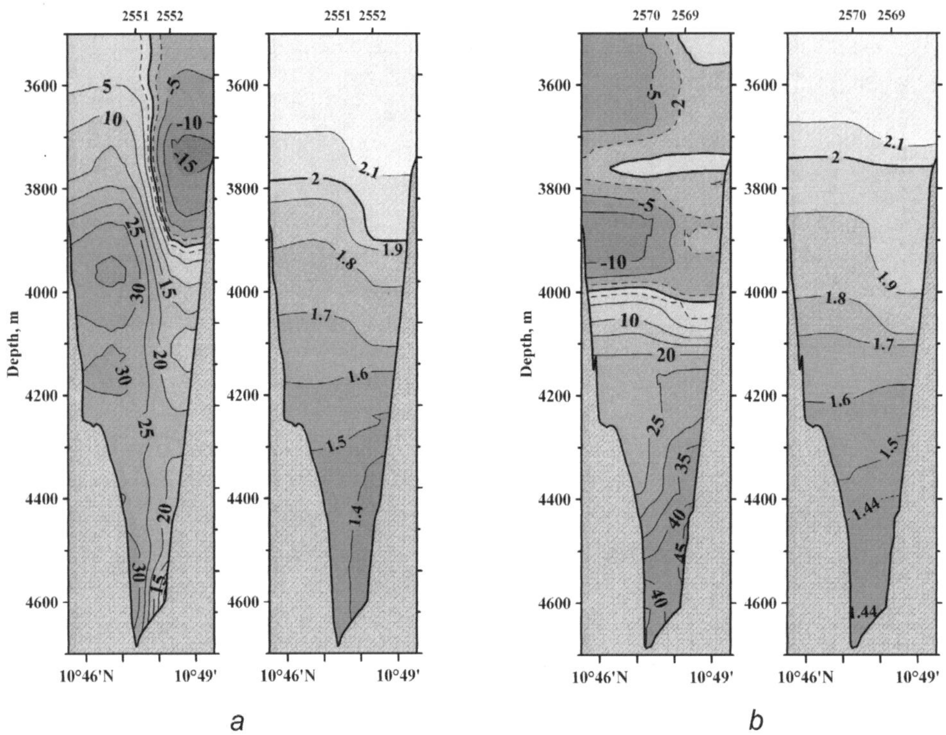

Figure 3. Structure of stratified flow in the northern channel of the Vema FZ. The cross-section along the longitude 41 01.0 W is shown: (a) 8 October 2014 and (b) 21 September 2015. Red color shows near-bottom flow directed to the east. Light blue color shows opposite overflow. Violet tones indicate the temperature stratification

In 2014, we occupied two stations over the sill of the northern channel placed along 41°01.0′ W longitude. The distance between these stations was approximately one mile. The width of the channel along the 3850 m isobath in this place is 7 km and the depth is up to 4690 m. The measurements showed that the minimum potential temperature at the bottom was 1.36°C, and the velocities of the bottom current in this channel reached 34 cm/s. The core of the maximum current velocity in the northern channel was displaced to the right relative to the direction of the flow, and the core of the minimum temperature was displaced to the left. The AABW transport was estimated to be 0.86 Sv. Thus, the total AABW transport through three abyssal channels of the Vema FZ was 1.20 Sv. This amount is more than five times of the transport of fresh water in the Amazon river.

In 2015, we repeated the measurements in the Vema FZ in order to observe the temporal variability of near-bottom stratified flows. It is interesting that the distribution of velocity and temperature changed essentially in northern channel compared to the data obtained in 2014.

Maximal velocity 46 m/s and minimal temperature 1.403°C were found in the deepest part of the northern channel (Figure 3b). This core of Antarctic Bottom Water was shifted to the northern wall in contrast with the flow picture from 2014 (see Figure 3a) which presents the pair of separated AABW jets displaced to the southern wall. The maximal AABW transport in the northern channel was estimated in 2015 to be less than 0.58 Sv. Such a decreasing after one year can be explained by location of intense AABW jet in the wide part of the northern channel in 2014. In general, this example illustrates high variability of intense abyssal flow affected by slight restoring buoyancy force in the presence of extremely weak stratification.

3. Modal characteristics of internal waves

Parameters of internal waves depend strongly on vertical structure of water density in the upper ocean layer, especially on the structure and stratification of the thermocline. Therefore the total density profile should be taken into account starting from the sea surface. Thus, we consider a mathematical model of internal gravity waves of small amplitude in a continuously stratified sea with constant depth h. We select the origin of the rectangular coordinate system (x, y, z) on the unperturbed free surface $z = 0$, so the z-axis is directed upwards. We assume that the water density in the state of rest depends only on the depth: $\rho_0 = \rho_0(z)$. Linearized equations and boundary conditions describing the dynamics of internal waves in dimensionless form and the approximation of the "rigid lid" are written as [10]

$$\mathbf{v}_t + \frac{1}{\rho_0}\nabla p = \mathbf{F}, \quad \rho_t + \rho_{0z} w = 0, \quad div\,\mathbf{v} = 0, \tag{1}$$

$$w = 0 \quad (z = 0,\ z = -h) \tag{2}$$

where dimensionless values are related to the dimensional values as

$$(\bar{x}, \bar{y}, \bar{z}, k) = h\,(x, y, z, \bar{k}), \qquad (\bar{t}, \sigma, f) = \sqrt{\frac{h}{g}}\,(t, \bar{\sigma}, \bar{f})\,.$$

$$(u, v, w) = \sqrt{gh}\,(\bar{u}, \bar{v}, \bar{w}), \qquad (\bar{\rho}_0, \bar{\rho}) = \bar{\rho}_0(0)\,(\rho_0, \rho), \qquad \bar{p} = \bar{\rho}_0(0)\,gh\,p.$$

Here, $\mathbf{v} = (u, v, w)$ is the velocity vector of wave perturbations; $\mathbf{F} = (fv, -fu, -\rho/\rho_0)$ is the mass force vector; p, ρ are wave perturbations of pressure and density, respectively; $\bar{f} = 2\Omega\,\sin\varphi$ is the Coriolis parameter; Ω is the angular velocity of the Earth; φ is the latitude; g is the acceleration due to gravity; σ is the frequency of wave oscillations; k is the horizontal wave number. We seek the solution of homogeneous problem (1), (2) as harmonic wave packets propagating horizontally with amplitudes depending on z,

$$(u, v, w, p, \rho) = \{U(z), V(z), W(z), P(z), R(z)\}\,\exp i(kx - \sigma t).$$

Substituting these functions into equations (1) and boundary conditions (2), we obtain the self-adjoint eigenvalue problem with spectral parameter k^2 at the given wave frequency σ,

$$\frac{d}{dz}\left(\rho_0 \frac{dW}{dz}\right) + k^2 \rho_0(z) \frac{N^2(z) - \sigma^2}{\sigma^2 - f^2} W = 0, \qquad W(0) = W(-1) = 0. \tag{3}$$

Here $N^2(z) = -g\rho_{0z}(z)/\rho_0(z)$ is the dimensionless Brunt — Väisälä frequency. Thus, according to the oscillation Sturm theorem, an infinite set of eigenvalues k_n^2 $(n = 1, 2, 3, ...)$ exists in the case $f < \sigma < \min N(z)$ considered here. These eigenvalues form an infinitely increasing monotonic series, and the complete orthogonal system of the corresponding eigenfunctions $W_n(z)$ (internal wave modes) exists [11].

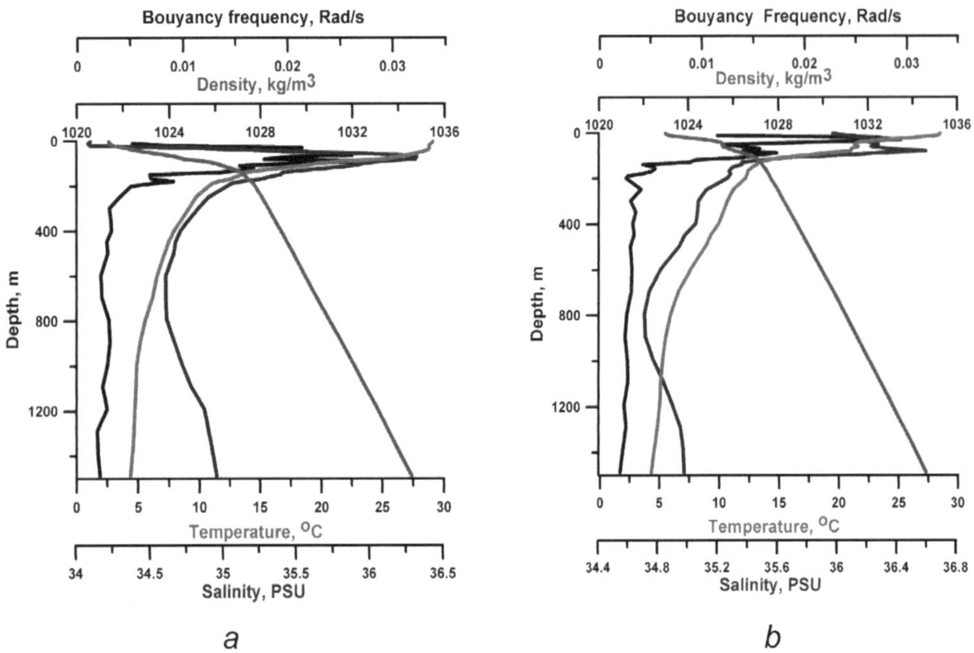

Figure 4. Termohaline characteristics measured in 2015: (a) station 2569 in the Vema FZ; (b) station 2577 in the Bogdanov fracture

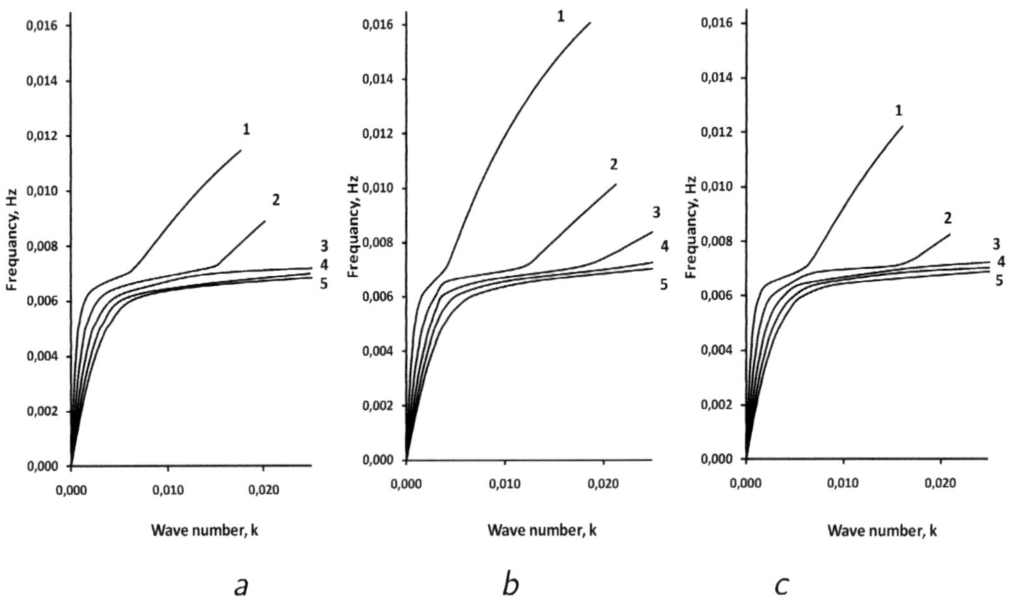

Figure 5. Dispersive curves: (a) station 2569 in the Vema FZ (2015); (b) station 2577 (the Bogdanov fracture, 2015); (c) station 2551 in the Vema FZ (2014)

The code based on finite elements method was used similarly to [12] by numerical solution of the spectral problem (3). We calculated modal characteristics of internal waves for the density profiles $\rho_0(z)$ obtained during the expeditions of RV *Akademik Sergey Vavilov* in 2014 and 2015. The Figure 4a illustrates the representative profiles of temperature, salinity, density and Brunt — Väisälä frequency obtained in the Vema FZ (station 2569 at 10°48.3 N, 41°01.0 W), and the Figure 4b shows CTD-data measured in the Bogdanov FZ (station 2577 at 07°09.6 N, 34°55.2 W). We note that the thickness of pycnocline at station 2569 is about 70 m. Station 2577 was occupied at a distance of about 200 km southeast of station 2569. The pycnocline there is thicker (approximately 120 m) due to the fresher upper layer.

Figures 5a and 5b show dispersion curves of modes 1-5 calculated by using the CTD-data from stations 2569 and 2577, respectively. In addition, Figure 5c shows dispersive curves calculated from the CTD-data measured in 2014 at station 2551, which was located in the same region in the Vema FZ as station 2569 in 2015. It is clear from this comparison that increased steepening (with respect to frequency) of dispersion curves at station 2577 occurs due to the difference between the density profiles in the upper layer at stations 2569 and 2577. This means that the pycnocline formed in the region of the Bogdanov FZ is an intense wave-guide for short-period internal waves. At the same time, stations 2569 and 2551 located at the same place in 2014 and 2015 show high similarity of dispersive curves despite high temporal variability of the local abyssal flows described in Section 2.

Acknowledgments

This work was supported by the grants of the Russian Foundation for Basic Research (project nos. 14-08-00107, 15-01-03942 and 15-31-53269). The expeditions were supported by the grants of the RFBR (project nos. 14-08-10002 and 15-08-10144). Analysis of data in the Vema FZ was supported by RNF grant 14-50-00095.

References

[1] Mantyla A W and Reid J L 1983 *Deep-Sea Res.* **30** 805–33
[2] Morozov E, Demidov A, Tarakanov R and Zenk W 2010 *Abyssal Channels in the Atlantic Ocean: Water Structure and Flows* (Dordrecht: Springer)
[3] Wüst G 1935 *Schichtung und Zirkulation des Atlantischen Ozeans. Die Stratosphäre. Wissenschaftliche Ergebnisse der Deutschen Atlantischen Expedition auf dem Forschungs- und Vermessungsschiff Meteor 1925-1927* (Berlin: Walter de Gruyter)
[4] Munk W H and Wunsch C 1998 *Deep-Sea Res.* **45** 1976–2009
[5] Morozov E G, Tarakanov R Yu, Lyapidevskii V Yu and Makarenko N I 2012 *Dokl. Earth Sci.* **446** 1211–14
[6] Gusev A V, Liapidevskii V Yu, Zatsepin A G and Nizov S S 2006 *Int. Conf. "Fluxes and Structures in Fluids"* (Moscow: IPMech RAS) pp 155–9
[7] Maslowe S A 1986 *Ann. Rev. Fluid Mech.* **18** 405–32
[8] Bacmeister J T and Pierrehumbert R T 1988 *J. Atmos. Sci.* **45** 63–80
[9] Morozov E G, Tarakanov R Yu and Makarenko N I 2015 *Oceanology* **55** 796–800
[10] LeBlond P H and Mysak L A 1978 *Waves in the Ocean* (Amsterdam: Elsevier)
[11] Yih C-S 1980 *Stratified Flows* (New York: Academic Press)
[12] Khartiev S M, Morozov E G, Grigorenko K S, Matishov D G, Solov'ev A N and Solov'eva A A 2013 *Dokl. Earth Sci.* **448** 70–3

Asymptotic theory of neutral stability curve of the Couette flow of vibrationally excited gas

Yu N Grigor'ev[1] **and I V Ershov**[2]

[1] Institute of Computational Technologies of SB RAS, Novosibirsk, 630090 Russia
[2] Department of Information Systems and Technologies, Novosibirsk State University of Architecture and Civil Engineering, Novosibirsk, 630008 Russia

E-mail: grigor@ict.nsc.ru, i_ershov@ngs.ru

Abstract. The asymptotic theory of neutral stability curve of the supersonic plane Couette flow of vibrationally excited gas is constructed. The system of two-temperature viscous gas dynamics equations was used as original mathematical model. Spectral problem for an eighth order linear system of ordinary differential equations was obtained from the system within framework of classical theory of linear stability. Transformations of the spectral problem universal for all shear flows were carried along the classical Dunn — Lin scheme. As a result the problem was reduced to secular algebraic equation with a characteristic division on "inviscid" and "viscous" parts which was solved numerically. The calculated neutral stability curves coincide in limits of 10 % with corresponding results of direct numerical solution of original spectral problem.

1. Introduction

The hydrodynamic stability theory used purely asymptotic methods during its formation. Application of analytical approaches allowed one to study some general mechanisms in hydrodynamic stability and turbulence transition. In particular, estimations of critical parameters of some flows important for practice were obtained with values close to real ones. Currently, asymptotic methods keep its role as independent source of theoretical results in spite of dominance of numerical studies in the flow stability problems. Primarily, it is of interest for investigation of optically active, thermally nonequilibrium and chemically reacting molecular gas flows in tracts of gas lasers, rocket engines and various technological plants. Studies in this domain require generalization of classical results [1] of the hydrodynamic stability theory to mediums and flows which were not considered earlier.

In the paper, constructing an asymptotic theory of neutral stability curve for the supersonic plane Couette flow of vibrationally excited gas is considered. It should be noted that earlier the corresponding spectral problem was not asymptotically investigated even for a perfect gas.

2. Statement of problem and basic equations

A problem of linear stability of a plane viscous Couette flow of a vibrationally excited molecular gas is considered. In the coordinate plane (x, y) the flow is bounded by two infinite parallel planes. It is assumed that the plane $y = 0$ is in a rest and the boundary $y = h$ moves uniformly in its own plane with a velocity U_0. The flow is described within the framework of two-temperature aerodynamics model [2, 3]. The quantities used for scaling are the channel width h, boundary velocity U_0, density ρ_0 and temperature T_0 of the stationary flow on the moving boundary of the channel, time $t_0 = L/U_0$, and pressure $p_0 = \rho_0 U_0^2$.

All parameters in the steady state flow are assumed depending only on the transverse coordinate y and vibrational temperature $T_{v,\,s}(y) = 0$. Original system of two-temperature aerodynamics equations is linearized near exact Navier — Stokes stationary solution

$$U_s(y) = y,\ T_s(y) = 1 + \frac{(\gamma-1)\mathrm{Pr}\mathrm{M}^2}{2}(1-y^2),\ \rho_s(y) = \frac{1}{T_s(y)},\ \ p_s(x_2) = \frac{1}{\gamma \mathrm{M}^2}. \quad (1)$$

which comply with following boundary conditions

$$U_s(0) = 0, \quad U_s(1) = 1, \quad \left.\frac{dT_s}{dy}\right|_{y=0} = 0, \quad T_s(1) = 1. \quad (2)$$

Here U_s is the velocity, ρ_s is the density, T_s is the static (translational) temperature of a gas, Pr, M are the Prandtl and Mach numbers in steady flow.

For linearized small-disturbance system $(\hat{u}_x,\ \hat{u}_y,\ \hat{\rho},\ \hat{T},\ \hat{T}_v,\ \hat{p})$ solutions in the form of travelling plane waves were considered:

$$\boldsymbol{q}(x,\,y,\,t) = \boldsymbol{q}(y)\,\mathrm{e}^{i\alpha(x-ct)}, \quad \boldsymbol{q}(x,\,y,\,t) = (\hat{u}_x,\ \hat{u}_y,\ \hat{\rho},\ \hat{T},\ \hat{T}_v,\ \hat{p}), \quad \boldsymbol{q}(y) = (u,\ \alpha v,\ \rho,\ \theta,\ \theta_v,\ p)\,,$$

where α is the wave number along periodic variable x, $c = c_r + ic_i$ is the complex phase velocity, i is the imaginary unity, $\boldsymbol{q}(y)$ is the vector of disturbance amplitudes. Substitution of $\boldsymbol{q}(x,\,y,\,t)$ in linearized system gives equations for small-disturbance amplitudes:

$$D\rho + \alpha\rho_s' v + \rho_s\sigma = 0, \quad (3)$$

$$\frac{1}{\mathrm{Re}}\triangle u - \rho_s Du - \alpha\rho_s v U_s' - i\alpha\varepsilon = 0, \quad (4)$$

$$\frac{\alpha}{\mathrm{Re}}\triangle v - \alpha\rho_s Dv - \varepsilon' = 0, \quad (5)$$

$$\frac{\gamma}{\mathrm{RePr}}\triangle\theta - \rho_s D\theta - \alpha\rho_s v T_s' - (\gamma-1)\sigma + \frac{2\gamma(\gamma-1)\,\mathrm{M}^2}{\mathrm{Re}}\left(u' + i\alpha^2 v\right)U_s' + \frac{\gamma_v\rho_s}{\tau}(\theta_v - \theta) = 0, \quad (6)$$

$$\frac{20\gamma\gamma_v}{33\mathrm{RePr}}\triangle\theta_v - \gamma_v\rho_s D\theta_v - \alpha\gamma_v\rho_s v T_s' - \frac{\gamma_v\rho_s}{\tau}(\theta_v - \theta) = 0, \quad \gamma\mathrm{M}^2 p = \rho_s\theta + \rho T_s. \quad (7)$$

In the equations (3) — (7), $\alpha_1 = \eta_b/\eta$ is the ratio of the bulk viscosity to the shear viscosity, γ is the adiabatic exponent, γ_v is a parameter characterizing a degree of nonequilibrium of a vibrational mode, τ is the characteristic relaxation time. Flow parameters are defined as follows: $\mathrm{Re} = \rho_0 h U_0/\eta$ and $\mathrm{M} = U_0/\sqrt{\gamma R T_0}$ are the Reynolds and Mach numbers, respectively, $\mathrm{Pr} = \eta c_V/\lambda$ is the Prandtl number; R is the gas constant, λ is the thermal conductivity coefficient. Hence,

$$D = i\alpha(U_s - c), \quad \sigma = \alpha(v' + iu), \quad \varepsilon = p - \frac{\sigma}{\mathrm{Re}}\left(\alpha_1 + \frac{1}{3}\right), \quad \triangle = \frac{d^2}{dy^2} - \alpha^2\,;$$

here and below the primes mean a differentiation with respect to the variable y. The system (3) — (7) is of order eight and with homogeneous boundary conditions

$$u\big|_{y=0} = u\big|_{y=1} = v\big|_{y=0} = v\big|_{y=1} = \theta\big|_{y=0} = \theta\big|_{y=1} = \theta_v\big|_{y=0} = \theta_v\big|_{y=1} = 0 \quad (8)$$

it defines a spectral problem, in which the eigenvalues are the complex phase velocities of perturbations $c = c_r + ic_i$, and the Mach number M and wave number α are parameters.

The asymptotic solutions of the system (3) — (7) for high Reynolds numbers Re are constructed in the form of a perturbation series

$$\boldsymbol{q}(y) = \boldsymbol{q}_0(y) + \mathrm{Re}^{-1}\boldsymbol{q}_1(y) + \ldots .$$

In the zeroth-order approximation one obtains a system of equations for inviscid perturbations. In the work [3] it has been shown that this system is reduced to the second order linear equation for pressure perturbation. Thus, the zeroth-order approximation allows us to find only two of the linearly independent solutions. One has to determine the remaining six solutions by direct consideration of the complete system.

3. Asymptotics of inviscid and viscous of solutions

Two linearly independent inviscid solutions of equation for pressure perturbation

$$p'' - \frac{T_s}{(y-c)^2}\left(\frac{(y-c)^2}{T_s}\right)' p' - \alpha^2\left(1 - \frac{\mathrm{M}^{*\,2}\,(y-c)^2}{T_s}\right)p = 0 \tag{9}$$

were constructed by the Frobenius method [4] using an asymptotical expansion in neighborhood of regular singular point $y = c$. Here

$$\mathrm{M}^{*\,2} = m^2\mathrm{M}^2, \quad m^2 = m_r^2 + i\,m_i^2,$$

$$m_r^2 = \frac{R_1\,(1+\gamma_v) + \Delta^2}{R_1^2 + \Delta^2}, \quad m_i^2 = -\frac{\gamma_v}{\gamma}\frac{(\gamma-1)\Delta}{R_1^2+\Delta^2},$$

$$R_1 = 1 + (\gamma_v/\gamma)\,, \quad \Delta = \alpha\tau\,(y-c)\,.$$

The solutions have the form

$$p_1(y) = (y-c)^3 - \frac{3\,(y-c)^4 T'_{s,c}}{4T_{s,c}} + \frac{(y-c)^5}{10}\left(\alpha^2 - \frac{6\,T''_{s,c}}{T_{s,c}} + \frac{6\,T'^2_{s,c}}{T^2_{s,c}}\right) + \ldots,$$

$$p_2(y) = 1 + \frac{\alpha^2 T'_{s,c}\,p_1(y)}{6\,T_{s,c}}\,\mathrm{Ln}(y-c) -$$

$$-\frac{\alpha^2\,(y-c)^2}{2}\left[1 + \frac{\alpha^2\,(y-c)^2}{6}\left(\frac{\alpha^2}{2} - \frac{T''_{s,c}}{T_{s,c}} + \frac{T'^2_{s,c}}{12\,T_{s,c}} + \frac{m^2_{r,c}\,\mathrm{M}^2}{T_{s,c}}\right)\right] + \ldots\,, \tag{10}$$

where

$$\mathrm{Ln}(y-c) = \begin{cases} \ln\eta, & \eta > 0, \\ \ln|y-c| - i\pi, & \eta < 0. \end{cases}$$

The index "c" denotes values of variables at point $y = c$ (in critical layer) where the phase velocity is equal to velocity of undisturbed flow. Other components of inviscid solutions

$$\boldsymbol{V}_k(y) = [u_{ki}(y),\, v_{ki}(y),\, \theta_{ki}(y),\, \theta_{v,ki}(y)]\,, \quad k = 1, 2, \tag{11}$$

are expressed through obtained p_1, p_2.

To find the viscous solutions the system (3) — (7) using some simplifications was reduced to the form which is analogical to the Dunn — Lin "viscous" system [5]:

$$v' + iu - i\,(y-c)\,\frac{\theta}{T_s} = 0, \tag{12}$$

$$u''' - i\alpha\mathrm{Re}\,(y-c)\,\frac{u'}{T_s} = 0, \tag{13}$$

$$\theta'' - i\alpha\mathrm{PrRe}\,(y-c)\,\frac{\theta}{T_s} + \frac{\gamma_v\mathrm{PrRe}}{\gamma\tau\,T_s}\,(\theta_v - \theta) = 0, \tag{14}$$

$$\theta_v'' - \frac{33}{20}\,i\alpha\mathrm{PrRe}\,(y-c)\,\frac{\theta_v}{T_s} - \frac{33\mathrm{PrRe}}{20\gamma^2\tau\,T_s}\,(\theta_v - \theta) = 0. \tag{15}$$

For diatomic gases one can approximately suppose that $33/(20\gamma) \approx 1$. It allows us by summation and subtraction of Eqs (14), (15) to introduce the equations for auxiliary functions

$$\theta_+ = \theta + \theta_v\frac{\gamma_v}{\gamma}, \quad \theta_- = \theta - \theta_v.$$

Then the temperatures are correspondingly expressed by formulae

$$\theta = \frac{\gamma\,\theta_+ - \gamma_v\,\theta_-}{\gamma + \gamma_v}, \quad \theta_v = \frac{\gamma\,(\theta_+ + \theta_-)}{\gamma + \gamma_v}. \tag{16}$$

The momentum equation (13) and equations for functions θ_+, θ_- were reduced to the Airy equations. Their solutions were presented through the generalized Airy functions of the first and second orders $A_k(z,\,p)$ [6]. As a result, using (16) the solutions for temperatures were found. This allowed us to splinter the last equation from system (12) — (15) and reduce it in such a way to a sixth order system.

Linearly independent solutions for disturbances of transversal velocity are obtained by integration of Eq. (12) for two alternatives under $\theta = 0$, $u \neq 0$ and $u = 0$, $\theta \neq 0$. They also were expressed through the generalized Airy functions of the first and second orders.

As a result, it was shown that the linearly independent solutions of simplified system have the form:

$$\boldsymbol{V}_3(y) = \left[u_{1v}(y),\, v_{1v}^{(1)}(y),\, 0\right], \quad \boldsymbol{V}_4(y) = \left[0,\, v_{1v}^{(2)}(y),\, \theta_{1v}\right], \tag{17}$$

$$\boldsymbol{V}_5(y) = \left[u_{2v}(y),\, v_{2v}^{(1)}(y),\, 0\right], \quad \boldsymbol{V}_6(y) = \left[0,\, v_{2v}^{(2)}(y),\, \theta_{2v}\right]. \tag{18}$$

$$u_{1,\,2v}\,(y) = \frac{dY}{d\zeta}\left(\frac{dy}{dY}\right)^{3/2} A_{1,\,2}\,(i\zeta,\,1), \quad v_{1,\,2v}^{(1)}\,(y) = -i\left(\frac{dy}{dY}\right)^{5/2}\left(\frac{dY}{d\zeta}\right)^{2} A_{1,2}\,(i\zeta,\,2),$$

$$v_{1,\,2v}^{(2)}(y) = \frac{i(y-c)}{(\gamma+\gamma_v)\,T_s}\left(\frac{dY_\theta}{d\zeta_\theta}\right)\sqrt{\left(\frac{dy}{dY_\theta}\right)^3}\left[\gamma\,A_{1,2}\,(i\zeta_\theta,\,1) - \frac{\gamma_v}{\sqrt{2}}\,A_{1,2}\left(i\sqrt[3]{2}\,\zeta_\theta,\,1\right)\right],$$

$$\theta_{1,\,2v} = \frac{1}{\gamma+\gamma_v}\sqrt{\frac{dy}{dY_\theta}}\left[\gamma\,A_{1,\,2}\,(i\zeta_\theta) - \frac{\gamma_v}{\sqrt[3]{2}}\,A_{1,\,2}\left(i\sqrt[3]{2}\,\zeta_\theta\right)\right],$$

where

$$Y = \sqrt[3]{\left(\frac{3}{2}\int\limits_c^y \sqrt{\frac{t-c}{T_s}}\,dt\right)^2}, \quad \zeta = Y\,\sqrt[3]{\alpha\mathrm{Re}}, \quad Y_\theta = Y\,\sqrt[3]{\mathrm{Pr}}, \quad \zeta_\theta = \zeta\,\sqrt[3]{\mathrm{Pr}}.$$

4. Secular equation

For derivation of the secular equation it is necessary to require that a linear combination of independent solutions (11), (17), (18) satisfies the uniform boundary conditions (8)

$$c_1\,\boldsymbol{V}_1(0)+c_2\,\boldsymbol{V}_2(0)+c_3\,\boldsymbol{V}_3(0)+c_4\,\boldsymbol{V}_4(0)+c_5\,\boldsymbol{V}_5(0)+c_6\,\boldsymbol{V}_6(0)=0, \tag{19}$$

$$c_1\,\boldsymbol{V}_1(1)+c_2\,\boldsymbol{V}_2(1)+c_3\,\boldsymbol{V}_3(1)+c_4\,\boldsymbol{V}_4(1)+c_5\,\boldsymbol{V}_5(1)+c_6\,\boldsymbol{V}_6(1)=0. \tag{20}$$

Large values of arguments of "viscous" solutions on both boundaries $\zeta(y)\geq O(10)$, $y=0,\,1$ allows one to estimate it using the asymptotic formulae [6] for the generalized Airy functions

$$A_{\pm}(x,\,p)\simeq\frac{(\pm 1)^p}{2\sqrt{\pi}}\,x^{-(2p+1)}\mathrm{e}^{\pm\eta},$$

$$\eta=\frac{2}{3}\,x^{3/2},\quad A_1(x,\,p)\sim A_-(x,\,p),A_2(x,\,p)\sim A_+(x,\,p).$$

Normalization of solutions (17), (18) using its maximal values on the one of boundaries gives a possibility to replace exponentially small elements in the determinant of system (19), (20) by zeros. As a result, the determinant obtains a disperse form

$$\Delta=\begin{vmatrix} u_{1i}(0) & u_{2i}(0) & u_{1i}(0) & 0 & 0 & 0\\ v_{1i}(0) & v_{2i}(0) & v_{1v}^{(1)}(0) & v_{1v}^{(2)}(0) & 0 & 0\\ \theta_{1i}(0) & \theta_{2i}(0) & 0 & \theta_{1v}(0) & 0 & 0\\ u_{1i}(1) & u_{2i}(1) & 0 & 0 & u_{2v}(1) & 0\\ v_{1i}(1) & v_{2i}(1) & 0 & 0 & v_{2v}^{(1)}(1) & v_{2v}^{(2)}(1)\\ \theta_{1i}(1) & \theta_{2i}(1) & 0 & 0 & 0 & \theta_{2v}(1) \end{vmatrix}=0\,.$$

This transition unweighted essentially a derivation of a secular equation.

Use of the Airy functions allows us to represent the viscous part of secular equation through the tabulated Tietjens function $F(Z)$ and its derivative $F'(Z)$, which are usually applied in the traditional asymptotic theory of stability [1, 6]. The secular equation obtained by such a way has the form:

$$\Pi(0)=\frac{C_1p_1'(0)+p_2'(0)}{\alpha^2\left[C_1p_1(0)+p_2(0)\right]}=-c\,I\left[\frac{F(Z)+KJ_1(Z_\theta)}{1-IF(Z)}\right], \tag{21}$$

$$C_1=-\frac{v_{2i}(1)}{v_{1i}(1)},\;K=\frac{(\gamma-1)\,\mathrm{M}^2c^2}{T_s(0)}\left[1-\frac{\tau\gamma_v}{\gamma+\gamma_v-i\gamma\alpha\tau c}\right],\;I=\frac{3}{2}\frac{\sqrt{T_s}}{c^{3/2}}\int\limits_0^c\sqrt{\frac{t-c}{T_s(t)}}\,dt,$$

$$J_1=\frac{\left(\gamma\sqrt{2}-\gamma_v\right)\left(\widetilde{G}\left(Z_\theta\right)-\gamma_v\left(\gamma\sqrt{2}-\gamma_v\right)^{-1}\left(\sqrt[3]{2}-1\right)Z_\theta\right)}{\sqrt[6]{2}\left(\gamma\sqrt[3]{2}-\gamma_v\right)\left(1+\left(\sqrt[3]{2}-1\right)\left(\gamma\sqrt[3]{2}-\gamma_v\right)^{-1}Z_\theta^3\,\widetilde{G_\theta}\left(Z_\theta\right)\left[1-F\left(Z_\theta\right)\right]\right)},$$

$$\widetilde{G}\left(Z_\theta\right)=-\frac{A_1\left(-Z_\theta,\,1\right)}{Z_\theta A_1\left(-Z_\theta\right)},\quad Z_\theta=-\zeta_\theta.$$

5. Numerical calculations of neutral curves

The secular equation (21) has a characteristic structure, which coincides with analogical equations [5, 7]. The left "inviscid" part of Eq. (21) depends on the phase velocity c and wave number α. At the same time the right "viscous" part of Eq. (21) is expressed through tabulated functions of variable Z, also depending on the phase velocity c. Therefore, points on neutral stability curves $\mathrm{Re}\,(\alpha,\,\gamma_v,\,\mathrm{M})$ are calculated on plane $(\mathrm{Re},\,\alpha)$ in the next sequence. The value of the phase velocity c was set for fixed values of the Mach number M and the degree of vibrational nonequilibrium γ_v in interval $c = [0, 1]$ with step $\Delta c = 10^{-3}$. The integral I on the right-hand side of Eq. (21) was calculated by the Simpson formula. The real and imaginary parts of the right-hand side of Eq. (21) depend on the variable Z. The arrays of their values were calculated for $Z = [0, 10]$ using tables [5, 8]. For $Z = [10, 50]$, we used an asymptotic formula

$$\Pi(0) = -cqI\left\{1 + i + 2iqI + \left[\frac{5i}{2} + K\mathrm{Pr}^{-1/2}(1+i)\right]q\right\},$$

where $q = Z^{-3/2}/\sqrt{2}$. In both cases the table step $\Delta Z = 0.04$ [5] was picked out.

The arrays on the left-hand side of Eq. (21) for a fixed value of the phase velocity c were calculated for $\alpha = [1, 3]$ with a step $\Delta\alpha = 10^{-3}$. Calculated arrays on the right-hand and left-hand sides of Eq. (21) were compared up to obtaining coincide within the accuracy 10^{-8} if such a state could be achieved for a given value of c. Then calculation was repeated for the next value of the phase velocity c. As a result, we constructed arrays of values of wave numbers $\overline{\alpha}_c$, phase velocities $\overline{c}_c$, and of variable $\overline{Z}_c$, which correspond to points on neutral curve. Using formula

$$\mathrm{Re}_c(\overline{\alpha}_c) = \frac{\overline{Z}_c^{\,3}}{\overline{\alpha}_c}\left[\frac{3}{2}\int\limits_0^{\overline{c}_c}\sqrt{\frac{\overline{c}_c - y}{T_s(y)}}\,dy\right]^{-2},$$

the values of the Reynolds number Re_k on the neutral curve were calculated. The neutral curve for mode I corresponds to the values of the phase velocity in interval $\overline{c}_k \in (0.5, 1)$. At the same time the other points for $\overline{c}_k \in (0, 0.5)$ belong to the neutral stability curve of mode I.

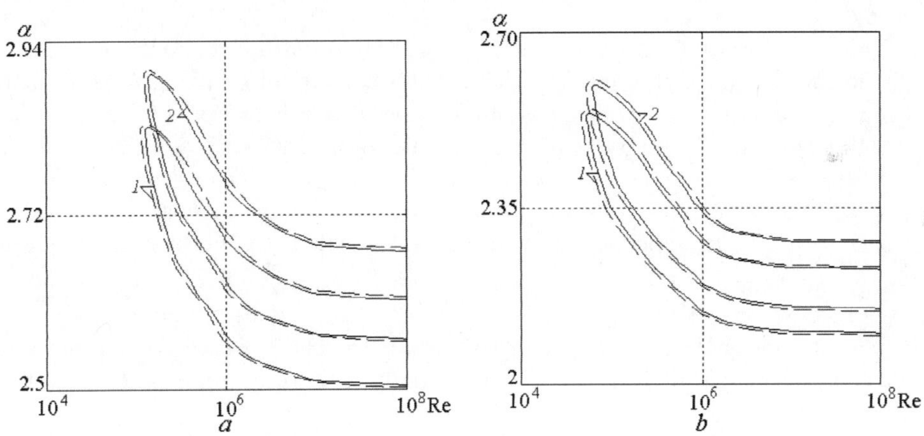

Figure 1. Neutral stability curves $\mathrm{Re}_c(\alpha_c)$ for modes I (*a*) and II (*b*) at M=3, $\gamma_v = 0$ (*1*) and $\gamma_v = 0.667$, $\tau = 1$ (*2*). The solid and dashed curves show the results of calculations by spectral problem (3) — (7) and the asymptotic theory, correspondingly.

Figure 1 shows in semi-logarithmic coordinates the results of calculations of neutral stability curves for the perfect and vibrationally excited gases at the Mach number M = 3. The curves plotted by dashed lines are derived from secular equation (21) of the asymptotic theory. The solid lines are presented for comparison with the corresponding results of numerical calculation of spectral problem (3) — (7). It is seen, that asymptotic curves are in satisfactory agreement with numerical calculation of origin spectral problem (3) — (7). In particular, dissipative effect of the vibrational relaxation is reflected clearly. The transition to the asymptotic theory extends slightly an instability area and reduces the critical Reynolds numbers.

Table 1. Numerical values Re_{cr} and α_{cr} for modes I and II.

	M = 3				M = 4			
				Mode I				
γ_v	Re^s_{cr}	α^s_{cr}	Re^a_{cr}	α^a_{cr}	Re^s_{cr}	α^s_{cr}	Re^a_{cr}	α^a_{cr}
0	123900	2.835	112000	2.842	82703	2.570	74800	2.576
0.667	137303	2.901	124100	2.910	91650	2.628	82800	2.639
				Mode II				
0	50060	2.546	44700	2.552	33415	2.307	29700	2.314
0.667	56061	2.603	50100	2.610	37421	2.359	33400	2.365

The table 1 allow one to compare data of numerical calculations of spectral problem (3) — (7) and asymptotic theory with respect to critical values of the Reynolds numbers Re_{cr} and wave numbers α_{cr}. It is seen that values Re^a_{cr} are approximately $9 \div 10$ percent less than the corresponding values Re^s_{cr}.

6. Conclusion

The asymptotic solutions of spectral problem of linear stability theory for the supersonic plane Couette flow of the vibrationally excited gas are constructed. The found solutions are universal for any two-dimensional perturbations of plane shear flows of molecular gases with one dedicated vibrational mode.

The asymptotic curves of neutral stability satisfactorily approximate results of direct numerical solution of spectral problem (3) — (7). These calculations once again confirm existence in a molecular gas of a dissipative effect, which is caused by relaxation of vibrational modes.

The essential point of the study is derivation and solution of secular equation for the case of compressible gas flow between two impermeable boundaries. When the excited vibrational mode is absent, a continuous transition to the compressible flow of a perfect gas takes place. As known, in such a case an asymptotic theory of neutral stability was not previously constructed.

Acknowledgments

This work was supported by the Russian Foundation for Basic Research (Grant No. 14-01-00274). The authors thank of Professor S A Gaponov for useful advices and discussion of this work results.

References

[1] Lin C C 1955 *The Theory of Hydrodynamic Stability* (Cambridge: Cambridge Univ. Press)
[2] Nagnibeda E A and Kustova E V 2009 *Nonequilibrium Reacting Gas Flows. Kinetic Theory of Transport and Relaxation Processes* (Berlin: Springer-Verlag)
[3] Grigor'ev Yu N and Ershov I V 2014 *J. Appl. Mech. Tech. Phys.* **55** 258
[4] Tricomi F G 1961 *Differential Equations* (New York: Blackie and Son Ltd)
[5] Reshotko E 1960 *Stability of the Compressible Laminar Boundary Layer PhD Thesis* (Pasadena: California Institute of Technology)
[6] Drazin P G and Reid W H 2004 *Hydrodynamic Stability* (Cambridge: Cambridge Univ. Press)
[7] Dunn D W and Lin C C 1955 *J. Aero. Sci.* **22** 455
[8] Miles J W 1960 *J. Fluid Mech.* **8** 593

Experimental measurements and visualisation of a viscous fluid flow in Y-branching modelling the common carotid artery bifurcation with MR and Doppler ultrasound velocimetry

N S Denisenko[1,6], **A P Chupakhin**[1,6], **A K Khe**[1,6], **A A Cherevko**[1,6], **A A Yanchenko**[1,6], **A A Tulupov**[2,6], **A V Boiko**[3], **A L Krivoshapkin**[4], **K Yu Orlov**[4], **M P Moshkin**[5] **and A E Akulov**[5]

[1] Lavrentyev Institute of Hydrodynamics, Novosibirsk, Russia
[2] International Tomography Center, Novosibirsk, Russia
[3] Khristianovich Institute of Theoretical and Applied Mechanics, Novosibirsk, Russia
[4] Meshalkin Novosibirsk Research Institute of Circulation Pathology, Novosibirsk, Russia
[5] Institute of Cytology and Genetics, Novosibirsk, Russia
[6] Novosibirsk State University, Novosibirsk, Russia

E-mail: raselkroy@gmail.com

Abstract. In our experiments, we investigate a flow of a viscous fluid in the model of the common carotid artery bifurcation. The studies are carried out using three hardware equipments: two magnetic resonance scanners by Philips and Bruker, and intravascular guidewire ComboWire. The flux is generated by a special pump CompuFlow that is designed to reproduce a flow similar to the one in the blood vessels. A verification of the obtained data is carried out. Conducted research shows the capabilities of the measurement instruments and reflects the character of fluid flow inside the model.

1. Introduction

One of the most important objects of research in haemodynamics is the brain circulatory system, since it manages all the functions of the body. From the point of view of mechanics of blood circulation, the brain is a separated haemodynamic system with four entrances and two outputs (blood enters the brain through the two internal carotid arteries and the two vertebral arteries, and its outflow occurs throughout the two jugular veins). The velocity of blood in brain arteries ranges from 20 cm/s to 150 cm/s. In the resting state, the brain consumes around 15% of the blood volume (750 ml/min) and about 20–25% of the oxygen obtained by breathing. In comparison with brain, in kidney at admission of 19% of the total blood flow, only about 7% of total oxygen volume is absorbed. Thus the brain is characterized by the elevated consumption of oxygen and consequently by the elevated consumption of energy [1].

Generally the pathological processes which cause cerebrovascular disturbance are characterised by the lesion of the cerebral and main arteries, the cerebral and jugular veins, and also the dural sinuses. There are various pathologies of vessels of the brain: thromboses, angulation and kinking, embolisms, luminal narrowing, arterial aneurysms, arteriovenous malformations. Therefore, a research of the cerebral haemodynamics in healthy state and in the presence of anomalies, is important challenge both from a theoretical point of view and from the point of view of medical applications.

Common carotid artery is a paired artery which originates in the chest region and rises almost vertically up along the neck and is divided into the external and internal carotid arteries at the top edge of the thyroid cartilage. The main function of the common carotid artery is the blood circulation of the brain, eyes and the larger part of the head. Carotid arteries provide about 70–85% of the blood flow to the brain. The length of the common carotid artery is 6–12 cm, the diameter is 7–11 mm, the internal carotid artery has the length of 10–11 cm and diameter of 5–8 mm, the external carotid artery is less in diameter in comparison with the internal one and almost immediately branches out. Because of the different kind of pathologies of the carotid arteries, the cerebral circulation is violated which leads to a decrease in the cerebral blood flow and disruption of metabolic processes (ischaemia).

An adequate modelling of the blood flow in the cerebral vessels is a fundamental problem, a solution of which allows us in particular to solve the problem of prediction and evaluation of quality of neurosurgical operations. For development and verification of mathematical models the experimental data are needed. The difficulty in obtaining these data *in vivo* leads to attempts to recreate and study *in vitro*. This work discusses an experimental research of the flow of a viscous fluid in the silicone model of the common carotid artery bifurcation.

One of the modern non-invasive methods of examining internal human organs is a magnetic resonance imaging (MRI). The scanner measures the electromagnetic response of atomic nuclei, most commonly the nuclei of hydrogen atoms, to their excitation by a particular combination of electromagnetic waves in a gradient magnetic field of a high strength. The results of the processing of these measurements can represent different values: the substance density, the velocity of molecule motion, the temperature, or the others. In particular, using the MRI one can investigate the flow of blood in vessels. The first description of the effects of flow on spin echoes was given by Hahn [2] many years before MRI became a widely accepted method to study the fluid flow in the vessels. A review of the flow effects in the MR imaging technique is given in a paper by Axel [3]. Determination of blood flow velocity by using the MRI as a non-invasive method has generated a great interest in medicine. To verify and determine the accuracy of the MRI methods a comparison with other reliable methods in model experiments is needed. A comparison of the MRI and the Doppler ultrasound techniques for measurement of the fluid flow velocity is conducted in [4]. The first experiments on estimation of the fluid velocity in the specially designed models with the help of the MR methods are described in [5]. The paper [6] presents the results of the MRI application for measurements of the flow rate in straight tubes. In addition the authors performed a comparison of MR and ultrasound methods on the abdominal aorta in volunteers which showed a good coincidence. In paper [7] an overview of MRI abilities for studying the flows of liquids is made. It is shown that MRI provides good possibilities for 3D visualisation of the flow and determination of the velocity field. MRI techniques are used to study the flow of gas and liquid in the channels of monolithic catalysts [8]. In the paper [9] it is shown that the nuclear magnetic resonance imaging can be used to study the flow of liquid and gas in the structured catalysts and reactors, allowing one to resolve the complexity of the flow.

Another method for measuring the fluid flow characteristics inside of the object is the use of a special intravascular guide wire ComboWire. This guide wire is used in neurosurgical endovascular operations for monitoring the velocity and pressure of blood flow inside the vessels. A feature of this sensor is the ability to measure the pressure inside the vessel, whereas the MR methods cannot measure this parameter. Velocity measurement is based on the ultrasound Doppler effect when ultrasonic waves reflect from the blood cells [10, 11]. Information about the dynamics of the flow inside the artery plays an important role. It is used by neurosurgeons for performing operations on treatment of various types of the vascular anomalies. The paper [12] presents the results of study on haemodynamics in 4 patients in the presence of an aneurysm. The measurements were performed with the guide wire ComboWire. The character of the flow inside the aneurysm and connecting arteries is shown. In the articles [13, 14, 15, 16] the results of measurements of the velocity and pressure with the guide wire ComboWire are used in computer modelling for obtaining additional information about the characteristics of the flow.

For creation of a flow similar to the flow in the artery, a special pump controlled by the computer is used. The pump is connected to the silicone model of carotid artery by the series of tubes. The measurements are performed using three hardware facilities: the intravascular guide wire and two different MR scanners.

Two series of experiments are performed. In the first series MR scanners are used to investigate the velocity profile of flow. In the second series the measurements are conducted with the intravascular guide wire and MR scanner, the local parameters—velocity and pressure in the flow—are examined.

2. Materials and methods

2.1. Model of the common carotid artery

The silicone model of the common carotid artery is designed by the Shelley Medical Imaging Technologies company [17] in accordance with the physiological parameters: the length of the common carotid artery is 90 mm, the diameter is 8 mm, the lengths of the internal and external carotid arteries are 60 mm, the diameters are 5.56 mm and 4.62 mm respectively. The channel which models carotid artery is made of elastic silicone block for modelling elastic vessels and external environment. The material used allows one to perform the measurements by using X-ray, magnetic resonance imaging (MRI), optical measurements of velocity fields of liquid or gas (PIV).

2.2. Experimental setup

The motions of blood is modelled by the motion of a viscous fluid in a closed system consisting of the model of carotid artery (Fig. 1, a), CompuFlow 1000 MR pump (Fig. 1, b) and the system of connecting tubes (Fig. 1, c), providing communication between the pump and the model.

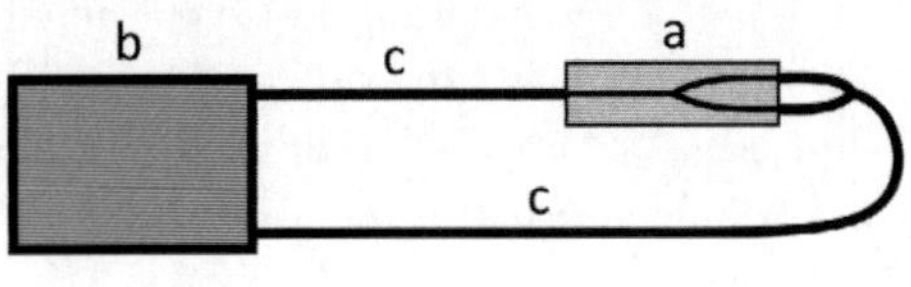

Figure 1. Experimental setup.

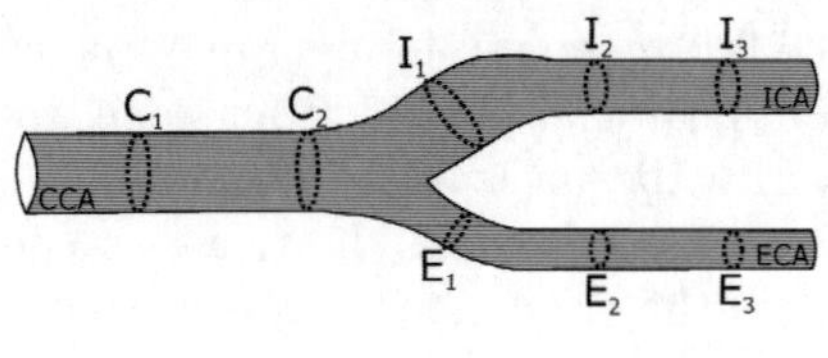

Figure 2. The model of carotid artery with marked places of measurements.

2.3. CompuFlow 1000 MR pump

Fluid motion is generated by the special pump CompuFlow 1000 MR developed by Shelley Medical Imaging Technologies. This pump is able to generate flow similar to the flow in blood vessels. CompuFlow 1000 MR has a number of default settings for dependencies between flow and time and also allows one to programme arbitrary dependence between flow and time in the range from 0.01 to 35 ml/s with an accuracy of $\pm 3\%$. The liquid used in the experiment has the same mechanical properties as blood: viscosity $\mu = 0.004$ Pa $\cdot$ s, density $\rho = 1000$ kg/m^3.

2.4. Magnetic resonance imaging

The method used to determine the flow rate is based on measuring the phase shift of the MR signal caused by liquid motion. As a result of the scanner's post-processing we get the distribution of the normal component of the velocity vector of liquid in a given cross section. The spatial resolution with which the velocity field can be obtained directly depends on magnetic field of the scanner: the stronger is the field the higher is the spatial resolution. The measurements of the liquid velocity field in the carotid artery model are performed with two devices:

- Philips Achieva with the field strength 1.5 T (designed for diagnostic studies of human).
- Bruker BioSpec 117/16 USR with the field strength 11.7 T (designed for studies of small animals in biomedical science, for biomedical and preclinical researches).

The main difference of Bruker scanner from Philips, except it's higher resolution, is that it is able to monitor the flow dynamics directly during the study with the help of method Flash-ToF-2D-Saturation, which allows one to track the movement of the magnetized transverse and longitudinal liquid layers (Fig. 3).

Using of two devices of different classes allows one to perform mutual verification of the measured values.

2.5. ComboWire guide wire

The intravascular ComboWire guide wire is designed for simultaneous measurement of the blood velocity and the pressure in blood vessels. The guide wire was developed by Volcano Corporation. The velocity is measured with the help of Doppler method with the frequency of 12 MHz, the pressure is measured with piezoelectric method. The measured data are recorded with the frequency of 200 Hz. Velocity and pressure sensors are mounted on the end of the guide wire which is 0.36 mm in diameter and 185 cm in length. The ComboWire guide wire is used together with the ComboMap computer system which directly processes the measured data.

2.6. Investigated flow regimes

The purpose of the measurements is to study liquid flow in the carotid bifurcation model under conditions that are close to ones in a living organism. For studying the flow we use the sequence of the flow regimes "from simple to complicated". This sequence allows at first to reveal effects that appear in transition from some simple flow regime to a more complicated one and secondly to verify the measurement method. Such a choice is caused also by the complexity of the measurement devices: the transition from one flow regime to the more complicated one often requires the correction of measurement parameters of the devices.

From the flow regimes chosen for study let us point out the next ones:

- Fluid flows with a constant rate: $5, 10, 15$ ml/s.
- Fluid flows in which the next periodic dependencies between flow rate and time are realized: $A+B\sin(\pi t), A+B\sin(2\pi t), A+B\sin(4\pi t), A+B\sin(2\pi t)+C\sin(4\pi t)$ ml/s. Here A, B, C are some constants varied in experiments. Constant A takes values from 0 to 10 ml/s, amplitudes B, C are chosen from the range of 1.25 to 15.
- Dependence between flow rate in carotid and time measured experimentally during neurosurgical operation [10].
- Dependence between flow rate and time corresponding to blood flow rate in carotid averaged over the big and representative group of patients.

Measurements are made on the straight section before bifurcation, in bifurcation area and daughter branches of carotid. In Fig. 2 measurement places are marked with dotted lines.

3. Results of the measurements

3.1. MRI

In the measurement of the velocity field there are high-frequency perturbations with amplitude ~15–20% superimposed on the main flow. Additional assumptions allow one to filter out the noise. Thus, in the measurement of the velocity profile along the straight section of the common carotid artery we can assume that the flow is axisymmetrical. This assumption allows us to perform the averaging of the velocity profile by angle:

$$\overline{F}_{2n}(r) = \frac{\sum\limits_{k=0}^{2n-1} F(r\cos(\pi\frac{k}{n}), r\sin(\pi\frac{k}{n}))}{2n}, \quad (1)$$

where F is measured velocity values, $2n$ is the number of the sectors by which averaging is performed, R is the vessel radius, $r \in (0, R)$.

In Fig. 3 the dynamics of the movement of the magnetized liquid layers in the area of the carotid artery bifurcation for the constant flow rate 5 ml/s is shown. In Fig. 4 averaged by angle velocity profiles along the straight section of the common carotid artery for the constant flow rate 5 ml/s measured by the Bruker and the Philips scanners are shown.

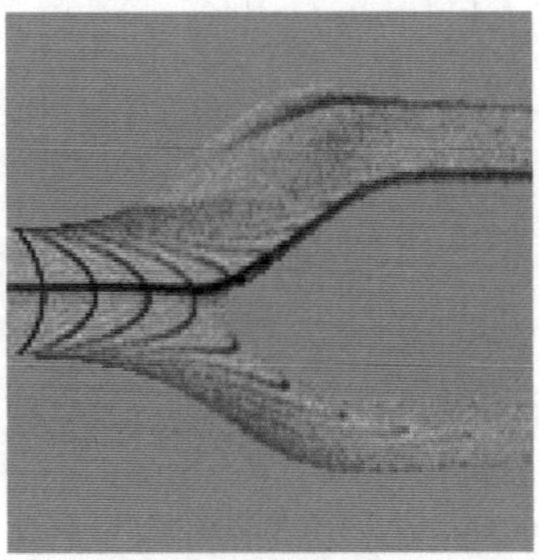

Figure 3. The movement of the magnetized liquid layers in the area of the carotid artery bifurcation for the constant flow rate 5 ml/s.

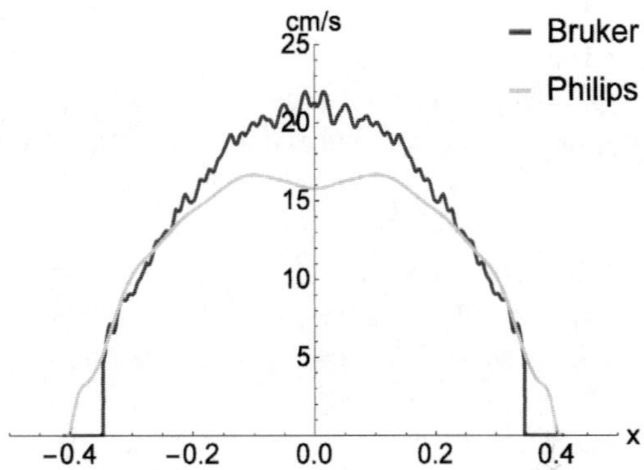

Figure 4. The velocity profiles along the straight section of the common carotid artery averaged by angle for the constant flow rate 5 ml/s.

Another stage of the measurement verification is a comparison of the measured and generated by the pump flow rates. For each moment of time at which the measurement is performed, the flow rate through a given cross section is calculated by formula

$$\int_{\Sigma} (\vec{v} \cdot \vec{n})\, d\sigma,$$

where Σ is the cross section of the vessel in which the measurement is performed, $\vec{v}$ is the liquid velocity, $\vec{n}$ is the vector normal to the cross section, $d\sigma$ is an element of the cross section. Note that the scanner measures the normal component of liquid velocity for a given cross section, equal to $\vec{v} \cdot \vec{n}$.

In Fig. 5 the results of measurements of flow with the flow rate $0.1 + 10\sin(2\pi t)$ ml/s in the straight section of the carotid artery are presented. In Fig. 6 the diametrical velocity profiles at the time moments t_1, t_2, t_3 are shown.

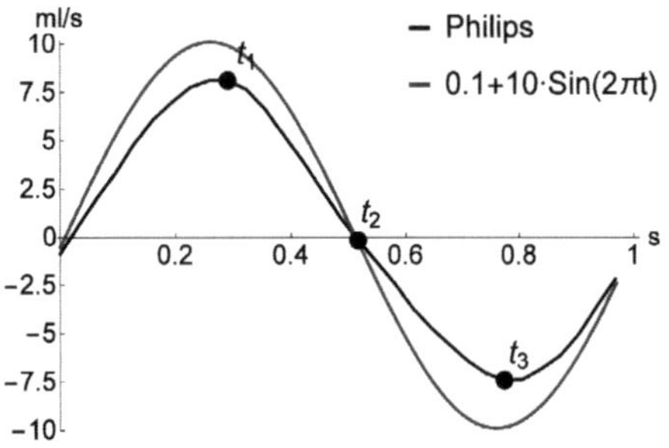

Figure 5. Generated by the pump and measured by the scanner dependencies between fluid flow rate $(0.1 + 10\sin(2\pi t)$ ml/s) and time. Dots indicate the time moments at which the averaged diametrical velocity profiles are calculated.

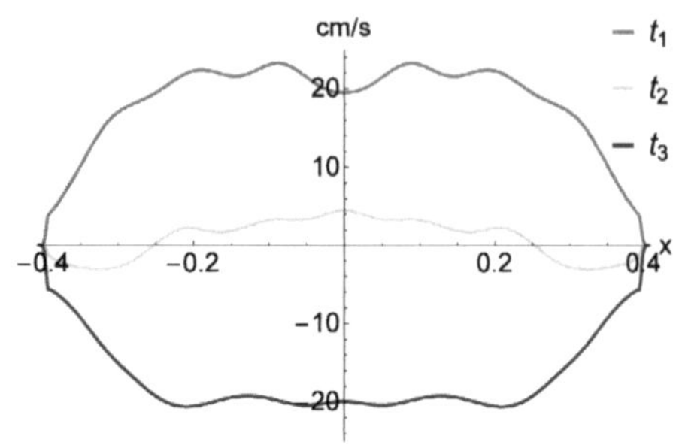

Figure 6. The averaged diametrical velocity profiles at the moments t_1, t_2, t_3.

The calculated fluid flow rate and the given constant flow rate (generated by the pump) differ by 1–3% for both devices.

During the experimental sessions about 100 measurements on the Philips scanner and 90 measurements on the Bruker scanner were performed.

3.2. The ComboWire guide wire

During the measurement the ComboMap system obtains data from ComboWire guide wire about velocity spectrum and builds an envelope around it. The value of the spectrum envelope is accepted as the value of the fluid velocity.

In Fig. 7 the velocity values measured by ComboWire guide wire and Philips scanner in the straight section before the bifurcation for the given flow rate typical for the carotid artery are presented. In Fig. 8 VP–diagrams ("velocity–pressure"), which visualize the relationship between pressure and velocity in the flow, are presented. In Fig. 9 measurements of the fluid flow typical for the carotid artery in cross section $\mathrm{C}_1, \mathrm{I}_1, \mathrm{I}_3, \mathrm{E}_1$ and E_3 (Fig. 2) are shown.

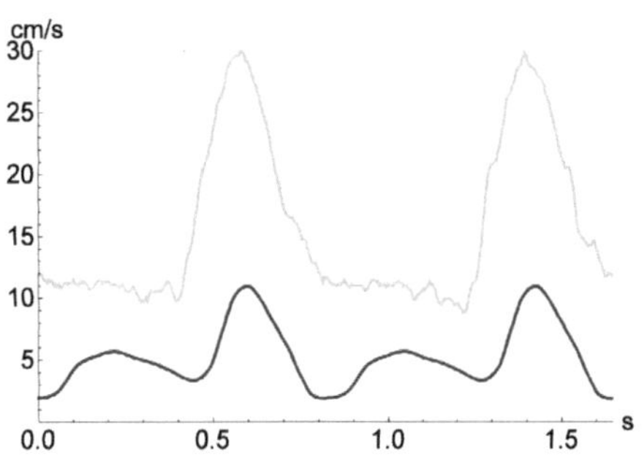

Figure 7. Velocity dependence on time measured by the ComboWire guide wire and Philips scanner.

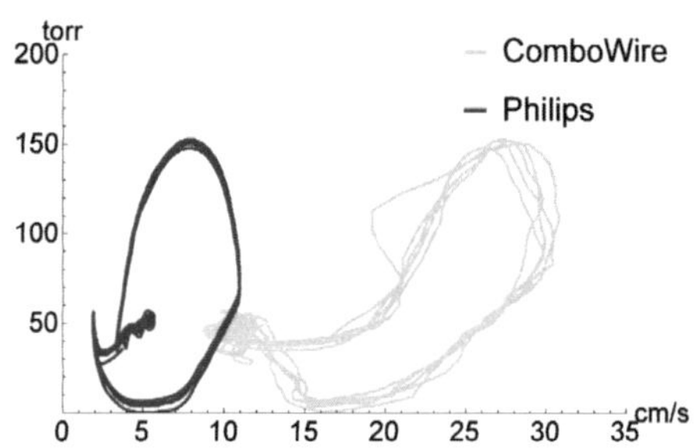

Figure 8. "Velocity–pressure" phase diagrams measured by the ComboWire guide wire and Philips scanner.

12 periodic flow regimes were investigated and 96 measurements were performed with ComboWire guide wire in total. The average duration of one measurement is 2–3 minutes.

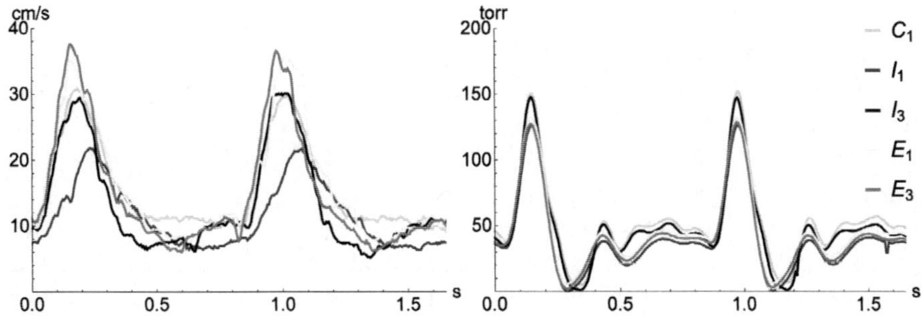

Figure 9. The velocity and the pressure measured by the ComboWire guide wire in different cross sections.

For four flow regimes corresponding measurements by Philips scanner were performed.

4. Discussion

The velocity profiles measured by the Philips and the Bruker scanners coincide quite well, except for the area where the liquid moves with the highest velocity (Fig. 4).

The calculated fluid flow rate and the given constant flow rate (generated by the pump) differ by 1–3% for both devices. Thus, under these conditions the Philips scanner gives results comparable to the results of the Bruker one.

The areas near the bifurcation, in which liquid magnetized layers do not propagate are stagnation areas (Fig. 3). From a medical point of view, such areas are potentially dangerous because blood properties significantly depend on its movement velocity and in a stagnant area a thrombus may occur.

Averaged over cross section velocity values measured by the Philips scanner differ in 1.5–2 times from the values measured by the ComboWire guide wire (Fig. 7). This difference is caused by the fact that the ComboWire guide wire is situated in the centre of the blood vessel and measures the local velocity, whereas the comparison is performed to the averaged over the cross section velocity measured by the MR scanner. The dependence between the measured average and maximum velocities is similar to the dependence between average and maximum velocities for steady Poiseuille flow.

The pressure measurement is important for further studies of liquid dynamics and external environment. Dependence between pressure and velocity defined in such way (Fig. 8) can be used for constructing the state equation which characterises the rheology of the composite system “fluid flow – elastic environment”. The measured pressures are used in computer modelling as boundary conditions.

The ComboWire guide wire flow investigations show that the maximum values of the velocity in cross section I_1 are $\sim$30% less than the maximum values of the velocity in C_1, maximum velocity values in sections C_1 and E_2 are greater than the maximum value of the velocity in common carotid artery by $\sim$20% and $\sim$25% respectively (Fig. 9). The maximum and minimum pressure values for each flow generated by the pump almost do not change throughout the model of the carotid artery bifurcation, in contrast to the velocity values.

When the pump is set up for the periodic flow rate $A + B\sin(N\pi t)$ ml/s with almost zero flow rate at the period ($0 \leq A \leq 1$) the experimental setup reproduces the defined flow not accurately enough: the difference between the maximum values of the measured and defined ones can reach 25% (Fig. 5).

During the experiment it has been found that the pressure sensor reacts to light and environment temperature — the zero level of the measured pressure shifts. Therefore, it is necessary to maintain a constant level of lighting and constant liquid temperature.

Performed investigations demonstrate the capabilities of the measuring devices and show the character of fluid flow inside the model.

5. Conclusion

The result of the experimental studies is qualitative and quantitative information about the dynamics of fluid flow inside the model of the carotid artery for different flow regimes.

Averaged over cross section velocity values measured by the Philips scanner differ in 1.5–2 times from the values measured by the ComboWire guide wire. This difference is caused by the fact that the ComboWire guide wire is situated in the centre of the blood vessel and measures the local velocity, whereas the comparison is performed to the averaged over the cross section velocity measured by the MR scanner.

The velocity profiles measured by the Philips and the Bruker devices coincide quite well, except for the area where the liquid moves with the highest velocity. The calculated fluid flow rate and the given constant flow rate (generated by the pump) differ by 1–3% for both MR scanner. So, under these conditions the Philips scanner gives the results comparable to the results of the Bruker one.

Thus, the measurement equipment consisting of the intravascular ComboWire guide wire, CompuFlow 1000 MR pump and two Philips and Bruker MR scanners can be used to obtain reliable and consistent data in the investigation of fluid flow in the vessel model.

Acknowledgments

ComboWire measurements were supported by the Russian Foundation for Basic Research (project No. 14-01-00036). MRI measurements were supported by the Russian Science Foundation (project No. 14-35-00020).

References

[1] Caro C, Pedley T, Schroter R, Seed W 1978 *Oxford University Press* New York
[2] Hahn E L 1950 *J. Phys. Rev.* **80** 580
[3] Axel L 1984 *Amer. J. Radiol.* **143** 1157
[4] Matsuda T, Shimizu K, Sakurai T, Fujita A, Ohara H, Okamura, Hashimoto S, Tamaki S and Kawai C 1987 *J. Radiology* **162** 857
[5] Snaar J, Guilfoyle D, van Duk P and Bohning D 1986 *Fifth Annual Meeting of Society of Magnetic Resonance in Medicine, Monreal* 76
[6] Meier D, Maier S and Bosiger P 1988 *J. Magn. Res. in Med.* **8** 25
[7] Bonn D, Rodts S, Groenink M, Rafai S, Shahidzadeh-Bonn N and Coussot P 2008 *Ann. Rev. of Fluid Mech.* **40** 209
[8] Koptyug I, Altobelli S, Fukushima E, Matveev A and Sagdeev R 2000 *J. Magn. Reson.* **147** 36
[9] Koptyug I, Ilyina L, Matveev A, Sagdeev R, Parmon V and Altobelli S 2001 *J.Cat. Tod.* **69** 385
[10] Chupakhin A, Cherevko A, Khe A, Telegina N, Krivoshapkin A, Orlov K, Panarin V, and Baranov V 2012 *Circul. Pathol. Cardiac Surgery* **4** 27
[11] Boiko A et al. 2015 *Suppl. 1 (Book of Abstracts ESMRMB 2015) - Edinburgh* **28** 384
[12] Fernsa S, Schneidersa J, Siebesb M, van den Berga R, van Bavelb E and Majoiea C 2010 *Amer. J. Neurorad.* **31** 324
[13] Vorobtsova N, Yanchenko A, Cherevko A, Chupakhin A, Krivoshapkin A, Orlov K, Panarin V, Baranov V 2013 *Russ. J. Numer. Anal. Math. Modelling* **28** 505
[14] Yanchenko A, Cherevko A, Chupakhin A, Krivoshapkin A, Orlov K 2014 *Russ. J. Numer. Anal. Math. Modelling* **29** 307
[15] Cebral J, Castro M, Burgess J, Pergolizzi R, Sheridan M and Putman 2005 *Amer. J. Neurorad.* **26** 2550
[16] Castro M, Putman C and Cebral J 2006 *Amer. J. Neurorad.* **106** 2061
[17] Smith R, Rutt B, Fox A, Rankin R and Holdsworth D 1996 *J. Acad. Radiol.* C **11** 898

On the phase shift in the Kuzmak–Whitham ansatz for nonlinear waves

S Yu Dobrokhotov [1,2]

[1]A.Ishlinsky Institute for Problems in Mechanics of RAS, Russia
[2]Moscow Institute of Physics and Technology (Technical University), Russia

E-mail: dobr@ipmnet.ru

Abstract. We consider one-phase (formal) asymptotic solutions in the Kuzmak–Whitham form for evolutionary nonlinear equations. In this case, the leading asymptotic expansion term has the form $X(S(x,t)/\varepsilon + \Phi(x,t), A(x,t), x, t) + O(\varepsilon)$, where ε is a small parameter and the phase $S(x,t)$ and slowly changing parameters $A(x,t)$ are to be found from the system of averaged Whitham equations. The equation for the phase shift $\Phi(x,t)$ is appearing by studying the *second-order* correction to the leading term. The corresponding procedure for finding the phase shift is then nonuniform with respect to the transition to a linear (and weakly nonlinear) case. We formulate the general conjecture (checked for some examples), which essentially follows from papers by R.Haberman and collaborators, is that if one incorporates the phase shift $\Phi(x,t)$ into the phase and adjust A by setting $S \to \tilde{S}(x,t,\varepsilon) = S + \varepsilon\Phi + O(h2)$, $A \to \tilde{A}(x,t,\varepsilon) = A + \varepsilon a + O(\varepsilon^2)$, then the new functions $\tilde{S}(x,t,\varepsilon)$ and $\tilde{A}(x,t,\varepsilon)$ become solutions of the Cauchy problem for the *same* Whitham system but with modified initial conditions. These functions completely determine the leading asymptotic term in the Whitham method.

1. The Kuzmak–Whitham ansatz

The Kuzmak–Whitham ansatz describes a class of special (formal) asymptotic solutions for a broad class of nonlinear evolution equations, including equations with variable coefficients. Recall some ideas leading to these asymptotics. The plane waves $u = A\cos(\omega t + k \cdot x + \Phi)$ are important solutions of linear evolution equations (or systems of equations). Here ω is the frequency, $k = (k_1, \ldots, k_n)$ is the wave vector, A is the amplitude (which is a vector in the vector case A), and the number Φ is a phase correction. The counterpart of this solution in the nonlinear case has the form

$$u = X(\omega t + k \cdot x + \Phi, A), \tag{1}$$

where $X(\theta, A)$ is a smooth 2π-periodic function (or vector function) of the scalar variable θ depending n the parameter(s) A. Usually the existence of a solution of this type implies that ω, k, and the parameters A (which are a nonlinear analog of the amplitude) satisfy some relation of the form $\omega = \Omega(k, A)$, which is known as the *dispersion relation.* In the linear case, Ω is independent of A. In many important physical examples, the functions $X(\theta, A)$ can be expressed via elliptic functions, e.g., the function cn, and the functions (1) are known as *cnoidal waves* in this case.

The exact solution (1) can be generalized in the following way. One can assume that the parameters ω, k, A are smooth functions slowly depending on t and $x = (x_1, \ldots, x_n)$. We characterize this slow dependence by a small parameter ε and set $\omega = \omega(\varepsilon t, \varepsilon x)$, $k = k(\varepsilon t, \varepsilon x)$,

$A = A(\varepsilon t, \varepsilon x)$, and $\Phi = \Phi(\varepsilon t, \varepsilon x)$. It is convenient to use a general phase function $S(\varepsilon t, \varepsilon x)$ instead of $\omega(\varepsilon t, \varepsilon x)t + k(\varepsilon t, \varepsilon x) \cdot x$, introduce the variables $t' = \varepsilon t$ and $x' = \varepsilon x$, and set $u = Y(\frac{S(t',x')}{\varepsilon} + \Phi(t', x'), A(t', x'))$. It is also convenient in what follows to omit the prime on the new variables, include small corrections related to the small parameter ε in the solution, and finally write

$$u = Y\Big(\frac{S(t,x)}{\varepsilon} + \Phi(t,x), t, x, \varepsilon\Big), \tag{2}$$

$$Y = X(\theta, A(t,x), t, x) + \varepsilon Y^1(\theta, t, x) + \varepsilon^2 Y^2(\theta, t, x) + \dots, \quad \theta = \frac{S(t,x)}{\varepsilon} + \Phi(t,x). \tag{3}$$

Here the functions X and Y^j are smooth in all arguments and 2π-periodic in θ. The right-hand side of (2) is called the *Kuzmak–Whitham ansatz*. It describes asymptotic solutions of many important nonlinear problems in mathematical and theoretical physics and appears first in Kuzmak's paper [1] for ordinary differential equation (in this case, the variables x disappear) and in Whitham's paper [2] in the case of partial differential equations. Let us briefly recall the meaning of this ansatz.

2. The Whitham equations

Formally, one can represent the partial differential equation under study in the form

$$\mathcal{L}(t, x, Q(\varepsilon\frac{\partial}{\partial t})u, R(\varepsilon\nabla)u, \varepsilon) = 0, \tag{4}$$

where L is a vector nonlinear (pseudo)differential operator acting on the vector function u, and $Q(\varepsilon\frac{\partial}{\partial t})$ and $R(\varepsilon\nabla)$ are some (pseudo)differential) operators acting on the vector function u and giving some new vector function. The main assumption about the last equation is that the ordinary (pseudo)differential equation with fixed (frozen) variables (t, x),

$$\mathcal{L}(t, x, Q\big(\omega(t,x)\big)\frac{\partial}{\partial\theta})X, R\big(k(t,x)\big)\frac{\partial}{\partial\theta})X, 0) = 0, \quad \omega(t,x) = \frac{\partial S}{\partial t}, \quad k(t,x) = \nabla S, \tag{5}$$

possesses a family of smooth solutions $X(\theta, A, t, x)$ 2π-periodic in θ and also depending on a scalar or vector parameter A. Just as in the case of constant coefficients, the existence of 2π-periodic solutions implies a dispersion relation, which may include dependence on the slow variables (t, x),

$$\omega = \Omega(k, A, t, x). \tag{6}$$

Actually, formula (2) gives an asymptotic series, and the function

$$u_0 = X\Big(\frac{S(t,x)}{\varepsilon} + \Phi(t,x), A(t,x), t, x\Big)$$

is its leading term. As a rule, one can efficiently describe only the leading term X and probably the first correction. *To construct the leading term, one should find the function X, the equations for the phase $S(x,t)$, the parameter(s) A, and the phase shift Φ.*

A scheme for constructing these objects was suggested in [3]. As we mentioned above, to find the function X and the dispersion equation (6), one has to solve equation (5). To derive the equation(s) for the function(s) $A(x)$, one has to consider the equation for the first correction Y^1, which is a linear inhomogeneous ordinary (pseudo)differential equation for Y^1 with respect to the variable θ varying on the circle $\mathbf{R} \bmod 2\pi$,

$$L(\omega, k, t, x, X, \frac{\partial}{\partial\theta})Y^1 = F_1(X, \omega, k, A, \Phi, t, x), \tag{7}$$

where F is a smooth function depending on X, its θ-derivatives, ω, k, A, Φ, and their derivatives with respect to t and x. Note that determination of F_1 involves two terms of the expansion. The effective variable in this equation is θ, and the variables t, x serve as parameters. The requirement of the existence of 2π-periodic solutions Y^1 of this equation is equivalent to the orthogonality conditions

$$\int_0^{2\pi} F_1(X, \omega, k, A, \Phi, t, x)\bar{Z}(\theta, t, x)d\theta, \tag{8}$$

where Z runs over the solutions of the equation $L^*Z = 0$ with the operator $L^*(\omega, k, t, x, X, \frac{\partial}{\partial\theta})$ adjoint to the operator L. Usually the dimension of the kernel of the operator L^* (as well as of the operator L) is not less than the number m of parameters $A = (A_1, \ldots, A_m)$, and the orthogonality conditions, together with the dispersion relation, give a system of equations determining $A_0(x,t) = S(t,x), A_1(t,x), \ldots, A_m(t,x)$,

$$A_t = G(\nabla A_0, \ldots, \nabla A_m, t, x), \tag{9}$$

where G is a smooth vector function with $m+1$ components, $G = (\Omega(k(t,x), A, t, x), G_1, \ldots, G_m)$. Note that the choice of the functions (parameters) A (including A_0) is not unique and is very important: a good choice of the parametrization of the function X may strongly simplify the form of (9). We refer to (9) as the *Whitham equations.* This type of systems originally appeared in the paper [2] by Whitham, who used averaging of conservation laws and variational principle for their derivation. The equivalence of this approach to the orthogonality conditions was shown in [4]. Now it becomes clear that the Whitham are very important objects in mathematical physics, they appear in many important problems and have very interesting properties. For instance they inherit the integrability properties of original system including the case of multi-phase asymptotics (see [5, 6, 7]).

3. Example: The nonlinear Klein-Gordon equation

In this case,

$$\varepsilon^2(u_{tt} - c^2(t,x)\triangle u) + V_u(u,t,x) = 0, \quad V = \frac{1}{2}q^2(t,x)u^2 + \lambda r(t,x)u^4 \tag{10}$$

Here $c(t,x), q(t,x), r(t,x)$ are smooth positive functions, and λ is a parameter. The Whitham equations for the phase S and the scalar function A are (see the specific forms of $X(\theta, A, t, x)$ and $\Omega(A,t,x)$ in [10])

$$S_t^2 - c^2(x,t)(\nabla S)^2 = \Omega(A,t,x), \quad \frac{\partial}{\partial t}\Big[\frac{A}{\Omega(A,t,x)}S_t\Big] - c^2(x,t)\Big\langle\nabla, \frac{A}{\Omega(A,t,x)}\nabla S\Big\rangle = 0,$$

4. The phase shift

The Whitham equations do not completely determine the leading term u_0 of the asymptotic solution (2); to this end, one needs to construct an equation for the phase shift $\Phi(t,x)$. But the derivation of the equation for Φ is not the same for the weakly and strongly nonlinear cases; moreover, even the order of this equation with respect to time t is not the same in these two cases. In the weakly nonlinear case (and also in the linear case, parameter $\lambda = \varepsilon$ in example (10)), Eq. (5) is linear, the frequency Ω in the dispersion relation (6) is independent of A, the operators L and L^* in (7) are independent of X, and the dimension of the kernels of the operators L and L^* is equal to $m+1$. Thus, in this case the orthogonality conditions (8) give an equation of the first order in t for the phase shift $\Phi(t,x)$,

$$\Phi_t = \tilde{\Omega}(\nabla\theta, \nabla A_0, \ldots, \nabla A_0, t, x), \tag{11}$$

where $\tilde{\Omega}$ is a smooth function of all of its arguments and is linear in θ. In the strongly nonlinear case ($\lambda = 1$ in example (10)), the situation is dramatically different in that the dimension of kernels of L and L^* is equal to m (the number of components of A), and the equation for the phase shift Φ follows from the analysis of the equation for *the second correction* Y^2. Thus, it includes some characteristics of the first correction Y^1, and its derivation is much more complicated than that of the Whitham equations. A method for this derivation was presented in [4]. Unfortunately, the final equation for the phase shift Φ in [4] includes an arithmetical mistake, which has been corrected in [8]. In contrast to the weakly nonlinear case, the phase shift now satisfies an equation of the *second order in time* t, which has a form

$$\Phi_{tt} = \mathcal{K}(\Phi_t, \nabla\Phi, A, t, x), \tag{12}$$

where $\mathcal{K}$ is a smooth function of all of its arguments. Thus, we see that the Kuzmak–Whitham type asymptotic solution (2) is not uniform with respect to passage from the weakly nonlinear case to the strongly nonlinear case. Moreover, there exists a "topological" obstruction to this passage: the dimension of the kernel of the adjoint variational operator L^* experiences a jump. Note that the concept of "weak nonlinearity" and "strong nonlinearity" is more or less artificial; the original equation becomes weakly nonlinear for a sufficiently small solution. Thus, the obstruction does not permit us to construct an asymptotic solution decaying outside some compact set describing the wave train propagation. Indeed, the head and tail of this solution are in the framework of the weakly nonlinear situation, and the "main part" is in the framework of the strongly nonlinear situation. Also, there exists an intermediate case, where one should match these two cases which also gives a lot of troubles.

5. "Phase Shift Principle" for the Kuzmak-Whitham anzats

The beautiful observation made in the paper [8] is that the equation for the phase shift for the nonlinear Klein–Gordon equation can be represented in the form of a system of equations for the vector $(\Phi, \delta A_1, \ldots, \delta A_n)$, and this system is none other than the linearization of the Whitham equations (9). The same fact is true for the Korteweg–de Vries equation. Using this fact, we can include the phase shift $\Phi(t, x)$ in the phase S and the corrections $\varepsilon a_j(t, x)$ in the functions A_j, consider these functions $\tilde{S}(t, x, \varepsilon)$ and $\tilde{A}(t, x, \varepsilon)$ as solutions of *the same Whitham equation* (9), and substitute these functions into the Kuzmak–Whitham ansatz

$$\tilde{S}(t, x, \varepsilon) = S(t, x) + \varepsilon\Phi(t, x), \quad \tilde{A}(t, x, \varepsilon) = A(t, x) + \varepsilon a(t, x), \tag{13}$$

$$u = X(\frac{\tilde{S}(t, x)}{\varepsilon}, \tilde{A}(t, x, \varepsilon), t, x) + \varepsilon\tilde{Y}^1(\frac{\tilde{S}(t, x)}{\varepsilon}, t, x) + \varepsilon^2\tilde{Y}^2(\frac{\tilde{S}(t, x)}{\varepsilon}, t, x) + \ldots. \tag{14}$$

The reconstruction (3), (14) *of the Kuzmak–Whitham ansatz is now uniform with respect to the passage from weak nonlinearity to strong nonlinearity and does not need any new objects to construct the leading term of the rapidly oscillating solution.*

Now we can state the main result (conjecture) of this paper. We only have a proof of this conjecture for the nonlinear Klein–Gordon equation and the Korteweg–de Vries equation (see [9, 10] and also [11]); thus, we state it in the form of

Phase Shift Principle for the Kuzmak–Witham Ansatz:
Assume that one constructs a (formal) asymptotic Kuzmak–Whitham solution (2), (3) *of some evolution equation* (4) *with functions* $S(t, x)$, $\Phi(t, x)$, *and* $A(t, x)$ *satisfying the Whitham equations* (9) *and the corresponding equation for the phase shift. Then there exist functions* $\tilde{S}(t, x, \varepsilon) = S(t, x) + \varepsilon\Phi(t, x) + O(\varepsilon^2)$ *and* $\tilde{A}(t, x, \varepsilon) = A(t, x) + O(\varepsilon)$ *satisfying the same Whitham equations* (9) *such that the asymptotic expansions* (2)–(3) *and* (13)–(14) *define the same function* u.

Journal of Physics: Conference Series **722** (2016) 012014 doi:10.1088/1742-6596/722/1/012014

Recall that Eqs. (13)–(14) describe some special class of asymptotic solutions. To define them, one should add certain additional conditions, for example, add the initial data $A|_{t=0} = A^0(x)$ for the Whitham equations (9), which fix the leading term X of the asymptotics for $t = 0$. From the viewpoint of the original equations (4), one should represent the function $X|_{t=0}$ in the prescribed form, but one should also choose the first correction $Y^1|_{t=0}$ in the prescribed form. (The corresponding formulas for the nonlinear Klein–Gordon and Korteweg–de Vries equations can be found in [9, 10].) The reconstruction (13)–(14) of the Kuzmak–Whitham ansatz is similar to Poincaré–Bogolyubov ideas in the averaging method concerning the construction of asymptotic solutions in the new dependent variables, which are changed at each step of the asymptotic procedure (see [12],[13]). Finally, note that the representation (13)–(14) shows that, owing to the presence of the phase shift $\Phi(t, x)$, the corrections $O(\varepsilon)$ to the initial data in the form of the Kuzmak–Whitham ansatz can strongly change the asymptotic solution in bounded time $t = O(1)$ (see Fig.1). This does not mean that the Kuzmak–Whitham-type solution is not stable, but this only means that, to achieve stability,the change in the initial data should be of the order of $O(h^{1+\kappa})$, $\kappa > 0$. This effect is well known as "orbital stability" in celestial mechanics [14]. One can find a more comprehensive discussion in [9, 10].

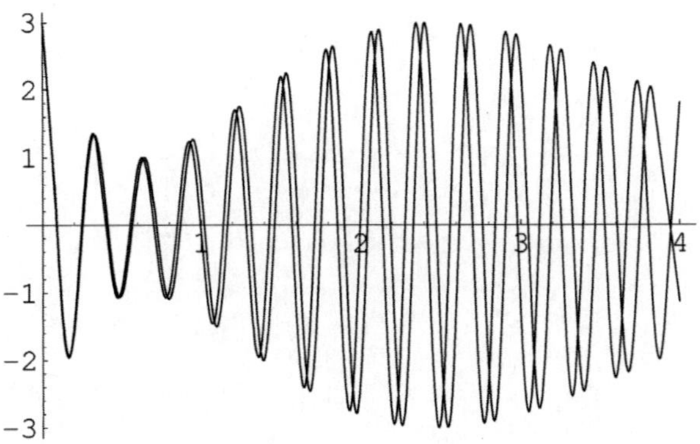

Figure 1. The influence of the phase shift on the asymptotics describing the wave train propagation

6. Some open problems

I think that it is of interest to check (or prove) the *Phase Shift Principle* for other equations where the Kuzmak–Witham ansatz applies. These include

1. Rapidly oscillating asymptotic solutions of the Navier–Stokes equations constructed in [15].
2. The integrable partial differential equations considered in [16].
3. The Benjamin–Ono equation, which is actually a pseudodifferential equation [17].

At last, it seems that the *Phase Shift Principle* should also work for the asymptotics of soliton-like solutions, and it is of interest to check it for the solutions of the equations considered in [18, 19].

Acknowledgments

The author are grateful to D. Minenkov and V. Nazaikinskii for useful discussion. This research was supported by RFBR grant no. 14-01-00521.

Journal of Physics: Conference Series **722** (2016) 012014 doi:10.1088/1742-6596/722/1/012014

References

[1] Kuzmak G N 1959 *J. Appl. Math. Mech.* **23** 730–44
[2] Whitham G B 1965 *Proc. Roy. Soc. London A* **283** 238–61
[3] Luke J C 1966 *Proc. Roy. Soc. London A* **292** 403–12
[4] Dobrokhotov S Y and Maslov V P 1981 *J. Sov. Math.* **16** 1433–87
[5] Flashka H, Forest M G and McLaughlin D W 1980 *Comm. Pure. Appl. Math.* **33** 739–84
[6] Dubrovin B A and Novikov S P 1989 *Russ. Math. Surveys* **44**(6) 35–124
[7] Tsarev S P 1991 *Mathematics of the USSR-Izvestiya* **37**(2) 397–419
[8] Haberman R 1991 *SIAM J. Appl. Math.* **51** 1638–52
[9] Dobrokhotov S Y and Minenkov D S 2010 *Regul. Chaotic Dyn.* **15**(2-3) 285–99
[10] Dobrokhotov S Y and Minenkov D S 2011 *Theor. Math. Phys.* **166**(3) 350–65
[11] Maltsev A Ya 2008 *Funct. Anal. Appl.* **42**(2) 103–15
[12] Bogolujbov N N and Mitropolski Yu A 1961 *Asymptotic Methods in the Theory of Non-Linear Oscillations* (New York: Gordon and Breach)
[13] Neishtadt A I 1984 *J. Appl. Math. Mech.* **48**(2) 133–9
[14] Arnold V I, Kozlov V V and Neishtadt A I 2006 *Mathematical Aspects of Classical and Celestial Mechanics* ed 3 (Berlin: Springer)
[15] Maslov V P 1986 *Russian Math. Surveys* **41**(6) 23–42
[16] Krichever I M 1988 *Funct. Anal. Appl.* **22**(3) 200–13
[17] Dobrokhotov S Y and Krichever I M 1991 *Math. Notes* **49**(6) 583–94
[18] Grimshow R 1979 *Proc. R. Soc. Lond. A* **358** 359–85
[19] Maslov V P and Omel'yanov G A 1981 *Russian Math. Surveys* **36**(3) 73–149

About the possibility of creating a stable transonic region in supersonic flow in the channel

V Zamuraev and A Kalinina

Khristianovich Institute of Theoretical and Applied Mechanics of SB RAS, Russia

E-mail: zamuraev@itam.nsc.ru, kalinina@itam.nsc.ru

Abstract. The work is devoted to the mathematical modeling of the influence of transversal jet and the near-wall energy sources on the shock wave structure of supersonic flow in channel with variable cross section. Stable regimes with the region of transonic velocities are obtained. Their stability is confirmed by the width of the corridor of the input power in the region of existence of the regime modes.

1. Introduction

The reduction of total pressure losses in combustion chamber is one of the main issues that defines the effectiveness of high-speed ramjet [1]. The technologies designed to implement reliability of ignition and stability of combustion also lead to total pressure losses. The gas-dynamic and kinetic methods are applied to solve the problem. In particular, jets for flow control are currently widely used. The injection of oncoming jet or co-current jet is considered [2]. A rotating detonation wave created with a jet [3] is also used. Subsurface sources of energy are of interest as well [4]. All the possible ways, active and passive, are considered as shown in the review section of the paper [5]. These methods either require a significant investment of energy, or do not give a stable result. In [1] a new method of combustion chamber starting is proposed. It enables the transition to supersonic flow in the combustion chamber by a jet and distributed fuel injection into the chamber. Experimental data confirm its effectiveness. The proposed work presents an analytical and numerical analysis of control of supersonic flow into a channel with both the jet and the surface energy supply to identify relatively stable regimes with an extended transonic region.

2. Mathematical formulation

Numerical simulation is carried out for unsteady flow in a channel of variable cross section. Plane channel consists of a nozzle with sound speed flow at the inlet, the section being of constant cross section into which the jet flows and near-wall energy supply is carried out (the Mach number at the entrance to this section is approximately equal to $\mathrm{M} = 2$). The expanding section ends with a constant cross section. Two-dimensional unsteady Euler equations in a dimensionless form for an ideal gas with a constant ratio of the specific heats γ are solved. They can be used to solve such problems, as it is shown, for example, in [6]. All lengths are related to the size d of the input section of the channel. The ratio of this size to the sound speed at the entrance of the channel is used as the time scale. The velocity components relate to the critical speed in the throat of the nozzle, the pressure and temperature relate to the corresponding values in the input hole. On the right border "soft" boundary conditions are used; impermeability conditions are used on the walls of the channel. The jet is supplied from a gas generator with a set pressure, and the temperature is maintained equal to its value at the entrance to the channel. Impulse energy supply is made instantly, but the gas density and its velocity are not changed. The gas energy density in the energy supply zones increases (correspondingly, the temperature and pressure increase). In the source model used the power density is proportional to the local density of the gas. The intensity of the supplied energy is determined from the condition of equivalence of the energy released during the combustion of hydrogen at a temperature of 1000 K with the excess air ratio α in the range of 2.6 to 10. Numerical integration of the differential equations is carried out in the intervals between the moments of energy supply. For the numerical solution to the problem the finite-volume scheme that reduces the total variation is used.

3. The parameters of the simulated system.

The calculations were performed for a plane channel with variable cross section. The shape and basic sizes of the channel are shown in the upper part of Figure 1, a fragment of the first section of channel constant cross section with the positions of the transverse jet and the zones of energy release indicated by rectangles, is shown in the lower part.

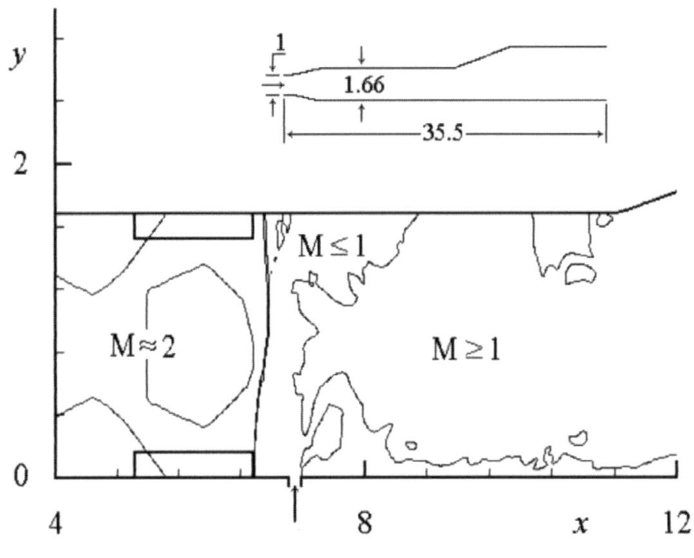

Figure 1. The channel scheme. The fragment of the Mach numbers distribution before the energy input in the zones.

The Mach number at the entrance to the channel is assumed to be $M = 1$. The nozzle creates in the first part of constant cross section (width is equal 1.66) a stream with Mach number $M \approx 2$ that corresponds to the conditions in the path of high-speed ramjet. At the exit of the channel, the Mach number varies from 2.4 to 2.9. The jet of width $\Delta x = 0.104$ is injected into the part with constant cross section. The hole size is selected from the following requirement: the jet energy should be negligibly small in comparison to the energy of the entire flow in the channel. The pressure in the gas generator, which forms the jet, is $p = 2$, that is significantly greater than the pressure in the undisturbed flow. The magnitude of pressure in the gas generator is based on the requirement of approximate equality of pressure drop between the gas generator and the excess pressure, which can create an instantaneous thermal pulse in the channel. The thermal

Journal of Physics: Conference Series **722** (2016) 012015 doi:10.1088/1742-6596/722/1/012015

pulse energy corresponds to the instantaneous reaction in the hydrogen–air mixture with the range of values of the excess air factor $2.6 < \alpha < 10$. This energy is supplied upstream from the jet in two subsurface zones which are placed opposite each other. The length of the zones of 1.7 along the channel is greatly larger than the perpendicular size of 0.16. As can be seen from Figure 1, the distance from the zones to the jet is comparable to the characteristic size of the zones along the channel axis, whereby the intensity of the pressure pulse will decrease slightly to the moment of its convective drift to the jet site. The jet flow leads to a compression of gas in front of it. The system can serve as a very approximate physical model of the experiment [1]. The energy supply period $\Delta t = 2.16$. The value of the period is chosen from the condition that the gas portion heated by an energy pulse has left the energy supply zone to the next energy supply. At the same time almost every portion of gas passing through the energy supply zone should have received energy.

4. The numerical simulation results

The impact of the considered surface energy supply only (without jet) on the flow does not lead to the formation of the extensive area with transonic speeds localized in the constant cross section. It is caused by demolition of the heated gas in the widening part of the channel (channel locking is not taking place). The jet without heat pulses does not cause locking of channel. The Mach number distribution in Figure 1 is obtained when the gas flow in the channel is preliminary heated to 1000 K (this corresponds to $\alpha = 2.9$)). This periodic solution is shown at a time immediately before the supply of energy. The flow with $M \approx 2$ is braked to subsonic speed ($M \approx 0.8$) in the shock wave, which in the period examined is carried out by the stream beyond the zone of energy release. Further subsonic flow is accelerated to the speed of sound, and then broadens and becomes supersonic. The formation of modes is shown in Figure 2 (*1* denotes the coordinate x_{sh} of the shock wave propagating upstream; t is the time, *2* and *3* are the wall coordinates of the shock wave caused by the jet in the absence of energy supply, *4* is the boundary of the energy supply zone). The front moves upstream, but remains in the constant cross section of the channel.

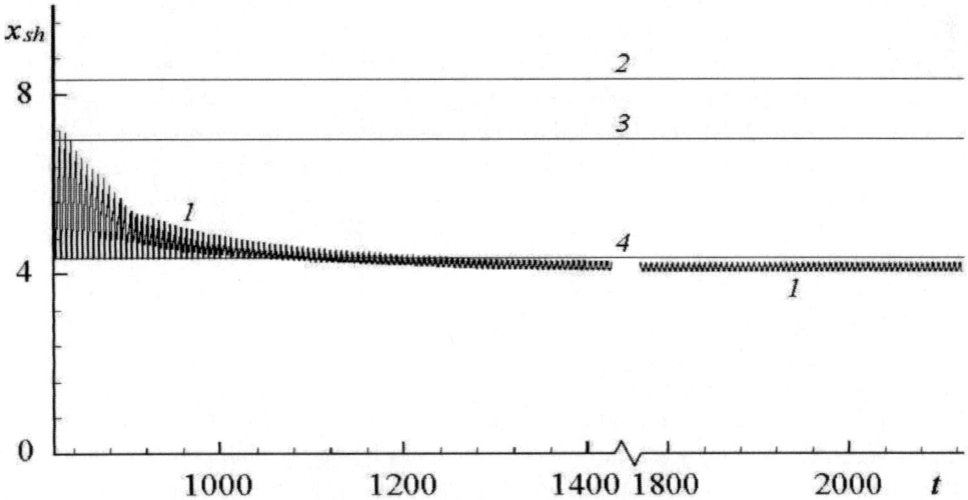

Figure 2. The process of formation of mode with transonic region (*1* is the coordinate x_{sh} of the shock wave propagating upstream; t is the time, *2* and *3* are the the wall coordinates of the shock wave caused by the jet in the absence of energy supply, *4* is the boundary of the energy supply zone).

At values of excess air factor $\alpha < 2.9$ the combined action of the sources and the jet leads to the locking of channel; for $\alpha > 4$ the flow remains supersonic. The interval $2.9 \leq \alpha \leq 4$ is quite wide. The significant interval width means the stability of this mode in relation to the fluctuating of parameters. Thus, the combined effect of surface energy supply and the jet leads to the formation of subareas with transonic velocities with the length along the channel which is comparative to the size of the channel inlet. This creates favorable conditions for effective supply and combustion of fuel in a widening channel section. The Table 1 shows the coordinates restricting the channel subsonic region (x_{up} and x_{sk} are the minimum and maximum coordinates of the subsonic region along the center line; $< x_{up} >$ and $< x_{sk} >$ are the minimum and maximum coordinates of the subsonic region averaged over the cross section of the channel), and corresponding sizes of the subsonic area depending on the excess air factor α.

Table 1. The coordinates of the boundaries of the subsonic region and its size along the center line and the values averaged over the cross section of the channel depending on the excess air factor α.

α	x_{up}	x_{sk}	Δx	$< x_{up} >$	$< x_{sk} >$	$< \Delta x >$
2.9	6.744	7.222	0.478	6.754	7.902	1.148
3.0	6.804	7.265	0.461	6.815	7.872	1.057
3.5	7.279	7.311	0.032	7.113	7.689	0.576
4.0	7.524	7.560	0.036	7.786	8.096	0.310

The values given correspond to the minimum values of the width of this region because they are obtained at the moment of time before the supply of energy. With decrease of excess air ratio the transonic speed region becomes wider. The ability to create such regimes is ensured by the use of negative feedback mechanism on the pressure. Its implementation is shown in Figure 3, which demonstrates the dependence of the pressure at the exit of the jet from the hole (curve *1*) and the corresponding jet consumption m (curve *2*).

These dependencies have an oscillatory character and are in opposite phase. When the high pressure area which was created by surface energy sources is drifting into the region of the inflow jet, the pressure difference decreases which leads to decrease of the flow rate in the jet and vice versa. This mechanism works quite effectively, if the pressure difference generated by the energy release, in the order of magnitude, is comparable to the pressure drop between the gas generator (which implies the jet) and the flow in the channel. Figure 4 demonstrates the chart of "normalized power N" – "period Δt of energy supply." Power N is normalized to the product $3\rho a^3 S$, where S is the cross section area of constant cross-section, ρ and a are the density and sound velocity at the channel inlet. Curves *1–5* and the dashed horizontal line correspond to the transversal source of energy. These curves are described in [7]. The dashed lines correspond to an analytical assessment for the transverse sources corridor of values when a disturbance moving upstream stopped in a channel part of constant cross section. The dot-dash line demonstrates the same assessment for the near-wall sources. The squares and diamonds present a numerical solution without locking for transverse and near-wall sources, respectively, the triangles (with a base at the bottom) and circles correspond to the solution with locking. An important result revealed by the chart is a very narrow corridor formed by lines *1–2*, which are obtained by numerical calculations for the transverse source, and the almost complete absence of similar corridor for near-wall sources. The small width of the corridors means that a small perturbation will result in either locking of channel or demolition of a shock wave downstream. The solid lines with triangles in Figure 4 marks the boundaries of this corridor for the joint

Journal of Physics: Conference Series **722** (2016) 012015 doi:10.1088/1742-6596/722/1/012015

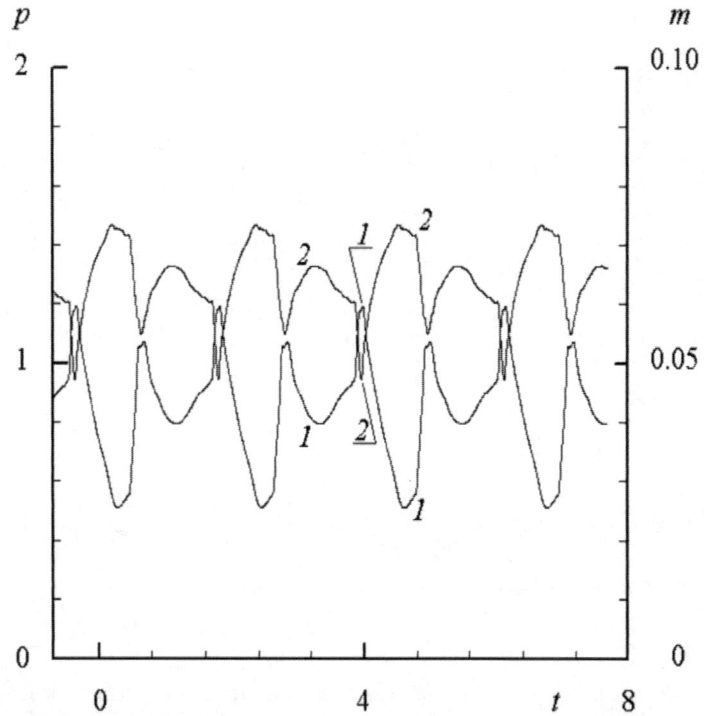

Figure 3. The dependence of the pressure *1* and the gas consumption from gas generator *2* at the exit of the jet from the hole.

action of the jet and the near-wall surface sources which is obtained numerically. The diagram shows that its width is almost 40% of the lower limit of the corridor. A sufficiently large width of the corridor means that the regime will be more resistant to disturbances than the regimes with energy sources only. This is provided by the mechanism of negative feedback on pressure. The mechanism of interaction between the flow and heat sources is a positive argument for the justification of the method for starting the combustion process in the combustion chamber. This mechanism explains why the intensification of combustion in the experiment [1] does not lead to locking up the channel, but forms a continuous mode of combustion.

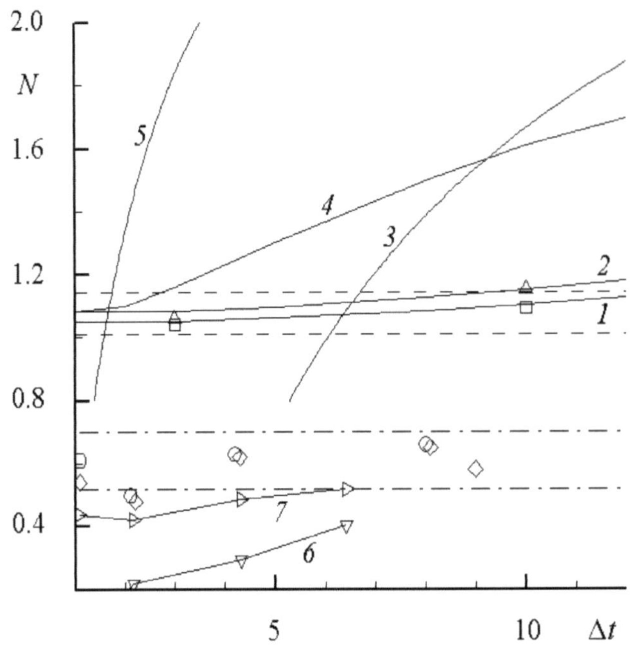

Figure 4. The mode chart (- - - - - analytical assessment for the transverse sources corridor of values when a disturbance, moving upstream, stopped in a channel part of constant cross section; — · — – analytical assessment for the near-wall sources corridor of values when a disturbance, moving upstream, stopped in a channel part of constant cross section; □ and ◇ represent a numerical solution without locking for transverse and near-wall sources, respectively; △ (with a base at the bottom) and ○ correspond to the solution with locking; —— with △ marks the boundaries of this corridor for the joint action of the jet and the near-wall surface source).

5. Conclusion

Thus, the possibility of realization of the stable modes with transonic region in supersonic flow is shown. Stability is achieved under the condition of approximate equality of pressure drop between the gas generator and the maximum additional pressure resulting from the near-wall energy. The significant wide corridor for input power in the mode chart diagram confirms the regime stability. Such modes may be used as the initial conditions for a realization of combustion in the combustion chambers.

Acknowledgments

Work performed under the project of SB RAS III.22.6.2 and with the financial support of the RFBR (grant No. 14-08-00820). Authors gratefully acknowledge Christopher Hearsey and Valeriya Hearsey for their assistance.

References

[1] Tretyakov P and Zabaykin V 2015 *Russian Congress on fundamental problems of theoretical and applied mechanics* ed Akhmetov D (Kazan) p 3778
[2] Guryleva N, Ivankin M, Lapinsky D and Timoshenko V 2012 *TsAGI Science Journal* **43** 753–74
[3] Frolov S, Dubrovsky A and Ivanov V 2012 *Russian Journal of Physical Chemistry B: Focus on Physics* **6** 276–88
[4] Zhao Z, Li J, Zheng J, Cui Y D and Knoo C 2015 *AIAA Journal* **53** 1336–48
[5] Zhuravskaya T and Levin V 2015 *Fluid Dynamic* **50** 283–93

[6] Zhuravskaya T and Levin V 2012 *Fluid Dynamics* **47** 793–801
[7] Zamuraev V and Kalinina A 2015 *Proc. VIII Lavrentyev readings on mathematics, mechanics and physics* (Novosibirsk: Lavrentyev Institute of Hydrodynamics) p 111

On the construction of heat wave in symmetric case

A L Kazakov and A A Lempert

Matrosov Institute for System Dynamics and Control Theory SB RAS, Irkutsk, Russia

E-mail: kazakov@icc.ru, lempert@icc.ru

Abstract. A nonlinear second-order parabolic equation with two variables is considered. Under additional conditions, this equation can be interpreted as the porous medium equation in case of dependence of the unknown function on two variables: time and distance from the origin. The equation has a wide variety of applications in continuum mechanics, for example, it is applicable for mathematical modeling of filtration of ideal polytropic gas in porous media or heat conduction. The authors deal with a special solutions which are usually called heat waves. A special feature of such solution is that it consists of two continuously joined solutions. The first of them is trivial and the second one is nonnegative. The heat wave solution can have discontinuous derivatives on the line of joint which is called the front of heat wave, i.e. smoothness of the solution, generally speaking, is broken. The most natural problem which has such solutions is the so-called "the Sakharov problem of the initiation of a heat wave". New solutions of the problem in the form of multiple power series for physical variables are constructed. The coefficients of the series are obtained from tridiagonal systems of linear algebraic equations. Herewith, the elements of matrices of this systems depend on the matrix order and the condition of the diagonal dominance is not fulfilled. The recurrent formulas for the coefficients are suggested.

1. Introduction

In this paper we consider the nonlinear porous medium equation [1] having the form

$$T_t = T\Delta T + \frac{1}{\sigma}(\nabla T)^2. \tag{1}$$

This differential equation represents two fundamental laws of physics: the Fourier law for heat conductivity [2] and the Darcy law for filtering [3] provided the dependencies of the heat conductivity coefficient on temperature and of the filtration coefficient on the gas density which are exponential.

In the case of one independent variable eq. (1) can be represented as

$$u_t = uu_{\rho\rho} + \frac{1}{\sigma}u_\rho^2 + \frac{\nu}{\rho}uu_\rho. \tag{2}$$

Here $u(t,\rho)$ is the unknown function; t is the time, $\rho = \sqrt{x_1^2 + \ldots + x_{\nu+1}^2}$ is the spatial coordinate; $\sigma > 0$ is a constant characterizing the properties of the medium; the parameter ν is equal to 0,1 or 2 in the cases of flat, axial or central symmetry, respectively.

Eq. (1) is a second-order nonlinear equation of parabolic type [4]. Note that (1) can be considered as an independent mathematical object (with no connection to applications), in which case the constants σ, ν may admit any real values (except $\sigma = 0$).

The heat waves (wave filtering) propagating on cold (zero temperature) background with a finite velocity are one of the interesting classes of solutions of the heat equation. Geometrically the solution of heat-wave type solution consists of two surfaces (perturbed solution $u(t,\bar{x}) > 0$ and zero temperature background $u \equiv 0$), which are continuously joined along some sufficiently smooth line $x = b(t)$, called the front. In the linear case such solutions are apparently known since the time of J. Fourier.

A simple example of the heat wave for a linear heat equation $u_t = u_{xx}$ is given by

$$u(t,x) = \begin{cases} \exp\left(-x/\sqrt{2}\right)\sin\left(t - x/\sqrt{2}\right), & 0 \le x < t\sqrt{2}, \\ 0, & x \ge t\sqrt{2}, \end{cases}$$

where $0 \le t \le \pi/2$. It is easy to see that the front here is the line $x = t\sqrt{2}$.

Note that in this case the question of uniqueness is not addressed: for the unique solvability we need to assume an additional condition (for example, to determine the derivative with respect to the spatial coordinate at $x = 0$, i.e. to consider the Cauchy problem).

In relation to Eq. (1) the terms "heat wave" and the "wave filter", have been previously used as well. For example, in the papers [5, 6] the composite solution:

$$u_*(t,\bar{x}) = \begin{cases} u(t,\bar{x}) > 0, & a(t, x_2, \ldots, x_n) > x_1, \\ 0, & x_1 \ge a(t, x_2, \ldots, x_n), \end{cases}$$

where $a_t(0, x_2, \ldots, x_n) > 0$ was understood in this sense. For the problems discussed in the paper, such definition is not quite convenient, because it assumes the specification of the front in the form of a sufficiently smooth function. So, we define the heat wave in the following (more general) way.

Definition 1. Let $u(t,\bar{x})$ be a continuous, non-negative function defined for $t \in [t_*, t^*)$, $\bar{x} \in X \subseteq \mathbb{R}^n$, with a compact simply connected support $\operatorname{supp} u = \overline{D}$, where $D = \{(t,\bar{x}) |\ u(t,\bar{x}) > 0\}$.

We call the function $u(t,\bar{x})$ a *heat wave*, if it is

(i) twice continuously differentiable in D with respect to the spatial variables $\bar{x}$ and continuously differentiable with respect to the time t;

(ii) satisfies Eq. (1) in D;

(iii) the domain D has the property: if $t_* \le t_1 < t_2 < t^*$, then $D(t_1) \subset D(t_2)$, where $D(t_i)$ is the projection of the section of D by the hyperplane $t = t_i$, $i = 1, 2$ to $\mathbb{R}^n$.

In the case when the function $u(t,\bar{x})$ is analytical in D, we consider *the analytical heat wave*. The boundary $\Gamma = \overline{D} \setminus D$ of D will be called the front of a heat wave or a zero temperature front.

Since the function $u \equiv 0$ obviously satisfies (2), the heat wave is a classical (smooth) solution of Eq. (1) everywhere in the domain of definition, except perhaps for a set Γ, where the derivatives (but not unknown function) are allowed to be discontinuous.

The simplest example of the solution of heat-wave type for Eq. (1) in the case of a plane symmetry and when $\nu = 0$, $\rho = x$ is the piecewise linear function

$$u(t,x) = \begin{cases} \alpha_1 t - \sqrt{\sigma\alpha_1}x, & x < b(t) = \alpha_1 t/\sqrt{\sigma\alpha_1}, \\ 0, & x \ge \alpha_1 t/\sqrt{\sigma\alpha_1}, \ \alpha_1 = const > 0. \end{cases}$$

It was Ya.B. Zeldovich, who first obtained solutions of (1), in the form of a heat wave for nonlinear heat conductivity problem [7]. In [3] similar results were obtained for filtration problems by G.I. Barenblatt. In the paper [8] by O.A. Oleinik and co-authors the boundary problems, in which the speed of propagation of the front filter is supposed to be finite, are studied in abstract functional spaces. Heat waves in the class of piecewise-analytic functions were first considered by A.F. Sidorov [9, 10]. The method of power series [11] for the porous medium equation with special boundary conditions is applied in [12].

The present paper discusses the solution of heat-wave type for Eq. (1). These solutions are constructed in the class of analytic functions (analytical heat wave) as power series with coefficients, which are recurrently determined by the tridiagonal systems of linear algebraic equations [13]. At the same time, unlike previously published papers of the A.F. Sidorov's scientific school [6, 10, 14, 15], the solution is constructed in an explicit form. Until now similar results for Eq. (2) were obtained only in the case of $\nu = 0$ [5] and $\nu = 1$ [15].

The constructed power series are compared with the exact solutions of the equation, which are obtained by the methods of the group analysis [16, 17] of differential equations. The comparison shows that the first coefficients of the Taylor series are the same, which confirms the correctness of the results.

2. Boundary condition and existence theorem

Let for Eq. (2) the following boundary condition

$$u(t, \rho)|_{\rho=\nu R} = f(t), \ f(0) = 0, \ f'(0) > 0. \tag{3}$$

This is be given the most natural problem, in which the solutions described in the previous section arise. In the Russian scientific literature it is called "The Sakharov problem of the initiation of a heat wave".

Note that (3) contains only one boundary condition for second-order equations.

Nevertheless due to the presence of degeneration that is associated with the vanishing factor to the highest derivative, Eq. (2) has the specific properties characteristic to first-order equations. So, for the problem (2), (3) we have the following theorem of existence and uniqueness of solutions.

Theorem 1. Suppose that $f = f(t)$ is analytical in some neighborhood of $t = 0$. Then the problem (2), (3) has a unique analytical solution in a full neighborhood of $t = 0$, $\rho = \nu R$, if the sign $u_\rho|_{t=0,\ \rho=\nu R}$ is choosen.

The proof of it is very cumbersome. So, here we present just the general scheme.

(i) We make some nondegenerate analytic changes of variables (A–D) that allow to bring the problem (2), (3) to a special (characteristic) form.
(A) The first substitution allows to transforms not yet known front of the heat wave in the coordinate axis. This adds to the problem an extra unknown function and an additional boundary condition (u is equal to zero on the front).
(B) The second substitution is a special analog of the hodograph transformation and changes the roles of the unknown function u and one of the independent variables. It allows to get rid of the unknown function b and one of the boundary conditions.
(C) Next substitution brings the curve $u = f(t)$ to the coordinate axis.
(D) The last substitution is a partial decomposition of the unknown function in a Taylor series, and the boundary condition is introduced into the equation.

(ii) The solution of transformed problem is constructed in the form of the characteristic power series of one variable, whose coefficients are uniquely determined (if we initiate the sign of the first one).

(iii) For the solution of the transformed problem we construct a majorizing series.

(iv) We prove the convergence of the majorizing series by constructing one more majorant.

3. Heat wave construction

We construct a solution of the problem (2), (3) in the form of a multiple power series

$$u = \sum_{l,m=0}^{\infty} u_{l,m} \frac{t^l}{l!} \frac{(\rho - \nu R)^m}{m!}, \quad u_{l,m} = \frac{\partial^{l+m} u(t,\rho)}{\partial t^l \partial \rho^m}\Big|_{t=0,\ \rho=\nu R}. \tag{4}$$

Using the condition (3) it is possible to find the coefficients $u_{l,0}$, $l = 0, 1, 2, \ldots$. In fact, by the assumption of the theorem 1, for the function $f(t)$ in some neighborhood of $t = 0$ we have the expansion

$$f(t) = \sum_{l=0}^{\infty} f_l \frac{t^l}{l!},$$

where $f_l = f^{(l)}(0)$. From (3) and (4) there follows the equality

$$u(t,\rho)|_{\rho=\nu R} = \sum_{l=0}^{\infty} u_{l,0} \frac{t^l}{l!} = \sum_{l=0}^{\infty} f_l \frac{t^l}{l!}.$$

We can see, that $u_{l,0} = f_l$, in particular, $u_{0,0} = u(0, \nu R) = f_0 = 0$, $u_{1,0} = f_1 = f'(0) > 0$. The remaining coefficients of the series (4) are found by induction on the total differentiation order $l + m$.

Let us assume in (2) that $t = 0, \rho = \nu R$. Taking into account that the values $u_{0,0} = 0$, $u_{1,0} > 0$ have already been found, we obtain the quadratic equation for $u_{0,1}$, which shows that $u_{0,1}$ is defined in two ways by the formula $u_{0,1} = \pm\sqrt{\sigma f_1}$. Thus, the base of induction is found.

Suppose now that all the coefficients up to order n are found, i.e. $m+l = 1, \ldots, n$. Then, the coefficients of the order $n+1$ are found by solving a system of linear algebraic equations (SLAE) with a tridiagonal matrix, where the main diagonal elements are $a_i = -(i + 2/\sigma)\, u_{0,1}$, $i = 0, \ldots, n$, the superdiagonal elements are $b_j = -j u_{1,0} < 0$, $j = n, \ldots, 1$ and the subdiagonal elements are equal to one [15]. Obviously, in general, such matrix is not diagonally dominant.

The resolution of the SLAE will be described below. But first we need to introduce some additional sequences

$$\lambda_0 = 1, \quad \lambda_1 = a_n, \quad \lambda_{k+1} = a_{n-k}\lambda_k - b_k \lambda_{k-1};$$

$$\eta_0 = 1, \quad \eta_1 = a_0, \quad \eta_{k+1} = a_k \eta_k - b_{n+1-k}\eta_{k-1}; \quad k = 1, \ldots, n.$$

By induction on k we can show that all $\lambda_k \neq 0, \eta_k \neq 0$ (regardless of the sign of $u_{0,1}$).

Let $|A|$ be the determinant of the matrix SLAE. Then $|A| = \eta_{n+1} = \lambda_{n+1}$.

In view of the introduced notations, $n+1$ order coefficients of the series (4) are determined by the following formulas

$$u_{n,1} = \frac{1}{|A|}\lambda_n (L_{n,0} - f_{n+1}) + \frac{1}{|A|}\sum_{j=2}^{n+1} \eta_0 \lambda_{n+1-j} \prod_{l=j-1}^{1} (-b_{n+1-l}) L_{n+1-j,j-1},$$

$$u_{n-1,2} = -\frac{1}{|A|}\eta_0 \lambda_{n-1}(L_{n,0} - f_{n+1}) + \frac{1}{|A|}\eta_1 \lambda_{n-1} L_{n-1,1} +$$
$$+ \frac{1}{|A|}\sum_{j=3}^{n+1} \eta_1 \lambda_{n+1-j} \prod_{l=j-1}^{2} (-b_{n+1-l}) L_{n+1-j,j-1}.$$

$$u_{n+1-k,k} = \frac{1}{|A|}\eta_0 \lambda_{n+1-k} (-1)^{k-1} (L_{n,0} - f_{n+1}) + \tag{5}$$

$$+\frac{1}{|A|}\sum_{j=2}^{k}\eta_{j-1}\lambda_{n+1-k}(-1)^{k-j}L_{n+1-j,j-1}+$$

$$+\frac{1}{|A|}\sum_{j=k+1}^{n+1}\eta_{k-1}\lambda_{n+1-j}\prod_{l=j-1}^{k}(-b_{n+1-l})L_{n+1-j,j-1},\ k=3,\ldots,n-1,$$

$$u_{1,n}=\frac{1}{|A|}\eta_0\lambda_1(-1)^{n-1}(L_{n,0}-f_{n+1})+$$

$$+\frac{1}{|A|}\sum_{j=2}^{n}\eta_{j-1}\lambda_1(-1)^{n-j}L_{n+1-j,j-1}-\frac{1}{|A|}\eta_{n-1}\lambda_0 b_1 L_{0,n},$$

$$u_{0,n+1}=\frac{1}{|A|}\eta_0\lambda_0(-1)^n(L_{n,0}-f_{n+1})+\frac{1}{|A|}\sum_{j=2}^{n+1}\eta_{j-1}\lambda_0(-1)^{n+1-j}L_{n+1-j,j-1}.$$

Here $L_{n-k,k}$ are known by the induction hypothesis (since it depends on the coefficients of the series of the order less than n):

$$L_{n-k,k}=\sum_{\substack{i=0\\ i+j\neq 0}}^{n-k}\sum_{\substack{j=0\\ i+j\neq 1}}^{k}C_{n-k}^{i}C_{k}^{j}u_{i,j}u_{n-k-i,k-j+2}+$$

$$+\frac{1}{\sigma}\sum_{\substack{i=0\\ i+j\neq 0}}^{n-k}\sum_{\substack{j=0\\ i+j\neq n}}^{k}C_{n-k}^{i}C_{k}^{j}u_{i,j+1}u_{n-k-i,k-j+1}+$$

$$+\nu\sum_{i=0}^{n-k}\sum_{j=0}^{k}C_{n-k}^{i}C_{k}^{j}\left[\sum_{l=0}^{j}C_{j}^{l}u_{i,l}\frac{(-1)^{j-l}(j-l)!}{R^{j-l+1}}\right]u_{n-k-i,k-j+1}.$$

Thus, the series (4) is constructed. Its convergence follows from Theorem 1, namely, from the existence and uniqueness of an analytic solution of the problem (2), (3) with the choice of the sign of $u_{0,1}$

The above construction of the series allows to prove the following statement. But first we make a shift of the spatial coordinate so that the boundary condition is set at the zero value of the new coordinate r. Let $r=\rho-\nu R$.

Theorem 2. Under the conditions of the theorem 1 the problem

$$(r+\nu R)^2u_t=u\Big[\nu(r+\nu R)u_r+(r+\nu R)^2u_{rr}\Big]+\frac{1}{\sigma}(r+\nu R)^2u_r^2,$$

$$u(t,r)|_{r=0}=f(t),\ u(t,r)|_{t=0}=0, \tag{6}$$

has piecewise analytical solution which, if $\nu=0,1,2,\ldots$, is an analytical heat wave in the neighborhood of $t=0$, $r=0$. The choice of the direction of wave motion (to the singular point $r=\nu R$ or from it) ensures the uniqueness.

Proof. From the theorem 1 and the results of this section there follows the existence of two piecewise analytical solutions of the boundary problem (2), (3). The choice of the sign of $u_{0,1}$ ensures the uniqueness. We show now that this implies the validity of the above statement.

As shown in the beginning of this section $u_{1,0}=f_1>0$, $u_{0,1}=\pm\sqrt{\sigma f_1}$. Hence, we have that for the solution $u=u_+(t,r)$ corresponding to a positive value of $u_{0,1}$, there is a line $r=b_+(t)$, $b_+(0)=0$, $b'_+(0)<0$, in the plane of the variables t,r on which the condition $u_+|_{r=b_+(t)}=0$ is satisfied in a neighborhood of $t = 0, r = 0$, if $t > 0$. For the solution $u = u_-(t, r)$ corresponding to a negative value of $u_{0,1}$, there is a line $r = b_-(t)$, $b_-(0) = 0$, $b'_-(0) > 0$, in the plane of the variables t, r on which the analogous condition $u_-|_{r=b_-(t)} = 0$ is satisfied in a neighborhood of $t = 0, r = 0$, if $t > 0$.

Since $b'_+(0) < 0$, $u_{1,0} > 0$, the solution $u = u_+(t, r)$ allows us to find the heat wave

$$u(t,x) = \begin{cases} u_+ > 0, & b_+(t) < r \leq 0, \\ 0, & r \leq b_+(t), \end{cases}$$

which is moving into the inner area towards the singular point $r = \nu R$.

On the other hand, $b'_-(0) > 0$, $u_{1,0} > 0$ and the solution $u = u_-(t, r)$ allows to determine the heat wave

$$u(t,x) = \begin{cases} u_- > 0, & b_-(t) > r \geq 0, \\ 0, & r \geq b_-(t), \end{cases}$$

which is moving to the external area.

It is easy to see that as $b_\pm(0) = 0$, then in both cases the condition $u|_{t=0} = 0$ is satisfied and also on the zero front we have discontinuities of the derivatives. The value of t^* (see the definition) is determined by the radius of convergence for the series (4). So, the constructed composite solutions, indeed, fall under the definition of an analytical heat wave. The choice of the sign of $u_{0,1}$ (which, recall, provides the uniqueness) is equivalent to the choice of the front of the heat wave motion direction. This proves the theorem.

4. Comparison

As already noted, one of the main results of the present study is that for the first time the authors constructed solutions of the heat wave initiation problem by given boundary conditions in a constructive way. In this section, the constructed series are compared with the exact solutions. We further assume that $\nu \neq 0$, as the case $\nu = 0$ was considered previously in [5].

Let us obtain the first coefficients of the series. Already obtained in the previous section

$$u_{0,0} = 0, \; u_{l,0} = f_l, \; l = 1, 2, 3, ..., \; u_{0,1} = \pm\sqrt{\sigma f_1},$$

Using (5), we can show that

$$u_{1,1} = \frac{\pm(1 + 2/\sigma)R\sqrt{\sigma f_1} f_2 - 2f_1^2}{(3 + 4/\sigma)Rf_1}, u_{0,2} = \frac{Rf_2 \mp 3f_1\sqrt{\sigma f_1}}{(3 + 4/\sigma)Rf_1}. \tag{7}$$

Note that if the boundary conditions are set in the form (3), then the parameter ν specifies the point where the condition is set. At the same time ν vanishes in the formulas for the coefficients.

The coefficients of the series (4) are compared with the coefficients of Taylor series for the exact solutions of Eq. (2) in a neighborhood of zero.

$$u = \pm\frac{\rho^2}{C \mp \alpha t} \mp \frac{C^{\beta-1}(\nu R)^2}{(C \mp \alpha t)^\beta}, \tag{8}$$

where $\alpha = 2 + 2\nu + 4/\sigma$, $\beta = (2 + 2\nu)/\alpha$; $C > 0$ is an arbitrary constant. The upper sign corresponds to the heat wave that moves from $\rho = R\nu$ in the direction of increasing of the coordinate ρ. The lower one corresponds to the movement in the direction of decreasing of ρ (to the singular point $\rho = 0$). This solution was obtained by using the group theory method [17].

We obtain the coefficients of Taylor expansion for the exact solution 8 by a direct differentiation. For definiteness assume $C = 1, \nu = 1, R = 1$ and consider the movement of the heat wave in the the outer area.

$$\begin{gathered} u_{1,0} = f_1 = \alpha(1-\beta) = \frac{4}{\sigma},\ u_{0,1} = -2, \\ u_{2,0} = f_2 = -\frac{16(3\sigma+2)}{\sigma^2},\ u_{1,1} = \frac{8(\sigma+1)}{\sigma},\ u_{0,2} = -2. \end{gathered} \tag{9}$$

It is not difficult to make sure by substituting the absolute values that formula (7) takes the form of (9). It should be observed that the high-order coefficients of the constructed series coincide with the respective Taylor expansion coefficients of (8) as well.

Note, that proposed power series is applicable for verification of numerical results [6, 18].

5. Conclusion

Summarizing the research, note that the boundary value problem with singularity for a nonlinear parabolic equation is considered. This equation becomes the one-dimensional porous medium equation in the case the parameter ν is a non-negative integer. At the same time it can be interpreted as a problem of the initiation of the heat wave by the boundary condition, which is specified in the sphere in $\mathbb{R}^{\nu+1}$.

For the considered problem the existence and uniqueness theorem in the class of analytical functions is proved. This theorem is an analogue of the classic Cauchy-Kovalevskaya theorem. Authors construct solution in the form of multiple series with respect to powers of the physical variables. Its coefficients are determined by solving the tridiagonal systems of linear algebraic equations without diagonal dominance. A comparison of the constructed series coefficients and the coefficients of Taylor series for exact solutions showed their identity.

Acknowledgments

The authors are grateful to P. Kuznetsov for collaboration in consideration of some particular problems of this paper.

The reported study was particulary funded by RFBR according to the research projects No. 16-01-00608 and No. 16-31-00291.

References

[1] Vazquez J 2007 *The Porous Medium Equation: Mathematical Theory* (Oxford: Clarendon Press)
[2] Evans L 2010 *Partial Differential Equations* (American Mathematical Society)
[3] Barenblatt G, Entov V and Ryzhik V 1990 *Theory of Fluid Flow of Fluids Through Natural Rocks* (Dordrecht/Boston/London: Kluwer Academic Publ.)
[4] Ladyzhenskaya O, Solonnikov V and Ural'tseva N 1968 *Linear and Quasilinear Equations of Parabolic Type* (American Mathematical Society)
[5] Kazakov A and Lempert A 2013 *J. Appl. Mech. Tech. Phys.* **54** 97–105
[6] Kazakov A and Spevak L 2013 *Applied Mathematical Modelling* **37** 6918–28
[7] Zeldovich Y and Raizer Y 1966 *Physics of Shock Waves and High-Temperature Hydrodynamic Phenomena* vol 2 (New York: Academic Press)
[8] Oleinik O, Kalashnikov A and Chzou Y 1958 *Izv. Akad. Nauk SSSR* **22** 667–704
[9] Vasin V and Sidorov A 1983 *Izvestiya Vysshikh Uchebnykh Zavedenii. Matematika* **7** 13–26
[10] Sidorov A 1985 *Dokl. Akad. Nauk SSSR* **280** 47–51
[11] Teshukov V 1980 *Journal of Applied Mechanics and Technical Physics* **21** 261–7
[12] Furtak-Cole E, Telyakovskiy A and Cooper C 2013 *AGU Fall Meeting Abstracts*
[13] Kazakov A and Lempert A 2011 *J. Appl. Mech. Tech. Phys.* **52** 352–62
[14] Filimonov M 2003 *Differential Equations* **39** 844–52
[15] Kazakov A and Kuznetsov P 2014 *J. Appl. Ind. Math.* **8** 227–35
[16] Bluman G and Kumei S 1989 *Symmetries and differential equations* (New York: Springer)

[17] Andreev V, Kaptsov O, Pukhnachov V and Rodionov A 2010 *Application of Group-Theoretical Methods in Hydrodynamics* (New York: Springer)
[18] Duque J C M, Almeida R M P and Antontsev S 2014 *Applied Mathematics and Computation* **235** 137–47

Two layer asymptotic model for the wave propagation in the presence of vorticity

M Yu Kazakova[1,2] and P Noble[2]

[1] Lavrentyev Institute of Hydrodynamics, Novosibirsk, Russia
[2] Institut de mathematiques de Toulouse, Toulouse, France

E-mail: maria.kazakova@math.univ-toulouse.fr

Abstract. In the present study, we consider the system of two layers of the immiscible constant density fluids which are modeled by the full Euler equations. The domain of the flow is infinite in the horizontal directions and delimited above by a free surface. Bottom topography is taken into account. This is a simple model of the wave propagation in the ocean where the upper layer corresponds to the (thin) layer of fluid above the thermocline whereas the lower layer is under the thermocline. Though even this simple framework is computationally too expensive and mathematically too complicated to describe efficiently propagation of waves in the ocean. Modeling assumption such as shallowness, vanishing vorticity and hydrostatic pressure are usually made to get the bi-layer shallow water models that are mathematically more manageable. Though, they cannot describe correctly the propagation of both internal and free surface waves and dispersive/non hydrostatic must be added. Our goal is to consider the regime of medium to large vorticities in shallow water flow. We present the derivation of the model for internal and surface wave propagation in the case of constant and general vorticities in each layer. The model reduces to the classical Green-Naghdi equations in the case of vanishing vorticities.

1. Introduction

We consider two-layer flow of invisid incompressible and immiscible fluids of constant densities ρ_1, ρ_2, for the stable stratification $\rho_1 > \rho_2$. The flow area is bounded above by a free surface $z = \zeta_1(\vec{x}, t)$ and below by nonuniform bottom $z = -h_{2_0} + \zeta_2(\vec{x}, t)$. Hereafter quantities with indexes '1' are addressed to the upper layer, and indexes '2' to the lower.

The governing equations is given by Euler equations for each layer (here $V_i = (\vec{u}_i(t, \vec{x}, z), w_i(t, \vec{x}, z))$, $\vec{u}_i(t, \vec{x}, z) = (u_i(t, \vec{x}, z), v_i(t, \vec{x}, z))$ and p_1, p_2 denote the velocities and pressure fields, respectively):

$$\begin{cases} (V_i)_t + V_i \cdot \nabla_{\vec{x},z} V_i = -\dfrac{1}{\rho_i} \nabla_{\vec{x},z} p_i - g e_z, \\ \nabla_{\vec{x},z} \cdot V_i = 0, \end{cases} \qquad (i = 1, 2)$$

which are complemented with kinematic boundary conditions:

$$\begin{cases} \zeta_{1_t} + \vec{u}_1\big|_{z=h_0^1+\zeta_1} \cdot \nabla\zeta_1 &= w_1\big|_{z=h_1^0+\zeta_1}, \\ \zeta_{2_t} + \vec{u}_1\big|_{z=\zeta_2} \cdot \nabla\zeta_2 &= w_1\big|_{z=\zeta_2}, \\ \zeta_{2_t} + \vec{u}_2\big|_{z=\zeta_2} \cdot \nabla\zeta_2 &= w_2\big|_{z=\zeta_2}, \\ \vec{u}_2\big|_{z=-h_{2_0}+b(x)} \cdot \nabla(b) &= w_2\big|_{z=-h_{2_0}+b(x)}, \end{cases}$$

and dynamic boundary conditions:

$$p_1\big|_{z=h_{1_0}+\zeta_1} = 0, \quad p_1\big|_{z=\zeta_2} = p_2\big|_{\zeta_2}.$$

It is well known that under assumption of shallowness, vanishing vorticity and hydrostatic pressure the model can be reduced to the bi-layer shallow water models which are mathematically more manageable. Though, they cannot describe correctly the propagation of both internal and free surface waves since they do not take into account dispersive/non hydrostatic effects. In [1, 2] these effects are added but still under the hypothesis of vanishing or small vorticity.

The goal of the present study is deriving a bi-layer model in the regime of medium to large vorticities with taking into account dispersive terms and non hydrostatic pressure components. For that purpose we follow the strategy found in [3, 4]. We reduce the full two-layer Euler system describing the evolution of the flow by a non-dimensional system of averaged equations in each fluid layer that describes the evolution of the fluids height, the horizontal momentum in each layer and other averaged quantities (vorticities, internal energy). This system is exact but too complex to work with, because it is required a closure relations for the 'Reynolds tensors' and pressure gradients for each layer (we are following here the terminology using in [3]).

As no irrotationality assumptions are made, to find a closure we follow the strategy for the one layer found in [3] and introduce shear velocities representing the horizontal vorticities contribution to the horizontal velocities (which is zero for rotational case). The 'Reynolds tensors' and pressure gradients can be represented in terms of the shear velocities. By using vorticity and incompressible equations, we found an equation for shear velocities, and therefore our system become closed. The case of constant vorticities is considered as the first step of demonstration of the model applicability. The next step is to consider the case of general vorticities.

2. Asymptotic analysis

2.1. Dimensionless form of the averaged equations

For non-dimensionalization of the equations we use the following dimensionless parameters:

$$\rho = \frac{\rho_1}{\rho_2} > 1, \quad \varepsilon_1 = \frac{a_1}{h_{1_0}}, \quad \varepsilon_2 = \frac{a_2}{h_{1_0}}, \quad \mu = \frac{h_{1_0}^2}{L^2}, \quad \delta = \frac{h_{1_0}}{h_{2_0}}, \quad \beta = \frac{\beta}{h_{1_0}}.$$

To form these parameters several characteristic quantities should be specified: the typical amplitudes a_1 of the surface waves and a_2 of the interface waves, the typical amplitude β of the bottom variations and the typical depths h_{1_0} and h_{2_0} of two layers together with the typical wave length L.

Parameters ε_1 and ε_2 are so-called nonlinearity parameters. The parameter μ is the shallowness, or dispersion, parameter, that defines which wave scale is into account.

We are interested in shallow water flows and therefore we assume that shallowness parameter is small, $\mu << 1$, but no smallness assumption is made for nonlinearity parameters $\varepsilon_1, \varepsilon_2$, hence we allow for the large-amplitude waves.

In terms of dimensionless variables Euler equations take form:

$$\begin{cases} (V_i)_t + \varepsilon_i V_i \cdot \nabla_{\vec{x},z} V_i = -\dfrac{R_i}{\varepsilon_i M}(\nabla_{\vec{x},z} p_i + e_z), \\ \nabla_{\vec{x},z} \cdot V_i = 0, \end{cases} \tag{1}$$

where $M = (0, 0, \mu)^\top$, $e_z = (0, 0, 1)^\top$, $R_1 = 1$, $R_2 = \rho$, coupled with dimensionless boundary conditions. Hereinafter unless indicated otherwise $i = 1, 2$, that is corresponding to the each layer.

Let us introduce the notations of the vertical averaging of the horizontal velocity components, here $h_1 = 1 + \varepsilon_1\zeta_1 - \varepsilon_2\zeta_2$ and $h_2 = \varepsilon_2\zeta_2 + \frac{1}{\delta} - \beta b(x)$ are changing layer depth:

$$\overline{u}_1 = \frac{1}{h_1}\int_{\varepsilon_2\zeta_2}^{1+\varepsilon_1\zeta_1} \vec{u}_1 \, dz, \quad \overline{u}_2 = \frac{1}{h_2}\int_{-\frac{1}{\delta}+\beta b(x)}^{\varepsilon_2\zeta_2} \vec{u}_2 \, dz.$$

Now we can construct decomposition of horizontal velocity field given by:

$$\vec{u}_1 = \overline{u}_1 + \sqrt{\mu}\vec{u}_1^*, \quad \vec{u}_2 = \overline{u}_2 + \sqrt{\mu}\vec{u}_2^*. \tag{2}$$

Integration of system (1) using boundary conditions leads to the system of averaging equations on changing depths h_1, h_2 and average velocities (implying decomposition (2)):

$$\begin{cases} (h_i)_t + \varepsilon_i \nabla \cdot (h_i \overline{u}_i) = 0, \\ (h_i\overline{u}_i)_t + \varepsilon_i \nabla \cdot (h_1 \overline{u}_i \otimes \overline{u}_i) + \varepsilon_i \mu \nabla \cdot \left(\displaystyle\int_{A_i}^{B_i} \vec{u}_i^* \otimes \vec{u}_i^* dz \right) + \dfrac{R_i}{\varepsilon_i} \displaystyle\int_{A_i}^{B_i} \nabla p_i dz = 0, \\ (w_i)_t + \varepsilon_i (u_i \nabla \cdot) w_i + \varepsilon_i w_i (w_i)_z = -\dfrac{1}{\mu \varepsilon_i}(R_i (p_i)_z + 1), \end{cases} \tag{3}$$

it should be noted that $\nabla = (\partial_x, \partial_y)$ here is two-dimensional operator, and $A_1 = \varepsilon_2\zeta_2$, $B_1 = 1 + \varepsilon_1\zeta_1$, $A_2 = -\frac{1}{\delta} + \beta b(x)$, $B_2 = \varepsilon_2\zeta_2$ denote the layer boundaries.

The system (3) is exact, but it requires the closure for 'Reynolds tensors' $\int_{A_i}^{B_i} \vec{u}_i^* \otimes \vec{u}_i^* dz$ and pressure terms $\frac{R_i}{\varepsilon_i}\int_{A_i}^{B_i} \nabla p_i$ in terms of $\overline{u}_i$, h_i. The closure is obtained within the framework of procedure introduced in [3]. For this purpose we now introduce vorticities $\Omega_i = \nabla \times u_i$ that can be represented in accordance with velocity fields decomposition as:

$$\Omega_i = \begin{pmatrix} w_{iy} - v_{iz} \\ u_{iz} - w_{ix} \\ v_{ix} - u_{iy} \end{pmatrix} = \frac{\varepsilon_i\sqrt{gh_{1_0}}}{h_{1_0}}\sqrt{\mu}\begin{pmatrix} -\tilde{v}_{iz}^* + \sqrt{\mu}\tilde{w}_{iy} \\ \tilde{u}_{iz}^* - \sqrt{\mu}\tilde{w}_{ix} \\ \nabla^\perp \cdot \vec{\tilde{u}}_i \end{pmatrix} \equiv \frac{\varepsilon_i\sqrt{gh_{1_0}}}{h_{1_0}}\sqrt{\mu}\begin{pmatrix} \vec{\omega}_h^i \\ \omega_v^i \end{pmatrix}, \tag{4}$$

the tildes here correspond to the dimensionless functions, orthogonal operator $\nabla^\perp = (-\partial_y, \partial_x)$.

2.2. Velocity fields decompositions

Now we can find a representation for the velocity fields in terms of vortisities. Indeed, integration of the discontinuity equations yields to the expressions for the vertical velocities components in terms of horizontal velocities:

$$\begin{aligned} w_1 &= -\nabla \cdot [\overline{u}_1(z - \varepsilon_2\zeta_2)] - \sqrt{\mu}\nabla \cdot \int_{\varepsilon_2\zeta_2}^{z} \vec{u}_1^* dz + \frac{\varepsilon_2}{\varepsilon_1}\zeta_{2_t}, \\ w_2 &= -\nabla \cdot \left[\overline{u}_2\left(z - \beta b(x) + \frac{1}{\delta}\right)\right] - \sqrt{\mu}\nabla \cdot \int_{\beta b(x) - \frac{1}{\delta}}^{z} \vec{u}_2^* dz, \end{aligned} \tag{5}$$

boundary conditions is implied.

Definition of the vorticities (4) give us an equations for the components of horizontal velocity fields:

$$\partial_z \vec{u}_i^* = \sqrt{\mu}\nabla w_i - (\vec{\omega}_h^i)^{\perp}.$$

Integrating it with taking into account expressions for vertical velocities (5),and introduce the operators for the sake of simplicity:

$$T_1 V = \int_z^{1+\varepsilon_1\zeta_1} \nabla\nabla\cdot V, \quad T_2 V = \int_z^{\varepsilon_2\zeta_2} \nabla\nabla\cdot V,$$
$$T_1^* V = (T_1 V)^*, \quad T_2^* V = (T_2 V)^*,$$

we found:

$$(1-\mu T_1^*)\vec{u}_1^* = \vec{u}_{1sh}^* + \sqrt{\mu}\left(T_1^*\overline{u}_1 - \left(\frac{\varepsilon_2}{\varepsilon_1}\nabla\zeta_{2_t}(1+\varepsilon_1\zeta_1 - z)\right)^*\right),$$
$$(1-\mu T_2^*)\vec{u}_2^* = \vec{u}_{2sh}^* + \sqrt{\mu}\left(T_2^*\overline{u}_2\right).$$

Here we introduce shear velocities representing the horizontal vorticities contribution to the horizontal velocities:

$$\vec{u}_{1sh} = \int_z^{1+\varepsilon_1\zeta_1} (\vec{\omega}_h^1)^{\perp}, \quad \vec{u}_{2sh} = \int_z^{\varepsilon_2\zeta_2} (\vec{\omega}_h^2)^{\perp}. \tag{6}$$

Then together with the velocity field decompositions (2) we found the following first order approximations:

$$u_1 = \overline{u}_1 + \sqrt{\mu}u_{1sh}^* + O(\mu), \quad u_2 = \overline{u}_2 + \sqrt{\mu}u_{2sh}^* + O(\mu),$$

and the second order approximations:

$$u_1 = \overline{u}_1 + \sqrt{\mu}u_{1sh}^* + \mu\left(T_1^*\overline{u}_1 - \left(\tfrac{\varepsilon_2}{\varepsilon_1}\zeta_{2t,x}(1+\varepsilon_1\zeta_1 - z)\right)^*\right) + \mu^{\frac{3}{2}}u_{1sh}^* + O(\mu^2),$$
$$u_2 = \overline{u}_2 + \sqrt{\mu}u_{2sh}^* + \mu\left(T_2^*\overline{u}_2\right) + \mu^{\frac{3}{2}}u_{2sh}^* + O(\mu^2).$$

2.3. Equations for the shear velocities

It is required to find an equations for shear velocities to close the system. For this purpose we use the dimensionless vorticity equations:

$$\omega_t^i + \varepsilon_i(\vec{u}_i\nabla_{x,z})\omega^i = \varepsilon_i\omega_h^i\nabla u_i + \frac{\varepsilon_i}{\sqrt{\mu}}\omega_v^i(u_i)_z.$$

Recalling the definition (6) of u_{1sh}, we integrate this equation for the upper layer with respect to z (we use here vectorial identity):

$$\partial_t u_{1sh} + \varepsilon_1\overline{u}_1\cdot\nabla u_{1sh} + \varepsilon_1 u_{1sh}\cdot\nabla\overline{u}_1 - (\varepsilon_1\nabla\cdot(\overline{u}_1(z-\varepsilon_2\zeta_2)) - \varepsilon_2\zeta_{2t})\partial_z u_{1sh} = O(\varepsilon_1\sqrt{\mu}).$$

We found the equation for the average component $\overline{u_{1sh}}$ by integration the equation obtained above. Substraction one from another gives us the following equation for u_{1sh}^*:

$$\partial_t u_{1sh}^* + \varepsilon_1\overline{u}_1\cdot\nabla u_{1sh}^* + \varepsilon_1 u_{1sh}^*\cdot\nabla\overline{u}_1 - (\varepsilon_1\nabla\cdot(\overline{u}_1(z-\varepsilon_2\zeta_2)) - \varepsilon_2\zeta_{2t})\partial_z u_{1sh}^* = O(\varepsilon_1\sqrt{\mu}) \tag{7}$$

Analogously, we obtain the equation for the shear velocity of lower layer:

$$\partial_t u_{2sh}^* + \varepsilon_2\overline{u}_2\cdot\nabla u_{2sh}^* + \varepsilon_2 u_{2sh}^*\cdot\nabla\overline{u_2} - (\varepsilon_2\nabla\cdot(\overline{u}_2(z-\beta b(x)+\frac{1}{\delta}))\partial_z u_{2sh}^* = O(\varepsilon_2\sqrt{\mu}). \tag{8}$$

It is should be noted that the order $O(\varepsilon_i\sqrt{\mu})$ of equations obtained in this section is enough to derive the models of the order $O(\mu)$ in the case of general vorticities. But to obtain models of higher order the more precise equations should be found.

2.4. Pressure fields decompositions

The pressure fields representation in terms of the vorticities is obtained by integrating vertical component of Euler equations (3). Implying dynamic boundary condition we have:

$$\frac{1}{\varepsilon_1}\nabla p_1 = \nabla\int_z^{1+\varepsilon_1\zeta_1}\left(-\frac{1}{\varepsilon_1}\partial_z p_1\right)dz = \nabla\zeta_1 + \mu\nabla\int_z^{1+\varepsilon_1\zeta_1}(w_{1t} + \varepsilon_1(u_1\nabla\cdot)w_1 + \varepsilon_1 w_1 w_{1z})\,dz,$$

$$\frac{\rho}{\varepsilon_2}\nabla p_2 = \nabla\left(\int_z^{\varepsilon_2\zeta_2}\left(-\frac{\rho}{\varepsilon_2}\partial_z p_2\right)dz + \frac{\rho}{\varepsilon_2}p_1\big|_{\varepsilon_2\zeta_2}\right) =$$
$$= \nabla\zeta_2 + \mu\nabla\int_z^{\varepsilon_2\zeta_2}(w_{2t} + \varepsilon_2(u_2\nabla\cdot)w_2 + \varepsilon_2 w_2 w_{2z})\,dz + \nabla\left(\frac{\rho}{\varepsilon_2}p_1\big|_{\varepsilon_2\zeta_2}\right).$$

The first terms on the right-hand side in both representations related to the hydrostatic pressure. An expansion for non-hydrostatic terms we will find in the next section in the case of constant vorticities, and further for general vorticity case.

3. Constant vorticities 1d case

As the first step of demonstration of the model applicability let us consider the case of constant vorticities. For the sake of simplicity the calculations are performed in the one dimensional case $x \in \mathbb{R}^1$.

In this case the only one component of each vorticities is non-zero, we assume that these components remain constant for all time:

$$\Omega_i = \begin{pmatrix} 0 \\ \omega_i \\ 0 \end{pmatrix}, \quad \omega_i = \partial_z\vec{u}_i^* - \sqrt{\mu}\partial_x w_i \equiv const.$$

To define the shear velocities we use (6) with $(\omega_h^i)^\perp = -(\omega_i, 0)^\top$:

$$u_{ish} = \int_{A_i}^{B_i}(\omega_h^i)^\perp dz = -\omega_i(B_i - z - \frac{h_i}{2}), \tag{9}$$

A_i, B_i are boundaries of the layers defined above. The vertical velocities, therefore, is defined as:

$$\begin{aligned} w_1 &= \partial_x(\overline{u}_1(z - \varepsilon_2\zeta_2)) + \frac{\sqrt{\mu}}{2}\partial_x(\omega_1(z - \varepsilon_2\zeta_2)(1 + \varepsilon_1\zeta_1 - z)) + \frac{\varepsilon_2}{\varepsilon_1}\zeta_{2_t}, \\ w_2 &= \partial_x(\overline{u}_2(z - \beta b(x) + \frac{1}{\delta})) + \frac{\sqrt{\mu}}{2}\partial_x(\omega_2(z - \varepsilon_2\zeta_2)(\beta b(x) - \frac{1}{\delta} - z)). \end{aligned} \tag{10}$$

3.1. 'Reynolds tensor' contributions for the constant vorticity case

Using velocity fields decomposition (2) and (9) we define 'Reynold tensor' contribution with the second order for the upper layer:

$$\varepsilon_1\mu\partial_x\int_{\varepsilon_2\zeta_2}^{1+\varepsilon_1\zeta_1}|u_1^*|^2 dz = \varepsilon_1\mu\partial_x\int_{\varepsilon_2\zeta_2}^{1+\varepsilon_1\zeta_1}|u_{1sh}^*|^2 + 2\varepsilon_1\mu^{\frac{3}{2}}\partial_x\int_{\varepsilon_2\zeta_2}^{1+\varepsilon_1\zeta_1}u_{1sh}^* T_1\overline{u}_1 - 2\varepsilon_2\mu^{\frac{3}{2}}$$
$$\partial_x\left(\zeta_{2_t}\int_{\varepsilon_2\zeta_2}^{1+\varepsilon_1\zeta_1}u_{1sh}^*(1 + \varepsilon_1\zeta 1 - z)dz\right) =$$
$$\frac{\varepsilon_1\mu}{12}\omega_1^2(h_1^3)_x - \frac{\varepsilon_1\mu^{\frac{3}{2}}}{12}\omega_1\partial_x\left(h_1^3(h_1\partial_x^2\overline{u}_1 - 4\varepsilon_2\partial_x\zeta_2\partial_x\overline{u}_1 - 2\varepsilon_2\partial_x^2\zeta_2\overline{u}_1)\right) + \frac{\varepsilon_1\mu^{\frac{3}{2}}}{6}\omega_1\partial_x(h_1^3\zeta_{2_t}) \tag{11}$$

and for the lower layer:

$$\varepsilon_2\mu\partial_x\int_{-\frac{1}{\delta}+\beta b(x)}^{\varepsilon_2\zeta_2}|u_2^*|^2dz =$$

$$\frac{\varepsilon_2\mu}{12}\omega_2^2(h_2^3)_x - \frac{\varepsilon_2\mu^{\frac{3}{2}}}{12}\omega_2\partial_x\left(h_2^3(h_2\partial_x^2\overline{u}_2 - 4\beta b'(x)\partial_x\overline{u}_2 - 2\beta b''(x)\overline{u}_2)\right). \quad (12)$$

These expressions are required closures for 'Reynold tensor' as long as they depends only on h_i, $\overline{u}_i$ and on the other known task parameters such as a bottom parametrization $b(x)$ and constant vorticities ω_i.

3.2. Pressure contribution

To compute the pressure contributions we represent the vertical velocity (10) in the following form:

$$w_1 = f_1(x,t) + zg_1(x,t), \quad w_2 = f_2(x,t) + zg_2(x,t),$$

here $f_i(x,t)$, $g_i(t,x)$ are defined as:

$$f_1(x,t) = \frac{\varepsilon_2}{\varepsilon_1}\zeta_{2_t} + \varepsilon_2\partial_x(\overline{u}_1\zeta_2) - \frac{\sqrt{\mu}}{2}\omega_1(\varepsilon_1\varepsilon_2\partial_x(\zeta_1\zeta_2) + \varepsilon_2\zeta_{2_x}),$$

$$g_1(x,t) = \frac{\sqrt{\mu}}{2}\omega_1\partial_x(\varepsilon_1\zeta_1 + \varepsilon_2\zeta_2) - \partial_x\overline{u}_1,$$

$$f_2(x,t) = \beta\partial_x(\overline{u}_2 b(x)) - \frac{1}{\delta}\partial_x\overline{u}_2 - \frac{\sqrt{\mu}}{2}\omega_2(\varepsilon_2\beta\partial_x(\zeta_2 b(x)) - \frac{1}{\delta}\varepsilon_2\zeta_{2_x}),$$

$$g_2(x,t) = \frac{\sqrt{\mu}}{2}\omega_2\partial_x(\varepsilon_2\zeta_2 + \beta b(x)) - \partial_x\overline{u}_2.$$

With such notations integrals take form:

$$\frac{1}{\varepsilon_1}\int_{\varepsilon_2\zeta_2}^{1+\varepsilon_1\zeta_1}\nabla p_1 = h_1\zeta_{1_x} + \mu\left(\varepsilon_2\zeta_{2_x}h_1\left(\mathcal{D}_1 f_1 + \varepsilon_1 f_1 g_1 + \frac{1}{2}(\mathcal{D}_1 g_1 + \varepsilon_1 g_1^2)(1+\varepsilon_1\zeta_1+\varepsilon_2\zeta_2)\right) + \right.$$
$$\partial_x\left(\frac{h_1^2}{2}\left(\mathcal{D}_1 f_1 + \varepsilon_2\zeta_2\mathcal{D}_1 g_1 + \varepsilon_1 f_1 g_1 + \varepsilon_1\varepsilon_2\zeta_2 g_1^2\right) + \frac{h_1^3}{3}\left(\mathcal{D}_1 g_1 + \varepsilon_1 g_1^2\right)\right)\Bigg)$$
$$- \mu^{\frac{3}{2}}\omega_1\varepsilon_1\left(\frac{h_1^3}{12}\varepsilon_2\zeta_{2_x}\partial_x^2\overline{u}_1 + \partial_x(\frac{h_1^3}{12}(\partial_x f_1 + (1+\varepsilon_1\zeta_1 - \frac{h_1}{2})\partial_x g_1) + \frac{h_1^4}{24}\partial_x g_1)\right),$$

$$\frac{1}{\varepsilon_2}\int_{-\frac{1}{\delta}+\beta b(x)}^{\varepsilon_2\zeta_2}\nabla p_2 = h_2\zeta_{2_x} + \mu\left(\beta b'(x)h_2\left(\mathcal{D}_2 f_2 + \varepsilon_2 f_2 g_2 + \frac{1}{2}(\mathcal{D}_2 g_2 + \varepsilon_2 g_2^2)(\varepsilon_2\zeta_2 - \frac{1}{\delta} + \beta b(x))\right)\right.$$
$$\partial_x\left(\frac{h_2^2}{2}\left(\mathcal{D}_2 f_2 - (\frac{1}{\delta} - \beta b(x))\mathcal{D}_2 g_2 + \varepsilon_2 f_2 g_2 - \varepsilon_2(\frac{1}{\delta} - \beta b(x))g_2^2\right) + \frac{h_2^3}{3}(\mathcal{D}_2 g_2 + \varepsilon_2 g_2^2)\right)\Bigg)$$
$$- \mu^{\frac{3}{2}}\omega_2\varepsilon_2\partial_x\left(\frac{h_1^3}{12}\beta b'(x)\partial_x^2\overline{u}_2 + \partial_x(\frac{h_2^3}{12}(\partial_x f_2 + (\varepsilon_2\zeta_2 - \frac{h_2}{2})\partial_x g_2) + \frac{h_2^4}{24}\partial_x g_2)\right),$$

with operators are given by:

$$\mathcal{D}_1 f = \partial_t f + \varepsilon_1\overline{u}_1\partial_x f, \quad \mathcal{D}_2 f = \partial_t f + \varepsilon_2\overline{u}_2\partial_x f.$$

Thus, summing 'Reynolds tensor' and pressure contributions, we finally obtained a model with first order of approximation:

$$\begin{cases} h_{1_t} + \varepsilon_1(h_1\overline{u}_1)_x = 0, \\ (h_1\overline{u}_1)_t + \varepsilon_1(h_1\overline{u}_1^2)_x + \dfrac{\mu\varepsilon_1}{12}\omega_1(h_1^3)_x + h_1\zeta_{1_x} + \dfrac{1}{6}\mu h_1 \\ \big(3\varepsilon_1\varepsilon_2\overline{u}_1^2\mathcal{G}(h_1,\zeta_1,\zeta_2) + 2h_1^2(\varepsilon_1\overline{u}_{1_x}\overline{u}_{1_{xx}} - \overline{u}_{1_{txx}}) + \overline{u}_1\mathcal{T}_1(\overline{u}_1) + 6\varepsilon_2\zeta_{1_x}(\zeta_{2_{tt}} + \varepsilon_1\zeta_{2_x}\overline{u}_{1_t}) + 3h_1\mathcal{T}_2(\overline{u}_1)\big) = 0 \\ h_{2_t} + \varepsilon_2(h_2\overline{u}_2)_x = 0, \\ (h_2\overline{u}_2)_t + \varepsilon_2(h_2\overline{u}_2^2)_x + \dfrac{\mu\varepsilon_2}{12}\omega_2(h_2^3)_x + h_2\zeta_{2_x} + \displaystyle\int_{-\frac{1}{\delta}+\beta b(x)}^{\varepsilon_2\zeta_2} \nabla(\frac{\rho}{\varepsilon_2}p_1\big|_{\varepsilon_2\zeta_2}) + \dfrac{1}{6}\mu h_2 \\ \big(3\beta\varepsilon_2\overline{u}_2^2\mathcal{B}(h_2,b) + \varepsilon_2\overline{u}_2\mathcal{R}_1(\overline{u}_2) + 6\beta b'(\beta b' + h_{2_x})\overline{u}_{2_t} + h_2\mathcal{R}_2(\overline{u}_2)\big) = 0. \end{cases}$$

In italic here we denote following operators:

$$\mathcal{T}_1(V) = -6\varepsilon_1 h_{1_x}(\varepsilon_1\zeta_{1_x}\partial_x V + h_1\partial_{xx}V) + \partial_x V(6\varepsilon_1^3(\zeta_{1_x})^2) + 9\varepsilon_1\varepsilon_2 h_1\zeta_{2_{xx}}) - \\ 2\varepsilon_1 h_1^2\partial_{xxx}V + 12\varepsilon_1\varepsilon_2\zeta_{1_x}\zeta_{2_{tx}} + 6\varepsilon_2 h_1\zeta_{2_{txx}},$$

$$\mathcal{T}_2(V) = 2\varepsilon_1^2(\partial_x V)^2\zeta_{1_x} + \varepsilon_2\zeta_{2_{xx}}\partial_t V - 2h_{1_x}\partial_{tx}V + 2\varepsilon_2\partial_x V\zeta_{2_{tx}} + \varepsilon_2\zeta_{2_{ttx}},$$

$$\mathcal{R}_1(V) = (6\beta^2(b')^2 + 6\beta b' h_{2_x} + 9h_2\beta b'')\partial_x V - 6h_2h_{2_x}\partial_x^2 V - 2h_2^2\partial_x^3 V,$$

$$\mathcal{R}_2(V) = 6\varepsilon_2(\beta b' + h_{2_x})(\partial_x V)^2 + 3\beta b''\partial_t V - 6h_{2_x}\partial_{tx}V + h_2(2\varepsilon_2\partial_x V\partial_{xx}V - 2\partial_{txx}V),$$

$$\mathcal{G}(h_1,\zeta_1,\zeta_2) = 2\varepsilon_1\zeta_{1_x}\zeta_{2_{xx}} + h_1\zeta_{2_{xxx}},$$

$$\mathcal{B}(h_2,b) = 2(\beta b' + h_{2_x})b'' + h_2 b'''.$$

The boundary term is defined as:

$$\int_{-\frac{1}{\delta}+\beta b(x)}^{\varepsilon_2\zeta_2} \nabla(\frac{\rho}{\varepsilon_2}p_1\big|_{\varepsilon_2\zeta_2}) = \rho\frac{\varepsilon_1}{\varepsilon_2}\partial_x\mu\left((\mathcal{D}_1(A_1) + \varepsilon_1 A_1B_1)h_1 + (\mathcal{D}_1(B_1) + \varepsilon_1 B_1^2)h_1(1 + \varepsilon_1\zeta_1 + \varepsilon_2\zeta_2)\right) \\ + \frac{\rho}{\varepsilon_2}h_{1x} + \rho\frac{\varepsilon_1}{\varepsilon_2}\mu^{3/2}\partial_x\left(\frac{h_1^3}{12}\omega_1 B_{1x}\right) \quad (13)$$

The model of the second order can be written directly from the contribution calculated above with additional terms in the boundary term.

4. 1d Equations with general vorticity

In this section we derive the generalized Green-Naghdi equations in the presence of general vorticities in each layer. Pressure fields contributions can be calculated explicitly in the first order but it requires the closure in the second order. In contrast, 'Reynolds tensor' contribution can not be computed even in the first order and we should found an equation for component of the tensor. This implies that we limit ourselves to first order model to simplify computations. Now vorticities $\omega_i = \omega_i(t,x,z)$ depend on space and time.

4.1. Computation of the 'Reynolds tensor' contribution

In the first order the 'Reynolds tensor' contribution can be written as:

$$\varepsilon_1\mu\partial_x\int_{\varepsilon_2\zeta_2}^{1+\varepsilon_1\zeta_1}|u_1^*|^2dz = \varepsilon_1\mu\partial_x\int_{\varepsilon_2\zeta_2}^{1+\varepsilon_1\zeta_1}|u_{1sh}^*|^2dz + O(\mu),$$

$$\varepsilon_2\mu\partial_x\int_{-\frac{1}{\delta}+\beta b(x)}^{\varepsilon_2\zeta_2}|u_2^*|^2dz=\varepsilon_2\mu\partial_x\int_{-\frac{1}{\delta}+\beta b(x)}^{\varepsilon_2\zeta_2}|u_{2sh}^*|^2dz+O(\mu).$$

To compute this contribution it is necessary to find equations for E^1, E^2:

$$E^1=\int_{\varepsilon_2\zeta_2}^{1+\varepsilon_1\zeta_1}|u_{1sh}^*|^2dz,\quad E^2=\int_{-\frac{1}{\delta}+\beta b(x)}^{\varepsilon_2\zeta_2}|u_{2sh}^*|^2dz.$$

For this purpose we will use equations for the shear velocities (7), (8) obtained above. Required equations are found directly by multiplying each equations by u_{ish}, respectively, and integrating in z:

$$E_t^1+\varepsilon_1\overline{u}_1E_x^1+3\varepsilon_1\overline{u}_{1x}E^1=0,$$
$$E_t^2+\varepsilon_2\overline{u}_2E_x^2+3\varepsilon_2\overline{u}_{2x}E^2=0.$$

In such a way we found a closure for 'Reynolds tensor' by adding additional equations to our model.

4.2. Pressure contribution

In the first order vorticities are not included in the pressure contribution, it allows us to compute the integrals explicitly. Moreover, it is clear that the result will coincide with the first order terms without vorticities calculated already in the case of constant vorticities.

Therefore the one dimensional Green-Naghdi equations in the presence of general vorticities in the first order have the following form:

$$\begin{cases}
h_{1_t}+\varepsilon_1(h_1\overline{u}_1)_x=0,\\
(h_1\overline{u}_1)_t+\varepsilon_1(h_1\overline{u}_1^2)_x+\mu\varepsilon_1E^1+h_1\zeta_{1_x}+\\
h_1\zeta_{1_x}+\frac{1}{6}\mu h_1\\
\left(3\varepsilon_1\varepsilon_2\overline{u}_1^2\mathcal{G}(h_1,\zeta_1,\zeta_2)+2h_1^2(\varepsilon_1\overline{u}_{1_x}\overline{u}_{1_{xx}}-\overline{u}_{1_{txx}})+\overline{u}_1\mathcal{T}_1(\overline{u}_1)+6\varepsilon_2\zeta_{1_x}(\zeta_{2tt}+\varepsilon_1\zeta_{2_x}\overline{u}_{1_t})+3h_1\mathcal{T}_2(\overline{u}_1)\right)=0\\
E_t^1+\varepsilon_1\overline{u}_1E_x^1+3\varepsilon_1\overline{u}_{1x}E^1=0,\\
h_{2_t}+\varepsilon_2(h_2\overline{u}_2)_x=0,\\
(h_2\overline{u}_2)_t+\varepsilon_2(h_2\overline{u}_2^2)_x+\mu\varepsilon_2E^2+h_2\zeta_{2_x}+\int_{-\frac{1}{\delta}+\beta b(x)}^{\varepsilon_2\zeta_2}\nabla(\frac{\rho}{\varepsilon_2}p_1|_{\varepsilon_2\zeta_2})+\\
\frac{1}{6}\mu h_2\left(3\beta\varepsilon_2\overline{u}_2^2\mathcal{B}(h_2,b)+\varepsilon_2\overline{u}_2\mathcal{R}_1(\overline{u}_2)+6\beta b'(\beta b'+h_{2_x})\overline{u}_{2_t}+h_2\mathcal{R}_2(\overline{u}_2)\right)=0,\\
E_t^2+\varepsilon_2\overline{u}_2E_x^2+3\varepsilon_2\overline{u}_{2x}E^2=0.
\end{cases}$$

The operators in italic are defined earlier. And boundary term is the same as (13) unless it has the first order.

5. Conclusion

We have derived here several fully nonlinear models for the wave propagation in two-layer flow in the presence of vorticity. Introducing of vorticities in each layer allow us to close the system and correct dispersive terms by taking into account rotational part of horizontal velocity fields. The presented models are derived in the one dimensional case for constant vorticities in each layer with the first and second order of approximation and for general vorticities case with first order.

The natural perspective of this work is to derived the second order model for the general vorticities case in order to correct non-hydrostatic part with additional terms. The other direction of the research is to perform the numerical simulation for obtained models.

References

[1] Barros R, Gavrilyuk S L and Teshukov V M 2007 *Stud. Appl. Math.* **119**(3) 191-211
[2] Duchene V 2010 *SIAM J. Math. Anal.* **42, 5** 2229-60
[3] Castro A and Lannes D 2014 *J. Fluid Mech.* **759** 642–75
[4] Richard G L and Gavrilyuk S L 2015 *J. Fluid Mech.* **773** 49-74

The impact of unilateral vibrations on aerodynamic characteristics of airfoils in transonic flow

V Zamuraev and A Kalinina
Khristianovich Institute of Theoretical and Applied Mechanics of SB RAS, Russia
E-mail: zamuraev@itam.nsc.ru, kalinina@itam.nsc.ru

Abstract. The work is devoted to the mathematical modeling of the influence of forced vibrations of a surface element on one side of the airfoil on the shock-wave structure of transonic flow around. The influence of parameters of oscillations on the airfoil wave drag and the lift force were qualitatively and quantitatively investigated for constant maximum velocity amplitude, which is close in magnitude to the sound velocity in the incoming flow, and for a wide range of frequencies. The arising of additional lift force is shown.

1. Introduction

The oscillations of the airfoil surface element at transonic streamline is possible to examine from different points of view. In the first place is necessary to understand the buffeting of aircraft [1, 2]. From another perspective, the forced oscillations as a means of streamline control is of interest. The author's study which concerns the management of transonic flow around airfoils with the aid of pulsed-periodic energy supply has shown significant favourable changes in the aerodynamic characteristics of profiles. Generalization of these results is presented in [3]. The initial results of the study of the impact of high-frequency oscillations of an airfoil surface element on transonic streamline were presented in [4], and the similarity of shock wave structure with the case of energy supply allows a conclusion to be drawn about the possibility of achieving similar effects. Transonic flow around an airfoil even without oscillations is a complex problem because it involves the viscous and thermal phenomena which are characterized by complex physical models [5]. The introduction of oscillations into the formulation greatly increases the computational complexity of the problem. Direct numerical simulation of the oscillations based on the Navier – Stokes equations contains certain difficulties, which manifest themselves as numerical oscillations. These oscillations are completely impossible to suppress even by non-reflecting boundary conditions [1]. As a result the search of regimes with the occurrence of buffeting was performed using the SST-turbulence model within ANSYS CFX [1, 2]. The present paper examines the oscillations of a small portion of the surface, an order of magnitude less than the chord of the profile. This issue was partially explored for high frequency symmetrical fluctuations of the symmetric profile which is streamlined by transonic flow at zero angle of attack [4]. Unlike of [4] here we consider the case of oscillations of the surface element on one side of the airfoil in a wide range of frequencies. The studied case is an airfoil NACA 0012 which is streamlined at zero angle of attack with the Mach number in the incoming flow of $M = 0.85$ (there is no flow separation). Numerical simulation is based on the solution of two-dimensional unsteady Euler equations with ooscillations of the surface element. This formulation allows us to estimate the maximum values of all effects, because consideration of the dissipative phenomena reduces the magnitude of the oscillation effect in the flow in comparison with a case without these phenomena.

2. Mathematical formulation

As a mathematical model to describe planar unsteady flow of non-viscous gas the Euler equations in conservative form are used. The system of equations is supplemented with boundary conditions on boundaries of two-dimensional rectangular area of a rectangle with an inner border corresponding to the contour of the considered airfoil. On the left, top and bottom boundaries of this area undisturbed flow conditions are used. On the right boundary "soft" conditions are used. The condition of impermeability is used on the airfoil contour. On the part of the contour with the boundaries x_1 and x_2 the geometry changes according to the following dependence $y = f_0(x)$:

$$f(x,t) = f_0(x) + A\sin(2\pi t/\Delta t)\sin(\pi(x - x_1)/(x_2 - x_1)).$$

Here all the spatial sizes are related to the chord length of the profile b; the x-coordinate is directed along the chord, y is the transverse coordinate; the profile is localized at $0 \leq x \leq 1$. Time t and the oscillation period Δt are measured in units of b/a_∞, where a_∞ is the sound velocity in the incoming flow. In detail the problem formulation is presented in [4]. For the numerical solution of the formulated problem the finite-volume scheme is used which reduces the total variation with taking into account the fluctuations of the surface profile.

3. The results

The results were obtained for airfoil NACA 0012 which is streamlined at zero attack angle by the ideal gas with the heat ratio $\gamma = 1.4$ and the Mach number $M_\infty = 0.85$. The oscillations amplitude and the period were varied for fixed localization of the moving surface element and its size $x_1 = 0.600$, $\Delta x = x_2 - x_1 = 0.096$, respectively. In Figure 1 for the relation $A/\Delta t = 0.2$ the dependence of the wave drag coefficients C_x (curve *1*) and the lifting force C_y (curve *2*) from the period are presented.
A given value of the ratio $A/\Delta t = 0.2$ corresponds to the maximum amplitude of the velocity oscillations which has a ratio to the speed of sound of incoming flow of 1.26. The character of the C_y dependence changes significantly with Δt. A rather sharp maximum at $\Delta t = 0.0075$ and a minimum at $\Delta t = 0.035$ are observed. The ratio of the oscillations amplitude of the profile element to the period specifies the amplitude of the velocity oscillations. For all points of the curves in Figure 1 the ratio $A/\Delta t = 0.2$ is characterized by the same value. Therefore, the flow gets the same energy for each period. However, its distribution in the streamline region depends on the period value. For small values of the period (high frequency oscillations) each portion of gas gets energy repeatedly, and this process leads to increasing of C_y. As a result the quasi-stationary regime of the airfoil streamline is observed. Such a regime takes place, in particular, for the period $\Delta t = 0.0075$. In this case, the reciprocal of the homochronicity number which is calculated along the length of oscillating region of the profile [3], consists of

$$1/Ho = 1/(\Delta t/(\Delta x/u)) = 11 \gg 1.$$

This value indicates the number of energy portions which is received by one portion of gas, and its magnitude corresponds to the quasi-stationary regime of the flow (here u is the velocity component along the x-axis). However, for very small values of the profile surface displacement, the spatial dispersion of energy is significant. The amplitude of the surface displacement increases with the increase of the oscillation period in accordance with condition $A/\Delta t = const$.

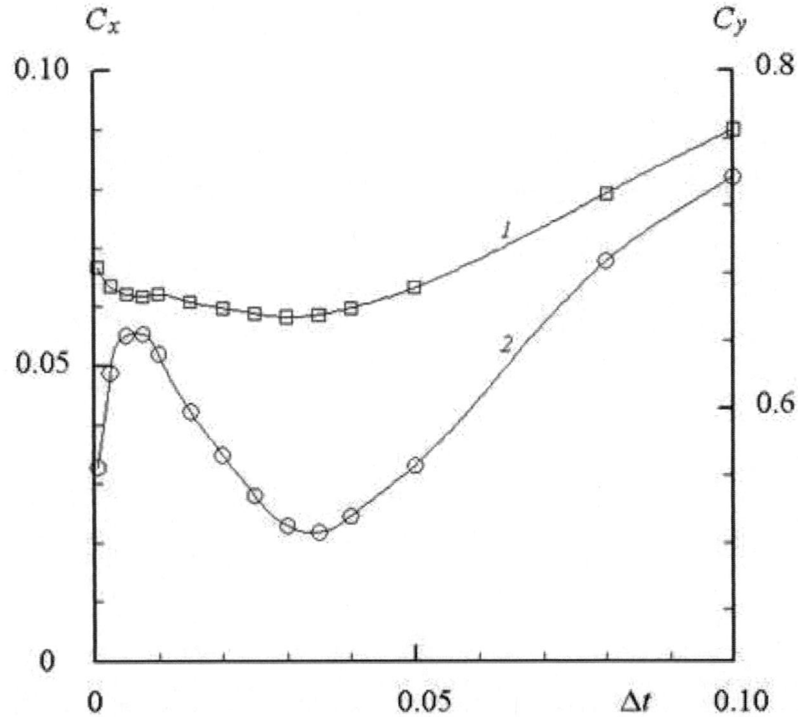

Figure 1. The wave drag coefficient C_x and the lifting force coefficient C_y depending on the period Δt at $A/\Delta t = 0.2$.

The absolute value of the spatial attenuation decreases with increasing of the magnitude A, but simultaneously reduces the energy which is received by the portion of gas when that portion passes the oscillations zone. The first peak on the graph arises due to these two competing processes. The closed shock wave which is localized on the profile side with the undisturbed surface goes on until the trailing edge. On the opposite side (perturbed surface) is displaced upstream. The shift of the closed shock waves is qualitatively similar to the case of unilateral energy supply [3]. The distributions of Mach number M along the contour profile in different phases of oscillation at $\Delta t = 0.0075$ is shown in Figure 2 for various oscillations phases. For comparison, the dashed line shows the distribution M in the absence of oscillations. Streams flowing down from both sides of the profile are separated by a contact discontinuity. The pressure on both sides of the contact discontinuity must be the same (as well as the Mach number and other values characterized by the discontinuity solution). After closing the shock wave on the bottom (disturbed) side of the profile the flow becomes subsonic. It is inhibited when approaching the trailing edge, and the pressure increases. To balance the pressures at the trailing edge, the closing shock wave on the top (undisturbed) surface should be displaced downstream. In the present case, it stops at the trailing edge. Vertical dashed lines near the x-axis in Figure 2 indicate the zone of oscillations on the profile. The change of the Mach number M within the period happen in essence in this zone too. The smooth nature of the distribution after the peak indicates a quasi-stationary mode.
For $\Delta t > 0.01$ the character of the streamline is changed. The traveling wave extends from the oscillating zone. In Figure 3, this wave is shown after half of the period (curves *1* and *5* for phase 0 and π respectively). The closing shock wave on the upper side of the profile is not displaced to the trailing edge; on the underside the offset of the closing shock wave is less

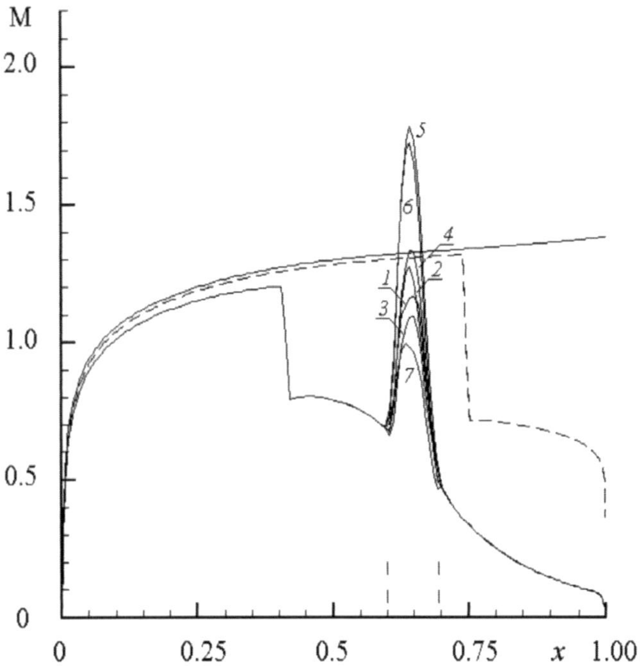

Figure 2. The Mach number at $\Delta t = 0.0075$ (*1* – phase 0, *2* – $\pi/4$, *3* – $\pi/2$, *4* – $3\pi/4$, *5* – π, *6* – $5\pi/4$, *7* – $7\pi/4$).

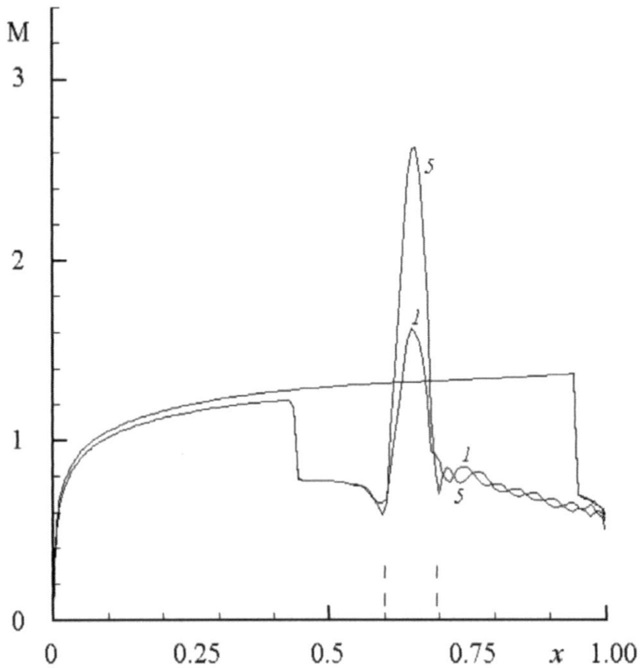

Figure 3. The Mach number at $\Delta t = 0.0075$ (*1* – phase 0, *5* – π).

than the data from Figure 2. All these factors lead to a significant reduction in additional lift force. With further increase of the period Δt the oscillation amplitude A increases, and, as a result, the spatial and temporal scale of the emerging wave structures increases. In Figure 4 and Figure 5 for the period $\Delta t = 0.08$ the formation of wave structures at the maximum deviation of the surface profile outside (Figure 4) and inside (Figure 5) is shown. In the first case the shock wave departs from the surface of the profile, and behind it a rarefaction wave runs; in the

Journal of Physics: Conference Series **722** (2016) 012018 doi:10.1088/1742-6596/722/1/012018

second case, the region of rarefaction wave after the shock wave which detaches from the profile is characterized by the more extending area.

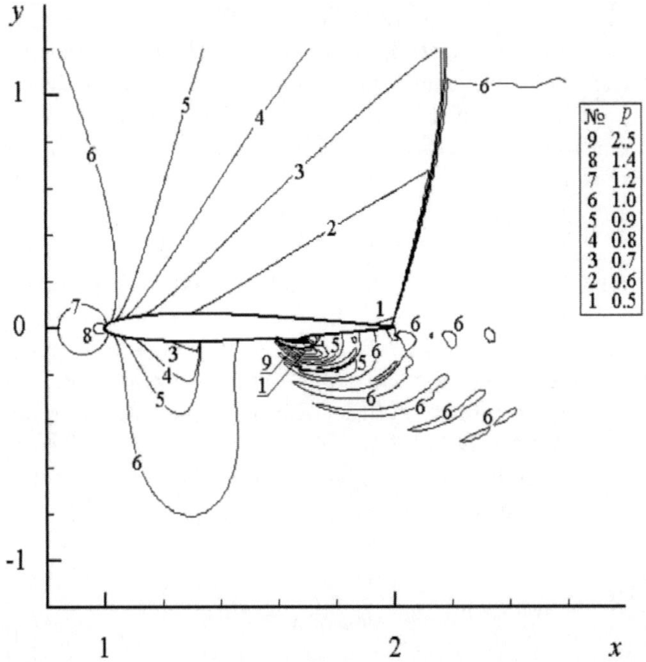

Figure 4. The pressure distribution at $\Delta t = 0.08$ and phase $\pi/2$.

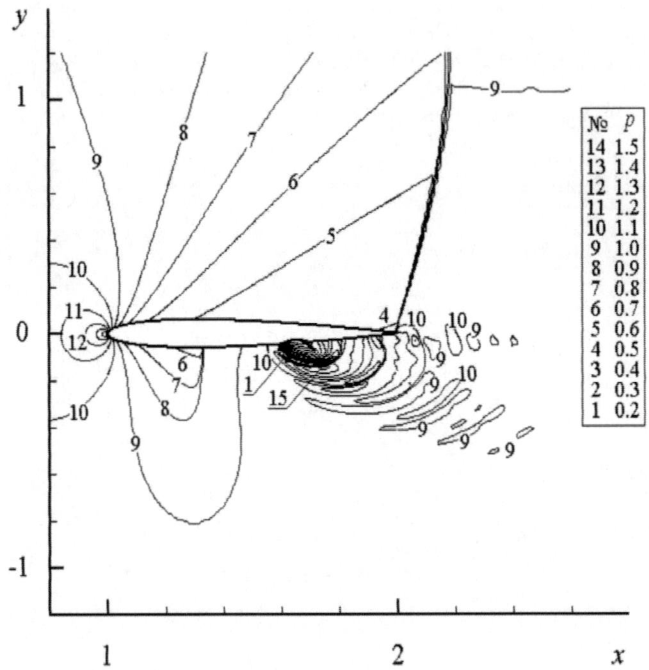

Figure 5. The pressure distribution at $\Delta t = 0.08$ and phase $3\pi/2$.

In the considered case, the ratio between the period of oscillation, the distance to the trailing edge and the flow parameters has such magnitudes that three "semi-cylindrical" shock waves appear which are localized between the oscillation surface and the trailing edge, and high pressure for the last wave leads to a shift of the closing shock wave on the upper side of the profile to the trailing edge. The effect is mostly caused by the fact that with increasing amplitude of the oscillation the value of the absolute attenuation decreases. As a result the lifting force is growing again, which explains the behavior of C_y in Figure 1 after the minimum. Neither in this case nor for other values of the oscillation period for the selected values of the velocity amplitude the formation of vortex structures with their detaching from the trailing edge or flow separation are observed. This is the fundamental difference between the force impact from the surface and the energy supply to the flow, which is sometimes characterized by the formation of the vortex structures due to the instability of the thermal track.

4. Conclusion

Thus, it was found that when unilateral force action on transonic airfoil (forced oscillations of a surface element) takes place, the airfoil aerodynamic characteristics have a non-monotonic dependence on frequency. The study has revealed the occurrence of the additional lifting force. For a constant value of the velocity amplitude which is close to the speed of the incoming flow, the dependence of lift force coefficient on the period of the oscillations is nonmonotonic. The relationship of these effects with the transformation of shock-wave structure of the flow is shown. Shift of the shock waves is qualitatively similar to the case of unilateral energy supply. The observed effects may be used to control the flow around airfoils at the corresponding regimes.

Acknowledgments

Authors gratefully acknowledge Christopher Hearsey and Valeriya Hearsey for their assistance.

References

[1] Lipanov A, Lipatov I and Karskanov S 2015 *Proc. of Institute of mechanics* (Izhevsk: Institute of mechanics) pp 164–79
[2] Lipatov I and Vinh T 2014 *Scientific Bulletin of Moscow state technical University of civil aviation* **200** 133
[3] Aulchenko S, Zamuraev V and Kalinina A 2011 *J. of Techn. Phys.* **56** 1552–61
[4] Aulchenko S, Zamuraev V and Kalinina A 2014 *Journal of Engineering Physics and Thermophysics* **87** 635–47
[5] Bogdanov A and Diesperov V 2015 *Russian Congress on fundamental problems of theoretical and applied mechanics* ed Akhmetov D (Kazan) p 485

Effect of the mode composition disturbances in a high-enthalpy wind tunnel on wave processes in the hypersonic viscous shock layer

S V Kirilovskiy[1,2], T V Poplavskaya[1,2] and I S Tsyryulnikov[1,2]

[1]Khristianovich Institute of Theoretical and Applied Mechanics SB RAS, 630090 Novosibirsk, Russia
[2]Novosibirsk State University, 630090, Novosibirsk, Russia

Email: kirilov@itam.nsc.ru, popla@itam.nsc.ru

Abstract. A numerical and experimental study of the spectral distribution of disturbances hot-shot wind tunnel was performed. It was found that disturbances are the superposition of fast acoustic waves propagating along axial direction and oblique slow acoustic waves which have less amplitude. Numerical simulation of disturbances evolution in the nozzle of the wind tunnel was carried out by solving 2D Navier–Stokes equations using the ANSYS Fluent program package. Experiments were carried out in the IT-302M high-enthalpy hotshot wind tunnel ITAM SB RAS.

1. Introduction

Characteristics of the incoming flow disturbances, in particular the mode composition, is important for the objectives of the study of receptivity and disturbances evolution in supersonic and hypersonic shock layers. A viscous shock layer is always formed on the leading edges of hypersonic aircraft. The peculiarity of hypersonic shock layer is a closeness of boundary layer to the bow shock and considerable thickness of boundary layer. Disturbances generated in the shock layer propagate downstream and affect the laminar-turbulent transition in a hypersonic boundary layer of the whole aircraft (figure 1). Khristianovich Institute of Theoretical and Applied Mechanics SB RAS has considerable experience in experimental and numerical simulation of hypersonic instability in the viscous shock layer on the plate at the Mach numbers from 10 up to 21 [1-4].

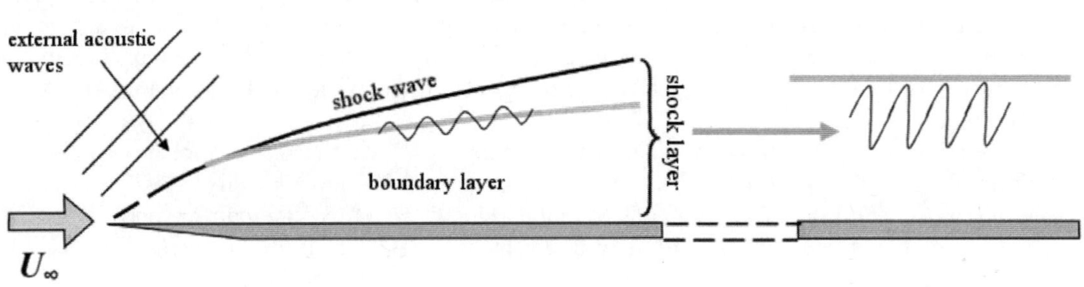

Figure 1. Viscous shock layer and evolution of the freestream flow disturbances

The transition in the compressible boundary layers has been the subject of research for many years, but only a limited number of existing experimental and numerical investigations have been conducted for the high-enthalpy hypersonic flows. Flight studies of gas flow stability are constructively complex and quite expensive. For this reason, numerical simulation and wind tunnel experiments are important. Experimental data on receptivity and evolution of disturbances in wind tunnels are widely used for constructing reliable theoretical models of the laminar-turbulent transition. However, it is difficult to estimate characteristics of disturbances without taking into account the processes in the nozzle channel of wind tunnel where the flow is formed. Therefore the free-stream processes (proceeding in the wind tunnel nozzle) have to be included in disturbance evolution simulations.

Thus the problem of disturbances evolution in hypersonic flow over plate is solved in three stages. In the beginning, the numerical simulation of flow in the nozzle of wind tunnel (figure 2) was performed and the flow parameters on the nozzle outlet were obtained. Then using the obtained data, the problem of flow over the plate in the test section of the wind tunnel was solved. Third stage was solving the transient problem of interaction between viscous shock layer and freestream plane acoustic waves. The knowledge of the mode composition of disturbances in the flow, i.e., separation of the field of disturbances of a supersonic flow into vortex, entropy, and acoustic modes, is important for refining the results of the problem of the perturbations evolution.

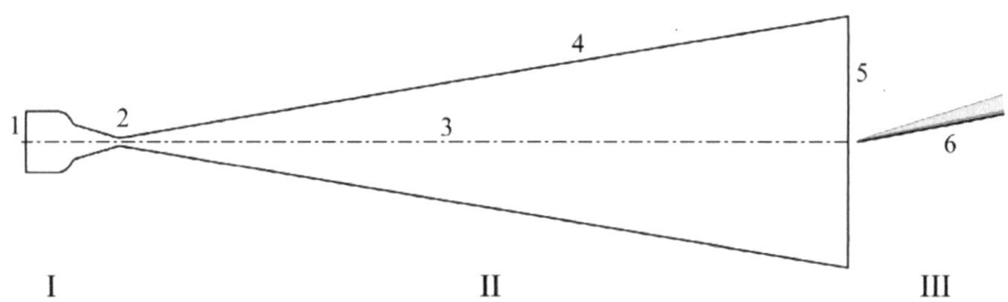

Figure 2. Sketch of the nozzle channel IT-302 M ITAM SB RAS: I – prechamber, II –conic nozzle, III – test chamber, 1 – inlet section, 2 – critical section, 3 – axis of symmetry, 4 – nozzle wall, 5 – output section, 6 – plate

The problem of determining the mode composition of disturbances in the nozzle channel of hotshot wind tunnel IT- 302M ITAM SB RAS is solved in this paper by using the numerical-experimental method.

2. Experiments in the wind tunnel

Experiments were carried out in the IT-302M high-enthalpy hotshot wind tunnel (Khristianovich Institute of Theoretical and Applied Mechanics, Siberian Branch, Russian Academy of Sciences) under conditions of air efflux from the nozzle at the following flow parameters: $M_\infty = 5.7$; $Re_{1\infty} = 8\times10^6$ m^{-1}. The gas was heated by an electric discharge in the prechamber of the wind tunnel. A high-speed digital camera (Phantom 310) operating at a frame rate of 120 kHz was used to obtain a sequence of the schlieren images of visualized flow in the test chamber. Then a sequence of frames was processed by the algorithm described below to determine the longitudinal phase velocity of the perturbation.

Figure 3 shows a plot of longitudinal phase velocity U_f of pressure disturbance propagation versus frequency, where the upper dashed straight line 1 corresponds to phase velocity $U_f=U_\infty+c_\infty$ of the fast acoustic mode, solid line 2 corresponds to phase velocity $U_f=U_\infty$ of the entropy and vortex modes, and lower dash–dot line 3 corresponds to phase velocity $U_f=U_\infty-c_\infty$ of the slow acoustic mode (where c_∞ is the velocity of sound in the incoming flow). As can be seen, the experimental points are placed close to the upper line, which indicates that the longitudinal phase velocity of pressure disturbances in the IT-302M wind tunnel corresponds to the phase velocity of fast acoustic waves.

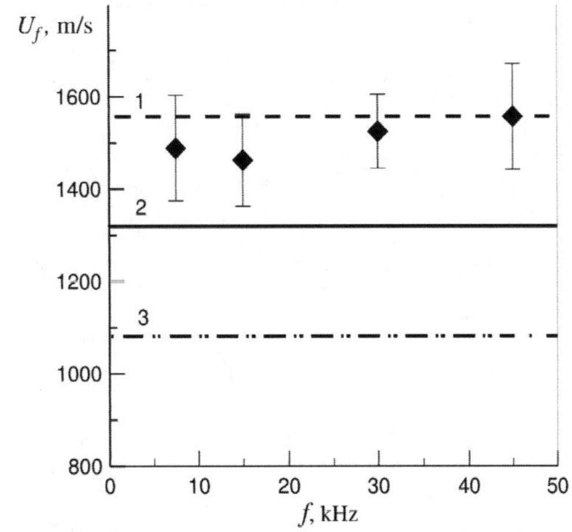

Figure 3. Frequency dependences of the longitudinal phase velocity U_f of pressure disturbances in the IT-302M wind tunnel (airflow, $M_\infty = 5.7$; $Re_{1\infty} = 8\times10^6\ m^{-1}$): 1 – fast acoustic mode, 2 –entropy-vortex mode, and 3 – slow acoustic mode; points – experimental data [3]

3. Numerical simulations

Numerical simulation of disturbances evolution in the nozzle of the wind tunnel IT-302M was carried out by solving 2D Navier–Stokes equations in axisymmetric formulation with *k-ω* SST turbulence model using the ANSYS Fluent program package with a density-based solver applied to an implicit second-order accuracy (in space) scheme with Roe-FDS convective flow splitting and explicit 4-stage Runge–Kutta scheme in time. The thermal conductivity and viscosity of a working gas were set by a formula from the kinetic theory and by Sutherland's law, respectively, and the heat capacity was taken to be a function of temperature.

Computational domain was constructed geometrically similar to the real channel of the wind tunnel IT-302M (see figure 2). It was covered by a rectangular grid with clustering near the critical section and nozzle walls (a total of 600000 meshes). The values of total temperature and total pressure T_0=3000K, P_0=200bar in the prechamber of the wind tunnel were set at the left boundary of the computational domain (beginning of the nozzle channel). The solution on the right boundary (end of the nozzle channel) was extrapolated from inside the computational domain. Non-slip conditions and the constant wall temperature T_w = 300 K were set at the nozzle wall, and the symmetry condition was set at the lower boundary (the nozzle axis).

Disturbances in the prechamber appear due to the impact of electric discharge gas. In numerical simulations of these disturbances, the variables on the left boundary of the computational domain were set as a superposition of the steady stagnation parameters and a plane monochromatic acoustic wave in the form (with using user defined functions embedded in the computational code):

$$P_0^* = P_0 + \acute{P}_0 \cdot sin(-2\pi f_0 \cdot t)$$

$$T_0^* = T_0 + \frac{(\gamma - 1)}{\gamma}\frac{\acute{P}_0 \cdot \bar{T}_0}{\bar{P}_0} sin(-2\pi f_0 \cdot t)$$

P_0, T_0–total pressure and total temperature, $\acute{P}_0$–amplitude of total pressure disturbances, f_0 –frequency of disturbances. The total amplitude of the pressure pulsations set equal to $\acute{P}_0 = 0.03P_0$, in the present calculations and pulsation frequency f_0 = 40kHz.

Figure 4 shows instantaneous field of pressure fluctuations, normalized on local pressure values. It is seen that disturbances propagate both along the nozzle axis and turbulent boundary layer on the nozzle wall.

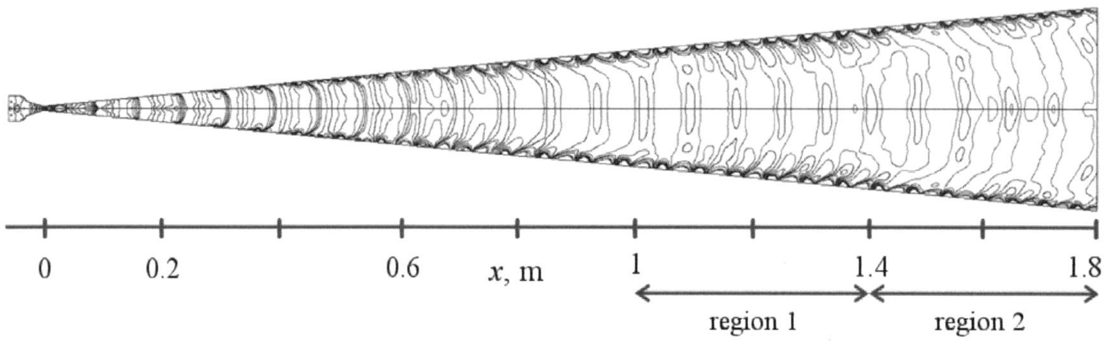

Figure 4. Instantaneous field of pressure fluctuations in the nozzle channel, normalized on local pressure values: $\acute{P}_0 = 0.03 \cdot P_0$, f_0=40kHz

4. Algorithm of longitudinal wave numbers determination

Determination of longitudinal wave numbers of disturbances k_x was performed using following algorithm shown in figure 5. Time series of static pressure values p(x,t) along x axis from the range 1-1.8m were processed by using the Fourier transform to obtain amplitude spectra $A_f(x,f)$ in each point along x axis from this range.

The original signal	⇨	Fourier transform of time series of signal	⇨	Filter for harmonics selection	⇨	Inverse Fourier transform	⇨	Fourier transform of spatial series of signal
p(x,t)		$A_f(x,f)$		$A_{f^*}(x,f^*) = \begin{cases} A_f(x,f^*) \\ 0, f \neq f^* \end{cases}$		A(f*,x,t)		A(f*,x,t*)

Figure 5. Algorithm of determination of longitudinal wave numbers of disturbances

Amplitude spectra of averaged pressure fluctuations over whole considered x range are shown in figure 6. It is seen that disturbances propagate both with main frequency f_0=40kHz and harmonic frequency f_1=80kHz in the nozzle channel.

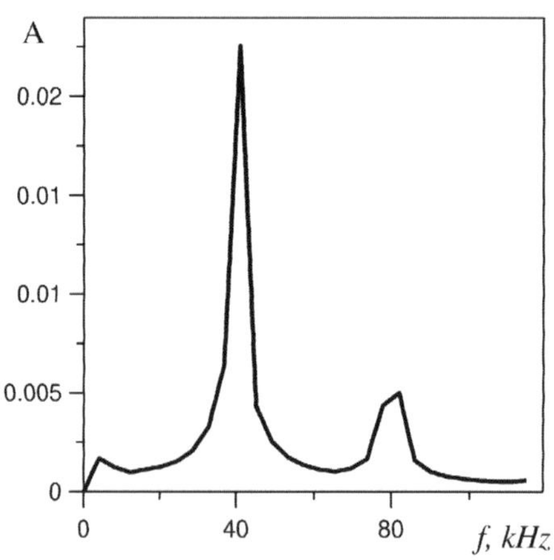

Figure 6. Amplitude spectra of pressure fluctuations averaged over whole considered x range

Then the filter was used to highlight the selected frequency f^*. The inverse Fourier transform is carried out after filtration procedures. This makes it possible to obtain the amplitude distribution A(f^*,x,t) along the axis for a single harmonic f^*. After this the expansion of the data on a longitudinal wave numbers k_x was carried out at arbitrary time moment t*. This makes it possible to identify the disturbances modes.

Similarly schlieren imaging sequence of frames was used in the processing of experimental data and the signal of brightness from the time was processed.

5. Results and discussion

Figure 7 shows the dependence of the pressure pulsation amplitude from the longitudinal wave number for both frequencies $f^*=f_0$=40 kHz (figure 7a) and 80 kHz (figure 7b) and for two regions of the nozzle axis.

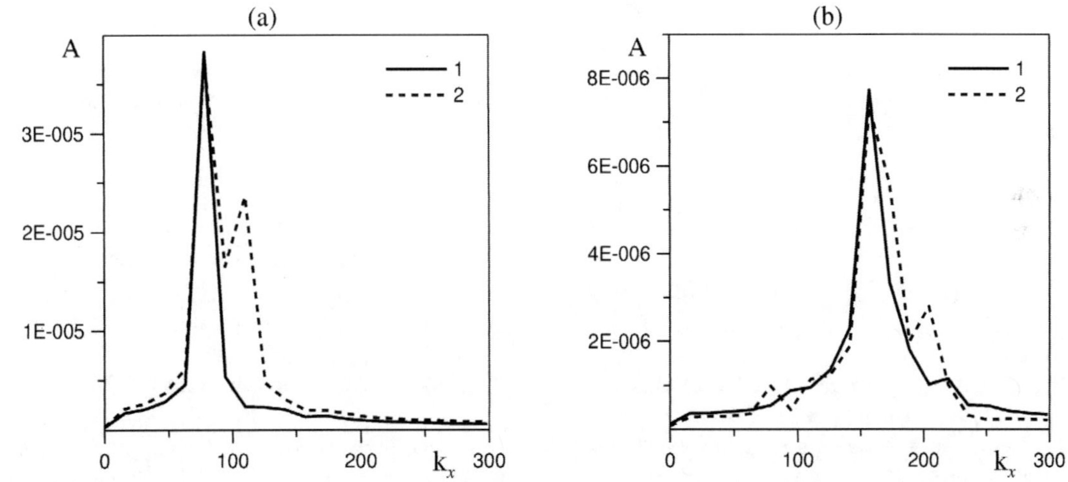

Figure 7. The spectral decomposition of the normalized amplitudes of the pressure pulsations at two regions along the axis of the nozzle on the longitudinal components of the wave vector: a – f_0 = 40 kHz, b – f_1 = 80 kHz, 1 – region corresponds to x from 1m to 1.4m, 2 – x from 1.4m to 1.8m

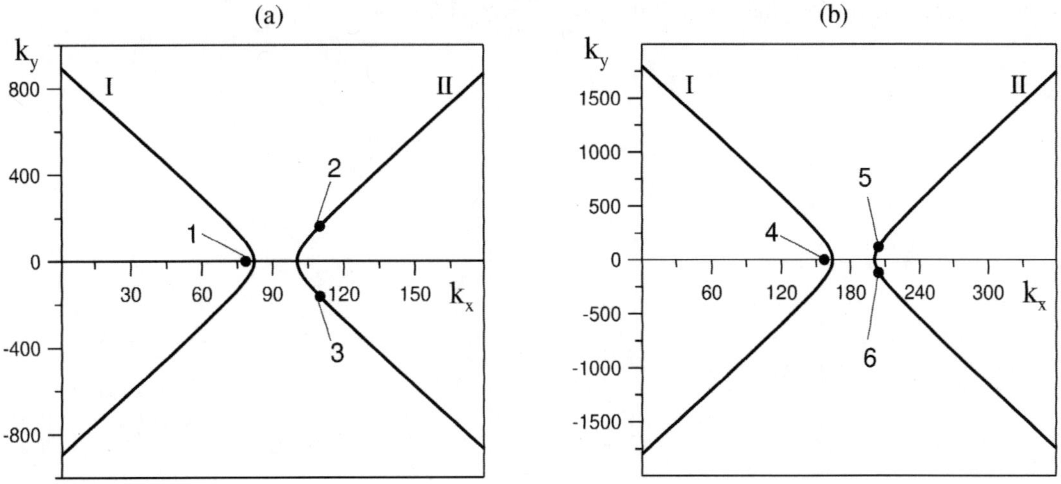

Figure 8. Wave diagrams for acoustic disturbances for f = 40 kHz –(a) and f = 80 kHz – (b): I – fast acoustic waves, II – slow, 1-6 – numerical wave numbers of disturbances in the nozzle channel

The region 1 corresponds to x from 1m to 1.4m, the region 2 corresponds x from 1.4m to 1.8m (the output of the nozzle channel). Figure 8 shows the wave diagrams for acoustic disturbances produced in accordance with the dispersion relations [5] for the frequency f_0=40 kHz (figure 8a) and f_1=80 kHz (figure 8b). Figure 7 shows that there is only one peak at the region 1 (line 1), and there are two peaks of amplitude at the region 2 (line 2). Figure 8 shows that the first peak corresponds to the fast acoustic wave propagating along the nozzle axis (points 1,4), and the second peak corresponds to the amplitude of slow acoustic wave with propagation angles $\beta = 60°$ for $f_0 = 40$ kHz (points 2,3) and $\beta = 30°$ for $f_1 = 80$ kHz (points 5,6). These slow acoustic waves are probably generated by the turbulent boundary layer on the walls of the nozzle as a result of its interaction with the fast mode perturbation.

6. Conclusion

The numerical and experimental investigations of the mode composition of disturbances in hotshot wind tunnel IT- 302M ITAM SB RAS were performed.

It was found that acoustic disturbances in the prechamber of wind tunnel IT-302M generate perturbations corresponding fast acoustic waves propagating along the nozzle axis. The fast acoustic disturbances propagating downstream lead to the appearance of inclined slow acoustic wave of smaller amplitude.

Acknowledgements

This work was supported by the Russian Foundation for Basic Research (Grant No. 16-08-00674) and by the Grant of the Government of the Russian Federation for Supporting Research Supervised by Leading Scientists (Agreement No. 14Z50.31.0019).

References

[1] Maslov A A, Mironov S G, Kudryavtsev A N, Poplavskaya T V and Tsyryulnikov I S 2010 *J. Fluid Mech.* **650** 81–118
[2] Maslov A A, Mironov S G, Poplavskaya T V and Tsyryulnikov I S 2010 *J. Appl. Mech. Tech. Phys.* **51** 482–8
[3] Kirilovskiy S V, Maslov A A, Poplavskaya T V and Tsyryulnikov I S 2015 *Technical Physics Letters* **41** 184–6
[4] Kirilovskiy S V, Maslov A A, Poplavskaya T V and Tsyryulnikov I S 2015 *Technical Physics* **60** 645–55
[5] McKenzie J F and Westphal K O 1968 *Phys. Fluids* **11** 2350–62

One-dimensional mixing layer model for a shear Hele-Shaw flow

P V Kovtunenko[1,2]

[1] Lavrentyev Institute of Hydrodynamics of SB RAS, 630090 Novosibirsk, Russia
[2] Novosibirsk State University, 630090 Novosibirsk, Russia

E-mail: pkovtunenko@gmail.com

Abstract. A shear flow of a viscosity-stratified fluid in a Hele-Shaw cell is considered. The long-wave approximation is applied to the governing equations. To describe the evolution of the mixing layer, a special flow with a three-layered structure is considered. A one-dimensional model is derived by averaging the motion equations over the cell width, taking into account the flow structure. For a stationary flow, solutions of motion equations are constructed. The influence of viscosity on the mixing layer evolution is investigated by performing a numerical experiment for a flow with different viscosities in the layers and for a flow with always zero viscosity. It is shown that viscosity has a significant influence on the flow evolution.

1. Introduction

The process of interaction of two fluids with different viscosities often appears to be unstable and leads to formation of different structures. In the case of a more viscous fluid displacing a less viscous one, a so-called viscous fingering occurs [1]. The classical mathematical model which describes evolution of the Saffman-Taylor instability in the Hele-Shaw cell is described in the papers [2, 3]. However, the need to incorporate inertial forces leads to development of more complex non-linear models [4], which makes it possible to consider flows with different velocities in the layers, when describing the formation and evolution of instabilities. In its turn, in shear flows the evolution of instabilities results in formation of such structures as mixing layers. The instabilities occur either on the interface between the fluids or on the free surface, which is then known as a turbulent bore. The relation of the vertical to horizontal characteristic flow size is usually small for those flows, which makes it possible to describe the instability evolution in the framework of a two-dimensional shallow water model, and additionally to consider mixing effects between the layers [5, 6]. In [7] a model describing the horizontal-shear flow was proposed, which allowed to generalize the concept of subcritical and supercritical flow. Further development of this model resulted in a method, which also takes into account the turbulent mixing effect and the mass transfer between the layers and is based on the multi-layered fluid theory. This method was successfully applied in works [8, 9] for describing evolution of a subsurface turbulent layer in plane-parallel flows. Using this method one can determine the size of the mixing area staying in the framework of a one-dimensional model, which simplifies the initially two-dimensional calculations drastically. The aim of this work is to derive equations for a three-layered shear flow in the Hele-Shaw cell under the assumption that the viscosity varies in the flow and to study the evolution of the mixing layer. The work is based on a model of a flow averaged through the gap with quadratic velocity and in the framework of the shallow water theory [9, 10]. The stationary solutions of the governing equations are sought and analyzed. It is shown that viscosity has a significant influence on the flow evolution.

2. Mathematical model

Gap-averaged shear flows of a weakly-compressible fluid in the Hele-shaw cell are described by the following system of equations [9, 10]:

$$\begin{aligned} &\rho_t + (u\rho)_x + (v\rho)_y = 0, \\ &(\rho u)_t + (\beta\rho u^2 + p)_x + (\beta\rho uv)_y = -\mu u, \\ &(\rho v)_t + (\beta\rho uv)_x + (\beta\rho v^2 + p)_y = -\mu v, \\ &(c\rho)_t + (uc\rho)_x + (vc\rho)_y = 0. \end{aligned} \tag{1}$$

Here $\rho(t,x,y)$ is the density, $u(t,x,y)$ and $v(t,x,y)$ are the horizontal velocity vector components, $p = p(t,x,y)$ is the pressure and $c = c(t,x,y)$ is the concentration, which is scaled such that if the flow divides into two layers with different properties, it is equal to unity in one of the fluid layers and zero in the other layer. The constant β comes from integration over the cell gap, and in general $\beta = 6/5$, but since it has a negligibly small effect on the flow evolution [4], in this work $\beta = 1$ is considered. The spatial configuration of the flow is shown in Figure 1. The cell gap, which stretches over the z-axis is considered to be small relative to the horizontal cell sizes. Motions of the fluid along the z-axis are restrained by solid plates.

Figure 1. The spatial configuration of the flow

To perform the long-wave approximation in the motion equations (1), one has to assume that the flow is essentially parallel and apply the scaling

$$t \to \varepsilon^{-1}t, \quad x \to \varepsilon^{-1}x, \quad v \to \varepsilon v, \quad \mu \to \varepsilon\mu$$

and then, discard the terms of power two (ε^2) for being negligibly small, where $\varepsilon = L_y/L_x \ll 1$ is the relation of characteristic size of the cell over the y-axis to the characteristic size over the x-axis. As a result, the following system of equations is derived:

$$\begin{aligned} &u_t + uu_x + vu_y + a\rho_x = -\mu u/\rho, \quad \rho_y = 0, \\ &c_t + uc_x + vc_y = 0, \quad \rho_t + (u\rho)_x + (v\rho)_y = 0. \end{aligned} \tag{2}$$

Moreover, a closing relation for the pressure is used to obtain the system (2). Namely, $p(\rho) = a\rho^2/2$, where $a^2 = c_0^2/\rho_0$, and the constants c_0 and ρ_0 determine the characteristic sound speed and density of the fluid.

In this work a flow between two solid plates parallel to the x-axis is considered. The plates are located on the levels $y = 0$ and $y = H$, and therefore the following additional impermeability conditions can be written:

$$v\big|_{y=0} = v\big|_{y=H} = 0.$$

In case when $\mu = 0$, a similar approach leads to the classical long-wave theory equations for an ideal fluid [12], if one excludes from consideration the equation for the concentration $c(t, x, y)$.

Furthermore, from the system (2), the conservation laws of momentum ρu and energy $E = (u^2 + a\rho)\rho/2$ can be derived:

$$\begin{aligned} &(\rho u)_t + (\rho u^2 + a\rho^2/2)_x + (\rho uv)_y = -\mu u, \\ &E_t + ((E + a\rho^2/2)u)_x + ((E + a\rho^2/2)v)_y = -\mu u^2. \end{aligned} \tag{3}$$

3. Multi-layered model

Hereinafter the problem is considered in the framework of a special flow structure [8]. Let the fluid flow into the cell from the left side. The shear flow consists of two layers with different viscosities and velocities. Then at some point in space (which will be further considered to be position $x = 0$) a mixing layer starts to develop, as shown in Figure 2.

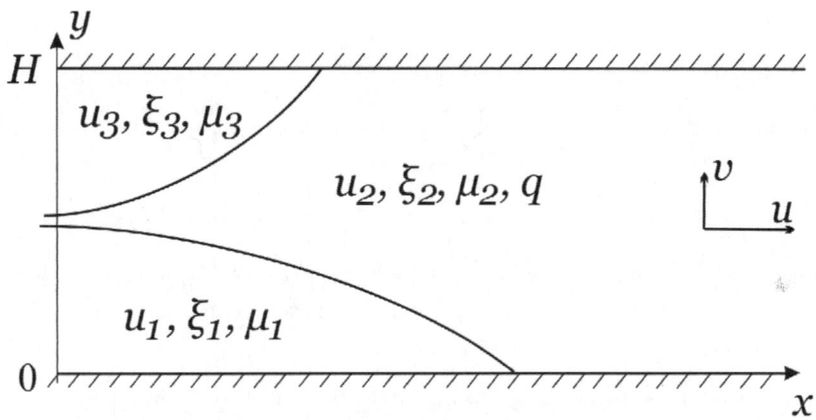

Figure 2. A three-layered flow in the Hele-Shaw cell

Hereinafter the index "1" corresponds to the lower flow layer, the index "2" corresponds to the mixing layer and the index "3" corresponds to the upper layer. The width of the i-th layer is ξ_i, velocity is u_i, and a constant viscosity μ_i is also given. The flow in the upper and lower layers is considered to be of the special class $u_y = 0$. Then the momentum equation in this layers can be written as:

$$u_{1t} + u_1 u_{1x} + a\rho_x = -\mu_1 u_1/\rho,$$

$$u_{3t} + u_3 u_{3x} + a\rho_x = -\mu_3 u_3/\rho.$$

Further on, the last equation of (2) and the equations (3) are averaged over the cell width

$y \in [0, H]$:

$$(\xi_1\rho)_t + (u_1\xi_1\rho)_x = -\sigma q\rho, \quad (\xi_2\rho)_t + (u_2\xi_2\rho)_x = 2\sigma q\rho, \quad (\xi_3\rho)_t + (u_3\xi_3\rho)_x = -\sigma q\rho,$$

$$Q_t + ((u_2^2 + q^2)\xi_2\rho + u_1^2\xi_1\rho + u_3^2\xi_3\rho + aH\rho^2/2)_x = -\mu_1\xi_1 u_1 - \mu_2\xi_2 u_2 - \mu_3\xi_3 u_3,$$

$$((u_1^2\xi_1 + (u_2^2 + q^2)\xi_2 + u_3^2\xi_3)\rho + aH\rho^2)_t + ((u_1^3\xi_1 + (u_2^2 + 3q^2)u_2\xi_2 + u_3^3\xi_3 + 2aQ)\rho)_x =$$
$$= -2(\mu_1\xi_1 u_1^2 + \mu_2\xi_2(u_2^2 + q^2) + \mu_3\xi_3 u_3^2) - \theta\rho q^3.$$

Here, the velocities in the mixing layer are described with the following quantities

$$u_2 = \frac{1}{\xi_2}\int\limits_{\xi_1}^{H-\xi_3} u\, dy, \quad q^2 = \frac{1}{\xi_2}\int\limits_{\xi_1}^{H-\xi_3} (u - u_2)^2\, dy,$$

where the total flow rate is $Q = \rho(u_1\xi_1 + u_2\xi_2 + u_3\xi_3)$, and the empirical constants σ, θ define the mass transfer and the energy dissipation in the flow, respectively. From these averaged equations, one can derive, as consequence, the following one-dimensional model of the flow:

$$\begin{aligned}
&(\xi_1\rho)_t + (u_1\xi_1\rho)_x = -\sigma q\rho, \quad (\xi_2\rho)_t + (u_2\xi_2\rho)_x = 2\sigma q\rho, \quad (\xi_3\rho)_t + (u_3\xi_3\rho)_x = -\sigma q\rho,\\
&u_{1t} + u_1 u_{1x} + a\rho_x = -\frac{\mu_1 u_1}{\rho}, \quad u_{3t} + u_3 u_{3x} + a\rho_x = -\frac{\mu_3 u_3}{\rho},\\
&u_{2t} + u_2 u_{2x} + \frac{(q^2\xi_2\rho)_x}{\xi_2\rho} + a\rho_x = -\frac{\mu_2 u_2}{\rho} + \frac{\sigma q}{\xi_2}\Big(u_1 - 2u_2 + u_3\Big),\\
&q_t + (u_2 q)_x = -\frac{\mu_2 q}{\rho} + \frac{\sigma}{2\xi_2}\Big((u_1 - u_2)^2 + (u_3 - u_2)^2 - (2 + \frac{\theta}{\sigma})q^2\Big).
\end{aligned} \tag{4}$$

In the process of flow evolution, the width of the upper or lower layer (for example, "3") can eventually become equal to zero. In this case, the flow becomes two-layered, and a similar approach leads to the model of a two-layered flow:

$$\begin{aligned}
&(\xi_1\rho)_t + (u_1\xi_1\rho)_x = -\sigma q\rho, \quad (\xi_2\rho)_t + (u_2\xi_2\rho)_x = \sigma q\rho,\\
&u_{1t} + u_1 u_{1x} + a\rho_x = -\frac{\mu_1 u_1}{\rho},\\
&u_{2t} + u_2 u_{2x} + \frac{(q^2\xi_2\rho)_x}{\xi_2\rho} + a\rho_x = -\frac{\mu_2 u_2}{\rho} + \frac{\sigma q}{\xi_2}(u_1 - u_2),\\
&q_t + (u_2 q)_x = -\frac{\mu_2 q}{\rho} + \frac{\sigma}{2\xi_2}\Big((u_1 - u_2)^2 - (1 + \frac{\theta}{\sigma})q^2\Big).
\end{aligned} \tag{5}$$

When both the upper and the lower layers vanish (their width becomes equal to zero), the model (4) is rewritten in a form which is close to equations describing the plane-parallel flows [13]:

$$\begin{aligned}
&\rho_t + (u_2\rho)_x = 0, \quad (u_2\rho)_t + ((u_2^2 + q^2)\rho + a\rho^2/2)_x = -\mu u_2,\\
&((u_2^2 + q^2)\rho + a\rho^2)_t + ((u_2^2 + 3q^2)u_2\rho + 2au_2\rho^2)_x = -2\mu_2(u_2^2 + q^2) - \theta h q^3.
\end{aligned} \tag{6}$$

For the sake of brevity, the characteristics of the two-layered model (5) are sought. The characteristics of the three- and one-layered models can be found using a similar approach. To do this, the model (5) is rewritten in the matrix form

$$\mathbf{U}_t + \mathbf{A}\mathbf{U}_x = \mathbf{F},$$

where $\mathbf{U} = (\rho, \xi_2, u1, u2, q)^{\mathrm{T}}$ is a vector of the unknown quantities,

$$\mathbf{F} = \left(0, \sigma q, -\frac{\mu_1 u_1}{\rho}, -\frac{\mu_2 u_2}{\rho} + \frac{\sigma q}{\xi_2}(u_1 - u_2), -\frac{\mu_2 q}{\rho} + \frac{\sigma}{2\xi_2}\left((u_1 - u_2)^2 - (1 + \frac{\theta}{\sigma})q^2\right)\right)^{\mathrm{T}}$$

is the right part of the equation, and the matrix $\mathbf{A}$ ca be written as follows:

$$\mathbf{A} = \begin{pmatrix} (\xi_1 u_1 + \xi_2 u_2)/H & \rho(u_2 - u_1)/H & \xi_1\rho/H & \xi_2\rho/H & 0 \\ \xi_1\xi_2(u_2 - u_1)/(H\rho) & (\xi_1 u_2 + \xi_2 u_1)/H & -\xi_1\xi_2/H & \xi_1\xi_2/H & 0 \\ a & 0 & u_1 & 0 & 0 \\ a + q^2/\rho & q^2/\xi_2 & 0 & u_2 & 2q \\ 0 & 0 & 0 & q & u_2 \end{pmatrix}.$$

The eigenvalues of the matrix $\mathbf{A}$ are sought from the characteristic equation:

$$\det(\mathbf{A} - \lambda\mathbf{E}) = (u_2 - \lambda)((u_2 - \lambda)^2 - 3q^2)\left((u_1 - \lambda)^2 - \frac{a\xi_1\rho}{H}\right) - \frac{a\xi_2\rho}{H}(u_1 - \lambda)^2 = 0.$$

It is obvious that this equation has a root $\lambda = u_2$ which corresponds to the characteristic $dx/dt = u_2$. It is also clearly seen then when the flow is close to the one-layered case, all other characteristics also become real.

4. Stationary solutions

Traveling waves can be investigated in a coordinate system moving with the same speed as the waves do. Therefore, to describe the mixing layer structure, stationary flow should be considered. Then the system (4) is rewritten as:

$$\begin{aligned} &(u_1\xi_1\rho)_x = -\sigma q\rho, \quad (u_2\xi_2\rho)_x = 2\sigma q\rho, \quad (u_3\xi_3\rho)_x = -\sigma q\rho, \\ &u_1 u_{1x} + a\rho_x = -\frac{\mu_1 u_1}{\rho}, \quad u_3 u_{3x} + a\rho_x = -\frac{\mu_3 u_3}{\rho}, \\ &u_2 u_{2x} + \frac{(q^2\xi_2\rho)_x}{\xi_2\rho} + a\rho_x = -\frac{\mu_2 u_2}{\rho} + \frac{\sigma q}{\xi_2}\left(u_1 - 2u_2 + u_3\right), \\ &(u_2 q)_x = -\frac{\mu_2 q}{\rho} + \frac{\sigma}{2\xi_2}\left((u_1 - u_2)^2 + (u_3 - u_2)^2 - (2 + \frac{\theta}{\sigma})q^2\right), \end{aligned} \tag{7}$$

and can be solved with respect to the derivative terms, which yields a system of ODEs:

$$\begin{aligned} &\rho' = \frac{G}{\Delta}, \quad u_1' = -\frac{\mu_1}{\rho} - \frac{a}{u_1}\rho', \quad u_3' = -\frac{\mu_3}{\rho} - \frac{a}{u_3}\rho', \\ &\xi_1' = -\frac{\sigma q}{u_1} - \frac{\xi_1}{u_1}u_1' - \frac{\xi_1}{\rho}\rho', \quad \xi_{3x} = -\frac{\sigma q}{u_3} - \frac{\xi_3}{u_3}u_3' - \frac{\xi_3}{\rho}\rho', \\ &u_2' = \frac{2\sigma q}{\xi_2} - \frac{u_2}{\xi_2\rho}\left(\xi_2\rho' - \rho\xi_1' - \rho\xi_3'\right), \quad \xi_2' = \frac{2\sigma q}{u_2} - \frac{\xi_2}{u_2}u_2' - \frac{\xi_2}{\rho}\rho', \\ &q' = -\frac{\mu_2 q}{\rho u_2} - \frac{q}{u_2}u_2' + \frac{\sigma}{2\xi_2 u_2}\left((u_1 - u_2)^2 + (u_3 - u_2)^2 - (2 + \frac{\theta}{\sigma})q^2\right), \end{aligned} \tag{8}$$

where

$$G = \frac{\mu_2\xi_2(u_2^2 - 2q^2)}{u_2(u_2^2 - 3q^2)} + \frac{\mu_1\xi_1}{u_1} + \frac{\mu_3\xi_3}{u_3} - \frac{\sigma q\rho}{u_1} - \frac{\sigma q\rho}{u_3} +$$
$$+ \frac{\sigma q\rho((u_1 - u_2)^2 + (u_3 - u_2)^2 + u_2(u_1 - 4u_2 + u_3) - (6 + \frac{\theta}{\sigma})q^2)}{u_2(u_2^2 - 3q^2)},$$

$$\Delta = H - \frac{a\xi_1\rho}{u_1^2} - \frac{a\xi_3\rho}{u_3^2} - \frac{a\xi_2\rho}{u_2^2 - 3q^2}.$$

The sign of the discriminant Δ indicates the flow type: subcritical when $\Delta < 0$, and supercritical when $\Delta > 0$.

The two- and singe-layered models (5) and (6) can also be rewritten in the form of systems of ODEs.

To calculate the initial values of the velocities, it is assumed that $\xi_2|_{x=0} \to 0$, and that the limits of u_2 and q exist. Then from the equations (7), the following relations can be derived:

$$q_0^2 = \frac{1}{2}u_{10}u_{20} - u_{20}^2 + \frac{1}{2}u_{20}u_{30}$$
$$u_{20} = \frac{(6+\theta/\sigma)(u_{10}+u_{30})}{4(4+\theta/\sigma)}\left(1 \pm \sqrt{1 - 16\frac{(4+\theta/\sigma)(u_{10}^2+u_{30}^2)}{(6+\theta/\sigma)^2(u_{10}+u_{30})^2}}\right), \tag{9}$$

where the index "0" denotes the values of the functions at $x = 0$. That root, which provides the relation $u_{10} \leq u_{20} \leq u_{30}$, is considered for the value u_{20}. This is necessary for the problem to be physically reasonable.

5. Numerical results

To investigate the viscous terms effects on the mixing layer evolution, a stationary flow model (8) is considered. This model can reduce to its two- or singe-layered version in the process of flow evolution.

The numerical experiment is performed by using a simple Euler method on a mesh with resolution $N = 5000$. The computational domain is $x \in [0, 12]$, the width of the cell $h = 1$, viscosity in the mixing layer is given by the formula $\mu_2 = \sqrt{\mu_1\mu_3}$. The initial conditions are: $u_1 = 0.3$, $u_3 = 0.75$, $\xi_1 = 0.599$, $\xi_3 = 0.399$, $\theta = 0.6$, $a = 1500$, the initial velocities u_2 and q are given by the formula (9). The parameter $\theta = 0.6$, and the parameter σ varies from 0.15 to 0.2.

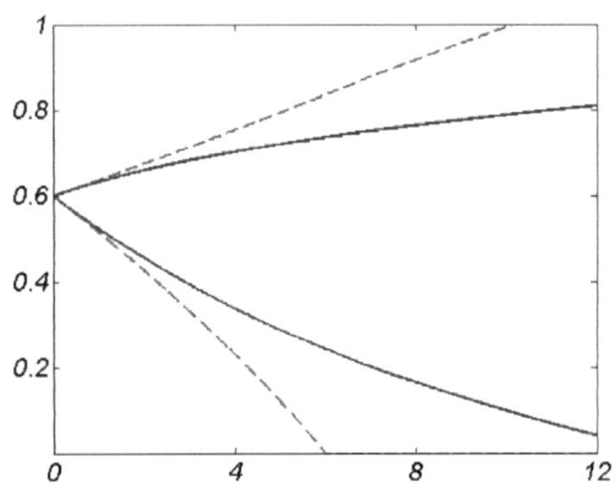

Figure 3. The mixing layer for the subcritical flow with viscosity (solid) and without viscosity (dotted)

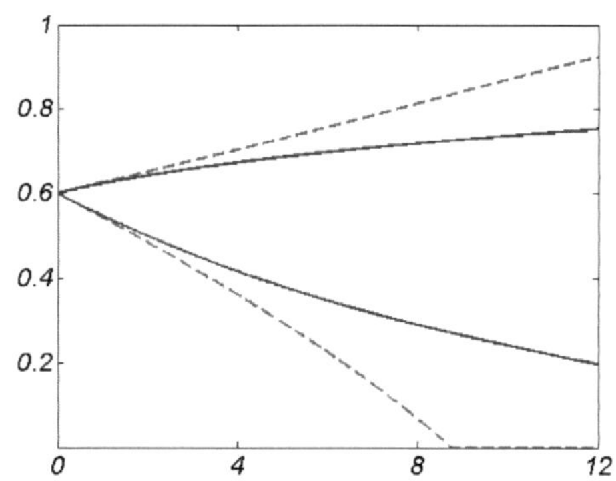

Figure 4. The mixing layer for the supercritical flow with viscosity (solid) and without viscosity (dotted)

The calculations for the subcritical flow are shown in Figure 3, and for the supercritical flow in Figure 4. The dotted lines show the modeling in assumption of a fluid with viscosity $\mu_1 = \mu_3 = 0$ and the solid lines show the modeling of a flow with different viscosities in the layers $\mu_1 = 0.5$ and $\mu_3 = 0.2$ The figures in both subcritical and supercritical case show that the viscous terms significantly slow the process of evolution of the mixing layer.

6. Conclusion

In the current work, a one-dimensional three-layered model of a shear flow in the Hele-Shaw cell is derived in the terms of long-wave approximation. Different viscosities of the fluid in the layers are assumed. A solution for the stationary flow was proposed and a numerical modeling was performed. The numerical results indicate that viscosity slows the process of evolution of the mixing layer significantly. Further work will be devoted to comparing the numerical results to the non-stationary flow solution and to the two-dimensional modeling of equations (1).

Acknowledgments

This work was supported by the Russian Science Foundation (Grant No.15-11-20013). The author is grateful to A. A. Chesnokov for discussions and helpful tips.

References

[1] Saffman P G and Taylor G 1958 *Proc. Roy. Soc.* A **245** 312-29
[2] Tan C T and Homsy G M 1986 *Phys. Fluids* **29** 3549-56
[3] Azaiez J and Singh B 2002 *Phys. Fluids* **14** 1559-71
[4] Gondret P and Rabaud M 1997 *Phys. Fluids* **9** 3267-74
[5] Jirka G H 2001 *J. Hydraul. Res.* **39** 567-73
[6] Rhoads B L and Sukhodolov A N 2008 *Water Resour. Res.* **44** DOI 10.1029/2007WR006634
[7] Chesnokov A A and Liapidevskii V Yu 2009 *J. Appl. Mech. Tech. Phys.* **50** 220-8
[8] Liapidevskii V Yu and Chesnokov A A 2014 *J. Appl. Mech. Tech. Phys.* **55** 299-310
[9] Chesnokov A A and Liapidevskii V Yu 2015 *ArXiv stored article* 1501.00366v2
[10] Chesnokov A A and Stepanova I V 2015 *Appl. Math. and Comput.* **265** 313-9
[11] Dias E O and Miranda J A 2011 *Phys. Rev.* **83** 066312
[12] Benney D J 1973 *Stud. Appl. Math.* **52** 45-50
[13] Teshukov V M 2007 *J. Appl. Mech. Tech. Phys.* **48** 303-9

The motion of elliptic cylinder under free surface

V K Kostikov[1,2] **and N I Makarenko**[1,2]

[1] Lavrentyev Institute of Hydrodynamics, Novosibirsk, Russia
[2] Novosibirsk State University, Novosibirsk, Russia

E-mail: makarenko@hydro.nsc.ru

Abstract. A problem on generation of unsteady nonlinear waves on the surface of an infinitely deep ideal fluid due to the motion of a submerged elliptical cylinder is considered. It is supposed that the cylinder can rotate in addition to translational two-dimensional motion. The initial formulation of the problem is reduced to an integrodifferential system of equations for the functions defining the free surface shape, the normal and tangential components of velocity on the free boundary. The small-time asymptotics of the solution is constructed in the case of the cylinder that moves with a constant acceleration from rest.

1. Introduction

In this paper we consider analytically the nonlinear problem on unsteady free surface motion of an ideal fluid generated by submerged elliptic cylinder. The initial stage of the flow forced by a circular cylinder under a free surface was studied in detail in [1, 2]. Unique solvability of a two-dimensional problem of motion of a circular cylinder under a free surface was established in the paper [3], where an asymptotics of the solution in terms of a small parameter characterizing the body size was justified. The small-time asymptotic solution was constructed in the case of translational motion of an elliptic cylinder in [4, 5]. The 3D-problem on the motion of a sphere under the free surface was considered in [6]. The formation of a cavity behind the submerged circular and elliptic cylinder moving with high acceleration was studied semi-analytically in [7, 8]. The problem under consideration was studied numerically in the papers [9]-[13]. Experimental results on unsteady motion of a cylinder near the free surface was presented in the paper [14].

Our analysis uses reduction of the Euler equations to the nonlinear boundary integral-differential system of equations for the wave elevation together with normal and tangential fluid velocities at a free surface. In this method the key role is played by the integral equation for a normal fluid velocity that discribes interaction of the cylinder with the free surface. This approach was also applied to the piston mechanism of the generation of tsunami-type waves due to a fast deformation of a compact bottom area [15].

2. Formulation of the problem

Two-dimensional potential flow of infinitely deep water is considered in the Cartesian coordinate system Oxy (see Figure 1). Initially the fluid is at rest. The line $y = 0$ corresponds to the undisturbed free surface. Non-dimensional variables are used below. The initial submersion depth h_0 of the cylinder center from undisturbed free surface is taken as the length scale of the problem and a characteristic speed of the body u_0 as the velocity scale of the flow. Free surface elevation $\Gamma(t)$ is described by the equation $y = \eta(x, t)$, where $\eta(x, 0) = 0$. The elliptic cylinder can rotate around its center. The non-dimensional coordinates of the center of the cylinder are given as $x = x_{cyl}(t)$ and $y = y_{cyl}(t)$, where $y_{cyl}(0) = -1$. In Figure 1, a and b are the non-dimensional semi-axes of the cylinder, $a \geq b$.

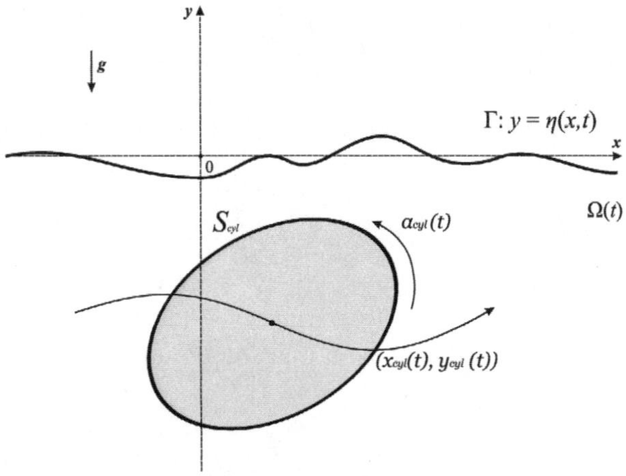

Figure 1. Scheme of motion.

Unsteady flow of incompressible fluid is described by the Euler equations for the velocity of the flow $\mathbf{u} = (U, V)$ and hydrodynamic pressure p

$$\begin{cases} U_t + UU_x + VU_y + p_x = 0, \\ V_t + UV_x + VV_y + p_y = -\lambda, \\ U_x + V_y = 0, \quad U_y - V_x = 0, \end{cases} \tag{1}$$

where $\lambda = gh_0/u_0^2$ is the square of the inverse Froude number and g is the gravity acceleration. The boundary conditions on the free surface have the form

$$\eta_t + U\eta_x = V, \quad p = 0 \qquad (y = \eta(x,t)). \tag{2}$$

and boundary condition on the cylinder surface $S_{cyl}(t)$ has the form

$$(\mathbf{u} - \mathbf{u_{cyl}}) \cdot \mathbf{n} = 0, \quad (x, y) \in S_{cyl}(t), \tag{3}$$

where $\mathbf{n}$ is the unit normal. Further we consider the translational motion of the cylinder together with rotation around its center. We assume that the coordinates of the cylinder center $(x_{cyl}(t), y_{cyl}(t))$ and the angle between the main axis of the cylinder and horizontal axis $\alpha(t)$ are known at each point of time t. We assume that the fluid is at rest at infinity, so we have $U, V, \eta \to 0$ as $|\mathbf{x}| \to \infty$. We assume also that the initial velocity field $\mathbf{u}_{|t=0} = \mathbf{u}_0$ satisfies the compatibility conditions

$$U_{0x} + V_{0y} = 0, \quad U_{0y} - V_{0x} = 0,$$
$$(\mathbf{u}_0 - \mathbf{u_{cyl}}(0)) \cdot \mathbf{n}_0 = 0, \quad (x, y) \in S_{cyl}(0).$$

These conditions are satisfied, in particular, when the cylinder starts moving smoothly from rest.

3. Integro-differential equations on the free surface

Let us introduce the tangential velocity $u(x,t) = (U + \eta_x V)_{|y=\eta}$ and normal velocity $v(x,t) = (V - \eta_x U)_{|y=\eta}$ of the fluid particles on the free surface $\Gamma(t)$. The equations (1)-(2) can be reduced to an equivalent system of one-dimensional equations for the unknown functions u, v, η:

$$\eta_t = v, \qquad u_t + \frac{1}{2}\frac{\partial}{\partial x}\left(\frac{u^2 - 2\eta_x uv - v^2}{1+\eta_x^2}\right) + \lambda\eta_x = 0. \tag{4}$$

The differential equations (4) should be considered together with the boundary integral equation

$$\pi v(x) + v.p.\int_{-\infty}^{\infty} A(x,s)\, v(s)\, ds = v.p.\int_{-\infty}^{+\infty} B(x,s)\, u(s)\, ds + v_{dip}(x), \tag{5}$$

which is the boundary equation for the analytic function $F(z,t) = U - iV$, where $z = x + iy$. The time t is not shown in kostik:IntegralBoundaryEquation because it plays the role of a parameter in this equation. The kernels A and B can be split as

$$A = A_f + r^2 A_r, \qquad B = B_f + r^2 B_r,$$

where $r = (a+b)/2$ and $c = \sqrt{a^2 - b^2}$ is proportional to the eccentricity of the body contour S_{cyl}. The kernels A_f and B_f depend on the deflection η as

$$A_f(x,s) + iB_f(x,s) = \frac{i[1 + i\eta_x(x)]}{x - s + i[\eta(x) - \eta(s)]}.$$

Note that the simplified version of the system (4)–(5) with truncated kernels $A = A_f$ and $B = B_f$ in (5) and $v_{dip} = 0$ corresponds to the model of hydroelastic waves (1)-(2) without any submerged body. Correspondingly, the kernels A_r and B_r describe the interaction of the body with free surface. To define these kernels, we introduce an auxiliary complex-valued function

$$\tau(z) = \begin{cases} \dfrac{1}{2}\left(z - z_{cyl} + \sqrt{(z - z_{cyl})^2 - c^2}\right), & -\dfrac{\pi}{2} < Arg(z) < \dfrac{\pi}{2}, \\[2ex] \dfrac{1}{2}\left(z - z_{cyl} - \sqrt{(z - z_{cyl})^2 - c^2}\right), & \dfrac{\pi}{2} < Arg(z) < \dfrac{3\pi}{2}, \end{cases}$$

where $z_{cyl}(t) = x_{cyl}(t) + iy_{cyl}(t)$. The function $\tau(z)$ presents analytic branches of inverse mapping by the Joukowski function $z(\tau) = z_{cyl} + \tau + c^2/(4\tau)$ which maps the exterior of the ellipse with semi-axes a and b onto the exterior $|\tau| \geqslant r$ of the circle of radius $r = (a+b)/2$. By using this mapping we can avoid the integrals along the body contour in kostik:IntegralBoundaryEquation by transforming them to the integrals along the interface Γ. The kernels A_r and B_r are given as

$$A_r(x,s) + iB_r(x,s) = \\ = \left(\frac{1}{\tau(s+i\eta(s)) - \overline{\tau_*(x+i\eta(x))}} + \frac{k}{\tau(s+i\eta(s)) - k\overline{\tau_*(x+i\eta(x))}}\right)\left(\frac{i}{\tau(x+i\eta(x))}\right)_x, \tag{6}$$

where bar denotes complex conjugate, and $\tau_* = r^2/\overline{\tau}$. Parameter $k = \frac{a-b}{a+b}$ signifies the deviation of the elliptic cylinder from the circular cylinder. The function v_{dip} in the right-hand side of (5) is given by

$$v_{dip}(x) = \Re\, 2\pi i r^2 \left[\frac{k e^{2i\alpha_0}\overline{z'_{cyl}(t)} - z'_{cyl}(t)}{\tau(x+i\eta(x))} + r^2 \frac{ik e^{2i\alpha_0}\alpha'_{cyl}(t)}{\tau^2(x+i\eta(x))}\right]_x$$

The velocity v_{dip} is a normal fluid velocity induced at Γ by conformal image of dipole concentrated at the origin $\tau = 0$ of auxiliary τ-plane. The integral equation (5) closing the system of differential equations (4) contains all information about interaction of the cylinder with the free surface. When parameter $k \to 0$ the kernels A_r and B_r and the function v_{dip} converge to the kernels of integral operators and integrated terms obtained for a circular cylinder [3]. The integro-differential system (4), (5) is equivalent to the original equations (1)–(3).

4. Small-time asymptotic solution

We consider the unsteady flow which starts from rest, $\eta(x,0) = u(x,0) = v(x,0) = 0$, and is caused by arbitrary two-dimensional motion of the cylinder with a constant acceleration. The solution of equations (4)-(5) is derived by the small-time expansion method (see [1]-[6] and [16]) in the form

$$\begin{aligned} \eta(x,t) &= t^2\eta_2(x) + t^3\eta_3(x) + t^4\eta_4(x) + \dots, \\ u(x,t) &= tu_1(x) + t^2u_2(x) + t^3u_3(x)\dots, \\ v(x,t) &= tv_1(x) + t^2v_2(x) + t^3v_3(x) + \dots. \end{aligned}$$

The coefficients η_n and u_n are evaluated explicitly by recursive formulae which follow from the differential equations (4) once the coefficients v_n are known. The integral equation (5) provides recursive equations for the coefficients $v_n(x)$ of the form

$$\pi v_n(x) + \int_{-\infty}^{+\infty} A_0(x,s)\, v_n(s)ds = \varphi_n(x), \quad (n \geqslant 1)$$

where the kernel A_0 consists of the lower-order terms of the kernels A and B with respect to the deflection η. Further simplifications are available by using a perturbation procedure with some appropriate small parameters. If the nondimensional semi-axes a and b are small, we obtain the non-linear version of dipole approximation for submerged elliptic cylinder. Combining this approach with the method of successive approximations one can obtain analytically explicit formulae for the leading-order solution.

We consider here wave patterns that are formed by an arbitrary plane motion of the cylinder. The trajectory of the center and angle of rotation can be taken in dimensionless form as:

$$z_{cyl}(t) = -i + t^2(\cos\theta + i\sin\theta), \qquad \alpha_{cyl}(t) = \alpha_0 + \varepsilon t^2.$$

The leading-order solution has the following coefficients for the free surface elevation $\eta(x,t)$:

$$\begin{aligned} \eta_2(x) = 2r^2\Big\{(\sin\theta + k\sin(\theta - 2\alpha_0))\,Q_x + (\cos\theta + k\cos(\theta - 2\alpha_0))\,P_x\Big\} + \\ + 2k\varepsilon r^4\Big\{\cos 2\alpha_0(QQ_x - PP_x) - \sin 2\alpha_0(PQ_x + QP_x)\Big\}, \quad (7) \end{aligned}$$

Here $P(x)$ is even and $Q(x)$ is odd functions of x. These functions are defined for $x \geqslant 0$ by the formulae

$$\begin{aligned} P(x) &= \frac{1}{2\sqrt{2}kr^2}\left(-\sqrt{2} + \sqrt{\sqrt{(x^2 - 4kr^2 - 1)^2 + 4x^2} - x^2 + 4kr^2 + 1}\right), \\ Q(x) &= \frac{1}{2\sqrt{2}kr^2}\left(\sqrt{2} - \sqrt{\sqrt{(x^2 - 4kr^2 - 1)^2 + 4x^2} + x^2 - 4kr^2 - 1}\right). \end{aligned}$$

5. Analysis of the solution

Figures 2-3 show typical shapes of the free surface corresponding to the solution (7) in the case of motion of elliptic cylinder with a semi-axes ratio $a/b = 2/1$ with rotation ($\lambda = 5$). One can see on the figure 2 that the formation of an asymetric splash jet is affected by rotation of the cylinder during its vertical submersion. Figure 3 demonstrates how the cylinder generates the travelling wave by it's horizontal accelerated motion together with rotation. Figure 4 shows the evolution of the free surface when the cylinder rotates without it's translation. One can see that free surface follows rotary motion of the body.

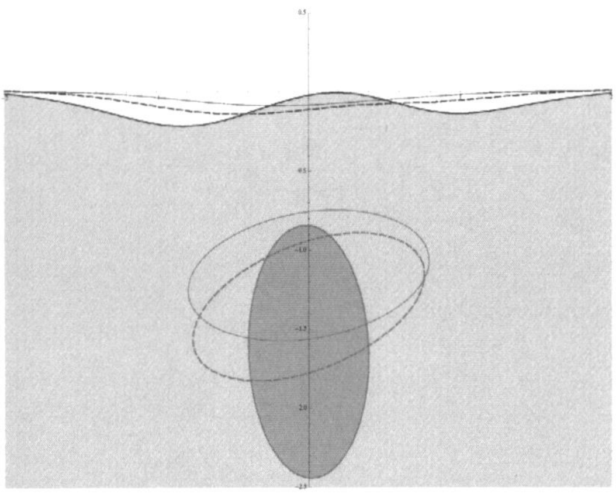

Figure 2. Submersion of elliptic cylinder with rotation at the time points $t = 0.4$ (thin line), $t = 0.6$ (dashed line) and $t = 0.8$ (thick solid line).

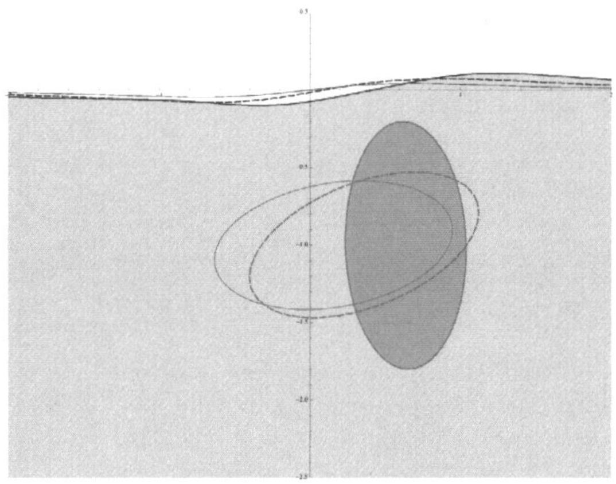

Figure 3. Horizontal motion of elliptic cylinder with rotation at the time points $t = 0.4$ (thin line), $t = 0.6$ (dashed line) and $t = 0.8$ (thick solid line).

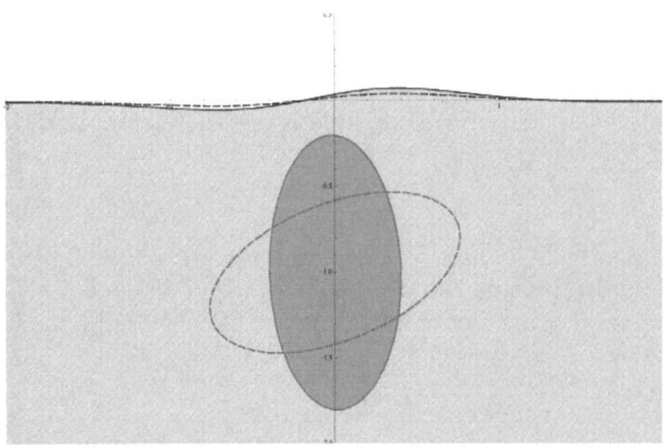

Figure 4. Rotation of elliptic cylinder without translational motion at the time points $t = 0.6$ (dashed line) and $t = 0.8$ (solid line).

Thus, the fully nonlinear integro-differential system on the free boundary was constructed for non-stationary motion of submerged body under the free surface. Small-time asymptotic solution was constructed in the case of uniformly accelerated elliptic cylinder that can rotate around its center. We can conclude from the analysis of this solution that the rotation of the cylinder causes considerable asymmetry of the wave patterns generated on the free surface.

Acknowledgments

This work was supported by the Russian Scientific Foundation (grant No. 15-11-20013).

References

[1] Tyvand P A and Miloh T 1995 *J. Fluid Mech.* **286** 67–101
[2] Tyvand P A and Miloh T 1995 *J. Fluid Mech.* **286** 103–16
[3] Makarenko N I 2003 *JOMAE Trans. of the ASME* **125**(1) 72–5
[4] Makarenko N I 2004 *Proc. 23rd Int. Conf. on Offshore Mech. and Arctic Eng. OMAE'04* (Vancouver: OMAE-51413)
[5] Makarenko N I and Kostikov V K 2013 *J. Appl. Mech. Tech. Phys.* **54**(3) 367–76
[6] Pyatkina E V 2003 *J. Appl. Mech. Tech. Phys.* **44**(1) 32–43
[7] Norkin M V 2012 *Comp. Math Math. Phys.* **51**(2) 292–301
[8] Norkin M V 2012 *J. Appl. Mech. Tech. Phys.* **53**(4) 532–9
[9] Haussling H J and Coleman R M 1979 *J. Fluid. Mech.* **92**(4) 767–81
[10] Terentiev A G, Afanasiev K E and Afanasieva M M 1988 *Proc. IUTAM Symp. Advanced Boundary Element Methods* pp 427–34
[11] Gorlov S I 1999 *J. Appl. Mech. Tech. Phys.* **40**(3) 393–8
[12] Wu G X and Taylor E 2003 *Ocean Eng.* **30** 387–400
[13] Zhu X, Faltinsen O and Hu C 2007 *JOMAE Trans. of the ASME* **129**(4) 253–64
[14] Greenhow M and Moyo S 1997 *Philos. Trans. Roy. Soc.* **355** 551–63
[15] Makarenko N I and Kostikov V K 2014 *Nat. Hazards Earth Syst. Sci.* **14** 751–56
[16] Karabut E A 1991 *Intern. Ser. Numer. Math.* **99** 215–24

Linear and nonlinear effects in detonation wave structure formation

S P Borisov[1,2] **and A N Kudryavtsev**[1,2]

[1] Khristianovich Institute of Theoretical and Applied Mechanics of SB RAS 630090 Novosibirsk, Russia

[2] Novosibirsk State University, 630090 Novosibirsk, Russia

E-mail: borisov@itam.nsc.ru, alex@itam.nsc.ru

Abstract. The role of linear and nonlinear effects in the process of formation of detonation wave structure is investigated using linear stability analysis and direct numerical simulation. A simple model with a one-step irreversible chemical reaction is considered. For linear stability computations, both the local iterative shooting procedure and the global Chebyshev pseudospectral method are employed. Numerical simulations of 1D pulsating instability are performed using a shock fitting approach based on a 5th order upwind-biased compact-difference discretization and a shock acceleration equation deduced from the Rankine–Hugoniot conditions. A shock capturing WENO scheme of the 5th order is used to simulate propagation of detonation wave in a plane channel. It is shown that the linear analysis predicts correctly the mode dominating on early stages of flow evolution and the size of detonation cells which emerge during these stages. Later, however, when a developed self-reproducing cellular structure forms, the cell size is approximately doubled due to nonlinear effects.

1. Introduction

Detonation is one of the most powerful, violent, and intriguing phenomena observed in nature and produced by people. In a good solid explosive, energy is converted at a rate of 10^{10} W per square centimeter of the detonation front so that a 20-m square detonation wave (DW) operates at a power equal to all the power the Earth receives from the Sun [1]. Traditionally, detonation studies were focused on producing powerful explosions as well as on their prevention and protection from them. However, the unique feature of detonative processes as very high density energy sources means that suitably controlled detonation fronts represent an attractive and perspective technology for application in many areas. In recent years, there was a significant interest in the use of detonation for development of novel propulsive devices such as standing DW engine [2], pulse detonation engine (PDE) [3], and continuously rotating DW engine [4]. In 2008 the first, 10-s flight of an aircraft powered by an experimental PDE was performed [5]. Moreover, a huge rate of heat release in DWs can also facilitate creation of microengines, entirely new microdevices, in the micrometer to millimeter range, able to produce mechanical work from chemical energy [6].

Nearly all self-sustaining detonations exhibit complicated unsteady 3D patterns known as a cellular or multifront structure [7, 8]. Not long after the discovery that the cellular structure is a widespread and common feature of propagating DWs, it was assumed that its emergence is caused by an instability of a plane detonation front with respect to transverse disturbances

[9]. Existence of such an instability (connected with the strong, exponential dependence of the chemical reaction rate on temperature) was shown, first, for the piecewise-constant basic flow behind the leading shock wave (i. e., in the limit of infinitely fast chemical heat release) [10] and then for the basic flow given by the Zeldovich–Neumann–Doring (ZND) solution of the 1D reactive Euler equations [11]. After a series of papers published by J.J. Erpenbeck in the mid-1960s [12, 13, 14], the linear analysis of DW instability has become a well-established theory. Since the early 1990s, there has been a renewed interest in this theory. The original techniques of Erpenbeck employing the Laplace transform has been replaced by a much simpler approach based on the method of normal modes [15, 16], and the stability of detonation for a simple chemical mechanism with one irreversible reaction has been investigated in a more systematic and nearly exhaustive manner (see the reviews [17, 18, 19] and the book [20]).

Nevertheless, the status of the linear theory of DW instability remains rather uncertain because, in a sharp contrast with the theory of hydrodynamic stability, the former wasn't ever confirmed directly by experiment. A better understanding of links between the stability theory and mechanisms of formation and self-reproduction of 3D DW structure might help us control detonation behavior in propulsion devices [19]. Relatively recently, a number of investigations were performed aimed at comparison of linear analysis predictions with data of direct numerical simulations in the 1D case [21, 22, 23, 24]. In the present paper, we investigate both the 1D pulsating instability and the 2D instability of a DW propagating in a plane channel trying to elucidate the role of linear and nonlinear effects in formation of the developed DW structure.

2. Governing equations and numerical techniques

A chemically reacting flow governed by the Euler equations for a perfect gas which undergoes a one-step irreversible reaction is considered:

$$\begin{aligned}
&\frac{\partial \rho}{\partial t} + \nabla\cdot\rho\mathbf{u} = 0, \\
&\frac{\partial \rho\mathbf{u}}{\partial t} + \nabla\cdot\rho\mathbf{u}\mathbf{u} + \nabla p = 0, \\
&\frac{\partial E}{\partial t} + \nabla\cdot(E+p)\,\mathbf{u} = 0, \qquad E = \frac{1}{2}\rho\,\mathbf{u}^2 + \frac{1}{\gamma-1}\frac{p}{\rho} - \lambda Q, \\
&\frac{\partial \rho\lambda}{\partial t} + \nabla\cdot\rho\lambda\mathbf{u} = \rho\,\dot{\omega}, \qquad \dot{\omega} = K\,(1-\lambda)\exp\left(-\frac{\rho E_a}{p}\right).
\end{aligned} \tag{1}$$

Here, the dependent variables are density ρ, velocity $\mathbf{u}$, pressure p, specific total energy E and reaction progress λ; t is time, $\gamma = 1.2$ is the ratio of specific heats, K is the predexponential factor, E_a is the activation energy, and Q is the specific heat release.

The 1D ZND solution of (1) consists of a leading plane nonreactive shock front followed by a chemical reaction zone [1, 20]. This structure is stationary in the frame of reference moving with the shock front speed D. The value of K can be specified in such a way that the reaction half-length, i.e., the distance between the shock front and the location where the reaction progress variable $\lambda = 1/2$, will be equal to unity for the ZND solution. Thereafter, all lengths are assumed to be normalized with the reaction half-length.

The ZND solution exists provided that $M = D/c_0 \geq M_{CJ}$ where c_0 is the sound speed ahead of the front and M_{CJ} is the Chapman–Jouguet (CJ) Mach number

$$M_{CJ} = \sqrt{\frac{(\gamma^2-1)Q}{2c_0^2}} + \sqrt{\frac{(\gamma^2-1)Q}{2c_0^2} + 1}. \tag{2}$$

The CJ detonation is a self-sustaining, freely propagating DW while *overdriven* detonations with $M > M_{CJ}$ can be considered as DWs driven by a piston moving with a constant velocity far behind the shock front. The overdrive parameter is defined as $f = (M/M_{CJ})^2$.

To investigate stability of a plane DW, small disturbances are added to the ZND solution, and the shape of shock front is also assumed to be distorted:

$$q(x,y,z) = \bar{q}(x) + \hat{q}(x)\, e^{\alpha t} e^{i(k_y y + k_z z)}, \qquad \psi(y,z,t) = 0 + \hat{\psi}\, e^{\alpha t} e^{i(k_y y + k_z z)}. \tag{3}$$

Here, q denotes any of flow variables, and the disturbed shape of shock front is given by the equation $x = \psi(y,z,t)$ where x is the coordinate normal to the undisturbed shock front, and y, z are two other Cartesian coordinates. The frame of reference is attached to the undisturbed shock front (located at $x = 0$), $\bar{q}(x)$ is the ZND solution, k_y and k_z transverse components of the disturbance wave number, $\alpha = \alpha_r + i\alpha_i$ is a complex number whose real part is the growth rate, the flow is unstable at $\alpha_r > 0$. It is worth noting that the axes y, z can be always orientated in such a way that one of the components k_y, k_z will vanish, so we will assume that $k_z = 0$.

The linearized reactive Euler equations along with the linearized Rankine–Hugoniot jump conditions at $x = 0$ and the radiation condition far from the shock front, at $x \to \infty$ results in an eigenvalue problem with respect to α. It is solved with either an iterative shooting procedure [15] or a pseudospectral Chebyshev method [25]. The latter provides an eigenvalue map and enables us to avoid time-consuming "carpet search" of initial guesses. Since the radiation condition for multidimensional disturbances leads to a nonlinear eigenvalue problem, non-reflective boundary conditions proposed in [26] are employed instead when solving the spectral problem.

A high-order shock-fitting approach is used to perform direct numerical simulation of evolution of 1D disturbances. The 1D reactive Euler equations are solved along with an equation for shock wave acceleration. The shock wave acceleration dD/dt can be expressed from the Rankine–Hugoniot conditions in terms of time derivatives of flow variables just behind the shock front, which, in turn, can be replaced using the Euler equations by their spatial derivatives. Since the spatial derivatives just behind, the shock front must be calculated using one-sided differences, it is preferable to use the Riemann invariant $\mathcal{L}$ corresponding to the characteristic running towards the shock front, which leads to the equation

$$\begin{aligned} \frac{dD}{dt} &= \frac{(\gamma+1)\,\mathcal{L}/2}{2\rho_0 D - \rho_s c_s \left[1 + (c_0/D)^2\right]}, \\ \mathcal{L} \equiv \frac{\partial p_s}{\partial t} - \rho_s c_s \frac{\partial u_s}{\partial t} &= -(u_s - D - c_s)\left(\frac{\partial p_s}{\partial x} - \rho_s c_s \frac{\partial u_s}{\partial x}\right) + (\gamma-1)\rho Q \dot{\omega}. \end{aligned} \tag{4}$$

Here, the subscripts 0 and s denote quantities ahead of and just behind the shock front, respectively. The spatial derivatives are approximated by the 5th-order upwind-biased compact differences [27], and the reactive Euler equations (1) along with (4) are integrated in time with the 5th-order, 6-stage SSP (Strong Stability-Preserving) Runge–Kutta scheme [28].

In the 2D case, numerical simulations are performed using the high-order shock-capturing finite-difference WENO (Weighted Essentially Non-Oscillatory) scheme [29].

3. Linear analysis for DW in a plane channel

In general, plane DWs are unstable over a wide range of the parameters Q, E_a and f. Figure 1 shows the eigenvalue maps at $Q = 50$, $E_a = 50$, $f = 1$ and different values of the transverse wave number k_y. There are 10 unstable eigenmodes at $k_y = 0$, however their number decreases as k_y grows so that the flow becomes stable if k_y exceeds a neutral wave number $k_y^N \approx 28$.

The growth rates of all these modes are shown in figure 2. For each value of k_y, there exists a maximally unstable mode. The dependence of the maximum growth rate on k_y is given by the envelope (shown as a bold solid line). For a DW traveling in a plane channel, boundary conditions for the transverse velocity v require $k_y = \pi n/H$, $n = 1, 2, 3...$, where H is the channel height.

If $H < \pi/k_y^N$ then the channel is too narrow for existence of unstable transverse disturbances, and formation of cellular DW structure can hardly be expected.

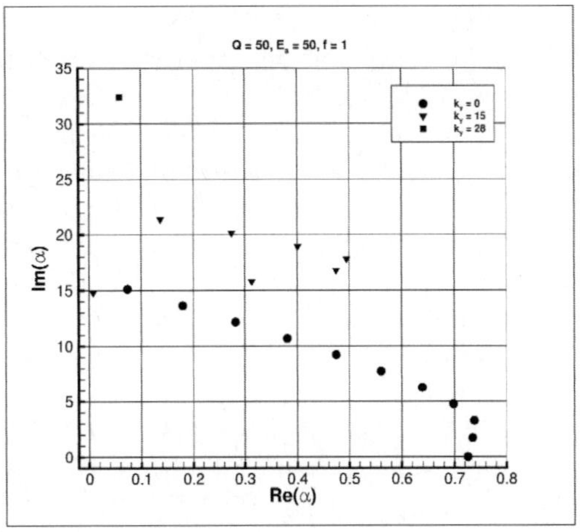

Figure 1. Eigenvalues at $k_y = 0, 15, 28$.

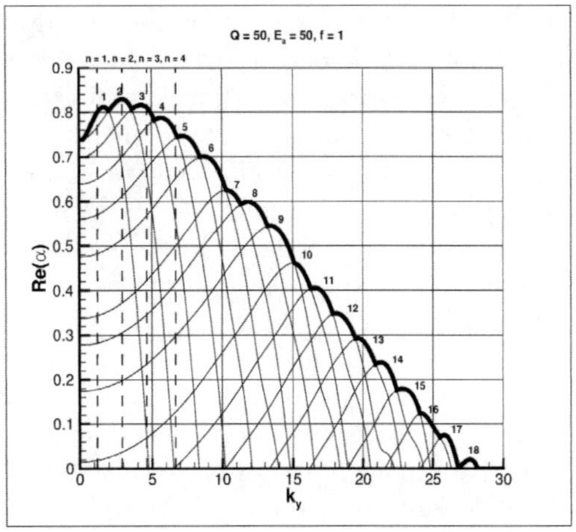

Figure 2. Growth rates as functions of k_y.

For wider channels, the growth rate is maximum for some value of $n = N$. It is natural to suppose [30] that the corresponding mode will dominate at early (at least) stages of DW structure evolution so that formation of the cellular structure containing N detonation *half-cells* (or $N/2$ entire cells) can be expected.

As an example, the number of detonation half-cells N and the cell size $a = 2H/N$ predicted by the linear theory at $Q = 50$, $E_a = 12.5$, $f = 1$ are shown in Figures 3 and 4, respectively.

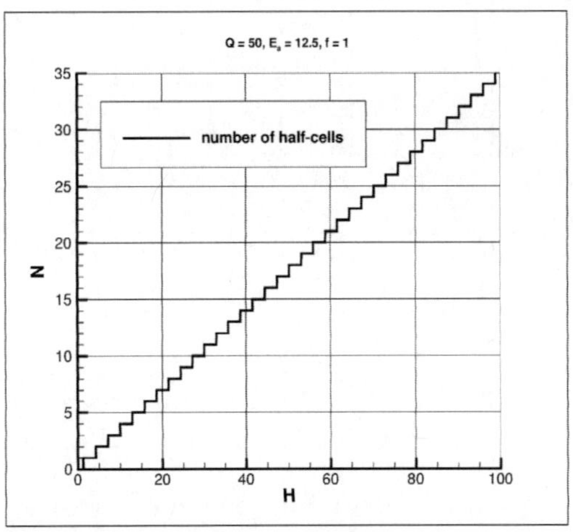

Figure 3. Predicted number of half-cells for DW in plane channel.

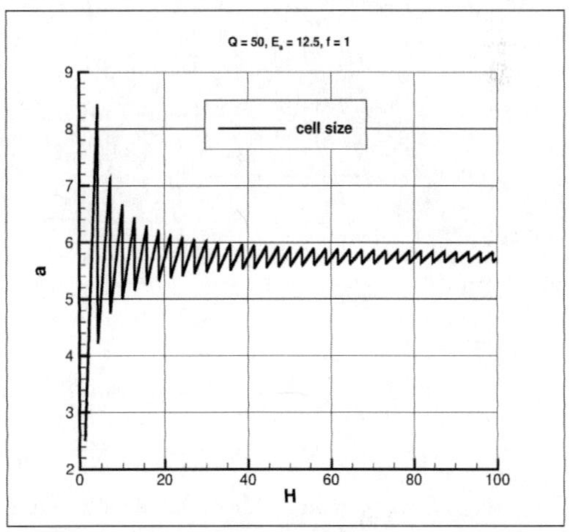

Figure 4. Predicted cell size for DW in plane channel.

It is seen that as H increases, the cell size also increases to fit boundary conditions and the number of cells does not change until a sudden transition to the next value of n happens. For very large channel heights the cell size approaches the limiting value $2\pi/k_y^M$ where k_y^M is the wave number corresponding to the maximum growth rate over all wave numbers (see Figure 2). Such behavior is in close agreement with that was observed in experiments.

4. Numerical simulations of pulsating 1D instability

Numerical simulations of the 1D instability cannot elucidate mechanisms governing formation of the cellular detonation structure, however they allow us to test predictions of the linear theory concerning growth rates of unstable disturbances, study mode competition phenomena and investigate the instability saturation and formation of nonlinear quasi-equilibrium states. In the present computations, the ZND solution with no imposed disturbances is taken as the initial condition. Nevertheless, as a result of accumulation of rounding errors, disturbances can emerge and, in unstable cases, grow.

The computations performed for $Q = 50$, $f = 1$ and varying values of E_a show that the behavior of disturbances agrees with the linear analysis. The disturbances grow at $E_a > E_a^N = 25.26$ and their numerical growth rate during the linear stage is very close to predicted one. After saturation, a stable limiting cycle is observed in the phase space $(D\,, dD/dt)$ if the activation energy slightly exceeds E_a^N. At larger values of E_a, the single limiting cycle gives way to a doubly periodic solution and, as the activation energy further increases, the system becomes more and more chaotic. These results agree very well with those in [24].

In Figures 5 and 6 the results of simulations performed at $Q = 50$, $E_a = 50$, $f = 1.4$ are compared with the linear analysis predictions. At these conditions, there are two unstable modes. As is seen, in numerical simulations the disturbances grow with the rate which is very close to the amplification coefficient predicted by the linear theory for the *faster growing* mode (Figure 5). This mode dominates for the entire stage of linear evolution. The spatial distribution of velocity disturbance during this stage nearly perfectly matches the linear eigenfuction of the more unstable mode (Figure 6).

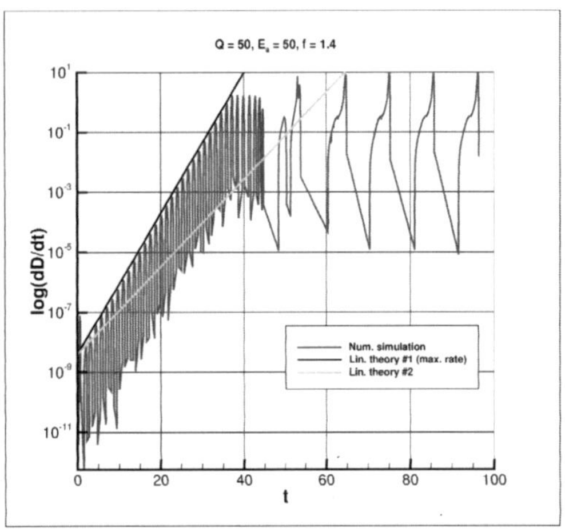

Figure 5. Time evolution of shock wave acceleration in 1D numerical simulations compared with linear theory predictions.

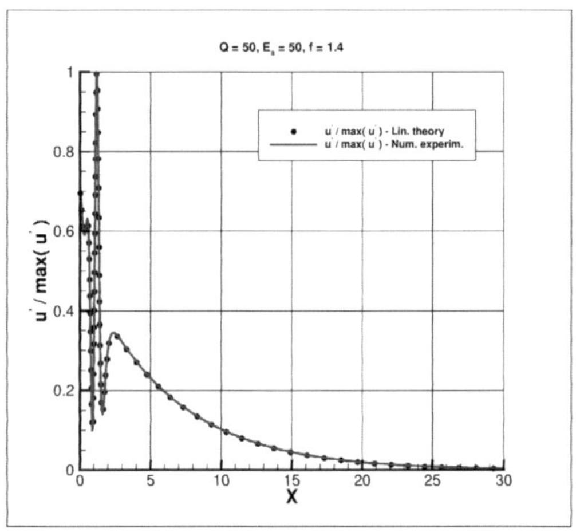

Figure 6. Spatial distribution of velocity disturbance at $t = 32$ in 1D numerical simulations compared with linear eigenfuction.

The period of DW oscillations during the stage of linear evolution is also very close to the period $2\pi/\alpha_i$ of the dominating mode. At a later moment it increases abruptly up to a value which is slightly larger than the period of the second, slower growing, mode and continues to increase permanently while the computation proceeds.

Journal of Physics: Conference Series **722** (2016) 012022 doi:10.1088/1742-6596/722/1/012022

5. Numerical simulation of formation of cellular detonation structure

To investigate the process of formation of the cellular structure, DW propagation in a plane channel of height $H = 100$ is simulated. The length of computational domain L_x is 80. A special techniques of periodic translations of the computational domain is used to keep the propagating DW inside the domain. The grid resolution is varied, the finest grid comprises of 2720×3400 cells; up to 80 cores of a multiprocessor cluster are used for the computations. A plane wave ZND solution with a superimposed long-wave (half-sine) disturbance of the transverse velocity is taken as the initial condition. The computations are performed at different values of the parameters Q, E_a, and f.

A numerically generated smoke foil resulted from a typical computation at $Q = 50$, $E_a = 12.5$, and $f = 1$ is shown in Figure 7 along with enlarged views of its leftmost and rightmost parts. At these parameters, the linear analysis predicts the existence of a single unstable mode. Its growth rate is maximum at the transverse wavenumber $k_y^M = 1.095$ which corresponds to the number of half-cells across the channel N between 34 and 35. As can be seen, detonation cells become visible approximately at $x = 180$. In spite of long-wave excitation, at first they are small. Later, however, their size increases and becomes roughly constant at $x \geq 350$. The cellular structure is far from completely regular so that both smaller and larger cells can be observed at any distance from the initial detonation front position.

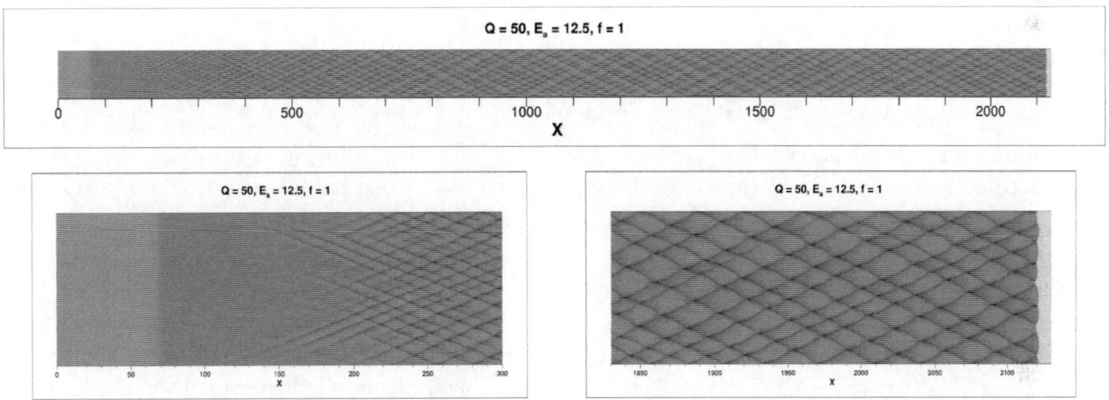

Figure 7. Numerical smoke foils for DW propagation in 2D channel, $H = 100$.

To characterize the process of formation and development of the cellular structure quantitatively, the number of cells as function of time is determined using the Fourier transform of the transverse velocity flowfield:

$$\hat{v}_k(x,t) = \frac{2}{H} \int_0^H v(x,y,t) \sin\left(\frac{k\pi y}{H}\right) dy. \tag{5}$$

The number of half-cells deduced from the wavenumber of Fourier harmonics with the maximum amplitude is shown in Figure 8. The line marked "max" is obtained by taking also the maximum of $|\hat{v}_k(x,t)|$ over the longitudinal coordinate x, whereas the second line ("ave") is built from the data averaged over some interval of x.

As is it seen just after formation of detonation cells their number is in good agreement with the one predicted by the linear theory. However, in a well-developed cellular structure the number of cell is approximately halved. This pairing process is obviously caused by nonlinear effects. Thus, it seems that the linear analysis predictions are valid only for early stages of formation of the cellular structure. It is worth noting that this conclusion disagrees with the results obtained in [31] where in numerical simulations of a weakly unstable detonation at $Q = 0.4$, $E_a = 50$, $f = 1$ the most unstable linear mode remains dominant even in the developed cellular regime.

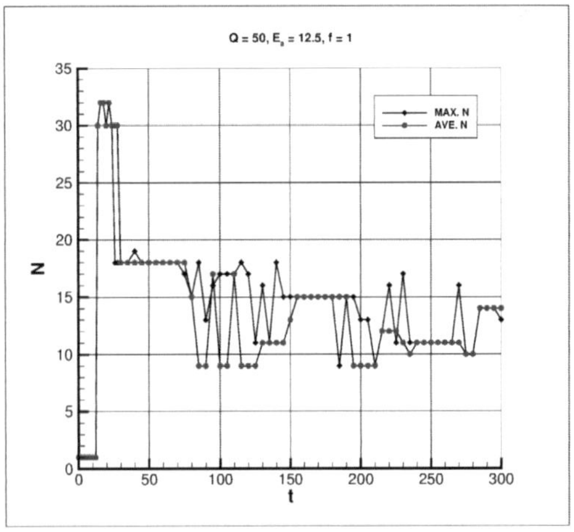

Figure 8. Time variation of number of half-cells for DW in plane channel.

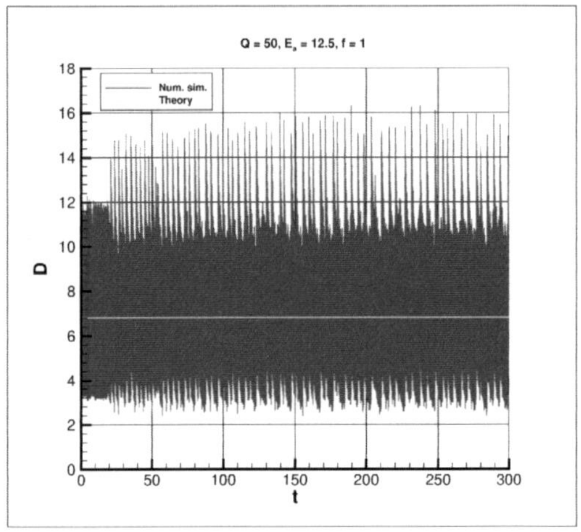

Figure 9. Time evolution of velocity of DW propagating in plane channel.

Another interesting feature of cellular DW propagation is the time evolution of the velocity of leading shock front shown in Figure 9. The theoretical CJ velocity is indicated by a white horizontal line. It can be seen that the averaged velocity obtained from the numerical simulation is slighter larger than its theoretical value predicted by the ZND model.

6. Conclusions

Numerical simulations confirm that the linear analysis predicts correctly the characteristics of disturbances which develop and dominate on early stages of flow evolution as well as the size of detonation cells emerging during these stages. It has been observed, however, that later, when a developed self-reproducing cellular structure forms, the cell size is approximately doubled — obviously due to nonlinear effects. A further exploration of non-linear mechanisms responsible for this "pairing" phenomenon can be considered as an interesting challenge for future research.

Acknowledgments

This work was supported by the Grant of the Government of the Russian Federation (Agreement No. 14.Z50.31.0019) for supporting research supervised by leading scientists.

References

[1] Fickett W and Davis W C 2004 *Detonation. Theory and Experiment* (Dover Publications)
[2] Sislian J P 2000 *Prog. Astronaut. Aeronaut.* **189** 823–89
[3] Roy G D, Frolov S M, Borisov A A and Netzer D W 2004 *Prog. Energy Comb. Sci.* **30**(6) 545–672
[4] Bykovsky F A, Zhdan S A and Vedernikov E F 2006 *Journal of Propulsion and Power* **22**(6) 1204–16
[5] Norris G 2008 *Aviation Week & Space Technology* **168**(7) 60
[6] Stewart D S 2005 *Mechanics of the 21st Century* (Springer Netherlands) pp 379–85
[7] Shchelkin K I and Troshin Y K 1965 *Gasdynamics of Combustion* (Mono Book Corp)
[8] Voitsekhovskii B V Mitrofanov V V and Topchiyan M E 1966 *Rep. The Structure of Detonation Front in Gases* (Wright Patterson AF Base) FTD-MTD-64-527
[9] Shchelkin K I 1959 *Soviet Physics JETP* **36(9)**(2) 416–20
[10] Zaidel R M 1961 *Doklady AN SSSR* in Russian **136**(5) 1142–5

[11] Pukhnachev V V 1963 *Doklady AN SSSR* in Russian **149**(4) 798–801
[12] Erpenbeck J J 1964 *Phys. Fluids* **7**(5) 684–96
[13] Erpenbeck J J 1965 *Phys. Fluids* **8**(6) 1192–3
[14] Erpenbeck J J 1966 *Phys. Fluids* **9**(7) 1293–306
[15] Lee H I and Stewart D S 1990 *J. Fluid Mech.* **216** 103–32
[16] Short M and Stewart D S 1998 *J. Fluid Mech.* **368** 229–62
[17] Clavin P 2004 *Chaos* **14**(3) 825–38
[18] Short M 2005 *CD Proc. of 20th Int. Colloquium on the Dynamics of Explosions and Reactive Systems* (Montreal) p 19
[19] Stewart D S and Kasimov A R 2006 *Journal of Propulsion and Power* **22**(6) 1230–44
[20] Lee J H S 2008 *The detonation phenomenon* (Cambridge Univ. Press)
[21] He L and Lee J H S 1995 *Phys. Fluids* **7**(5) 1151–8
[22] Daimon Y and Matsuo A 2003 *Phys. Fluids* **15**(1) 112–22
[23] Kasimov A R and Stewart D S 2004 *Phys. Fluids* **16**(10) 3566–78
[24] Henrick A K, Aslam T D and Powers J M 2006 *J. Comput. Phys.* **213** 311–29
[25] Tumin A 2007 *AIAA Journal* **45**(9) 2356–9
[26] Borisov S P and Kudryavtysev A N 2015 *CD Proc. of Int. Conf. "Advanced Mathematics, Computations and Applications 2015" AMCA* in Russian (Novosibirsk) pp 121–8
[27] Zhong X 1998 *J. Comput. Phys.* **144** 662–709
[28] Gottlieb S, Shu C-W and Tadmor E 2001 *SIAM Review* **43** 89–112
[29] Jiang G-S and Shu C-W 1996 *J. Comp. Phys.* **126** 202–28
[30] Kudryavtysev A N and Borisov S P 2012 *Proc. of Int. Conf. on Non-equilibrium Processes in Nozzles and Jets NPNJ* in Russian (Moscow: MAI) pp 230–2
[31] Taylor B D, Kasimov A R and Stewart D S 2009 *Combustion Theory and Modelling* **13**(6) 973-92

Mathematical modeling of formation, evolution and interaction of galaxies in cosmological context

I Kulikov[1,2], I Chernykh[1] and V Protasov[1,2]

[1] Institute of Computational Mathematics and Mathematical Geophysics SB RAS, Novosibirsk, Russia
[2] Novosibirsk State Technical University, Novosibirsk, Russia

E-mail: kulikov@ssd.sscc.ru

Abstract. The results of mathematical modeling of formation of galaxies in cosmological context with using of multiphase hydrodynamical model are presented in the paper. Mathematical model of the problem of cosmological modeling, numerical methods for solving the hyperbolical equations and brief description of parallel implementation of the software complex CosmoPhi are described in details. The results of numerical experiments of large-scale cosmological simulations are presented.

1. Introduction

The subject of study of modern astrophysics is the study of physical processes in the Universe, how had they influenced on self-organization and evolution of astrophysical objects, their further dynamics and interaction. Despite the great progress in observation we usually can see only static image, and then the model (or theory) of behavior of astrophysical objects before or after the observational moment is built based on the observational data and known physical processes. Essentiality of accounting of gravitational and magnetic fields, the complexity of reproducing of cosmic conditions in vitro impose significant restrictions on experimental studying of astrophysical objects. In fact, mathematical modeling – is the main, if not the only one, way for theoretical research of astrophysical processes and objects.

The evolution of cold dark matter described with collisionless N-body model in the context of cosmological simulation of clusters of galaxies (see for example [1]) within the standard cosmological model ΛCDM was computed. In the recent years series of global cosmological simulations were performed [2, 3, 4]. But to get on the next level, individual galaxies, such approach has significant restrictions due to inability to obtain some of elliptical galaxies. So it is necessary to simulate also baryonic matter in hydrodynamic model and carry out a mutual cosmological modeling like in the projects "Bolshoi" [5], "Illustris" [6, 7, 8] and "Eagle" [9, 10].

Refinement of cosmological model is done by not only using of multiphase model, but and sophistication of so-called subgrid physical model. Structural element of clusters of galaxiesis actually galaxies themselves. The problems of simulation of dynamics of galaxies could be divided by time of their dynamics. Since, the time of evolution of particular galaxy is up to several billion of years, while the interaction between individual galaxies needs hundreds of millions of years. In the field of research of collisions between galaxies the GALMER project of the Paris observatory – the database of numerical experiments [11] on collision between different types of galaxies – should be singled out. Subgrid processes (starburst [13], AGN [14], formation of supermassive double black holes [15, 16], chemokinetics [17]) are accelerated signifcantly in the problems of collision of galaxies [12], hence the explicit consideration of these processes in the mathematical model is required. The same consideration should be done in the problems of evolution of galaxies, for example, the separate proceedings [18] was devoted to the problem of consideration of different processes in the context of the problem of "galactic archaeology" of Milky Way.

The mathematical model and its implementation as software complex CosmoPhi for cosmological simulation of formation and collision of galaxies is described in the paper. The mathematical model is based on a joint solving of equations of multicomponent one-speed hydrodynamics, that describes gas component, and equations of the first momenta of Boltzmann collisionless equation, that describes collisionless components (stars and dark matter). The equations are written with consideration of cosmological expansion. The using of mathematical model written in form of hyperbolic equations allow us to use unified numerical method. This numerical method consist of special combination of Fluids-in-Cells method [19, 20], Godunov method [21] and piece-wise parabolic method on a local template [22, 23].

2. Mathematical model

The main ingredients of the cosmological simulation are collisionless dark matter and baryonic matter consisting of gas and stellar components, the dynamics of which is affected by dark energy. The dynamics of baryonic component is affected by the number of subgrid processes, such as chemokinetic and related processes of radiation and heating, and also processes of formation and destruction of stars.

2.1. Description of components of cosmological model

Redefined system of equations of gravitational gas dynamics describing baryonic gas component in comoving coordinate system with consideration of cosmological expansion is as follows:

$$\frac{\partial}{\partial t}\begin{pmatrix}\rho\\ \rho_i\\ \rho\vec{u}\\ \rho\varepsilon\\ \rho E\end{pmatrix} + \frac{1}{a}\nabla\cdot\begin{pmatrix}\rho\vec{u}\\ \rho_i\vec{u}\\ \rho\vec{u}\vec{u}\\ \rho\varepsilon\vec{u}\\ \rho E\vec{u}\end{pmatrix} = \begin{pmatrix}\mathcal{S}-\mathcal{D}\\ s_i+\mathcal{S}\frac{\rho_i}{\rho}-\mathcal{D}\frac{\rho_i}{\rho}\\ -\frac{1}{a}\nabla p - H\rho\vec{u} - \frac{\rho}{a^2}\nabla\Phi + \mathcal{S}\vec{v} - \mathcal{D}\vec{u}\\ -\frac{1}{a}(\gamma-1)\rho\varepsilon\nabla\cdot(\vec{u}) - 2H\rho\varepsilon + \Gamma - \Lambda + \varepsilon\frac{\mathcal{S}}{\rho} - \varepsilon\frac{\mathcal{D}}{\rho}\\ -\frac{1}{a}\nabla\cdot(p\vec{u}) - 2H\rho E - \frac{1}{a^2}(\rho\vec{u},\nabla\Phi) + \Gamma - \Lambda + \varepsilon\frac{\mathcal{S}}{\rho} - \varepsilon\frac{\mathcal{D}}{\rho}\end{pmatrix} \quad (1)$$

where $\rho = \sum_i \rho_i$ – density, ρ_i – density of each component of gaseous mixture, s_i – the speed of formation of i-th component of mixture, $\mathcal{S}$ – the speed of formation of supernovae, $\mathcal{D}$ – the speed of starburst, $\vec{u}$ – velocity, $p = \rho\varepsilon(\gamma-1)$ – pressure, $\rho\varepsilon$ – internal energy, γ – adiabatic index, $\rho E = \rho\varepsilon + \rho u^2/2$ – full mechanical energy, Φ – gravitational potential, Γ – heating function, Λ – cooling function, H – Hubble constant, $a = 1/(1+z)$ – the expansion parameter, computed from the Einstein equation:

$$\frac{da}{dt} = H\sqrt{\Omega_M\left(a^{-1}-1\right) + \Omega_\Lambda\left(a^2-1\right) + 1} \quad (2)$$

where $\Omega_M = \Omega_B + \Omega_D$ – the part of matter in the total mass, Ω_B – the part of visible baryonic matter (gas and stars), Ω_D – the part of dark matter, Ω_Λ – the part of dark energy. To describe the baryonic stellar component and dark matter we use collisonless hydrodynamical approach based on the equations for the first momenta of collisionless Boltzmann equation. It was successfully used for the problems of evolution and collision of galaxies [19, 24, 25]. In this approach we will use only diagonal tensor of dispersion of velocities because collisions almost play no role, hence, the off-diagonal elements of tensor could be neglected. The system of first momenta of collisionless Boltzmann equation in comoving coordinate system with consideration of cosmological expansion is as follows:

$$\frac{\partial}{\partial t}\begin{pmatrix} n \\ n\vec{v} \\ \Pi_{xx} \\ \Pi_{yy} \\ \Pi_{zz} \\ nW \end{pmatrix} + \frac{1}{a}\nabla\cdot\begin{pmatrix} n \\ n\vec{v} \\ \Pi_{xx} \\ \Pi_{yy} \\ \Pi_{zz} \\ nW \end{pmatrix} = \begin{pmatrix} \mathcal{D}-\mathcal{S} \\ -\frac{1}{a}\nabla\cdot(\Pi) - Hn\vec{v} - \frac{n}{a^2}\nabla\Phi + \mathcal{D}\vec{u} - \mathcal{S}\vec{v} \\ -\frac{2}{a}\Pi_{xx}\frac{\partial v_x}{\partial x} - 2H\Pi_{xx} + \varepsilon\frac{\mathcal{D}}{3\rho} - \varepsilon\frac{\mathcal{S}}{3\rho} \\ -\frac{2}{a}\Pi_{yy}\frac{\partial v_y}{\partial y} - 2H\Pi_{yy} + \varepsilon\frac{\mathcal{D}}{3\rho} - \varepsilon\frac{\mathcal{S}}{3\rho} \\ -\frac{2}{a}\Pi_{zz}\frac{\partial v_z}{\partial z} - 2H\Pi_{zz} + \varepsilon\frac{\mathcal{D}}{3\rho} - \varepsilon\frac{\mathcal{S}}{3\rho} \\ -\frac{1}{a}\nabla\cdot(\Pi\vec{v}) - 2HnW - \frac{1}{a^2}(n\vec{v},\nabla\Phi) + \varepsilon\frac{\mathcal{D}}{\rho} - \varepsilon\frac{\mathcal{S}}{\rho} \end{pmatrix} \quad (3)$$

where n – the density of stellar component, $\vec{u}$ – velocity, $\Pi_{\xi\xi}$ – diagonal tensor of dispersion of velocities, $nW = \frac{1}{2}(\Pi_{xx} + \Pi_{yy} + \Pi_{zz}) + \frac{nv^2}{2}$ – full mechanical energy of the stellar component. The system of equations is supplemented by the equation for the gravitational potential:

$$\triangle\Phi = 4\pi G(\rho + n - \rho_0) \quad (4)$$

where ρ_0 – average density of baryonic component of the dark matter.

2.2. Model of subgrid physics

The model of subgrid physics play a major role in cosmological problems. In our approach we will focus on the following, in our opinion, the most important physical processes.

2.2.1. Process of star formation and effect of the supernovae explosion Let us formulate the neccessary conditions for star formation proccess as follows[26]:

$$T < 10^4 K \qquad \nabla\cdot\vec{u} < 0 \qquad \rho > 1.64\frac{M_\odot}{pc^{-3}} \quad (5)$$

Then the star formation rate will be used in the following form [1]:

$$\Delta^{SFR} = \frac{dn}{dt} = \mathcal{C}\frac{\rho}{\tau_{dyn}} = \mathcal{C}\rho^{3/2}\sqrt{\frac{32G}{3\pi}} \quad (6)$$

where $\mathcal{C} = 0.03$ – the coefficient of efficiency of star formation. We use the collapse rate of stellar component in a result of supernovae explosion as follows [27]:

$$\Delta^{SN} = \frac{d\rho}{dt} = \beta\mathcal{C}\frac{n}{\tau_{dyn}} = \beta\mathcal{C}n^{3/2}\sqrt{\frac{32G}{3\pi}} \quad (7)$$

where the coefficient of explosion of young star is $\beta = 0.1$. Also with each explosion of supernova with mass equal to $M^{SN} = M_\odot$ the energy in an amount of $10^{51}erg$ is released, which will play the role of one of members of the function of gas heating

$$\Gamma_{SN} = 10^{51}\frac{M^{SN}}{M_\odot}erg \quad (8)$$

2.2.2. Cooling function Intergalactic gas, heated to a temperature $\sim 10^4 - 10^8 K$ while collision, will be cooled with time. The cooling rate of gas with a temperature more than $\sim 10^4$ is estimated in paper [28] and is used as follows:

$$\Lambda_{cooling} = 10^{-22}n^2\frac{erg}{cm^{-3}} \quad (9)$$

where n – the concentration of atomic hydrogen.

2.2.3. Primordial model of chemical kinetics Let us specially focus on the model of chemokinetics. We will treat chemical evolution of the basic forms of hydrogen and helium [29, 30, 31]. For this we will use the following 20 chemical reactions.

(1) Collisional ionization $H + e \longrightarrow H^+ + 2e$, the reaction rate (hear and further in cm^3/s) k_1 is given in table (1).

(2) Collisional recombination with radiation $H^+ + e \longrightarrow H + \gamma$, reaction rate k_2 in case with $T > 5500K$ is given in table (1), with lower temperature $k_2 = 3.92 \times 10^{-13} \times T^{-0.6353}$.

(3) Collisional ionization with radiation $H + e \longrightarrow H^- + \gamma$, reaction rate $k_3 = 6.77 \times 10^{-15} \times T^{0.8779}$.

(4) Dissociative electrone attachement $H^- + H \longrightarrow H_2 + e$, reaction rate k_4 in case with $T > 1160K$ is given in table (1), with a lower temperature $k_4 = 1.43 \times 10^{-9}$.

(5) Association with radiation $H + H^+ \longrightarrow H_2^+ + \gamma$, reaction rate in case with $T > 6700K$ $k_5 = 5.81 \times 10^{-16} \left(\frac{T}{56200}\right)^{-0.6657 \times \log_{10}(T/56200)}$, with a lower temperature $k_5 = 1.85 \times 10^{-23} \times T^{1.8}$.

(6) Recombination of hydrogen $H_2^+ + H \longrightarrow H_2 + H^+$, reaction rate $k_6 = 6 \times 10^{-10}$.

(7) Recombination of molecular hydrogen $H_2 + H^+ \longrightarrow H_2^+ + H$, reaction rate k_7 in case with $T > 350K$ is given in table (1), there is no reaction with a lower temperature.

(8) Collisional dissociation $H_2 + e \longrightarrow 2H + e$, reaction rate in case with $T > 350K$ $k_8 = 5.6 \times 10^{-11} \times \sqrt{T} \times \exp^{-102124/T}$, there is no reaction with a lower temperature.

(9) The first reaction of collisional detachment of electrons $H^- + e \longrightarrow H + 2e$, reaction rate k_9 in case with $T > 45K$ is given in table (1), there is no reaction with a lower temperature.

(10) The second reaction of collisional detachment of electrons $H^- + H \longrightarrow 2H + e$, reaction rate k_{10} in case with $T > 45K$ is given in table (1), with a lower temperature $k_{10} = 2.56 \times 10^{-9} \times T^{1.78186}$.

(11) Neutralization $H^- + H^+ \longrightarrow 2H + \gamma$, reaction rate $k_{11} = 6.5 \times 10^{-9} \times T^{-1/2}$.

(12) Collisional association $H^- + H^+ \longrightarrow H_2^+ + e$, reaction rate $k_{12} = 10^{-8} \times T^{-0.4}$.

(13) Dissociative recombination $H_2^+ + e \longrightarrow 2H + \gamma$, reaction rate with a temperature $T > 617K$ $k_{13} = 1.32 \times 10^{-6} \times T^{-0.76}$, with a lower temperature $k_{13} = 10^{-8}$.

(14) Molecular neutralization $H_2^+ + H^- \longrightarrow H + H_2$, reaction rate $k_{14} = 5 \times 10^{-7} (100 \times T)^{-0.5}$.

(15) Collisional molecular association $3H \longrightarrow H_2 + H$, reaction rate with a temperature $T > 300K$ $k_{15} = 1.3 \times 10^{-32} (T/300)^{-1}$, with a lower temperature $k_{15} = 1.3 \times 10^{-32} (T/300)^{-0.38}$.

(16) Collisional molecular dissociation $H_2 + H \longrightarrow 3H$, reaction rate $k_{16} = 5 \times 10^{-7} (100T)^{-0.5} \frac{1.0671 \times 10^{-10} \times T^{2.012}}{\exp^{4.463/T} (1 + 0.2472 \times T)^{3.512}}$.

(17) Collisional ionization $He + e \longrightarrow He^+ + 2e$, reaction rate k_{17} in case with $T > 900K$ is given in table (1), there is no reaction with a lower temperature.

(18) Collisional recombination with radiation $He^+ + e \longrightarrow He + \gamma$, reaction rate with a temperature $T > 900K$ $k_{18} = 3.92 \times 10^{-12} \times T^{-0.6353} + 1.54 \times 10^{-9} \times T^{-1.5} \frac{1 + \frac{0.3}{e^{8.0993/T}}}{e^{40.4966/T}}$, with a lower temperature $k_{18} = 3.92 \times 10^{-13} T^{-0.6353}$.

(19) Collisional ionization $He^+ + e \longrightarrow He^{++} + 2e$, reaction rate k_{19} in case with a temperature $T > 900K$ is given in table (1), there is no reaction with a lower temperature.

(20) Collisional molecular dissociation $He^{++} + e \longrightarrow He^+ + \gamma$, reaction rate $k_{20} = 3.36 \times 10^{-10} \times T^{-0.5} \frac{(T/1000)^{-0.2}}{1 + (T/10^6)^{0.7}}$

Cooling functions realted to reactions are given in appendix. We will calculate effective adiabatic index from the equation:

$$\gamma = \frac{5n_H + 5n_{He} + 5n_e + 7n_{H_2}}{3n_H + 3n_{He} + 3n_e + 5n_{H_2}} \tag{10}$$

where n_H – concentration of atomic hydrogen, n_{He} – atomic helium, n_e – electrons, n_{H_2} – molecular hydrogen. We will use the classical initial distribution of elements in the Universe [31] in the numerical experiments.

3. Numerical method

The original numerical method based on the combination of operator splitting approach, Godunov method, Roe scheme and piece-wise parabolic reconstruction on a local template [22, 23] is used to solve the hydrodynamical equations. The method based on Fast Fourier Transform is used to solve the Poisson equation. At the eulerian stage of the numerical method the equations without consideration of advective members and function, that simulate "subgrid" processes, are solved. To approximate spatial derivatives the solution of the linearized Riemann problem is used. For this purpose the averaging of density and pressure (in the case of collisionless component we use diagonal member of the tensor of dispersion of velocities) is made on all of the interfaces between cells (L – is the designation of the left cell, R – is the designation of the right cell) with using of the following formulae:

$$\rho = \frac{\rho_L^{3/2} + \rho_R^{3/2}}{\sqrt{\rho_L} + \sqrt{\rho_R}} \qquad p = \frac{p_L\sqrt{\rho_L} + p_R\sqrt{\rho_R}}{\sqrt{\rho_L} + \sqrt{\rho_R}}$$

We assume that solution in treated cells is piece-wise parabolic function. The procedure of construction of local parabolas is described in classic work [32] in details. Omitting the detailed calculations of the development of scheme with using of the Godunov method (the numerical method and its verification are described in [33] in details) the solution of the Riemann problem at the eulerian stage with consideration of cosmological expansion is:

$$U = \frac{u_L(-\frac{\lambda\tau}{a}) + u_R(\frac{\lambda\tau}{a})}{2} + \frac{p_L(-\frac{\lambda\tau}{a}) - p_R(\frac{\lambda\tau}{a})}{2}\sqrt{\frac{(\sqrt{\rho_L} + \sqrt{\rho_R})^2}{\frac{\gamma_L+\gamma_R}{2}(\rho_L^{3/2} + \rho_R^{3/2})(p_L\sqrt{\rho_L} + p_R\sqrt{\rho_R})}}$$

$$P = \frac{p_L(-\frac{\lambda\tau}{a}) + p_R(\frac{\lambda\tau}{a})}{2} + \frac{u_L(-\frac{\lambda\tau}{a}) - u_R(\frac{\lambda\tau}{a})}{2}\sqrt{\frac{\frac{\gamma_L+\gamma_R}{2}(\rho_L^{3/2} + \rho_R^{3/2})(p_L\sqrt{\rho_L} + p_R\sqrt{\rho_R})}{(\sqrt{\rho_L} + \sqrt{\rho_R})^2}},$$

where

$$\lambda = \sqrt{\frac{\frac{\gamma_L+\gamma_R}{2}(p_L\sqrt{\rho_L} + p_R\sqrt{\rho_R})}{\rho_L^{3/2} + \rho_R^{3/2}}}$$

$$q_L(-\nu t) = q_i^R - \frac{\nu t}{2h}\left(\triangle q_i - q_i^6\left(1 - \frac{2\nu t}{3h}\right)\right) \qquad q_R(\nu t) = q_i^L + \frac{\nu t}{2h}\left(\triangle q_i + q_i^6\left(1 - \frac{2\nu t}{3h}\right)\right)$$

At the lagrangian stage the advection of hydrodynamical quantities takes place and the equations has a following form:

$$\frac{\partial f}{\partial t} + \frac{1}{a}\nabla \cdot (f\vec{v}) = 0$$

where f is symbolic notation of the density ρ, n, momentum $\rho\vec{u}, n\vec{v}$, density of the full mechanical $\rho E, nW$ or internal $\rho\epsilon, \Pi_{\xi\xi}$ energy of the gas. To solve the equation we use the same approach,

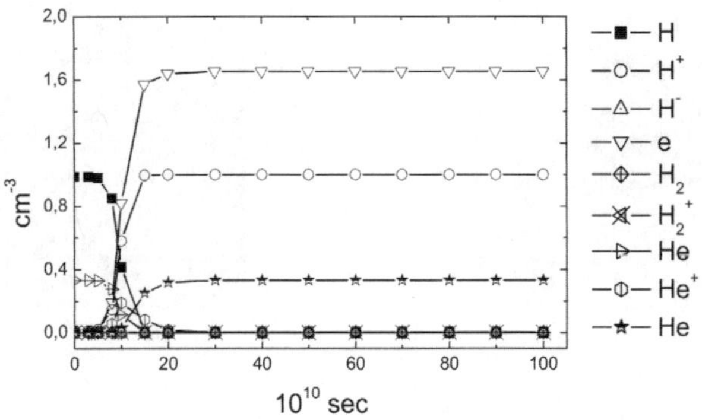

Figure 1. Behaviour of concentration during the simulation.

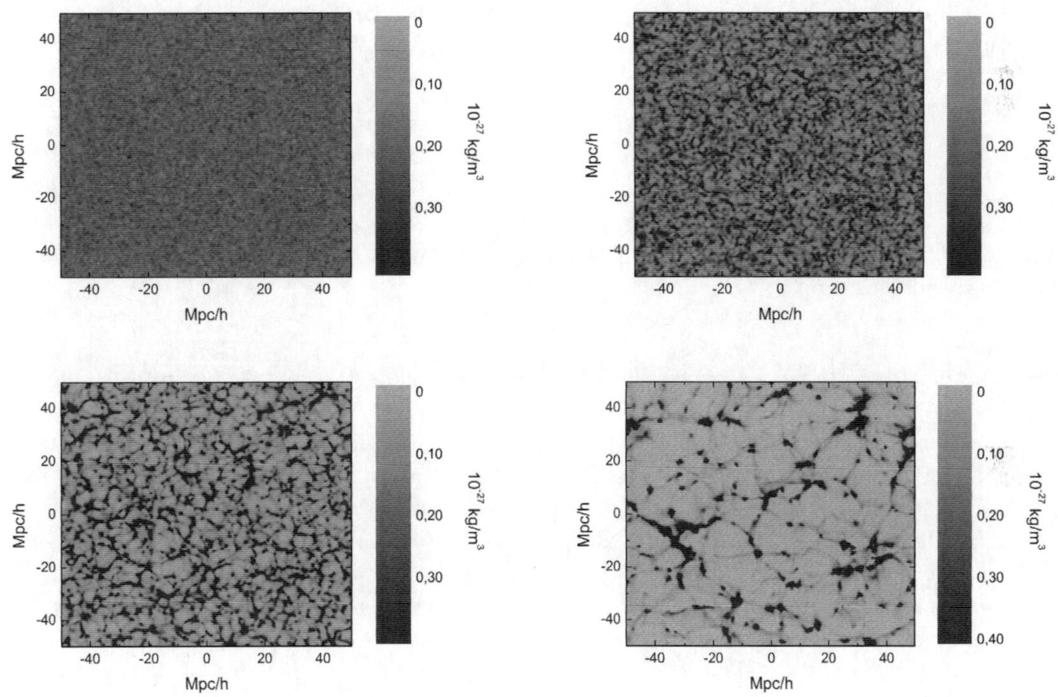

Figure 2. The density of dark matter in $z = 2$ (top left), $z = 1$ (top right), $z = 0.4$ (bottom left), $z = 0$ (bottom right).

that is used at the eulerian stage. To compute the flow $F = f\vec{v}$ with $\lambda = |\vec{v}|$ the following formula is used:

$$F = v \times \begin{cases} f_L(-\frac{\lambda\tau}{a}), v \geq 0 \\ f_R(\frac{\lambda\tau}{a}), v < 0 \end{cases} \qquad v = \frac{v_L\sqrt{\rho_L} + v_R\sqrt{\rho_R}}{\sqrt{\rho_L} + \sqrt{\rho_R}}$$

where $f_L(-\frac{\lambda\tau}{a})$ $f_R(\frac{\lambda\tau}{a})$ – piece-wise parabolic function for the quantity f. To construct the piece-wise parabolic solution the similar procedure is used. At the final stage of solving of the

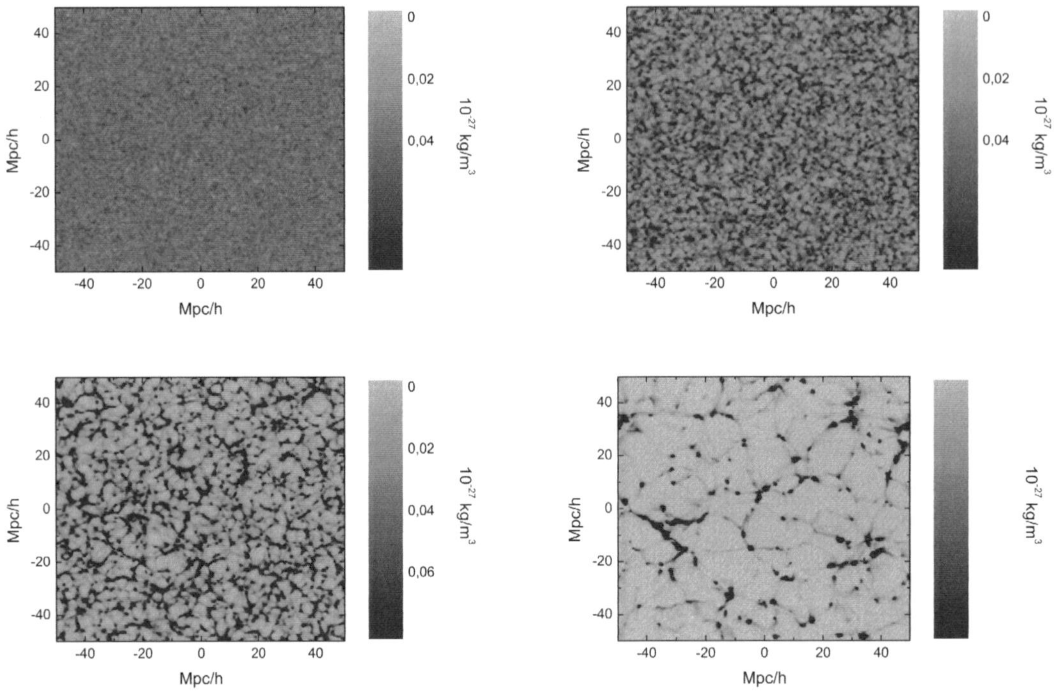

Figure 3. The density of baryon matter in $z = 2$ (top left), $z = 1$ (top right), $z = 0.4$ (bottom left), $z = 0$ (bottom right).

hydrodynamical equations the correction of solution is done. At the gas-vacuum interface [34] the correction of velocity takes place, in the remaining region the internal energy is corrected, that guarantees nondecreasing entropy [35]. Such modification provides detailed balance of energies and guarantees nondecreasing entropy. Method of solving of the Poisson equation is based on Fast Fourier Transform. Geometrical decomposition of computational domain is in a basis of parallel implementation. Unified numerical method allow us to use the same scheme of geometrical decomposition, that leads us to more simple and effective implementation. The numerical experiments on the study of performance of the CosmoPhi code on the cluster RSC PetaStream were carried out. As a result, 134-fold acceleration (strong scalability) on one accelerator Intel Xeon Phi, 77% of efficiency (weak scalability) with using of 224 accelerators Intel Xeon Phi (more than 50 000 logical cores), and 40% of peak performance with using of scalar computations on the accelerator Intel Xeon Phi, were achieved. Using of simulation of behaviour of programm implementation [36] on such architecture had shown 80% of efficiency on up to 983040 cores, that corresponds to exaflop computational level.

4. Results of cosmological simulation

From the moment of time, corresponding to $z = 99$, we will deal with expanding cubic region with the lenght of cube $L = 100/h$ Mpc $= 3 \times 10^{23}/h$ m and periodic border conditions on each dimension. As typical value of the density we will get $\rho = 1.88 \times 10^{-26} h^2$ kg/m^3. Portion of the dark energy $\Omega_\Lambda = 0.73$, dark matter $\Omega_D = 0.226$, visible baryonic matter $\Omega_B = 0.044$ (in the beginning a lack of stars is presumed). The temperature of the gas component is $T = 10^4$ K. The Hubble constant is $H = 67.8$ km/s/Mpc. To set the initial conditions the small fluctuations with uniform density distribution are used. To set the random fluctuations the normal distribution

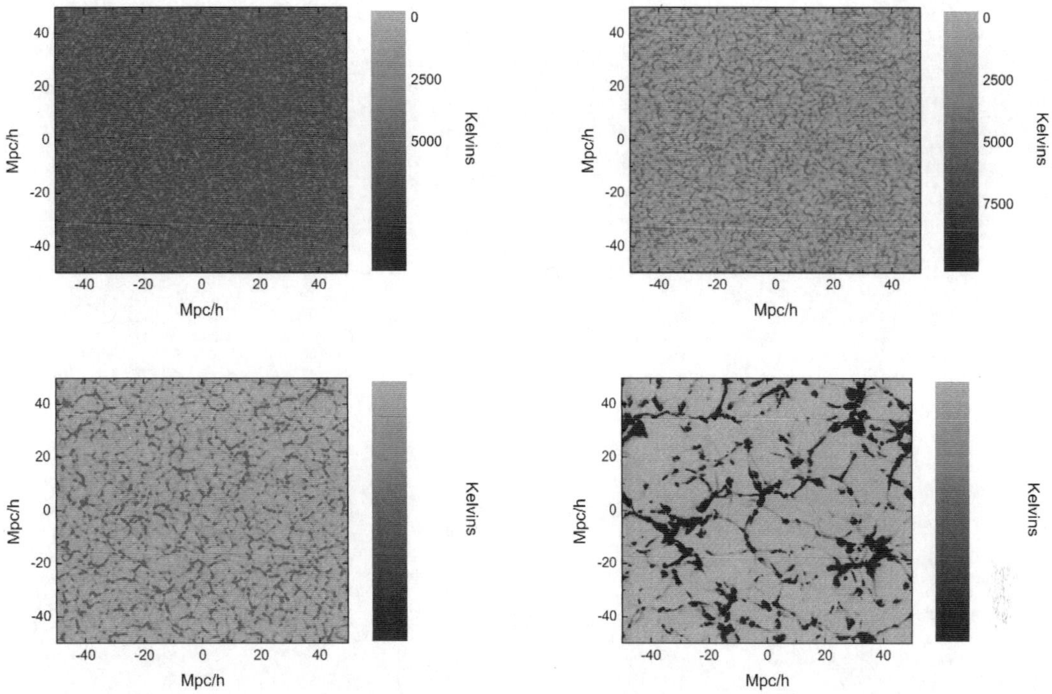

Figure 4. The temperature of gas in $z = 2$ (top left), $z = 1$ (top right), $z = 0.4$ (bottom left), $z = 0$ (bottom right).

with the range, corresponding to energetic spectrum (equation (25) in the paper [37]), is formed. Then the inverse Fourier transform is performed.

To study the need of using of equilibrium chemokinetics in problems of cosmological simulation the typical concentrations of different forms of hydrogen and gelium [29, 30, 31] with a temperature $T = 10^4$ K were taken. Then, with using of software package ChemPAK [38] developed earlier, the numerical experiment, which had shown that all of the chemical reactions happens for about a ten of thousands of years, was carried out. The model of equilibrium chemokinetics, with consideration that typical timestep in problems of cosmological simulation is equal to one million of years, was chosen and implemented.

Formation of large-scale cosmological structures – filaments, pancakes, clusters of galaxies, voids – within two-phase multicomponent hydrodynamical model with consideration of cosmological expansion and subgrid processes was simulated. These large-scale objects are formed as a result of evolution of instability. Then individual cluster of galaxies was selected with using of zoom-in approach, approximate exponential profile of the density of gas and collisionless components was build and numerical experiment in the region containing only this cluster was carried out. As a result of the numerical experiment, qualitative compliance of structure of simulated and observed clusters, quantitative compliance of masses of simulated and observed galaxies and distances between them, were shown.

Also the behavior of starburst process as a result of cosmological simulation was studied. Simulated value of fraction of the stars in the total mass corresponds to the observed value. Also the process of active starburst begins with $z = 2$, that corresponds to the observed data.

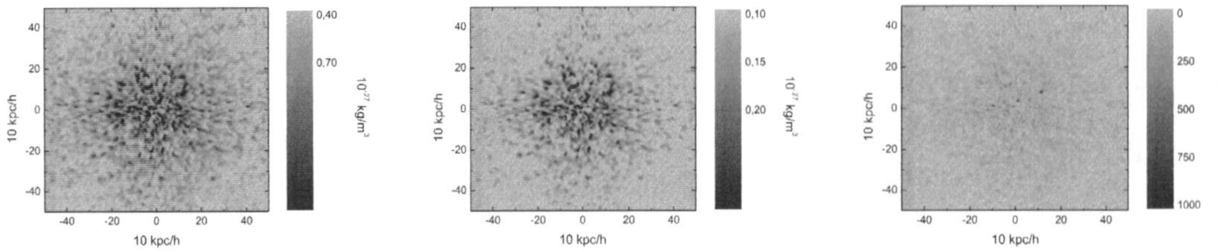

Figure 5. The density of dark matter (left) and baryon matter (center), temperature of gas (right).

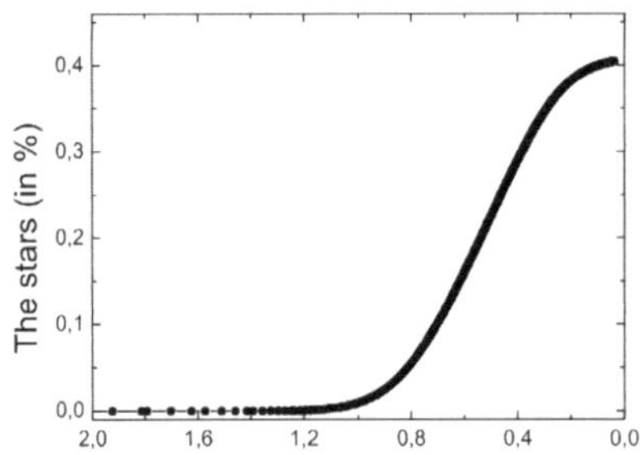

Figure 6. Starburst behaviour during the simulation.

5. Conclusion

The new hydrodynamical model of formation of galaxies in cosmological context within multiphase hydrodynamical model, based on the joint solving of equations of multicomponent gas dynamics, first momenta of collisionless Boltzmann equation and Poisson equation for the potential, with consideration of cosmological expansion and subgrid processes, is shown in paper. The mathematical model of problem of cosmological simulation and numerical methods for solving of hyperbolical equations are described in details, also parallel implementation as software complex CosmoPhi is given briefly in the paper. The using of such model allowed us to formulate the unified numerical method for solving of hyperbolical equations. This numerical method is implemented effectively on a heterogeneous supercomputer equipped with accelerators Intel Xeon Phi. As a result, 134-fold acceleration on one accelerator Intel Xeon Phi, 77% of efficiency with using of 224 accelerators Intel Xeon Phi, and 40% of peak performance with using of scalar computations on the accelerator Intel Xeon Phi, were achieved. The results of numerical experiments on large-scale cosmological simulation are presented.

Acknowledgments

The research was financially supported by RFBR within the scientific projects no. 15-31-20150 mol-a-ved and no. 15-01-00508, as well as by the grant of the President of the Russian Federation MK-6648.2015.9.

Appendix

Cooling function (here and further in ergs $\mathrm{cm}^{-3}\ \mathrm{s}^{-1}$ [29, 30, 31]) for collisional reactions:

$$Q_1^{col} = 7.50 \times 10^{-19} \left(1 + \sqrt{T/10^5}\right)^{-1} \exp^{-118349/T} \times n_e n_H$$

Table 1. Coefficients of rates of reactions in form $A(N) = A \times 10^{-N}$

N	k_1	k_2	k_4	k_7	k_9	k_{10}	k_{17}	k_{19}
A_0	-32.71	-28.61	-20.07	-24.24	-18.01	-20.37	-44.09	-68.71
A_1	13.53	-7.24(1)	2.28(1)	3.40	2.36	1.13	23.91	43.93
A_2	-5.73	-2.02(2)	3.59(2)	-3.89	-2.82(1)	-1.42(1)	-10.75	-18.48
A_3	1.56	-2.38(3)	-4.55(3)	2.04	1.62(2)	8.46(3)	3.05	4.71
A_4	-0.28	-3.21(4)	-3.10(4)	-5.41(1)	-3.36(2)	-1.43(3)	-5.68(1)	-7.69(1)
A_5	3.48(2)	1.42(5)	1.07(5)	8.41(2)	1.17(2)	2.01(4)	6.79(2)	8.11(2)
A_6	-2.63(3)	4.98(6)	-8.36(6)	-7.87(3)	-1.65(3)	8.66(5)	-5.01(3)	-5.32(3)
A_7	1.12(4)	5.75(7)	2.23(7)	4.13(4)	1.06(4)	-2.58(5)	2.06(4)	1.97(4)
A_8	-2.04(6)	-1.85(8)	0	-9.36(6)	-2.63(8)	2.45(8)	-3.64(6)	-3.16(6)

$$Q_2^{col} = 9.10 \times 10^{-27} \left(1 + \sqrt{T/10^5}\right)^{-1} T^{-0.1687} \exp^{-13179/T} \times n_e^2 n_{He}$$

$$Q_3^{col} = 5.54 \times 10^{-17} \left(1 + \sqrt{T/10^5}\right)^{-1} T^{-0.397} \exp^{-473638/T} \times n_e n_{He^+}$$

for reactions of ionizations:

$$Q_1^{ion} = 2 \times 10^{-11} \times k_1 n_e n_H \qquad Q_2^{ion} = 4 \times 10^{-11} \times k_{17} n_e n_{He} \qquad Q_3^{ion} = 9 \times 10^{-11} \times k_{19} n_e n_{He^+}$$

$$Q_4^{ion} = 5 \times 10^{-27} \left(1 + \sqrt{T/10^5}\right)^{-1} T^{-0.1687} \exp^{-55338/T} \times n_e^2 n_{He^+}$$

for reaction of recombination:

$$Q_1^{rec} = 8.71 \times 10^{-27} (T/1000)^{-0.2} \left(1 + \left(T/10^6\right)^{0.7}\right)^{-1} \times n_e n_{H^+}$$

$$Q_2^{rec} = 1.55 \times 10^{-26} T^{0.3647} \times n_e n_{He^+}$$

$$Q_3^{rec} = 1.24 \times 10^{-13} T^{-1.5} \left(1 + 0.3 \exp^{-94000/T}\right) \exp^{-94000/T} \times n_e n_{He^+}$$

$$Q_4^{rec} = 3.48 \times 10^{-26} \sqrt{T} (T/1000)^{-0.2} \left(1 + \left(T/10^6\right)^{0.7}\right)^{-1} \times n_e n_{He^{++}}$$

where n_H – concentration of atomic hydrogen, n_{He} – atomic helium, n_e – electrons, n_{H_2} – molecular hydrogen, n_{H^+} – ionized hydrogen, n_{He^+} – ionized helium, $n_{He^{++}}$ – twice ionized helium. We will write the speeds of chemical reaction in the following form $k = \exp^{\sum_{i=0}^{8} A_i \times \log^i(T)}$. Coefficients A_i for reactions will be written in the table (1).

References

[1] Borgani S and Kravtsov A 2011 *Advanced Science Letters* **4** 204–27
[2] Kim J, Park C, Richard Gott III J and Dubinski J 2009 *The Astrophysical Journal* **701**(2) 1547–59
[3] Springel V et al 2005 *Nature* **435** 629–36
[4] Teyssier R et al 2009 *Astronomy & Astrophysics* **497**(2) 335–41
[5] Klypin A, Trujillo-Gomez S and Primack J 2011 *The Astrophysical Journal* **740**(2,102) 1–17
[6] Genel S et al 2014 *Monthly Notices of the Royal Astronomical Society* **445**(1) 175–200
[7] Vogelsberger M et al 2014 *Monthly Notices of the Royal Astronomical Society* **444**(2) 1518–47
[8] Vogelsberger M et al 2014 *Nature* **509** 177–82
[9] Crain R et al 2015 *Monthly Notices of the Royal Astronomical Society* **450**(2) 1937–61
[10] Schaye J et al 2015 *Monthly Notices of the Royal Astronomical Society* **446**(1) 521–54
[11] Chilingarian I, Di Matteo P, Combes F, Melchior A and Semelin B 2010 *Astronomy & Astrophysics* **518**(61) 1 –14

[12] Tutukov A, Lazareva G and Kulikov I 2011 *Astronomy Reports* **55**(9) 770–83
[13] Schweizer F 2005 *Astrophysics and Space Science Library* **329** 143–52
[14] Sol Alonso M, Lambas D, Tissera P and Coldwell G 2007 *Monthly Notices of the Royal Astronomical Society* **375**(3) 1017–24
[15] Blecha L, Loeb A and Narayan R 2013 *Monthly Notices of the Royal Astronomical Society* **429**(3) 2594–616
[16] Rodriguez C, Taylor G, Zavala R, Pihlstrom Y and Peck A 2009 *The Astrophysical Journal* **697**(1) 37–44
[17] Combes F and Melchior A 2002 *Astrophysics and Space Science* **281**(1-2) 383–7
[18] Aoki W 2012 *Astronomical Society of the Pacific Conference Series* **458** 1–435
[19] Kulikov I 2014 *The Astrophysical Journal Supplements Series* **214**(12) 1–12
[20] Kulikov I M, Chernykh I G, Snytnikov A V, Glinskiy B M and Tutukov A V 2015 *Computer Physics Communications* **186** 71–80
[21] Godunov S K 1959 *Mathematichesky Sbornik* in Russian **47** 271–306
[22] Popov M and Ustyugov S 2007 *Computational Mathematics and Mathematical Physics* **47**(12) 1970–89
[23] Popov M and Ustyugov S 2008 *Computational Mathematics and Mathematical Physics* **48**(3) 477–99
[24] Mitchell N, Vorobyov E and Hensler G 2013 *Monthly Notices of the Royal Astronomical Society* **428** 2674–87
[25] Vorobyov E, Recchi S and Hensler G 2015 *Astronomy & Astrophysics* **579**(A9) 1–22
[26] Katz N, Weinberg D and Hernquist L 1996 *The Astrophysical Journal Supplement Series* **105** 19–35
[27] Springel V and Hernquist L 2003 *Monthly Notices of the Royal Astronomical Society* **339**(2) 289–311
[28] Sutherland R and Dopita M 1993 *The Astrophysical Journal Supplement Series* **88** 253–327
[29] Abel T, Anninos P, Zhang Y and Norman M 1997 *New Astronomy* **2**(3) 181–207
[30] Anninos P, Zhang Y, Abel T and Norman M 1997 *New Astronomy* **2**(3) 209–24
[31] Grassi T, Bovino S, Schleicher D, Prieto J, Seifried D, Simoncini E and Gianturco F 2014 *Monthly Notices of the Royal Astronomical Society* **439**(3) 2386–419
[32] Collela P and Woodward P R 1984 *Journal of Computational Physics* **54** 174–201
[33] Kulikov I and Vorobyov E 2016 *Journal of Computational Physics* **317** 318–46
[34] Vshivkov V, Lazareva G, Snytnikov A, Kulikov I and Tutukov A 2011 *Journal of Inverse and Ill-posed Problems* **19**(1) 151–66
[35] Godunov S and Kulikov I 2014 *Computational Mathematics and Mathematical Physics* **54**(6) 1012–24
[36] Podkorytov D, Rodionov A, Sokolova O and Yurgenson A 2010 *Lectures Notes of Computer Science* **6235** 247–50
[37] Eisenstein D and Hu W 1999 *The Astrophysical Journal* **511**(1) 5–15
[38] Chernykh I, Stoyanovskaya O and Zasypkina O 2009 *Chemical Product and Process Modeling* **4**(4) 1934–2659

Autowave process of the localized plastic deformation of high-chromium steel saturated with hydrogen

A V Bochkareva[1,2], S A Barannikova[1,3,4], Yu V Li[1], A G Lunev[1,2] and L B Zuev[1,3]

[1] Institute of Strength Physics and Materials Science, SB RAS, Tomsk, 634055, Russia
[2] National Research Tomsk Polytechnic University, Tomsk, 634050, Russia
[3] National Research Tomsk State University, Tomsk, 634050, Russia
[4] Tomsk State University of Architecture and Building, Tomsk, 634003, Russia

E-mail: avb@ispms.tsc.ru

Abstract. The deformation behavior of high-chromium stainless steel of sorbitic structure upon high-temperature tempering and of electrically saturated with hydrogen in the electrochemical cell during 12 hours is investigated. The stress-strain curves for each state were obtained. From the stress-strain curves, one can conclude that hydrogen markedly reduces the elongation to the fracture of specimen. Using double-exposed speckle photography method it was found that the plastic flow of the material is of a localized character. The pattern distribution of localized plastic flow domains at the linear hardening stage was investigated. Comparative study of autowave parameters was carried out for the tempered steel as well as the electrically saturated with hydrogen steel.

1. Introduction

In recent years, we have found experimentally for a range of metals and alloys that plastic deformation at all its stages is prone to localization [1,2]. The localization patterns observed are wave-like in character; the kind of wave depends on the acting law of work hardening, that is, the form of function $\theta(\varepsilon)$ [3]. There are only four of them: single deformation fronts move in the easy glide stage for single crystals and in a yield plateau for polycrystals; a system of equidistant mobile localized-strain centers in the linear hardening stage; spacio-periodic structure of plastic strain localization regions in the Taylor parabolic work hardening stage, and the structure of nonuniform mobile strain-localization regions peculiar to the final (prefracture) deformation stage.

These surveys of the localized-strain pattern features for the high-chromium stainless steel widely used in water-vapor, aquatic and acid environments, provide means of getting more information about material behavior being part of a critical structure. For example, a chemical reactor or oil and gas equipment made of high-chromium stainless steel operate in the presence of sever atmosphere of hydrogen. The presence of hydrogen (H) in solid solution of metals and alloys is related mainly to the small diameter of this element and its capacity to diffuse with certain ease in solid state. Different factors contribute to elevate or diminish the solubilization and/or diffusion of H in steels. The main ones are temperature, alloy composition, crystalline structure and substructure. Nevertheless, the presence of H in metals, particularly in steels, is not desired in most of the cases, since H alters considerably the mechanical–metallurgic properties of these materials with the possibility of fracture [4-8]. The main aim of this investigation was to elucidate the effect of dissolved hydrogen on the macroscopic plastic flow localization patterns in tensile strained stainless steel polycrystals.

2. Test material and investigation technique

The investigation was performed using specimens of high-chromium stainless steel (0.4%C–0.6%Si–0.55%Mn–12.5%Cr). The specimens were configured as dog bones with the gage section measuring 50×10×2 mm. The test pieces were cut out from sheet steel along the rolling direction. According to condition of supply, this steel has a ferrite-carbide structure; therefore, it was subjected to heat treatment [9, 11]. The specimens were quenched at T = 1320 K for 3 hours by fast air cooling to provide a martensitic high-strength structure with a small amount of chromium carbides. After that, in order to increase plasticity, the specimens were under high-temperature tempering from T = 870 K for 3 hours and furnace cooling, so that sorbitic structure (ferrite and carbides) was generated (figure 1).

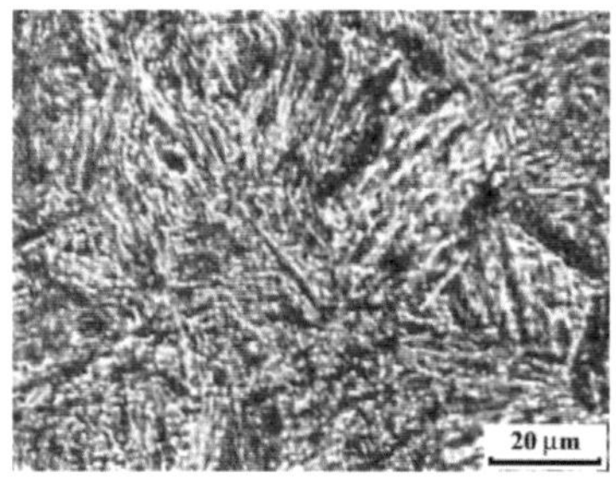

Figure 1. Microstructure of high-chromium steel in tempered state.

Prepared in this manner, the specimens were electrically saturated with hydrogen. The electrolytic hydrogenation of alloy samples was carried on for 12 and 24 hours under a controlled cathode potential of –600 mV relative to a reference electrode of silver chloride in the 0.1N solution of sulphuric acid with addition of 20 mg/L of thiourea [10]. The current-voltage curves were recorded using an IPC-Compact potentiostat. The time between hydrogen charging and tensile testing did not exceed 30 min and the testing time was less than 60 min. The specimens were subjected to uniaxial tension at room temperature at the rate of $6.67 \cdot 10^{-5} s^{-1}$ using a universal testing machine LFM-125 (Switzerland). The stress-strain curves for the material are shown in figure 2. The plastic flow curves plotted for the samples tested in tension were analyzed to single out a linear-hardening flow stage, whereas work hardening coefficient $\theta = const$. Concurrently, the displacement vector fields of points on the surface of the specimens were recorded by a double-exposure speckle detailed in [3]. An analysis of the spatial-periodical distribution of local elongation ε_{xx} of the specimen observed in the linear-hardening stage makes it possible to estimate the autowave spacing λ as well as velocity $V_{aw} = dX / dt$.

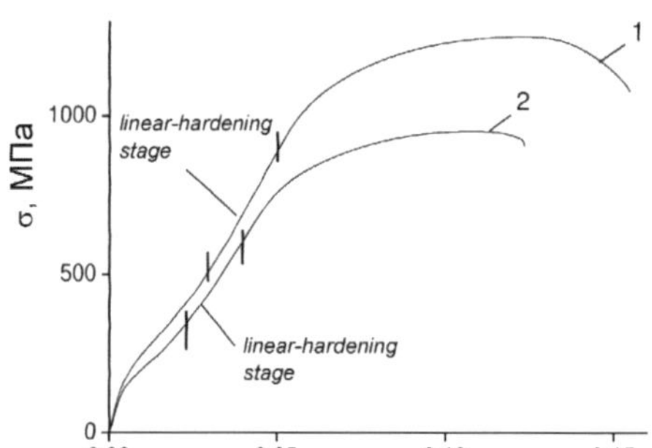

Figure 2. Stress-strain curves for high-chromium steel in the tempered state (1) and after electrolytic hydrogenation for 12 h (2).

The deformation diagrams for the alloy under review refer to the general type diagrams, therefore can be described by the Lüdwick equation [12]

$$s = s_0 + \theta e^n \quad (1)$$

where s is the true stress and can be calculated as $s = \sigma(1+\varepsilon)$, e is the true strain calculated as $e = \ln(1+\varepsilon)$, $s_0 \approx \sigma_y$ is the true critical share stress, $\theta = d\sigma/d\varepsilon$ is the work hardening coefficient expressed in MPa units, as it is a stress, and $n \le 1$ is the work hardening exponent.
In different portion of the stress-strain curve constructed in $\ln(s - s_0) - \ln e$ coordinates, the strain exponent has different values, but for the certain deformation stage, there are particular constant values of n and θ. For the linear work-hardening stage, n should be equal to 1.

3. Linear work-hardening stage and autowaves of plasticity

Hydrogen charging for 12 hours (2) results in remarkable changes of stress-strain curves as shown in figure 2. It is found that a 35 % decrease in the yield stress and 26 % decrease in the ultimate stress are observed for the counterpart subjected to hydrogenation for 12 hours (2) relative to the tempered samples of steel (1). From the obtained stress-strain curves, one can conclude that hydrogen markedly reduces the elongation to the fracture of specimen. The results obtained for the high-chromium steel electrically saturated with hydrogen for 24 hours [13] show the further decrease in mechanical properties of the material tested. It is found a 70 % decrease in the yield stress and 40 % decrease in the ultimate stress relative to the tempered samples of steel. The mechanical properties of the studied steel in different states are summarized in table 1.

One can see from the table 1 that the stress-strain curves of steel in all three states were found to include the linear work-hardening stage.

Table 1. Mechanical properties of stainless steel in different states.

	σ_y, MPa	σ_B, MPa	δ, %	θ, GPa	Linear work-hardening stage, $n \approx 1$	
					ε_{in}	ε_{tot}
Tempered	218	1251	12.9	19.3	0.03	0.054
12 h hydrogenation	141	922	9.9	14.7	0.024	0.037
24 h hydrogenation	73	798	4.7	14.6	0.047	0.058

The plastic-strain localization patterns for each of the stress-strain curves were studied by the double-exposure speckle photography. The displacement-field data for strained specimens were transformed into spatial distributions of local elongations ε_{xx}, using numerical differentiation as it shown in figure 3 for the tempered stainless steel.

So, if the distribution is spatial, the spatial period (λ) can be determined (figures 3 and 4). It should be noted that the spatial period of local elongation distribution significantly decreased after electrolytic hydrogenation of tempered steel specimen in the three electrode electrochemical cell. As a result, the localized-strain centers became clearly visible, and the coordinate of localized maximum can be determined.

As the local distributions of elongation component ε_{xx} in the linear-hardening stage in the tempered steel and in the steel saturated with hydrogen are of the periodic spatio-temporal structure and one have the sequence of these distributions in time, the positions of the strain centers as a function of the loading time can be plotted, as it shown in figure 5 for the high-chromium steel in the studied states. The sequence of coordinates for each of the localized plastic deformation domains in this stage is approximated by nearly parallel straight lines, where the slope of the curves enables the velocity of the domains to be estimated $V_{aw} = dX/dt$.

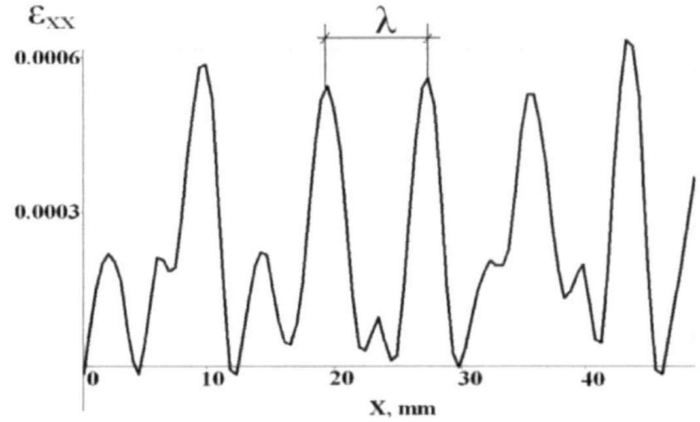

Figure 3. Distribution of local elongation component over tension axial: tempered steel in the linear-hardening stage

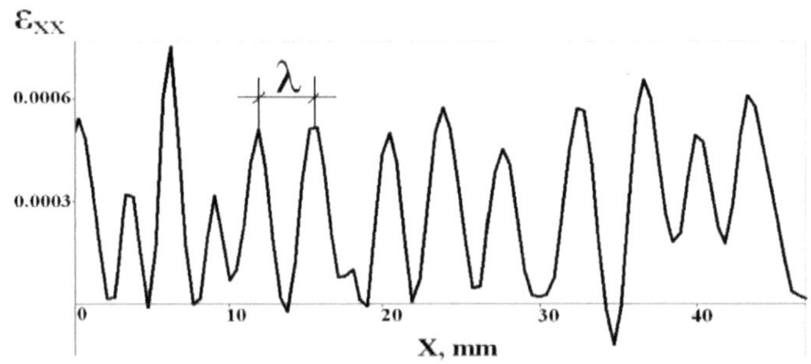

Figure 4. Distribution of local elongation component over tension axial: tempered steel saturated with hydrogen (12h) in the linear-hardening stage

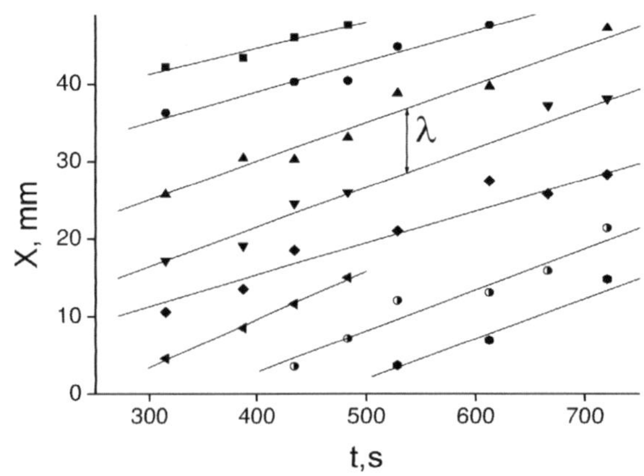

Figure 5. Mobile localized centres of local elongation component for the tempered high-chromium stainless steel.

Figure 4 shows the spatial distributions of local elongations ε_{xx}, for the steel saturated with hydrogen for 12 hours. Thus, all strain centers are seen to move uniformly at nearly the same velocities to generate the phase autowave [3, 13]. The autowave velocity V_{aw} was determined from the slopes of the

X(t) curves, whereas the autowave length (λ) was found from the curve spacing measured along the axis X. For the tempered specimen, the autowave length and autowave velocity are λ=6.8±1.5 mm and V_{aw}=(4.9±0.9)×10^{-5} m/s, correspondingly. As for the steel specimen after electrolytic hydrogenation, λ=4.6±0.5 mm and V_{aw}=(4.0±0.7)×10^{-5} m/s. The autowave velocity and length for the states studied are given in Table 2.

Table 2. The characteristics of localized-strain autowaves.

	λ, mm	$V_{aw}\times10^5$ m/s
Tempered	6.8±1.5	4.9±0.8
12 h hydrogenation	4.6±0.5	4.0±0.7
24 h hydrogenation	3.5±0.5	3.9±0.8

The comparing analysis of spatial distribution of local elongation for the specimen observed in the linear-hardening stage for the tempered state was conducted using the standard statistical processing data obtained such Student's t-criteria [15]. According to this, experimental results should be intercompared by the equation

$$t=(\bar{x}_1-\bar{x}_2)\cdot\left(\frac{(n_1-1)\cdot s_1^2+(n_2-1)\cdot s_2^2}{n_1+n_2-2}\right)^{-\frac{1}{2}}\cdot\left(\frac{n_1\cdot n_2}{n_1+n_2}\right)^{\frac{1}{2}} \tag{2}$$

where $\bar{x}_1,\bar{x}_2$ are the average values of two sampled populations, n_1,n_2 – the units in the sample, s_1,s_2 – the dispersion.

Using Student's t-criteria for the certain confidence level $\alpha=0.95$, one must compare $|t|$ calculated with the tabulated value of $t_{\alpha.f}$. If the calculated value exceeds the tabulated one, it means that two average values differ significantly.

The resultant quantity $|t|\geq t_{\alpha.f}$ is true for the value of the autowave length λ. This fact suggests that the average λ-values obtained for the tempered alloy and the hydrogenated one are quite different from each other. For the autowave velocity V_{aw} the $|t|\geq t_{\alpha.f}$ expression is not justified, so one can say that the velocity of autowave propagation is almost not change.

It is found experimentally that the motion rate of localization nuclei observed at the stage of linear work hardening is inversely proportional to the work hardening coefficient. This dependence is common to all the metals and alloys investigated [3, 16]. The shape of this dependence (V_{aw} ~ 1/θ) indicates that the wave processes in question differ in nature from the well-known plasticity waves (Kolsky waves) for which the dependence has the following form V_{pl} ~ √θ (here ρ is density). The distinction between the above two dependences leads one to conclude that a new type of wave processes is discovered. They are self-excited waves of the plastic flow. The new type waves will be generated in the deforming specimen, irrespectively, of electrolytic hydrogenation [10, 13, 14].

4. Conclusion

The deformation behavior of high-chromium stainless steel subjected to high-temperature tempering and saturated with hydrogen in three electrode electrochemical cell at controlled constant cathode potential for 12 and 24 hours was investigated. The mechanical properties of the steel in three states were detected. The hydrogen charging for 12 hours results in remarkable changes of yield stress, ultimate stress and total elongation of material under study. The stress-strain curves of the examined states of the stainless steel are characterized by the linear work-hardening stage with constant values of $n\approx1$ and θ. Because of this, phase autowave of plasticity occurs for steel in the states mentioned above.

Hydrogen not only provides the remarkable softening effect on the high-chromium stainless steel but it also initiates the significant reconstruction in the spatial scale for distribution and magnitude of plastic strain. The change in the microstructure of hydrogen charged tempered steel affects the stress-strain curves as well as the plastic strain localization patterns.

The comparison of the data for two conditions (tempered condition and after hydrogenation condition) of steel showed that the hydrogen significantly enhances localization and changes the quantitative parameters of the macroscopic plastic strain localization, particularly, the wave length of autowave plastic strain localization. The mechanism of hydrogen-stimulated plastic strain localization is still under discussion.

Acknowledgments
The work was performed in the frame of the Tomsk State University Academic D.I. Mendeleev Fund Program, of the Program of Fundamental Research of State Academies of Sciences for the period 2013-2020 yrs. and was partially supported by the Russian Foundation for Basic Research (Project No. 16-08-00385-a)

References

[1] Fressengeas C, Beaudoin A, Entemeyer D, Lebedkina T, Lebyodkin M and Taupin V 200 *Phys. Rev.* B **79** 014108
[2] Asharia A, Beaudoin A and Miller R 2008 *Math. Mech. Sol.* **13** 292–315
[3] Zuev L B and Barannikova S A 2014 *Int. J. Mech. Sci.* **88** 1–8
[4] Hudson R M and Stragand G L 1960 *Corros.* **16** 253–57
[5] Hirth J P 1980 *Met. Trans.* **11A** 861–90
[6] Fuchigami H, Minami H and Nagumo M, 2006 *Phil. Mag. Lett.* **86** 21–9
[7] Sofronis P, Liang Y and Aravas N 2001 *Eur. J. Mech. A Solid.* **20** 857–72
[8] Robertson I M 2001 Eng. *Frac. Mech.* **68** 671–92
[9] Pelleg J 2013 *Mechanical properties of materials* (New York, London: Springer–Dordrecht–Heidelberg) p 620
[10] Zuev L B, Barannikova S A , Nadezhkin M V and Mel'nichuk V A 2011 *Techn. Phys. Lett.* **37** 793–6
[11] Firger I V 1982 *Thermal treatment of allows. Reference book.* (Leningrad: Mashinostroyenie Publ.) p 304
[12] Trefilov V I, Moiseev V F, and E P Pechkovskii 1989 *Strain Hardening and Fracture of Polycrystalline Metals* (Kiev: Naukova Dumka)
[13] Bochkareva A V, Barannikova S A, Lunev A G, Li Yu V and Zuev L B 2015 *Proc. of Int. Conf. on Mechanical Engineering, Automation and Control Systems* (Tomsk: IEEE) pp 1–3
[14] Barannikova S A, Lunev AG, Nadezhkin M V and Zuev L B 2014 *Adv. Mater. Res.* **880** 42–7
[15] Hudson D J 1964 *Statistics* (Geneva: CERN)
[16] Barannikova S A 2004 *Tech. Phys. Lett.* **30** 338–40

Journal of Physics: Conference Series **722** (2016) 012025 doi:10.1088/1742-6596/722/1/012025

Numerical modeling of 1D heterogeneous combustion in porous media under free convection taking into account dependence of permeability on porosity

N A Lutsenko [1,2]

[1] Institute of Automation and Control Processes FEB RAS, Vladivostok, 690041, Russia
[2] Far Eastern Federal University, Vladivostok, 690950, Russia

E-mail: NickL@inbox.ru

Abstract. Using numerical experiment the one-dimensional unsteady process of heterogeneous combustion in porous object under free convection is considered when the dependence of permeability on porosity is taken into account. The combustion is due to exothermic reaction between the fuel in the solid porous medium and oxidizer contained in the gas flowing through the porous object. In the present work the process is considered under natural convection, i.e. when the flow rate and velocity of the gas at the inlet to the porous objects are unknown, but the gas pressure at object boundaries is known. The influence of changing of permeability due to the changing of porosity on the solution is investigated using original numerical method, which is based on a combination of explicit and implicit finite-difference schemes. It was shown that taking into account the dependence of permeability on porosity, which is described by some known equations, can significantly change the solution in one-dimensional case. The changing of permeability due to the changing of porosity leads to the speed increasing of both cocurrent and the countercurrent combustion waves, and to the temperature increasing in the combustion zone of countercurrent combustion wave.

1. Introduction

Heterogeneous combustion in porous media is quite common in nature. From the point of view of mechanics the porous media include peat, coal dumps, landfills, debris of destroyed buildings and so on. Heterogeneous combustion of porous objects may appear as a result of natural and man-made disasters: explosions at atomic and industrial facilities, destruction of buildings involving fires, underground explosions, peat fires, spontaneous self-ignition of solid waste dumps (landfills) and so on. In such porous objects the filtration combustion takes place under natural convection. In this case the flow rate of gas passing through the porous object is unknown and the flow rate of oxidant, which enters into the reaction zone in porous object, regulates itself. One type of heterogeneous combustion in porous media is smoldering as heterogeneous chemical reactions take place during smoldering.

There are a lot of publications devoted to both the solid porous media combustion and the gas combustion in filtration mode; general concepts and basic theory of filtration combustion were presented in the review paper [1]. Analytical investigations of counter-flow combustion of solid porous medium under free convection, when the gas and the combustion wave in porous object move in opposite directions, were carried out in [2]. The co-flow combustion of solid porous medium under free convection, when the gas moves in the same direction as the combustion wave, was investigated analytically in [3]. In [4] the ability of energy to be concentrated in front of a co-flow filtration combustion wave in a porous solid was analyzed. In [5] it was shown that reduction of the permeability in the combustion products can lead to hydrodynamic instability of combustion wave. In [6] a one-dimensional transient model of smoldering was presented and solved numerically. A novel computational model of smoldering combustion capable of predicting both the countercurrent and the cocurrent wave propagation was developed in [7]. A generalized pyrolysis model, Gpyro, for simulating the gasification of a variety of combustible solids encountered in fires was presented in [8]. In [9], [10] Gpyro was expanded to two-dimensional case. In [11] the one-dimensional unsteady process of heterogeneous combustion in porous object under free convection in the gravity field was investigated numerically using proposed mathematical model and an original numerical method. Two regimes of combustion wave propagation have been revealed – wave movement up the object (cocurrent burning) and down the object (countercurrent burning) – which differ significantly from each other by the degree of burnout of solid combustible material, the temperature in the combustion zone and the speed of combustion wave propagation. The details of the original numerical method were described in [12]. In [13] the effects of gravity field and pressure difference at the boundaries of the porous object on the appearance of stable heterogeneous combustion waves in object under free convection were investigated numerically.

This work is devoted to the numerical investigations of one-dimensional unsteady processes of heterogeneous combustion in porous object under free convection in the gravity field when the dependence of permeability on porosity is taken into account. The proposed mathematical model is briefly described, the model is similar to the one used in [11]–[13] and allows us to describe the processes both under free convection and forced filtration. The influence of the permeability changing due to the changing of porosity on the solution is investigated using numerical experiment.

2. Mathematical model and numerical method

Consider a homogeneous motionless porous object with a height H, which is bounded of impermeable non-heat-conducting side walls and opened on two opposite sides (at the top and at the bottom). In the porous medium the heterogeneous reactions take place that leads to the heat release. The cold gas can flow into the open walls of the porous object; the gas can flow through porous medium and flow out. Suppose that a solid porous substance consists of combustible and inert components, and at the same time the solid combustible material transforms into a gas in the reaction with gaseous oxidizer, so we have the following expression:

$$\text{Solid Fuel} + (\mu)\text{Oxidizer} \to (1+\mu)\text{Gaseous Product}, \quad (1)$$

where μ is the mass stoichiometric coefficient for oxidizer.

The model is based on the assumption of interacting interpenetrating continua [14] using the classical approaches of the theory of filtration combustion [1]–[3]. In the energy equation for the solid components not only heat generation is taken into account but also the thermal conductivity and the intensity of the interphase heat exchange which is assumed to be proportional to the difference of the phase temperatures at the considered point of the medium. In the energy equation for gas the thermal conductivity is not considered because of its smallness, and it is assumed that homogeneous reactions do not occur. For describing the dynamics of gas, the equation of momentum conservation for porous media is used, which is more correct than the classical Darcy's equation. The solid phase is assumed to be fixed, so the equation of motion for it degenerates. In the model the changes in volume and weight of the phases at their interaction are taken into account. Supposing that the diffusion of the oxidizer takes place, the perfect gas equation of state is valid. Combustion processes are described by one-step chemical reaction of first order with respect to both arguments (the mass concentration of oxidizer and the degree of conversion of the solid combustible component). As it was shown in [15], the allowance for the temperature dependence of gas viscosity in its motion through a porous heat-evolutional medium can change the solution both quantitatively and qualitatively, so we will assume that the dynamical viscosity of gas is determined by Sutherland's formula. Thus, the system of equations modeling the time-dependent gas flow in a porous object with zones of heterogeneous combustion is the following:

$$(\rho_{cf}c_{cf}+\rho_{ci}c_{ci})\frac{\partial T_c}{\partial t}=-\alpha(T_c-T_g)+Q\rho_{cf0}W+(1-a_g)\lambda_c\Delta T_c,$$

$$\rho_g c_{gp}(\frac{\partial T_g}{\partial t}+(\mathbf{v}_g\cdot\nabla)T_g)=\alpha(T_c-T_g),$$

$$\rho_g(1+\chi(1-a_g))(\frac{\partial \mathbf{v}_g}{\partial t}+(\mathbf{v}_g\cdot\nabla)\mathbf{v}_g)=-a_g\nabla p+\rho_g\mathbf{g}-a_g^2\frac{\mu_1}{k_1}\mathbf{v}_g-\rho_{cf0}W\mathbf{v}_g,$$

$$\frac{\partial\rho_g}{\partial t}+\nabla\cdot(\rho_g\mathbf{v}_g)=\rho_{cf0}W,\quad p=\rho_g RT_g/(a_g\cdot M),\tag{2}$$

$$\rho_g(\frac{\partial C}{\partial t}+(\mathbf{v}_g\cdot\nabla)C)=\nabla\cdot(\rho_g D_g\nabla C)-\mu\rho_{cf0}W-\rho_{cf0}WC,$$

$$D_g=D_{g0}(T_g/273)^b,\quad W=(1-\eta)Ck\exp(-E/(RT_c)),$$

$$\frac{\partial\eta}{\partial t}=W,\quad \rho_{cf}=(1-\eta)\rho_{cf0},\quad a_g=a_{g0}+a_{cf0}\eta,\quad \mu_1=c_{s1}\frac{T_g^{1.5}}{c_{s2}+T_g},$$

where a is the volume concentration, is the exponent in the expression for the diffusion coefficient, is the mass concentration of oxidizer, c is the specific heat, c_{s1} and c_{s2} are the constants in Sutherland's formula, D_g is the diffusion coefficient of gas, E is the activation energy, $\mathbf{g}$ is the gravity acceleration, k is the pre-exponential factor in the expression for the rate of reaction, k_1 is the permeability coefficient, is the molar mass of gas, p is the gas pressure, Q is the heat of reaction, R is the universal gas constant, t is the time, T is the temperature, $\mathbf{v}_g$ is the gas velocity, W is the rate of the chemical reaction, α is the constant determining the interphase heat transfer intensity, η is the degree of conversion of the combustible component of the solid medium, λ is the thermal conductivity, μ_1 is the dynamic viscosity of the gas, ρ is the effective density, χ is the coefficient, taking into account the inertial interaction of the phases in their relative motion [14]; subscripts: "0" denotes the initial moment, "c" denotes the condensed phase (solid medium), "i" denotes the inert component, "f" denotes the combustible component, "g" denotes the gas, "p" denotes values at constant pressure.

We assume that at the inlet of the porous object (at the lower open boundary where gas flows into the porous object) the gas pressure, the gas temperature and the mass concentration for the oxidizer are known. At the outlet (the upper open boundary where gas flows out the porous object) the pressure is known. The conditions of heat exchange at the inlet and outlet from the porous object and on the bounding impermeable walls are also known. A distinctive feature of the considered model is that the flow rate and gas velocity at the inlet to the porous object are unknown and have to be found from the solution of the problem. Thus, the boundary conditions for system (2) are as follows:

$$p|_{x\in G_1}=p_0(x),\quad \lambda\partial T_c/\partial n|_{x\in G_1}=\beta(T_{g0}-T_c|_{x\in G_1}),$$

$$T_g|_{x\in G_1}=T_{g0}\quad\text{and}\quad C|_{x\in G_1}=C_0\quad\text{if}\quad \mathbf{v}_g|_{x\in G_1}\cdot\mathbf{n}|_{x\in G_1}>0,\tag{3}$$

$$\partial T_g/\partial n|_{x\in G_1}=0\quad\text{and}\quad \partial C/\partial n|_{x\in G_1}=0\quad\text{if}\quad \mathbf{v}_g|_{x\in G_1}\cdot\mathbf{n}|_{x\in G_1}>0,$$

$$\partial T_c/\partial n|_{x\in G_2} = 0, \quad \partial T_g/\partial n|_{x\in G_2} = 0, \quad \mathbf{v}_g|_{x\in G_2} \cdot \mathbf{n}|_{x\in G_2} = 0,$$

where G_1 is the object boundary opened to the atmosphere, G_2 is the impermeable boundary of the object, $\mathbf{n}$ is the outward vector directed normally to G_1 or to G_2, β is the heat removal coefficient.

The permeability coefficient k_1 is usually assumed to be constant when modeling the filtration combustion in porous media. It is possible for the permeability of the combustion products to be equal to that of the initial porous substance, since permeability is determined by both the porosity and the structure of the porous medium. But typically, the combustion products are more permeable than the initial porous substance.

One of the earliest expressions for the dependence of permeability on porosity, called Kozeny equation, was proposed in [16]. A similar, but more detailed expression, called Ergun equation, was proposed in [17]. Using the Kozeny equation or neglecting the second-order term in the Ergun equation, we can write the following expression:

$$k_1 = k_{11}\frac{a_g^3}{(1-a_g)^2}, \tag{4}$$

where k_{11} is a constant. The Kozeny (or Kozeny-Carman) equation and the Ergun equation are widely used in filtration theory. But the following expression proposed in [18] have also been used in studying of some porous media:

$$k_1 = k_{10}\left(\frac{a_g}{a_{g0}}\right)^n, \tag{5}$$

where n is usually equal to 10.

In the present work we solve system (2) using equations (4) and (5) and compare the results with the case when the permeability coefficient is constant: $k_1 = k_{00}$. This investigation is carried out numerically by means of original computational algorithm which was proposed previously in [11] and is based on a combination of explicit and implicit finite-difference schemes. The details of the algorithm were described in [12]. It should be noted that used original algorithm is the development of another computational algorithm, which had previously been successfully used for the calculation of gas flows through porous objects with heat zones when pressure difference at the inlet and the outlet from the object is known [19]–[22].

3. Numerical investigation of 1D heterogeneous combustion taking into account the dependence of permeability on porosity

Consider the one-dimensional process of natural convection in vertical porous object, i.e. when the gas pressure at the object boundary opened to the atmosphere (at the top and at the bottom of porous object) is equal to the atmospheric pressure at assigned heights. Suppose that prior to the start time the gas flow in the object is absent and its temperature is equal to the ambient temperature T_{g0}. At initial time in the place of ignition, which is located either on the object bottom or on its top, the temperature of the solid phase reaches a value T_{c0} equal to or exceeding the self-ignition temperature T_{kr}, and burning is started. System (2) with boundary conditions (3) is solved in dimensionless variables which are introduced as follows: $\tilde{x} = x/H$, $\tilde{t} = t/t_*$, $\tilde{\mathbf{v}}_g = \mathbf{v}_g/v_*$, $\tilde{p} = p/p_*$, $\tilde{\rho}_g = \rho_g/\rho_*$, $\tilde{T}_c = T_c/T_*$, $\tilde{T}_g = T_g/T_*$, $\tilde{\rho}_f = \rho_f/\rho_{f0}$, $\tilde{W} = W/k$, $\tilde{D}_g = (T_*/273)^{-b}D_g/D_{g0}$; where subscript "*" denotes typical values or values at "normal" conditions; further in the text the tildes are removed.

In the calculations, unless otherwise stated, we use the following parameter values:

$$H = 10\,m, \quad t_* = 3600\,s, \quad v_* = 1\,m/s, \quad T_* = 300\,K, \quad p_* = 10^5\,Pa, \quad \rho_* = 1.2\,kg/m^3,$$

$$\rho_{cf0} = 1.1 \cdot 10^2 kg/m^3, \quad \rho_{ci} = 6.6 \cdot 10^2\, kg/m^3, \quad c_{cf} = 1.84 \cdot 10^3\, J/(kg\,K),$$
$$c_{ci} = 1.84 \cdot 10^3\, J/(kg\,K), \quad \alpha = 10^3\, J/(m^3\,K\,s), \quad c_{gp} = 10^3\, J/(kg\,K), \quad k_{00} = 10^{-8}\, m^2,$$
$$c_{s1} = 1.458 \cdot 10^{-6}\, kg/(m\,s\,K^{1/2}), \quad c_{s2} = 110.4\,K, \quad g = 9.8\,m/s^2, \quad \lambda_c = 1.2\,J/(m\,K\,s), \quad (6)$$
$$\beta = 10\,J/(m^2\,K\,s), \quad R = 8.31441\,J/(mole\,K), \quad M = 2.993 \cdot 10^{-2}\,kg/mole, \quad Q = 8 \cdot 10^6\,J/kg,$$
$$k = 3.16 \cdot 10^7 1/s \quad [23], \quad E = 110 \cdot 10^3\,J/mole \quad [23], \quad D_{g0} = 1.82 \cdot 10^{-5}\,m^2/s,$$
$$b = 1.724, \quad \mu = 2.667, \quad a_{g0} = 0.3, \quad a_{cf0} - 0.1, \quad \chi = 0.5,$$

and the dimensionless boundary conditions:

$$T_{g0} = 1 \quad \text{and} \quad C_0 = 0.23. \quad (7)$$

We select the coefficients k_{11} and k_{10} so that the initial permeability k_1 in these cases is equal to the constant permeability $k_{00} = 10^{-8}\,m^2$, which is also used for solving system (2).

We also assume that if the temperature of the solid medium below the temperature of self-ignition $T_{kr} = 1.1$, the exothermic reactions do not occur.

At first we consider the case when the ignition zone is located at the bottom of the porous object. For arising self-sustaining cocurrent combustion wave, the changing of permeability due to the changing of porosity does not lead to the temperature increasing in the combustion zone, but leads to the speed increasing of the combustion wave. This fact is clearly demonstrated in Fig. 1, which shows the distribution of solid medium temperature in the considered porous object at t = 700 at different permeability coefficient when the width of ignition zone located at the bottom of the object is equal to 1/10 and the initial ignition temperature T_{c0} is equal to 2.5.

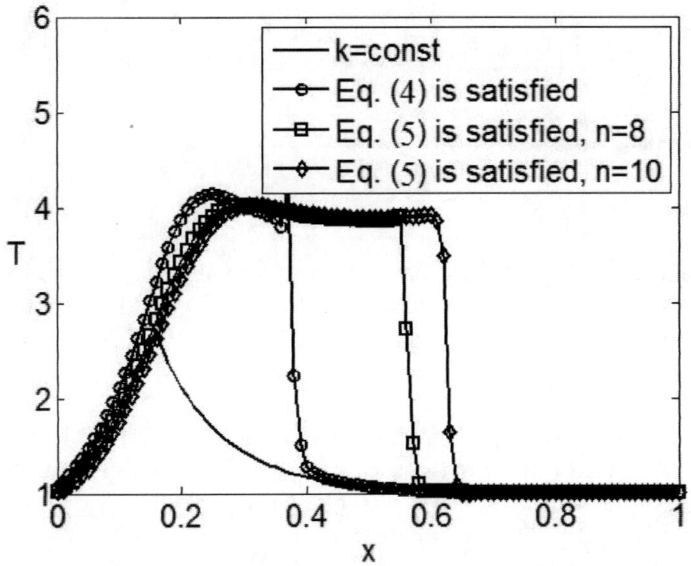

Figure 1. Distribution of the solid medium temperature in the porous object at t = 700 at different permeability coefficient when the width of ignition zone located at the bottom of the object is equal to 1/10 and the initial ignition temperature T_{c0} is equal to 2.5.

Combustion changes porosity, and if the permeability depends on porosity, the resistance of the porous medium decreases and gas velocity increases; this leads to increasing of the oxidizer supply, so the combustion wave propagates with a higher speed. Note that when we use equation (5) the permeability is modified more than when using equation (4). At the same time, when we use n equal to 10 in equation (5), the permeability is modified more than when using n equal to 8 in this equation. Thus, as can be seen from the Fig. 1, the more changes in permeability due to changes in porosity, the greater the speed of the combustion wave increases.

Consider the case when the ignition zone is located at the top of the porous object. For arising self-sustaining countercurrent combustion wave, the changing of permeability due to the changing of porosity leads to increasing both the temperature in the combustion zone and the speed of the combustion wave. This fact is clearly demonstrated in Fig. 2, which shows the distribution of solid medium temperature in considered porous object at t = 850 at different values of the permeability coefficient when the width of ignition zone located at the top of the object is equal to 1/10 and the initial ignition temperature T_{c0} is equal to 2.5 for the "direct" wave propagation, i.e. when combustion wave moves down.

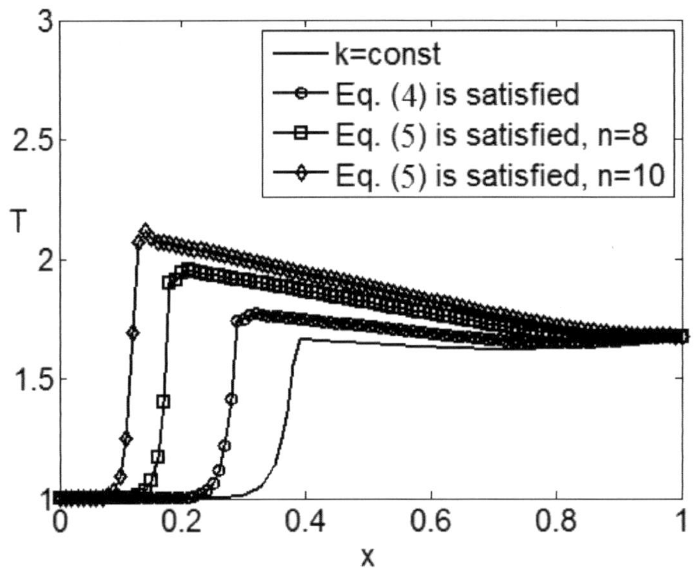

Figure 2. Distribution of the solid medium temperature in the porous object at t = 850 at different values of the permeability coefficient when the width of ignition zone located at the top of the object is equal to 1/10 and the initial ignition temperature T_{c0} is equal to 2.5 for "direct" wave propagation.

As for the countercurrent combustion wave, the rate of oxygen supply controls the process, increasing gas velocity due to decreasing resistance of the porous medium leads to increasing degree of conversion of the solid combustible component, so the temperature in the combustion zone increases with time. And the more changes in permeability due to changes in porosity, the greater the temperature and the speed of the combustion wave increase.

When the countercurrent combustion wave reaches the lower boundary of the object, it is reflected and starts to move up with reburning completely the remaining solid combustible substance. This reflected wave is cocurrent wave, so the propagation of the reflected wave is similar to the propagation of the cocurrent combustion wave which was considered earlier. Thus, in this case the changing of permeability due to the changing of porosity does not lead to the temperature increasing in the combustion zone, but leads to a significant speed increasing of the combustion wave. This fact is clearly demonstrated in Fig. 3, which shows the distribution of solid medium temperature in considered porous object at different times at different values of the permeability coefficient when the width of ignition zone located at the top of the object is equal to 1/10 and initial ignition temperature T_{c0} is equal to 2.5 for "reflected" wave propagation, i.e. when combustion wave moves up after the reflection from the bottom.

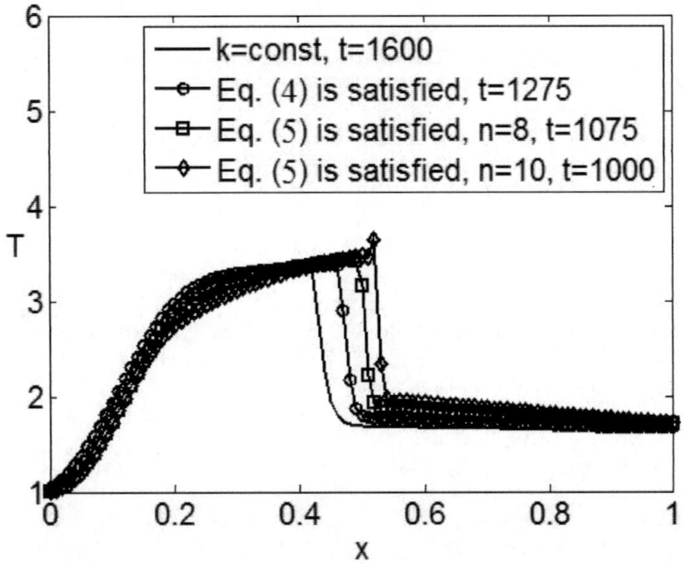

Figure 3. Distribution of the solid medium temperature in the porous object at different times at different values of the permeability coefficient when the width of ignition zone located at the top of the object is equal to 1/10 and the initial ignition temperature T_{c0} is equal to 2.5 for "reflected" wave propagation.

We can see again, that the more changes in permeability due to changes in porosity, the greater the speed of the combustion wave increases. It should be noted that the small temperature oscillation can be seen in the high-temperature zone of cocurrent combustion wave. When the temperature in cocurrent combustion wave is sufficiently high, oscillatory instabilities may occur for both constant and variable permeabilities, which is typical for kinetically controlled processes, but it is not discussed in this paper.

Thus, the considered dependence of permeability on porosity, which is described by equations (4) or (5), leads mainly to quantitative changes in solution of the system of equations (2) in one-dimensional case. Qualitative change is the increase of the temperature in the combustion zone with time in the countercurrent wave.

4. Conclusions

The one-dimensional unsteady process of heterogeneous combustion in porous object under free convection is considered when the dependence of permeability on porosity is taken into account. Using original computational algorithm, numerical investigations are carried out with some known equations which describe the changing of permeability due to the changing of porosity. It is shown that for both cocurrent and countercurrent self-sustaining combustion waves the changing of permeability due to the changing of porosity leads to the speed increasing of the combustion wave. For self-sustaining countercurrent combustion wave the changing of permeability due to the changing of porosity can also increase the temperature in the combustion zone with time.

Acknowledgments

The work was supported financially by the Russian Foundation for Basic Research (project 16-01-00103-a), the Far-Eastern Branch of the Russian Academy of Sciences (project 15-I-4-021). The author would like to acknowledge and thank Academician Vladimir Levin for his encouragement and valuable comments.

References

[1] Aldushin A P and Merzhanov A G 1988 The theory of filtration combustion: concept and state of research *Thermal Wave ropa ation in Hetero eneous Media* ed Y S Matros (Novosibirsk: Nauka) pp 9–52 (in Russian)
[2] Aldushin A P, Matkowsky J and Schult D A 1996 *om ust. Flame* **107** 151
[3] Aldushin A P, Matkowsky J and Schult DA 1997 *J. En . Math.* **31** 205
[4] Aldushin A P, Rumanov I E and Matkowsky J 1999 *om ust. Flame* **118** 76
[5] Aldushin A P and Matkowsky J 1998 *om ust. Sci. and Tech* **133**(4) 293
[6] Leach S V, Rein G, Ellzey J L, Ezekoye O A and Torero J L 2000 *om ust. Flame* **120** 346
[7] Rein G, Fernandez-Pello A C and Urban D L 2007 *roc. om ust. nst.* **31** 2677
[8] Lautenberger C and Fernandez-Pello C 2009 *Fire Safety J.* **44** 819
[9] Dodd A B, Lautenberger C and Fernandez-Pello A C 2009 *roc. om ust. nst.* **32** 2497
[10] Dodd A B, Lautenberger C and Fernandez-Pello A C 2012 *om ust. Flame* **159** 448
[11] Lutsenko N A 2013 *roc. om ust. nst.* **34** 2289
[12] Lutsenko N A and Sorokova S N 2014 *Advanced Materials Research* **880** 115
[13] Lutsenko N A and Levin V A 2014 *om ust. Sci. Technol.* **186** 1410
[14] Nigmatulin R I 1978 *rinciples of the Mechanics of Hetero eneous Media* (Moscow: Nauka) (in Russian)
[15] Lutsenko N A and Levin V A 2006 *J. En . hys. Thermophys.* **79** 33
[16] Kozeny J 1927 *Sitzun s erichte A ademie der Wissenschaften, Wien* **136**(2a) 271
[17] Ergun S 1952 *hem. En . ro .* **48**(2) 89
[18] Nikolaevskiy V N, Basniev K S, Gorbunov A T and Zotov G A 1970 *Mechanics of saturated porous media* (Moscow: Nedra) (in Russian)
[19] Levin V A and Lutsenko N A 2008 *Thermophys. Aeromech.* **15** 381
[20] Levin V A and Lutsenko N A 2010 *Math. Models omput. Simul.* **2** 635
[21] Levin V A and Lutsenko N A 2011 *Fluid Dyn.* **46** 826
[22] Lutsenko N A 2014 *nt. J. Heat Mass Transfer* **72** 602
[23] Sørum L, Grønli M G and Hustad J E 2001 *Fuel* **80** 1217

Lagrangian approach to modeling unsteady gas-liquid flow in a well

V Yu Liapidevskii[1,2] and V Tikhonov[3]

[1] Lavrentyev Institute of Hydrodynamics SB RAS, 630090 Novosibirsk, Russia
[2] Novosibirsk State University, 630090 Novosibirsk, Russia
[3] Weatherford, 125047 Moscow, Russia

E-mail: liapid@hydro.nsc.ru

Abstract. The purpose of this paper is to develop a numerical method of solving the problem of evolution of the finite gas volume that entered in a liquid flow at a set flow rate. The drift-flux model is used as gas-liquid mixture equations. The velocities of both phases, mixture and gas, are related by the Zuber-Findlay equation which coefficients depend on flow regime and gas void fraction. Lagrangian coordinates are used to simplify the initial equations. The numerical solution scheme is proposed. The numerical solution of the Riemann problem is verified by comparison with the exact self-similar solution. The model and numerical method efficiency is illustrated by examples of gas kick calculations in a vertical well.

1. Introduction

Managed Pressure Drilling (MPD) is one of the most advanced technologies of drilling wells in complex hydrogeological conditions. Bottomhole pressure is managed by applying back pressure in the annulus and changing the mud flow rate in the drill pipe at the surface. Sudden gas kicks from the formation are an issue while drilling [1]. Mathematical modeling is an effective method to assess the evolution of the gas-liquid bulk that has entered the well, as it is raised to the surface by the fluid flow, and to evaluate the kick intensity.

The most commonly used gas-liquid flow model is the Drift-Flux Model (DFM) known for over forty years [2]. It is instrumental to describe various flow regimes (bubbly flow, slug flow, etc.). The DFM considers mass conservation equations for each phase taken separately and the total momentum equation for the gas-liquid mixture in isothermal statement. The velocities of phases in the DFM are connected with the Zuber and Findlay equation [3], which parameters depend on the flow regime.

The Advection Upstream Splitting Method (AUSM) [4, 5] is used in most papers for the numerical DFM solution. However, when it is required to calculate a mixture flow rising from a thousand meters depth, the computing time becomes considerably high, which limits this method application for operative evaluation of gas kick to surface time and intensity. This is associated with a need to select quite a small integration time step due to the high sound speed in liquid. In practical problems aimed to evaluate the evolution of the gas-liquid bulk and gas kick intensity, liquid compressibility can be neglected. In this setting, it is convenient to use Lagrangian coordinates to simplify problem solution [6].

The paper proposes a Lagrangian approach to describe the evolution of the finite gas-phase volume placed in a vertical tube or an annular channel with a specified liquid flow rate. It considers dependence of the slip coefficients (coefficients of the Zuber-Findlay model) on the gas void fraction [7].

The basic assumptions and initial equations of a two-phase flow in Eulerian coordinates and their rearrangement to a divergent form and transfer to Lagrangian coordinates are described in Section 2. Section 3 contains an exact solution of the obtained equations for special case, and Section 4 describes the numerical scheme. The numerical scheme is verified by comparing the exact and numerical solutions in Section 5. Numerical examples of calculations of unsteady gas-liquid bulk flow rising from the well bottom with a surface are provided in Section 6. The conclusion contains the basic study results and further research lines aimed at model updating.

2. Governing equations

Let us consider the 1-D DFM equations in isothermal setting including two liquid and gas mass conservation equations and one gas-liquid mixture momentum equation [4], [6]:

$$\begin{aligned}&(\rho_l\alpha_l)_t+(\rho_l\alpha_l u)_x=\dot{m}_l\\&(\rho_g\alpha_g)_t+(\rho_g\alpha_g U)_x=\dot{m}_g\\&(\rho_l\alpha_l u+\rho_g\alpha_g U)_t+(\rho_l\alpha_l u^2+\rho_g\alpha_g U^2+p)_x=F,\end{aligned}\tag{1}$$

where u and U are the mean velocities of liquid and gas, respectively; p is the pressure; ρ, α and $\dot{m}$ are the density, the mean void fracture and the mass by unit of volume entering into the phase across the interface, respectively; index "l" is referred to the liquid and index "g" to the gas; $\alpha_l+\alpha_g=1$; $\dot{m}_l+\dot{m}_g=0$; F is the source term; t is the time; x is the space coordinate.

The source term is represented as a sum of gravity and viscous friction components:

$$F=-\rho_m(g\cos\theta+0.5fu_m|u_m|/D_h),\tag{2}$$

where $\rho_m=\rho_l\alpha_l+\rho_g\alpha_g$ and $u_m=u\alpha_l+U\alpha_g$ are the mean gas-liquid mixture density and velocity, respectively; θ is the angle of the inclination of the well; f is the friction coefficient; D_h is the hydraulic diameter.

Let one assume that the liquid is incompressible and the gas is ideal, then the pressure is

$$p=a_0^2\rho_g,\tag{3}$$

where a_0 is the sound speed in gas.

The closure equation of the system is the Zuber-Findlay equation [3] connecting the gas velocity with the mixture velocity:

$$U=K(\alpha_g)u_m+S(\alpha_g).\tag{4}$$

The values of coefficients K and S in equation (4) depend on the gas-liquid mixture flow regime, which, in turn, is connected to the gas void fraction $\alpha_g=1-\alpha_l$ [7]. The main flow regimes are bubbly, slug and annular flows [8] (figure 1). Dependencies of coefficients $K(\alpha_l)$ and $S(\alpha_l)$ derived from a series of runs (see details in [7]) are shown in figure 2. For the purpose of further transformations, we assume that empirical dependences $K(\alpha_l)$ and $S(\alpha_l)$ can be approximated using smooth curves.

Let one introduce new variables $\bar{\rho}=\alpha_g\rho_g/\rho_l$ and $c=\alpha_l/\bar{\rho}$ and rearrange the system (1) to a divergent form

$$\begin{aligned}
&\bar{\rho}_t + (\bar{\rho}U)_x = 0 \\
&(\bar{\rho}c)_t + (\bar{\rho}cU + A)_x = \dot{m}_l / \rho_l \\
&(\bar{\rho}\varphi)_t + (\bar{\rho}\varphi U + \Psi)_x = F / \rho_l,
\end{aligned} \tag{5}$$

where $A = (a-1)U - b$; $a = 1/K$; $b = S/K$; $\varphi = (\bar{\rho}cU + A + \bar{\rho}U)/\bar{\rho}$; $\Psi = A^2 / \bar{\rho}c + AU + p/\rho_l$.

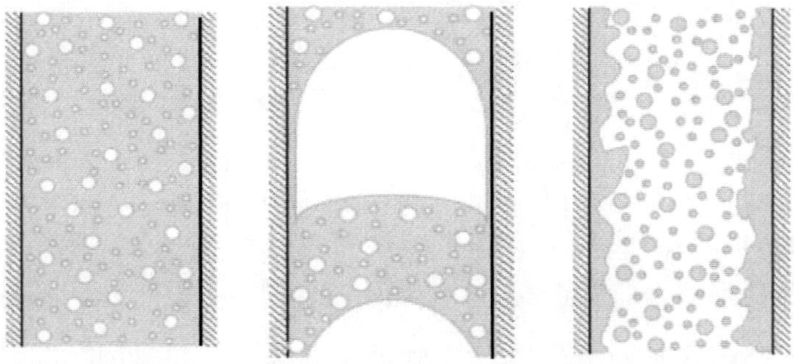

Figure 1. Flow patterns in a typical vertical gas-liquid flow

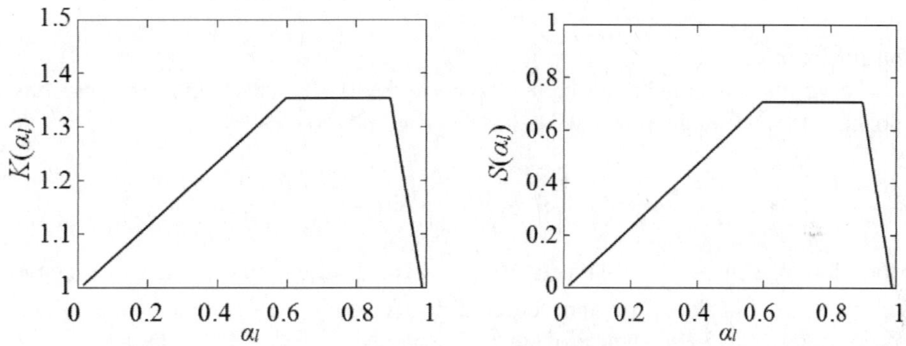

Figure 2. Dependences of the DFM parameters $K(\alpha_l)$ and $S(\alpha_l)$

The system (5) is written in relation to the variables $\bar{\rho}$, c, U. The functions A, φ, Ψ are expressed in terms of the above variables.

Let one introduce the Lagrangian coordinate ξ instead of the Eulerian coordinate, where $x(\xi,t)$ will be determined by solving the Cauchy problem [5]:

$$\mathrm{d}x/\mathrm{d}t = U(x,t), \quad x(\xi,0) = \xi.$$

Then, using the mass Lagrangian coordinate as an independent variable

$$q = \int_0^{\xi} \bar{\rho}(\xi,0)d\xi \tag{6}$$

and introducing the variable $v = 1/\bar{\rho}$, the system (5) is represented as:

$$\begin{aligned} &v_t - U_q = 0 \\ &c_t + A_q = \dot{m}_l\, v / \rho_l \\ &\varphi_t + \Psi_q = Fv / \rho_l . \end{aligned} \tag{7}$$

At that, the relation between the mass Lagrangian coordinate q and the Eulerian coordinate x is established by the differential equation $\partial x / \partial q = v$.

The characteristics of the system (7) are determined from the equation:

$$\Delta = \begin{vmatrix} -\lambda & 0 & -1 \\ A_v & A_c - \lambda & A_U \\ \Psi_v - \lambda\varphi_v & \Psi_c - \lambda\varphi_c & \Psi_U - \lambda\varphi_U \end{vmatrix} = 0, \tag{8}$$

which reduces to the solution of the cubic equation. If coefficients of the Zuber-Findlay model, $K = \text{const}$ and $S = \text{const}$, then the derivatives $A_v = 0$, $A_c = 0$. In this case, one of the system characteristics is $\lambda_0 = \mathrm{d}q / \mathrm{d}t = 0$. The respective Riemann invariant $W = (a-1)v + c$ will satisfy the equation $W_t = 0$. The other two roots of the set of Eqs. (8), λ^+ and λ^-, are determined from the quadratic equation. If gas void fractions α_g are rather high, close to 1, the characteristics λ^+ and λ^- become complex conjugate and the system (7) loses hyperbolicity. This constraint is specific of all known DFMs. To overcome it, the internal flow structure should be considered.

3. Riemann problem

In the general case, the system (7) can be solved numerically. However, in the special case, $F = 0$, $\dot{m}_l = 0$, a solution of the Cauchy problem with step-wise initial data:

$$\mathbf{V}_0(q) = (v_0(q), c_0(q), U_0(q)) = \begin{cases} \mathbf{V}_1, & q < 0, \\ \mathbf{V}_2, & q > 0 \end{cases} \tag{9}$$

will be in the class of self-similar solutions $\mathbf{V}(\zeta)$, where $\zeta = q/t$, and consist of a combination of contact discontinuities, shock waves and centered rarefaction waves (simple waves). This problem solution can be used to verify the numerical unsteady solution.

One begins with the case when $K = \text{const}$ and $S = \text{const}$.

It is assumed that $\mathrm{d}q/\mathrm{d}t = D$ is the discontinuity line of solution (7). Then the Rankine-Hugoniot conditions are:

$$\begin{aligned} &D[v] + [U] = 0 \\ &D[c] - [A] = 0 \\ &D[\varphi] - [\Psi] = 0. \end{aligned} \tag{10}$$

At contact discontinuities $D = 0$, and, as it follows from the first and the third equations (10):

$$[U] = 0, \quad [\Psi] = 0. \tag{11}$$

For shock waves $D \neq 0$ and the Riemann invariant is continuous, $[W] = 0$. Then, excluding variables D and c from the system (10), one can determine the dependence of shock wave velocity $U = U_s(v; \mathbf{V}_0)$ connecting states $\mathbf{V}_0$ and $\mathbf{V}$. There will be two shock transition curves in the plane (v, U), passing through point (v_0, U_0) and according to the chosen root of the quadratic equation connecting the variables v and U. For each set of characteristics λ^+ and λ^- determined by solution of equation (8) in the class of self-similar solutions $\mathbf{V}(\zeta)$

$$\zeta = \frac{\mathrm{d}q}{\mathrm{d}t} = \lambda^{\pm}. \tag{12}$$

Then, based on the first equation of the system (7), one gets the formula for simple waves:

$$\frac{\mathrm{d}U^{\pm}}{\mathrm{d}v} = -\zeta = -\lambda^{\pm}. \tag{13}$$

where $U^{\pm} = U_w^{\pm}(v;\mathbf{V}_0)$; $\lambda^{\pm} = \lambda^{\pm}(v,U^{\pm},c;\mathbf{V}_0)$; $c = c_0 + (a-1)(v_0 - v)$.

Thus, similar to two shock transition curves in the plane (v,U), there will be two curves corresponding to simple waves and also passing through the point (v_0,U_0).

In case of the arbitrary dependence $K(\alpha_l)$, $S(\alpha_l)$, the characteristics are determined from the cubic equation and can't be represented in an explicit form. To compare it to the results of the numerical unsteady calculations, one considers an example for definite functions $K(\alpha_l)$, $S(\alpha_l)$, shown in figure 2.

Let one assume that the wave directed to the left is simple. Then the dependences $U = U_w^1(v;\mathbf{V}_1)$ and $c = c_w^1(v;\mathbf{V}_1)$ are determined from the set of ordinary differential equations:

$$\frac{\mathrm{d}U}{\mathrm{d}v} = -\lambda_1, \quad \frac{\mathrm{d}c}{\mathrm{d}v} = \frac{\lambda_1 A_U - A_v}{A_c - \lambda_1} \tag{14}$$

with initial conditions $U|_{v=v_1} = U_1$, $c|_{v=v_1} = c_1$, where $\lambda_1 = \lambda_1(\mathbf{V})$ is the least (real) root of the equation (8).

This solution describes the admissible simple wave at $v > v_1$. One denotes by $v_3' > v_1$ the intermediate value v at the right boundary of the simple wave and considers the simple wave corresponding to the intermediate root λ_2 of the equation (8), $\lambda_1 < \lambda_2 < \lambda_3$. The dependencies $U = U_w^2(v;\mathbf{V}_3')$ and $c = c_w^2(v;\mathbf{V}_3')$ will be determined by solving the system of equations similar to the system (13) with formal replacement of root λ_1 with λ_2.

The simple wave of the 2nd family is admissible for the selected dependence $K(\alpha_l)$, if $v > v_3'$. It is supposed that the values K and S in the middle part of the above dependence are constant (figure 2). Let one take $v_3'' > v_2$ and assume that the obtained state $\mathbf{V}_3'$ is connected with state $\mathbf{V}_2$ via the shock transition described above. The discontinuity connecting $3''$ and 2 is stable at $v_2 > v_3''$. Then, setting $v_2 > v_3''$, determine $c_2 = c_3''$ and $U_2 = U_s^2(v_2;\mathbf{V}_3'')$.

The self-similar solution of the Riemann problem with the initial data (9) is built (figure 4).

4. Numerical solution

Godunov's method is used to solve the unsteady problem [9]. A uniform space mesh with step Δq is considered. One assumes mesh function values at the n-th step with respect to time as $\mathbf{V}_i^n$ and $\mathbf{V}_{i\pm 1/2}$, $i = 1:N$, where the lower integer index stands for the function value in the center of the i-th cell, and semi-integer indices $i \pm 1/2$ stand for mesh functions at cell boundaries. It is assumed that all mesh parameters and functions are constant inside each cell, i.e., there are function discontinuities at cell boundaries.

According to the Godunov's explicit scheme, the solution of the set of equations (7) at the (n+1)-th step with respect to time will be:

$$\begin{aligned} v_i^{n+1} &= v_i^n - (\Delta t / \Delta q)(U_{i-1/2} - U_{i+1/2}) \\ c_i^{n+1} &= c_i^n + (\Delta t / \Delta q)(A_{i-1/2} - A_{i+1/2}) + \Delta t \dot{m}_{li}^n v_i^n / \rho_l \quad , \\ \varphi_i^{n+1} &= \varphi_i^n + (\Delta t / \Delta q)(\Psi_{i-1/2} - \Psi_{i+1/2}) + \Delta t F_i^n v_i^n / \rho_l \end{aligned} \tag{15}$$

where Δt is the time step selected based on the scheme stability condition:

$$\Delta t = \min_{i=1(1)N} (C\Delta q / \lambda_i^+, C\Delta q / |\lambda_i^-|),$$

where C is the Courant number. (16)

Fluxes $U_{i-1/2}$, $A_{i-1/2}$, $\Psi_{i-1/2}$ can be determined by considering the equations (7) in an integral form. Two neighboring cells with numbers (i-1) and i are shown in figure 3. Assuming that the slope of each characteristic, λ^- and λ^+, within each of the cells remains constant, one integrates the third equation of the system (7) within the loop ABCD:

$$(\varphi_{i-1/2}^{-,n+1} - \varphi_{i-1}) |\lambda_{i-1}^-| + (\varphi_{i+1/2}^{+,n+1} - \varphi_i)\lambda^+ + \Psi_i - \Psi_{i-1} = 0 . \tag{17}$$

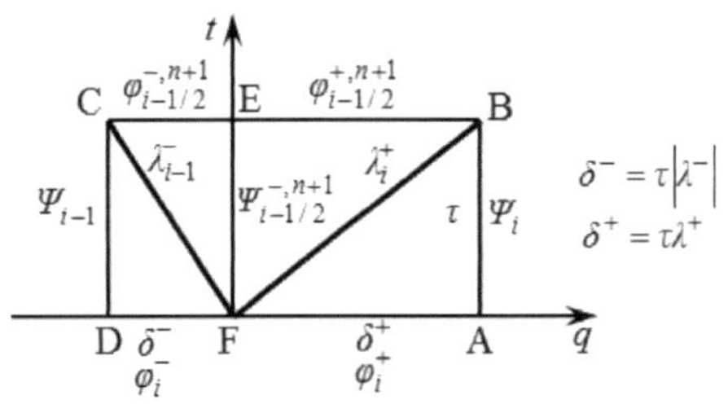

Figure 3. Integration by cells

The superscript standing for the variable value at the n-th step in respect of time is hereinafter omitted for simplicity.

Further, the function $\varphi_{i-1/2}^{\pm,n+1}$ should be expressed through $U_{i-1/2}^{n+1}$ and $v_{i-1/2}^{\pm,n+1}$. Let one determine the function $v_{i-1/2}^{\pm,n+1}$ by integrating the first equation of the system (7) within the loops ABEF and FECD (figure 3):

$$v_{i-1/2}^{-,n+1} = v_{i-1} + \frac{U_{i-1/2}^{n+1} - U_{i-1}}{|\lambda_{i-1}^-|}, \quad v_{i-1/2}^{+,n+1} = v_{i-1} - \frac{U_{i-1/2}^{n+1} - U_i}{\lambda^+} . \tag{18}$$

As a result of the above transformations, the flux $U_{i-1/2}^{n+1}$ is found:

$$U_{i-1/2}^{n+1} = \frac{(W_i + 1)\lambda_i^+ U_i - (W_{i-1} + 1)\lambda_{i-1}^- U_{i-1} + \Psi_{i-1} - \Psi_i + b_i U_i - b_{i-1} U_{i-1}}{(W_i + 1)\lambda_i^+ - (W_{i-1} + 1)\lambda_{i-1}^- + b_i - b_{i-1}} , \tag{19}$$

Using the known values $U_{i-1/2}^{n+1}$ it is easy to determine fluxes $A_{i-1/2}^{n+1}$ and $\Psi_{i-1/2}^{n+1}$ required to calculate the vector $\mathbf{V}_i^{k+1}$ by the formulae (15).

5. Numerical scheme verification

Numerical scheme verification included comparison of the unsteady solution of the Riemann problem with the exact (self-similar) solution described in Section 3.

The dependence $K(\alpha_l)$ shown in figure 2 was approximated by smooth function $K(\alpha_l) = 1.2 - 0.2/(1+\alpha_l)^{20}$, the value $S(\alpha_l)$ was assumed as being equal to zero. Initial data: $v_1 = 5.1$, $v_2 = 10$. Mesh dimension $N = 4000$.

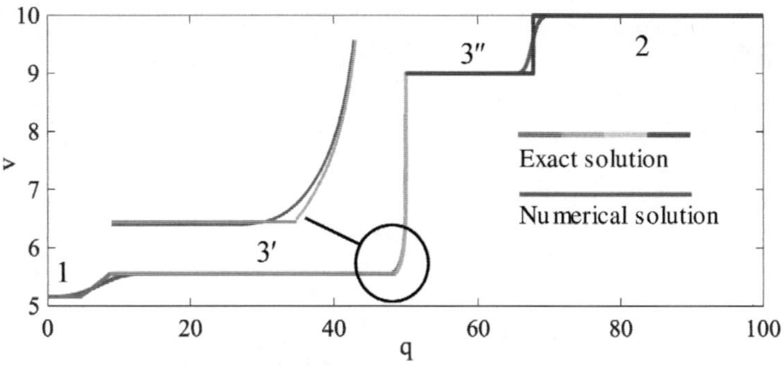

Figure 4. Numerical solution verification

Figure 4 shows solutions $v = v(q)$ at $t = 2$ sec. Obviously, both solutions are practically the same.

6. Modeling of gas-liquid bulk rising

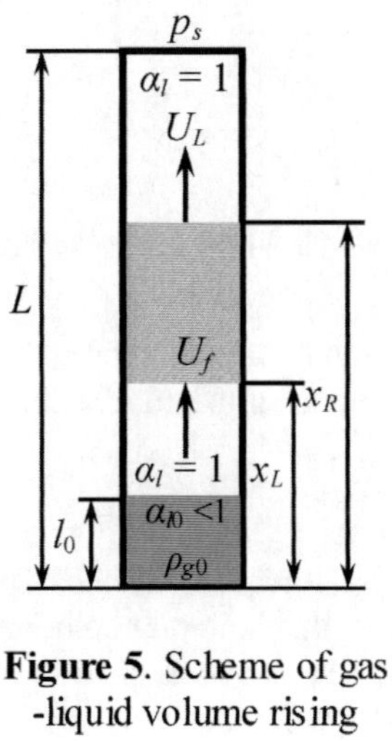

Figure 5. Scheme of gas -liquid volume rising

Let one return to the problem of gas-liquid bulk rising during a gas kick in a vertical well at the depth L (figure 5). It is assumed that the velocity of liquid circulating in the well, U_f, the initial gas-liquid bulk height, l_0, and the initial gas void fraction, α_{g0}, are specified. The surface pressure is p_s. The boundary conditions are as follows:

$$U(0) = \frac{dx_L}{dt} = U_f, \quad x_L(0) = 0$$
$$U(q_0) = \frac{dx_R}{dt} = U_R, \; \rho_L \frac{dU_R}{dt} + \frac{\partial p_L}{\partial x} = U_R, \; x_R(0) = L - l_0 \quad (20)$$

The boundary conditions (20) are approximated using the Godunov's method [8]:

$$U_{1/2}^{n+1} = U_f, \quad x_L^{n+1} = x_L^n + U_f \Delta t$$
$$U_{N+1/2}^{n+1} = U_{N+1/2}^n + \frac{\Delta t}{\rho_l}\left(F_l^n - \frac{p_s - p_N^n}{L - x_R^{n+1}}\right), \; x_R^{n+1} = x_R^n + U_{N+1/2}^n \Delta t \quad (21)$$

Fluxes $\Psi_{1/2}^{n+1}$, $A_{1/2}^{n+1}$, $\Psi_{N+1/2}^{n+1}$, $A_{N+1/2}^{n+1}$, required to calculate vectors $\mathbf{V}_{1/2}^{n+1}$ и $\mathbf{V}_{N+1/2}^{n+1}$ are determined based on the known values of functions $U_{1/2}^{n+1}$ и $U_{N+1/2}^{n+1}$ (21).

The following example with initial data is considered:

- Well depth = 4200 m;
- Hydraulic radius of the pipe = 0.2 m;
- Circulation rate of liquid = 1.2 m/sec;
- Friction coefficient = 0.004;
- Surface pressure = 1 atm;
- Initial height of the gas-liquid bulk= 70 m;
- Initial gas void fraction = 0.2.

Figures 6-9 show the gas-liquid bulk evolution in time $(t_m = 3800\text{sec})$. The curves at constant surface pressure p_s are highlighted in blue, and the curves at back pressure p_s =10 atm starting from $t > 0.8t_m$ are highlighted in green.

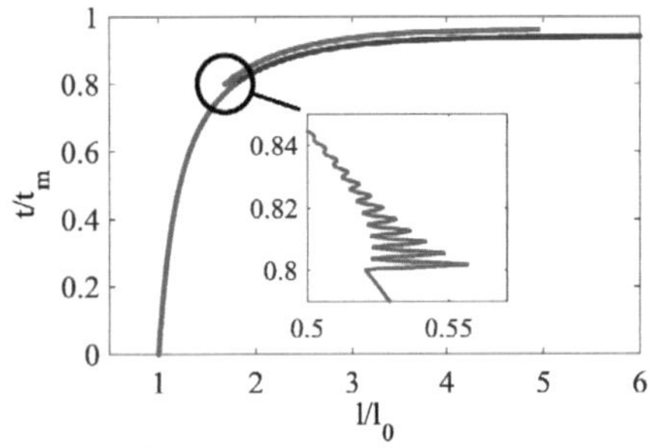

Figure 6. Changes of the relative gas-liquid bulk height

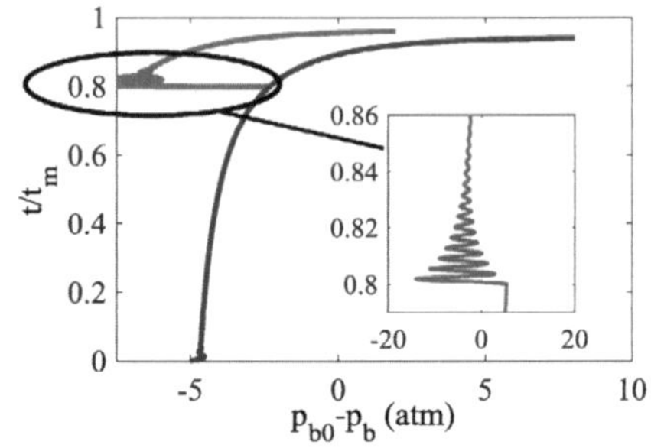

Figure 7. Changes of the bottomhole pressure increment

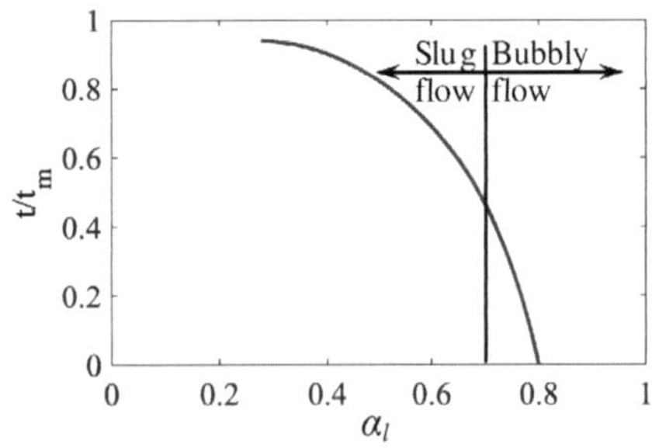

Figure 8. Changes of liquid fraction in the gas-liquid bulk

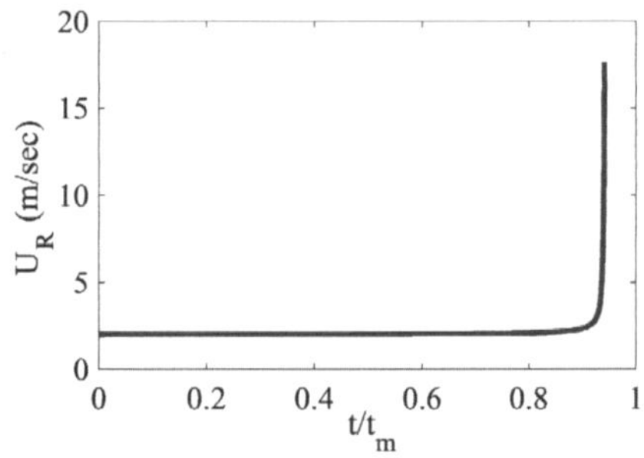

Figure 9. Changes of the top liquid cap velocity

According to the above processes, major parameter changes start as the mixture bulk approaches the surface. One notes an abrupt, six-time acceleration of the top liquid cap during mixture volume rising to the surface resembling champagne cork shooting out of the bottle.

7. Conclusion

A model and numerical integration scheme of unsteady gas-liquid flow in a well was developed. The numerical solution of the Riemann problem was verified by its comparison with the exact solution. The examples of changes in dimension, pressure, liquid fraction in the annulus and top liquid cap velocity during the gas-liquid bulk rising both at constant pressure out of the well and with applied back pressure were provided.

The Lagrangian coordinates permit to reduce the running time at calculation of unsteady gas-liquid flow by many tens of times;

Consideration of the internal flow structure (determination of a transition criterion to the annular flow at $\alpha_l << 1$), application of the real gas model, introduction of the second liquid phase, oil, etc., are defined as the further research lines.

Acknowledgments
The work is supported by the Russian Science Foundation (Project No. 15-11-20013).

References

[1] Hannegan D M and Mahmood A 2015 *Proc. of Offshore Technology Conference* (Houston, Texas: Curran Associates, Inc.) OTC-25723-MS
[2] Ishii M 1977 *Proc. of NATO Advanced Study Inst.* ed S Kakaç and F Mayinger (Washington, DC:

Hemisphere Publishing Corp.) 187
[3] Zuber N and Findlay J A 1965 *J. Heat Transfer* **87** 453–68
[4] Liou M-S 1996 *J. Comput. Phys.* **129** (2) 364–82
[5] Evje S and Fjelde 2003 K K *Computer and Fluids* **32** 1497–530
[6] Gavrilyuk S L and Fabre J 1996 *Int. J. Multiphase Flow* **22**(3) 453–60
[7] Varadarajan P A and Hammond P S 2015 *Int. J. Multiphase Flow* **73** 57–70
[8] Westende J M C van't 2008 *Droplets in annular-dispersed gas-liquid pipe flows* PhD thesis (University of Delft)
[9] Kulikovski A G, Pogorelov N V and Semenov A J 2001 *Mathematical Aspects of Numerical Solution of Hyperbolic Systems* (Moscow: Phizmathlit)

Nonlinear Waves: Theory and New Applications (Wave16) IOP Publishing
Journal of Physics: Conference Series **722** (2016) 012027 doi:10.1088/1742-6596/722/1/012027

Stratified flows over a complex relief

N I Makarenko[1,2], **J L Maltseva**[1,2] **and A A Cherevko**[1,2]
[1]Lavrentyev Institute of Hydrodynamics SB RAS, Novosibirsk 630090, Russia
[2]Novosibirsk State University, Novosibirsk 630090, Russia

E-mail: makarenko@hydro.nsc.ru

Abstract. A non-linear problem on steady non-homogeneous flows over an uneven bottom is considered. A semi-analytical model deals with asymptotic solutions of the Dubreil-Jacotin — Long equation of uniformly stratified fluid. Approximate solutions are constructed by the perturbation procedure combined with the Fourier method of modal expansion. Stationary wave patterns forced above the rough terrain of finite horizontal extension are calculated and analyzed.

1. Introduction

In this paper, a mathematical model of non-homogeneous fluid flows over complex topography is considered. Such stratified flows are of interest to meteorology in the context of study air currents above the mountain ranges [1, 2, 3]. It is well known that stationary wave patterns can occur in the lee of the terrain obstacles for appropriate upwind conditions [4]. In the theory of lee waves, exact analytic solutions are known only for simplest topographies such as single bell-shaped or semi-circular obstacle [5, 6]. Interference of lee waves over several obstacles has been studied experimentally in the papers [7, 8], and semi-analytically in the papers [9, 10, 11, 12, 13]. We develop an analytical approach involving von Mises transformation of the flow domain into the fixed one which provides efficient numerical calculation of the flow patterns over complex terrain.

2. Basic Equations

We consider a steady two-dimensional flow of a heavy inviscid incompressible fluid in the horizontal layer of finite depth affected by the gravity. A basic model involves the non-linear Euler equations

$$\rho\,(uu_x + vu_y) + p_x = 0, \quad \rho\,(uv_x + vv_y) + p_y = -\rho g, \tag{1}$$

$$u\rho_x + v\rho_y = 0, \qquad u_x + v_y = 0, \tag{2}$$

where ρ is the fluid density, (u, v) is the fluid velocity, p is the pressure and g is the gravity acceleration. It is supposed that the flow is bounded from above by a rigid lid $y = H$ where H is identified with the troposphere depth. The topography has the form $y = h(x)$, so the boundary conditions at the bottom and the top lid are

$$v - uh_x = 0\big|_{y=h(x)}, \qquad v = 0\big|_{y=H}. \tag{3}$$

It is supposed that the upstream flow is presented by the uniform current having constant speed U and the known density profile $\rho_\infty(y)$, so we have the condition

$$(u, v) \to (U, 0), \quad \rho \to \rho_\infty \qquad (x \to -\infty) \tag{4}$$

Introducing the stream function ψ by means of $u = \psi_y$, $v = -\psi_x$, we obtain the dependence $\rho = \rho(\psi)$ which is specified by the upstream condition (4) as $\rho(\psi) = \rho_\infty(\psi/U)$. It is assumed that the buoyancy frequency N defined by the formula $N^2(y) = -g\rho'_\infty(y)/\rho_\infty(y)$ is constant in the upstream flow. In this case, the density ρ_∞ depends exponentially on the height, $\rho_\infty(y) = \rho_0 \exp(-N^2 y/g)$ where ρ_0 is the reference density at $y = 0$. For this stratification, the basic dimensionless parameters are the Boussinesq parameter σ and the squared inverse *densimetric* Froude number λ defined by the formulae

$$\sigma = \frac{N^2 H}{g}, \qquad \lambda = \frac{\sigma g H}{U^2}.$$

The parameter σ determines the slope of the density profile at upstream flow, and parameter λ qualifies the sub- or super-criticality of this flow. In addition, we use dimensionless height of topography $\alpha = a/H$ as small parameter by applying the perturbation method. Selecting the height H as the length scale and upwind speed U as the velocity scale, dimensionless variables are introduced as follows:

$$(x, y, h) = H\,(\bar{x}, \bar{y}, \bar{h}), \quad \psi = UH\,\bar{\psi}.$$

In these variables, the shape of the bottom topography rewrites as $\bar{y} = \alpha\bar{h}(\bar{x})$, and upstream density profile takes the dimensionless form $\rho_\infty/\rho_0 = \exp(-\sigma\bar{y})$.

Eliminating the pressure p from the momentum equations (1) reduces the system (1)–(3) to the equivalent boundary value problem for the non-linear Dubreil – Jacotin — Long equation [1] which has the dimensionless formulation

$$\Delta\psi + \lambda\,(\psi - y) = \frac{1}{2}\,\sigma\left(|\nabla\psi|^2 - 1\right) \qquad (\alpha h(x) < y < 1) \tag{5}$$

$$\psi = 0|_{y=\alpha h(x)}, \qquad \psi = 1|_{y=1} \tag{6}$$

Here $\Delta = \partial_x^2 + \partial_y^2$ is the Laplace operator, and the bar is dropped in notation of variables. The upstream condition takes the form

$$\psi(x, y) \to y \qquad (x \to -\infty). \tag{7}$$

3. Constructing analytic solution

The uneven bottom topography essentially complicates solving the problem (5)–(7) even in the case of the simplest bottom shape $h(x)$. Therefore we involve the von Mises independent variables (x, ψ) in order to transform original flow domain to the unit strip in the (x, ψ)– plane. Now we seek the streamlines in the form $y = Y(x, \psi)$ so the equation (5) takes a more complicated form

$$-\frac{\partial}{\partial x}\frac{Y_x}{Y_\psi} + \frac{1}{2}\frac{\partial}{\partial\psi}\frac{1 + Y_x^2}{Y_\psi^2} = \lambda\,(Y - \psi) + \frac{1}{2}\,\sigma\left(\frac{1 + Y_x^2}{Y_\psi^2} - 1\right) \qquad (0 < \psi < 1) \tag{8}$$

but the boundary conditions (6) simplify as

$$Y(x, 0) = \alpha h(x), \qquad Y(x, 1) = 1. \tag{9}$$

The upstream condition (7) reduces to

$$Y(x,\psi) \to \psi \qquad (x \to -\infty) \tag{10}$$

The advantage of the problem (8)–(10) is a formulation of equations in rectangular domain with the exact topographic condition at the bottom line $\psi = 0$. However, the limitation appears while the von Mises transformation suggests all the streamlines to be not overhanging. Thus, the nonlinear model (8)–(10) may serve for simulating the topographic flows over a gently shaped relief having relatively small height of peaks. Thus, we are looking for the solution Y by the perturbation method with a small parameter α

$$Y(x,\psi) = \psi + \alpha w_0(x,\psi) + \alpha^2 w_1(x,\psi) + ...$$

Substituting this power series into (8)–(9) leads to the recursive set of equations

$$L(\sigma,\lambda)\, w_n = f_n \quad (0 < \psi < 1; \quad n = 0, 1, 2, ...), \tag{11}$$

$$w_0(x,0) = h(x), \;\; w_0(x,1) = 0; \quad w_n(x,0) = w_n(x,1) = 0 \;\; (n = 1, 2, ...) \tag{12}$$

Here $L(\sigma,\lambda)$ is the linear second-order differential operator acting by the formula

$$L(\sigma,\lambda)\, w = w_{xx} + w_{yy} - \sigma w_\psi + \lambda w,$$

and the non-linearity realizes by the recursive operators $f_n = f_n(w_0, ..., w_{n-1})$. Particulary, we have

$$f_0 = 0, \qquad f_1(w_0) = (w_{0x} w_{0\psi})_x + \frac{1}{2}\,(w_{0x}^2 + 3w_{0\psi}^2)_\psi - \frac{1}{2}\,\sigma\,(w_{0x}^2 + 3w_{0\psi}^2).$$

Further, we apply a version of the Fourier method using the eigenfunctions

$$\varphi_n(x,\psi) = e^{ikx} e^{\sigma\psi/2} \sin \pi n\psi \qquad (n = 1, 2, 3, ...)$$

which satisfy the equation $L(\sigma,\lambda_n)\varphi_n = 0$ with the eigenvalues $\lambda_n = \pi^2 n^2 + k^2 + \sigma^2/4$. The leading-order solution w_0 has the form

$$w_0(x,\psi) = e^{\sigma\psi/2} \left\{ y(\psi)\, h(x) + \sum_{n=1}^{\infty} w_0^{(n)}(x) \sin \pi n\psi \right\} \tag{13}$$

where the function

$$y(\psi) = \frac{\sin k_0\,(1-\psi)}{\sin k_0}, \qquad k_0 = \sqrt{\lambda - \sigma^2/4} \tag{14}$$

presents the hydrostatic mode, and the infinite series in (13) involves normal modes of the waves generated by topography. More precisely, for upstream flow with a given value of the Froude number λ taken from the interval

$$\pi^2 m^2 + \frac{1}{4}\sigma^2 < \lambda < \pi^2 (m+1)^2 + \frac{1}{4}\sigma^2$$

we obtain a m-modal lee wave solution having the coefficients

$$w_0^{(n)}(x) = -\frac{c_n}{k_n} \int_{-\infty}^{x} \sin k_n(x-s)\, h''(\xi)\, d\xi \qquad (n = 1, ..., m)$$

with the *oscillation* wave-number $k_n = \sqrt{\lambda - \sigma^2/4 - \pi^2 n^2}$, and

$$w_0^{(n)}(x) = \frac{c_n}{2\kappa_n} \int_{-\infty}^{+\infty} e^{-\kappa_n |x-s|} h''(\xi)\, d\xi \qquad (n = m+1, m+2, ...)$$

with the *attenuation* wave-number $\kappa_n = \sqrt{\pi^2 n^2 + \sigma^2/4 - \lambda}$. Here c_n are the known Fourier coefficients of the hydrostatic modal function $y(\psi)$ from (14) calculated with respect to the harmonic basis $\{\sin \pi n \psi\}_{n=1}^{\infty}$. The constructed solution w_0 satisfies automatically no wave upstream condition $w_0 \to 0$ as $x \to -\infty$, and it has the lee-wave downstream behavior

$$w_0(x, \psi) \sim e^{\sigma\psi/2} \sum_{n=1}^{m} \left(a_j^{(0)} \sin k_n x + b_j^{(0)} \cos k_n x \right) \sin \pi n \psi$$

as $x \to +\infty$. The constant wave amplitudes $a_j^{(0)}$ and $b_j^{(0)}$ depend here on the upstream flow parameters as well as on the topography shape. The non-linear correction to the leading-order solution w_0 can be found by constructing the function w_1 via a similar analytical procedure which involves the right-hand nonlinearity term $f_1(w_0)$ calculated by the first step of this iterative method.

4. Numerical calculations

Wolfram Mathematica [14] was used by semi-analytic calculations of the flows for a wide class of the bottom obstacles. Symbolic computer algebra was used to represent a truncated solution (13), which provides fast convergence. Numerical simulations used a series of ten basic harmonics for the series presenting the functions w_0 and w_1. Computations involved the discretization with 50 discrete points in the horizontal direction x and 10 points in ψ. The calculation of the non-linear second-order solution used 6-order interpolation splines for the leading-order discrete linear solution. Multiple series of calculations were carried out on the computer cluster at Novosibirsk State University.

Preliminary simulations used the first-order linear solution [11] and the second-order non-linear solution [12]. Wave solutions calculated for the topography with a finite number of peaks showed that the linear theory overestimates the magnitude of the waves in the interference zone. Figure 1 shows the 1-mode stratified flow over the double-bumped topography with the fixed parameters $\alpha = 0.05$, $\lambda = 15$, $\sigma = 02$, but the distance between the bumps can vary. It is clear, that the wave patterns are very sensitive to this variation. However, the nonlinear correction smooths essentially the wave crests over the second obstacle only. This overestimating phenomenon of the linear theory has been pointed in the laboratory experiments [7] with two bell-shaped obstacles.

The calculations also found the fragmentation effect which occurs for near-field wave patterns forced by multi-bumped topography of finite extension. Experimental setup was considered in the paper [8] with sinusoidal topography of finite extension. The existence of sharp fronts separating the flow domains having different wave scales was discussed in our paper [13]. Namely, using the Fourier expansions allow to evaluate analytically the impact of modal decomposition on the fragmentation observed in numerical experiments. Figure 2 illustrates a 1-mode stratified flow over irregular-shaped obstacle with the parameters $\alpha = 0.16$, $\lambda = 15$, $\sigma = 0.1$. These dimensionless values of parameters correspond to an upstream wind speed of 10 m/s in the flow over the terrain having maximum peak height of 1600 m by the depth of troposphere 10 km. The sheared wind profile in Fig. 2 demonstrates the air jet over the peak with the maximum velocity about 50 m/s.

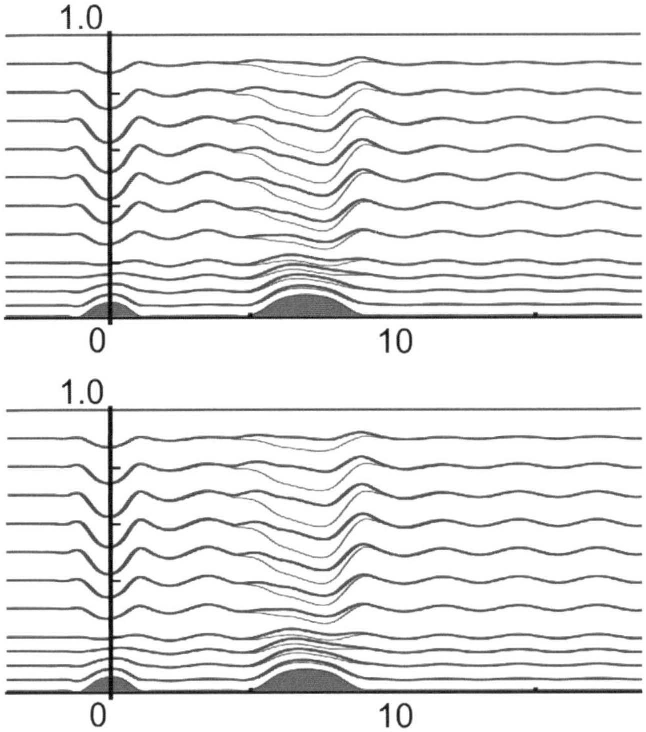

Figure 1. Stratified flow over two-bumped topography. Thin lines correspond to the linear model

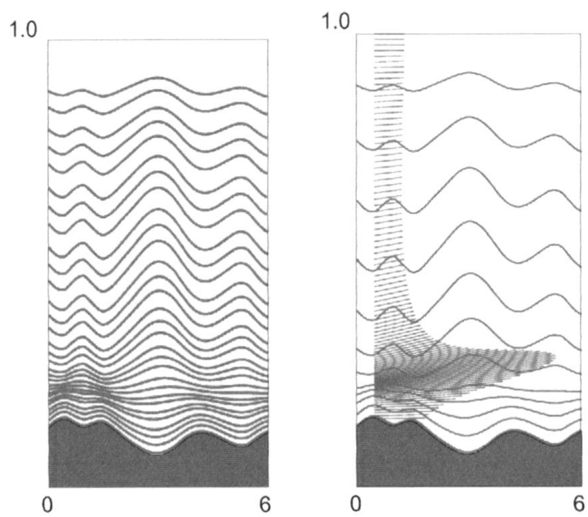

Figure 2. The streamlines (left) and the fluid velocity profile (right) in the stratified flow over irregular topography

Acknowledgments
The work was supported by the Russian Foundation of Basic Research (Grant No 15-01-03942).

References
[1] Yih C S 1980 *Stratified flows* (New York: Acad. Press)
[2] Baines P G 1998 *Topographic effects in stratified flows* (Cambridge: Cambridge Univ. Press)
[3] Grubisic V and Lewis J M 2004 *Bull. Am. Meteorol. Soc.* **85** 1127
[4] Long R R 1972 *Annu. Rev. Fluid Mech* **4** 69
[5] Kozhevnikov V N 1968 *Atmos. Ocean. Phys.* **4** 16–27
[6] Miles J W 1968 *J. Fluid Mech.* **32** 549
[7] Gyure B and Janosi I M 2003 *Dynam. Atmospher. Oceans* **37** 155
[8] Aguilar D, Sutherland B R and Muraki D J 2006 *Deep-Sea Res.* **53** 96
[9] Humi M 2009 *Nonlinear Process. Geophys.* **16** 533
[10] Muraki D J 2011 *J. Fluid Mech.* **681** 173
[11] Makarenko N I and Maltseva J L 2011 *Nat. Hazards Earth System Sci.* **11** 27
[12] Makarenko N I and Maltseva J L 2014 *J. Appl. Mech. Techn. Phys.* **55**(2) 311–7
[13] Makarenko N I, Maltseva J L and Cherevko A A 2015 *AIP Conf. Proc.* **1685** 090005-1
[14] *Mathematica, Version 9.0* 2012 (Champaign, IL: Wolfram Research, Inc.)

Synchronization of traveling waves in a dispersive system of weakly coupled equations

Z V Makridin[1,2] **and N I Makarenko**[1,2]

[1] Lavrentyev Institute of Hydrodynamics, Novosibirsk 630090, Russia
[2] Novosibirsk State University, Novosibirsk 630090, Russia

E-mail: `makarenko@hydro.nsc.ru`

Abstract. The system of weakly coupled differential equations describing traveling waves in dispersive media is considered. The Lyapunov — Schmidt construction is used to study the branching of cnoidal-type periodic solutions. The analysis of bifurcation equations uses the group symmetry and cosymmetry of original equations. Sufficient condition for existence of the phase-shifted modes of cnoidal waves is formulated in terms of the Pontryagin's function determined by the nonlinear perturbation terms.

1. Introduction

We study periodic solutions for weakly coupled nonlinear differential equations describing traveling waves in the multi-modal dispersive systems. For a coupled model, one often observes the phase shift arising among the modes of non-perturbed systems. In this context, the synchronization means the existence of phase-shifted solutions which can be constructed from decoupled modes by an appropriate perturbation procedure. Using the Lyapunov — Schmidt construction, we reduce the original problem to an equivalent implicit system of bifurcation equations. The asymptotic analysis of these equations results in the sufficient condition of synchronization which can be formulated via coupling nonlinear terms. In fact, this analytical condition reduces generic problem to the search for a simple root of a special Pontryagin type function which depends on the unknown phase shift c. We refer to the original paper [1] where this construction was suggested by the analysis of periodic solutions of planar systems close to the Hamiltonian ones. Bruno [2] considered the case of multiple roots of Pontryagin's function, and Malkin [3] extended the result known for periodic solutions of planar systems to the class of Lyapunov systems with n degrees of freedom. Similar existence condition involving Pontryagin's function was obtained in [4, 5] for solitary-wave solutions of the bimodal KdV-type systems. In present work, we use this bifurcation technique by the study of synchronized cnoidal wave solutions.

2. Statement of the problem

Let us consider the system of ordinary differential equations

$$\frac{d^2u}{dx^2} = \Phi_u(u, v, \varepsilon), \quad \frac{d^2v}{dx^2} = \Phi_v(u, v, \varepsilon), \tag{1}$$

Journal of Physics: Conference Series **722** (2016) 012028 doi:10.1088/1742-6596/722/1/012028

where $u(x)$, $v(x)$ are unknown scalar functions, ε is small parameter and the potential function Φ has the form

$$\Phi(u,v,\varepsilon)=\frac{1}{2}\left(u^2+v^2-u^3-v^3\right)+\varepsilon\Omega(u,v,\varepsilon).$$

For a simplicity, the perturbation term Ω is assumed to be analytic, and the following condition is also satisfied

$$\Omega(0,0,\varepsilon)=\Omega_u(0,0,\varepsilon)=\Omega_v(0,0,\varepsilon)=0.$$

Equations (1) arise for example by describing traveling waves in the system of two weakly coupled Korteweg — de Vries equations [5, 6]. For $\varepsilon=0$, decomposed system has a solution in the form of a cnoidal wave

$$u_0(x)=\alpha_2+(\alpha_3-\alpha_2)\,cn^2(rx;æ),\quad v_0(x)=u_0(x+c), \tag{2}$$

where the parameters r and æ are defined as

$$r=\frac{\sqrt{\alpha_3-\alpha_1}}{2},\quad æ^2=\frac{\alpha_3-\alpha_2}{\alpha_3-\alpha_1},$$

and parameters $\alpha_3>\alpha_2>\alpha_1$ are simple roots of the cubic polynomial $-u^3+u^2+2h=0$ with given constant h. The phase shift c is arbitrary for decoupled system at leading order in ε due to the translation invariance of the model equations. The problem is to find the values of c providing bifurcation of the solution by the perturbation in ε. It is well-known that the function u_0 formulated above approaches to the solitary wave solution $u_0(x)=\alpha_1+(\alpha_3-\alpha_1)\,ch^{-2}(rx)$ when $\alpha_2\to\alpha_1$. This case was studied in [4, 5].

3. Bifurcation of harmonic solutions

The problem under consideration is closely related to a problem of the perturbation of a limit cycle to the dynamical systems

$$\frac{du}{dx}=-H_v(u,v)+\varepsilon p(u,v,\varepsilon),\quad \frac{dv}{dx}=H_u(u,v)+\varepsilon q(u,v,\varepsilon) \tag{3}$$

being close to the Hamiltonian systems which were first studied in a two-dimensional case by Pontryagin [1]. Specifically, it was proved the following

Theorem 3.1 *Let C_0 be a closed curve whose points satisfy the equation $H(u,v)=h_0$ and functions H_u and H_v are not equal to zero simultaneously. Then there exist a closed curve C_h near C_0 whose points satisfy the equation $H(u,v)=h$ when $|h-h_0|$ is sufficiently small. Let*

$$\psi(h)=\iint\limits_D\left(p_{\varepsilon u}+q_{\varepsilon v}\right)dudv, \tag{4}$$

where D is the region inside a curve C_h. If $\psi'(h_0)=e\neq 0$ and $\psi(h_0)=0$ then there exist a unique limit cycle of (2) which depends continuosly on ε and it's characteristic number has the same sign as εe. Moreover $C_h\to C_0$ when $\varepsilon\to 0$.

At first step, we demonstrate how the theorem 3.1 can be extended to some simple class of multi-dimensional systems of non-linear oscillators having rationally independent basic frequencies. Let us consider the autonomous system of differential equations

$$\frac{dw}{dx}=\mathcal{A}w+f(w;\varepsilon) \tag{5}$$

where $w = (w_1, \ldots, w_n) \in \mathbf{R}^n$ is an unknown vector function with even number of components $n = 2s$, and the matrix $\mathcal{A}$ is the block-diagonal matrix

$$\mathcal{A} = \begin{pmatrix} \omega_1 J & & & \\ & \omega_2 J & & \\ & & \ddots & \\ & & & \omega_s J \end{pmatrix}, \qquad J = \begin{pmatrix} 0 & 1 \\ -1 & 0 \end{pmatrix}.$$

The frequencies $\omega_1, \ldots, \omega_s$ are assumed to be rationally independent, and the analytic vector function $f(w;\varepsilon) = (f_1, f_2, ..., f_n)$ is assumed to satisfy the condition $f(w, 0) = 0$. Let us note that the system (5) is linear at $\varepsilon = 0$, and it has in this case the Hamiltonian

$$H(w) = \frac{1}{2}\sum_{j=1}^{s} \omega_j (w_{2j-1}^2 + w_{2j}^2).$$

Therefore, the periodic 1-mode solution of (5) should have the form of harmonically oscillate solution

$$w_0(x) = \varrho_0 \left(\zeta e^{i\omega_1 x} + \overline{\zeta} e^{-i\omega_1 x}\right) \tag{6}$$

as $\varepsilon \to 0$. Here $\zeta = (1, i, 0, ..., 0)$ is the complex eigenvector of the real-valued matrix $\mathcal{A}$ corresponding to its eigenvalue $i\omega_1$, bar denotes complex conjugate and the real-valued amplitude parameter ϱ_0 depends on the constant level $H(w_0) = h_0$. Now we seek the solution w of a nonlinear system (5) to be periodic in x with an unknown period $T(\varepsilon) = 2\pi/\omega(\varepsilon)$ where the frequency $\omega(\varepsilon)$ is analytic in ε and $\omega(0) = \omega_1$. Changing independent variable $x = \omega(\varepsilon)\tau$, we consider an unknown solution in the form

$$w(\tau; \varepsilon, \omega, \varrho_0) = w_0(\tau; \varrho_0) + \varepsilon\theta(\tau; \varepsilon, \omega, \varrho_0).$$

By that way, the problem reduces to the finding 2π-periodic solution θ of equivalent operator equation

$$A\theta = R(\theta; \omega, \varepsilon) \tag{7}$$

where A is the linear differential operator acting by the formula

$$A\theta = \frac{d\theta}{d\tau} - \mathcal{C}\theta, \quad \mathcal{C} = \omega_1^{-1}\mathcal{A}, \tag{8}$$

and nonlinear mapping R is defined as follows:

$$R(\theta; \varepsilon, \omega) = \frac{1}{\varepsilon\omega_1}\Big\{(\omega_1 - \omega)\frac{dw_0}{d\tau} + \varepsilon(\omega_1 - \omega)\frac{d\theta}{d\tau} + f(w_0 + \varepsilon\theta; \varepsilon)\Big\}. \tag{9}$$

Let E and F be a Banach spaces of 2π-periodic real-valued functions $\theta(\tau)$ being smooth and continuous respectively, which can be presented by a uniformly convergent Fourier series

$$\theta(\tau) = \sum_{m=-\infty}^{+\infty} \theta^{(m)} e^{im\tau}, \quad \theta^{(m)} \in \mathbf{C}^n : \theta^{(-m)} = \bar{\theta}^{(m)}$$

having finite norms

$$\|\theta\|_E = \sum_{m=-\infty}^{\infty} (1 + |m|)|\theta^{(m)}|, \qquad \|\theta\|_F = \sum_{m=-\infty}^{\infty} |\theta^{(m)}|.$$

Lemma 3.2 *Let the frequencies $\omega_1, ..., \omega_s$ satisfy the inequalities $|\omega_1^{-1}\omega_j| < 2$ $(j = 2, \ldots, s)$. Then the non-homogeneous equation $Aw = g$ with a given vector function $g(\tau) \in F$ has a 2π-periodic solution $w(\tau) \in E$ if and only if the function g satisfies the orthogonality condition*

$$(g, \zeta e^{i\tau}) \stackrel{def}{=} \int_0^{2\pi} g(\tau) \cdot \zeta\, e^{i\tau}\, d\tau = 0 \tag{10}$$

where ζ is the complex eigenvector from (6) and the symbol $\cdot$ denotes Hermitian scalar product.

This lemma imply that operator $A : E \to F$ is the Fredholm operator. It has a two-dimensional null-space $Ker\, A$ generated by harmonic solutions of the form (6), so the Banach spaces E and F split into the direct sums $E = Ker\, A \oplus X$, $F = Y \oplus Im\, A$ where $X \subset E$ is closed infinite-dimensional subspace of E, and Y is the two-dimensional subspace of F. Thus, according to the Lyapunov — Schmidt construction [7]–[9], we can search for the small solution of the equation (7) in the form

$$\theta = \xi\varphi + \overline{\xi}\overline{\varphi} + \sigma, \qquad \xi \in \mathbf{C},\ \ \varphi(\tau) = \zeta e^{i\tau},\ \ \sigma \in X.$$

By that, the operator equation (7) reduces to equivalent bifurcation equation

$$QR(\xi\varphi + \overline{\xi}\,\overline{\varphi} + \sigma(\tau; \xi, \overline{\xi}, \varepsilon, \omega); \varepsilon, \omega) = 0 \tag{11}$$

where $Q : F \to Y$ is the projection associated with the orthogonality condition (10), and the nonlinear mapping $\sigma = \sigma(\tau; \xi, \overline{\xi}, \varepsilon, \omega)$ is specified by implicit equation

$$\sigma = \widetilde{A}^{-1}(I - Q)R(\xi\varphi + \overline{\xi}\,\overline{\varphi} + \sigma; \varepsilon, \omega) \tag{12}$$

where the isomorphism $\widetilde{A} : X \to Im\, A$ is generated by the restriction of the operator A. The system (11) at leading-order $\varepsilon = 0$ results in the pair of independent scalar relations

$$\varrho_0\, \omega'(0) + \Gamma(\varrho_0) = 0, \qquad \Psi(\varrho_0) = 0 \tag{13}$$

with smooth functions Γ and Ψ. More precisely, the function Ψ has the form

$$\Psi(\varrho_0) = \int_0^{2\pi} \left\{ \cos\tau\, \frac{\partial f_1}{\partial \varepsilon}(w_0; 0) - \sin\tau\, \frac{\partial f_2}{\partial \varepsilon}(w_0; 0) \right\} d\tau \tag{14}$$

with the components f_1, f_2 of the perturbation term f from (5), this is analog of the Pontryagin's function (4) for the ODE system (5). Let us remark that nonlinear bifurcation equation (11) is completely degenerated as far as a direct application of the implicit function theorem is concerned. However, the presence of group symmetry allows to reduce the number of parameters. Specifically, the kernel of the linear operator A is invariant with respect to the representation T_γ of the translation group $\tau \mapsto \tau + \gamma$. This representation acts on the null space $Ker\, A$ in accordance with the formula $T_\gamma\, \varphi(\tau) = e^{i\gamma}\, \varphi(\tau)$. Therefore, T_γ induces the representation of a compact group $SO(2)$ acting in a complex parametric plane $\xi \in \mathbf{C}$ by the formula $S_\gamma\, \xi = e^{i\gamma}\xi$. Thus, we can use the reduction theorem which was proved in [10]. According to this theorem, all solutions of the equations (11) can be presented at $\varepsilon = 0$ in the form

$$\theta_0 = T_\gamma \left\{ |\xi|(\varphi + \overline{\varphi}) + \widetilde{A}^{-1}(R_0) \right\} \tag{15}$$

with appropriate $\gamma \in [0, 2\pi]$. Thus, we can fix here the phase shift by choosing $\gamma = 0$ without loss of generality. In this case, the system (11) simplifies to the system

$$\begin{cases} |\xi|\Psi'(\varrho_0) + \Pi_1(\omega'(0), \varrho_0) + \varepsilon\chi_1(|\xi|; \varepsilon, \varrho_0, \omega) = 0 \\ |\xi|\Pi_2(\varrho_0, \omega'(0)) - \varrho_0\omega''(0) + \Pi_3(\omega'(0), \varrho_0) + \varepsilon\chi_2(|\xi|; \varepsilon, \varrho_0, \omega) = 0 \end{cases} \tag{16}$$

where explicit form of smooth functions Π_j $(j = 1, 2, 3)$ and χ_j $(j = 1, 2)$ is not essential for analysis. Note that the parameters ϱ_0 and $\omega'(0)$ have already known from the leading-order equations (13). Thus, if the condition $\Psi'(\varrho_0) \neq 0$ holds and the parameter ε is sufficiently small then we can apply the implicit function theorem in order to determine the magnitude $|\xi|$ and the parameter $\omega''(0)$ from the equations (16). Finally, we obtain the following result.

Theorem 3.3 *If ϱ_0 is a simple root of the function Ψ from (14) and ε is sufficiently small, then there exists $2\pi/\omega(\varepsilon)$-periodic solution to (5) of the form $w = w_0 + \varepsilon\theta$ such that $\omega(\varepsilon) \to \omega_1$ as $\varepsilon \to 0$.*

4. Bifurcation of cnoidal wave solutions

Let us return to the original system (1). Similarly, we can seek the periodic solution

$$u(x; \omega, c, \varepsilon) = u_0(x) + \varepsilon\, u_1(x; \omega, c, \varepsilon), \quad v(x; \omega, c, \varepsilon) = u_0(x + c) + \varepsilon\, v_1(x; \omega, c, \varepsilon) \tag{17}$$

having unknown period $T(\varepsilon) = 2K(æ)/r\omega(\varepsilon)$ such that $\omega(0) = 1$ where u_0 is the cnoidal-wave solution (2) and $K(æ)$ is the complete elliptic integral of the first kind. In the same way, this problem can be reduced by changing independent variable $x = \omega(\varepsilon)\tau$ to the problem of finding $2K/r$-periodic solution $w = (u_1, v_1)$ to the equivalent operator equation

$$Aw = R(w; \omega, \varepsilon) \tag{18}$$

which has the linear part

$$Aw = \Big(u_1{}'' + (3u_0 - 1)u_1,\ v_1{}'' + (3v_0 - 1)v_1\Big).$$

Accordingly, the nonlinear operator $R(w; \omega, \varepsilon) = (f_1(u_1, v_1; \omega, \varepsilon), f_2(u_1, v_1; \omega, \varepsilon))$ has the components

$$\begin{aligned} f_1 &= \varepsilon^{-1}(1 - \omega^2)(u_0'' + \varepsilon u_1'') - \frac{3}{2}\varepsilon u_1^2 + \Omega_u(u_0 + \varepsilon u_1, v_0 + \varepsilon v_1, \varepsilon) \\ f_2 &= \varepsilon^{-1}(1 - \omega^2)(v_0'' + \varepsilon v_1'') - \frac{3}{2}\varepsilon v_1^2 + \Omega_v(u_0 + \varepsilon u_1, v_0 + \varepsilon v_1, \varepsilon). \end{aligned} \tag{19}$$

Let $k \geq 1$ be an integer. Denote by $H^k_{2K/r}$ the class of $2K/r$-periodic function which belongs to the Sobolev space $W_2^k[0, 2K/r]$. Let us consider the linear differential equation

$$u'' + \big(3u_0(\tau + c) - 1\big)u = f(\tau) \tag{20}$$

with $2K/r$-periodic function f. Let G_c be the linear integral operator which is defined by following formula

$$\begin{aligned} (G_c f)(\tau) = &-u_0'(\tau + c)\int_0^{\tau+c} u_*(s)f(s - c)\,ds - u_*(\tau + c)\int_{\tau+c}^{2K/r} u_0'(s)f(s - c)\,ds - \\ &-\frac{r}{4KL}\,u_*(\tau + c)\int_{-2K/r}^{0} u_*(s)f(s - c)\,ds, \end{aligned} \tag{21}$$

where u_* is a non-periodic solution of homogeneous equation (20) with $c = 0$ which is given by the Liouville's formula

$$u_*(\tau) = u_0'(\tau) \int_0^{\tau} \frac{ds}{u_0'^2(s)}$$

and the constant L is defined via complete elliptic integrals [11].

Lemma 4.1 *Let $f \in H^k_{2K/r}$ and following condition holds*

$$\int_0^{2K/r} f(\tau)\, u_0'(\tau + c)\, d\tau = 0. \qquad (22)$$

Then $G_c f \in H^{k+2}_{2K/r}$ and inequality $\|G_c f\|_{H^{k+2}_{2K/r}} \le C \|f\|_{H^k_{2K/r}}$ is valid.

According to this lemma the general solution of nonhomogeneous equation (20) is of the form

$$u(\tau) = \xi u_0'(\tau + c) + G_c f(\tau), \quad \xi \in \mathbf{R}.$$

Let us return to the equation (18). The linear operator $A : H^{k+2}_{2K/r} \times H^{k+2}_{2K/r} \to H^k_{2K/r} \times H^k_{2K/r}$ is Fredholm and it has a two-dimensional kernel with the basis $e_1 = (u_0', 0)$, $e_2 = (0, v_0')$. The solvability conditions of nonhomogeneous equation $Aw = (f_1, f_2)$ reduces to that of two scalar equations of the form (20). It is important to note that the kernel invariance of linearized operator is crucial to the reduction of bifurcation equation for the system (5) considered above in the Section 3. In contrast, this property is not held for the original ODE system (1) even this system is also autonomuos. However, the group symmetry also plays a key role while the system (1) possesses the potential

$$l(w; \varepsilon) = \int_0^{T(\varepsilon)} \Big\{ \frac{1}{2} u'^2(x) + \frac{1}{2} v'^2(x) + \Phi(u(x), v(x), \varepsilon) \Big\}\, dx$$

which is invariant with respect to the translation group $T_\gamma w = w(x + \gamma)$ where $w = (u, v)$. In this case, the following relation holds $\big\langle \nabla_w l(w, \varepsilon), Xw \big\rangle = 0$ where $X = \partial_x$ is the infinitesimal operator of the translation group, and $\langle \cdot, \cdot \rangle$ denotes a scalar product in $L_2[0, T(\varepsilon)] \times L_2[0, T(\varepsilon)]$. The operator X here is the cosymmetry operator for the system (1) in the sense of the paper [12]. More generally [4], if a potential operator has the invariant Lagrangian under action of the Lie group, then cosymmetry is given by the Lie algebra of infinitesimal generators of the symmetry group factorized with respect to the isotropy subgroup of unperturbed solution (u_0, v_0). Yudovich [12, 13] explained by the presence of cosymmetry the branching of solution families near the known non-cosymmetric solution w_0 $(Xw_0 \ne 0)$. Therefore, the reduction theorem proved in [4] reduces a two-dimensional system of bifurcation equations for the system (1) to single scalar equation having the form

$$\int_0^{2K/r} \Big\{ \varepsilon^{-1}(1 - \omega^2)(u_0'' + \varepsilon u_1'') - \frac{3}{2}\varepsilon u_1^2 + \Omega_u(u_0 + \varepsilon u_1, v_0 + \varepsilon v_1, \varepsilon) \Big\}\, u_0'(\tau) d\tau = 0. \qquad (23)$$

Here the implicit mapping $(\xi_1, \xi_2; c, \omega) \to (u_1, v_1)$ is specified at leading order $\varepsilon = 0$ by the formulae

$$u_1 = \xi_1 u_0' + G_0 f_1^0, \quad v_1 = \xi_2 v_0' + G_c f_2^0 \qquad (24)$$

where the operator G_c acts by the formula (21), and the functions f_1^0, f_2^0 are defined by the formula (19) taken at the limit $\varepsilon = 0$. Taking into account these notations, we obtain the limit form of the equation (23) at $\varepsilon = 0$ which involves the Pontryagin's function as follows:

$$\widetilde{\Psi}(c) \stackrel{def}{=} \int_0^{2K/r} \Omega_u(u_0(\tau), u_0(\tau + c), 0)\, u_0'(\tau) d\tau = 0.$$

Using this fact and substituting explicit formulae (24) into equation (23) we arrive at the final form of the bifurcation equation

$$\widetilde{\Psi}'(c)(\xi_1 - \xi_2) + \nu(c) + \varepsilon\Theta(\xi_1, \xi_2; \omega, c, \varepsilon) = 0, \tag{25}$$

where explicit form of smooth functions ν and Θ is not essential for analysis. It is clear that equation (25) uniquely determines the parameter ξ_2 by the implicit function theorem when the condition $\widetilde{\Psi}'(c) \neq 0$ holds. Let us remark that the similar sufficient condition was obtained in [4, 5] in the limit case of solitary-wave solutions. In addition, it turns out that the stability of these solutions considered in the context of the time evolution model can be also checked in terms of the Pontryagin's function. Specifically, the stability of the bifurcate solution depends on a sign of $\widetilde{\Psi}'(c)$ in accordance with [5].

Acknowledgments

This work was partially supported by the Russian Scientific Foundation (Grant No. 15-11-20013).

References

[1] Pontryagin L S 1934 *Zh. Exp. Theor. Phys.* **4** 234–38
[2] Bruno A D 1990 *Math. Notes* **48** 1100–08
[3] Malkin I G 2004 *Methods of Lyapunov and Poincare in the theory of nonlinear oscillations* (Editorial: Moscow) (in Russian)
[4] Makarenko N I 1996 *Dokl. Math.* **53** 369–71
[5] Wright J D and Scheel A 2007 *Z. Angew. Math. Phys.* **58** 535–70
[6] Gear J A and Grimshaw R 1984 *Stud. Appl. Math.* **70** 235–58
[7] Vainberg M M and Trenogin V A 1974 *Theory of branching of solutions of non-linear equations* (Leyden: Noordhoff)
[8] Iooss G and Joseph D D 1980 *Elemenatry stability and bifurcation theory* (New York: Springer)
[9] Vanderbauwhede A 2009 *Encyclopedia of Complexity and Systems Science* (New York: Springer) 5299–315
[10] Loginov B V and Trenogin V A 1971 *Math. USSR Sb.* **14** 438–52
[11] Littman W 1957 *Comm. Pure Appl. Math.* **10** 241–70
[12] Yudovich V I 1991 *Math. Notes* **49** 540–45
[13] Kurakin L G and Yudovich V I 2000 *Sib. Math. J.* **41** 114–24

Journal of Physics: Conference Series **722** (2016) 012029 doi:10.1088/1742-6596/722/1/012029

Construction and investigation of 3D vessels net of the brain according to MRI data using the method of variation of scanning plane

A A Cherevko[1,2], G S Yankova[2], S V Maltseva[1,2], D V Parshin[1], A E Akulov[3], A K Khe[1,2] and A P Chupakhin[1,2]

[1] Lavrentyev Institute of Hydrodynamics of the Russian Academy of Sciences, Novosibirsk 630090, Russia

[2] Novosibirsk State University, Novosibirsk 630090, Russia

[3] Institute of Cytology and Genetics of the Russian Academy of Sciences, Novosibirsk 630090, Russia

E-mail: sv_maltseva@mail.ru

Abstract. The blood realizes the transport of substances, which are necessary for livelihoods, throughout the body. The assumption about the relationship genotype and structure of vasculature (in particular of brain) is natural. In the paper we consider models of vessel net for two genetic lines of laboratory mice. Vascular net obtained as a result of preprocessing MRI data. MRI scanning is realized using the method of variation of slope of scanning plane, i.e. by several sets of parallel planes specified by different normal vectors. The following special processing allowed to construct models of vessel nets without fragmentation. The purpose of the work is to compare the vascular network models of two different genetic lines of laboratory mice.

1. Introduction

Magnetic resonance imaging (MRI) is a non-destructive technique used to examine the inner structure of objects. The MRI is based on the Nuclear Magnetic Resonance (NMR) phenomenon, which is the rearrangement of magnetic moments of nonzero-spin nuclei in the external magnetic field. Employment of MRI in the investigation of bio-tissues and bio-fluids is based on NMR of hydrogen atoms. A specialized MRI technique called Magnetic Resonance Angiography (MRA) was developed to examine the circulatory system in living organisms.

Reconstructing the vasculature (and, in particular, that of the brain) of small laboratory animals is of interest by virtue of their wide using as objects of medical and biological research. However, this type of task is associated with the problem of insufficient sensitivity, even of high-field imaging devices, to the blood flow in cerebral blood vessels in non-enhanced imaging; images of some vessels appear fragmented. In the paper [1] suggested a constructive algorithm that allows for fragmentation-free reconstruction of models of complex vasculature without vessel fragmentation.

Method of variation of slope of scanning plane [1] allowing for reconstruction of geometrical models of vessel nets according to MRI data is briefly described in Section 2. On the basis of constructed models of vessel nets a comparison of the geometric configuration of blood net models for animals under investigation are carried out in Section 3. Results of measurement of angles between the vessels of the Willis' circle are shown and results of statistical analysis are presented. Results of numerical analysis of the blood flow in the constructed models are obtained using software ANSYS/CFX and presented in Section 4.

2. Method of variation of slope of the scanning plane

Let us formulate basic principles of investigation of the object using MRI. Object under investigation is scanned by the set of parallel planes given by the normal vector to this set. Elementary MRI investigation includes the following main stages.

(i) In accordance with specified set of scanning planes scanner discretizes the physical continuous three-dimensional space into elementary volumes — voxels.

(ii) MRI scanner registers summary NMR signal of hydrogen atoms being in the same voxel.

(iii) Voxels lying in the one scanning plane create image of the one section of the object.

Results of the scanning are saved by the scanner in the form of images of the section. Wherein the created image is grey-scaled so that level of grey is proportional to the level of NMR signal. Mathematically every this image is given by the numerical matrix and rows and columns of the matrix correspond to numeration of pixels in the image. Value of element of this matrix is defined by the level of NMR signal in the corresponding voxel.

Level of NMR signal in voxel, and consequently brightness of corresponding pixel of the image, depends on the scalar product (v_{ijk}, n), v_{ijk} — velocity of the blood flow in voxel (ijk), n — normal vector to the scanning plane. Dependence of the pixel brightness on the said scalar product is such that when it equals (or approximately equals) to zero, then scanner registers it as absence of the blood flow in voxel (ijk). This is true only if $v_{ijk} = 0$ and is false when $v_{ijk} \perp n$. The described feature of creation of image according to MRI data leads to the following. In models of vessel nets (created by such images) arise interrupts of vessels (vessels are fragmented) in voxels where $v_{ijk} \perp n$. Fragmented models of nets are unsuitable for hemodynamic computation and are limitedly suitable for analysis of their geometric features. Obviously there is necessity of creation of the method oriented to building non-fragmented models of vessel nets.

In the paper [1] method of MRI data processing was proposed, it is called method of variation of the slope of scanning plane, it allows to build non-fragmented models of vessel net. This method consists of carrying out scanning using several sets of parallel planes, recalculation of all available data to the unit grid domain and combination (in some way) of recalculated data. Used sets of planes are given by different normal vectors. Direct building of vessel net is performed by obtained combined data using software for image segmentation [2, 5].

3. Comparison of models of nets for different genetic lines

3.1. Description of genetic lines

As comparing genetic lines of laboratory mice in this paper, we consider the following two lines.

(i) Genetic line with the pathology: the line with a knock-out on the tumour necrosis factor (TNF), more details about this genetic line, see [7].

(ii) Control genetic line: C57Bl/6; the knock-out line have been obtained on the basis of this genetic line. Line C57Bl/6 is usually used as standard one for majority pharmacological preclinical research.

3.2. Building of models of vessel nets

Building of geometric models of vessel nets of investigated animals was realised using method of variation of the slope of scanning plane with five sets of scanning planes. Scanning planes are deviated from the axial one (i.e. from the standard scanning direction being the axis of symmetry of the object) on 15 and 30 degree in two orthogonal directions. Choice of these angles of deviation is caused by the following. When the deviation angle of the scanning plane from the axial one is increased, relation signal/noise is decreased and becomes unacceptably small when the angle is near 90 degree. In addition when scanning plane is turned on 45 degree or more, MRI scanner recalls coordinate axes leading to the turning of the image.

Non-fragmented three-dimensional models of vessel nets of all animals under investigation have been constructed using method of variation of the slope of the scanning plane with the usage of software ITK-Snap [2]. In Figure 1 you can see images of three-dimensional models of vessel nets for animals of genetic line with TNF knock-out, in Figure 2 — for animals of genetic line C57Bl/6.

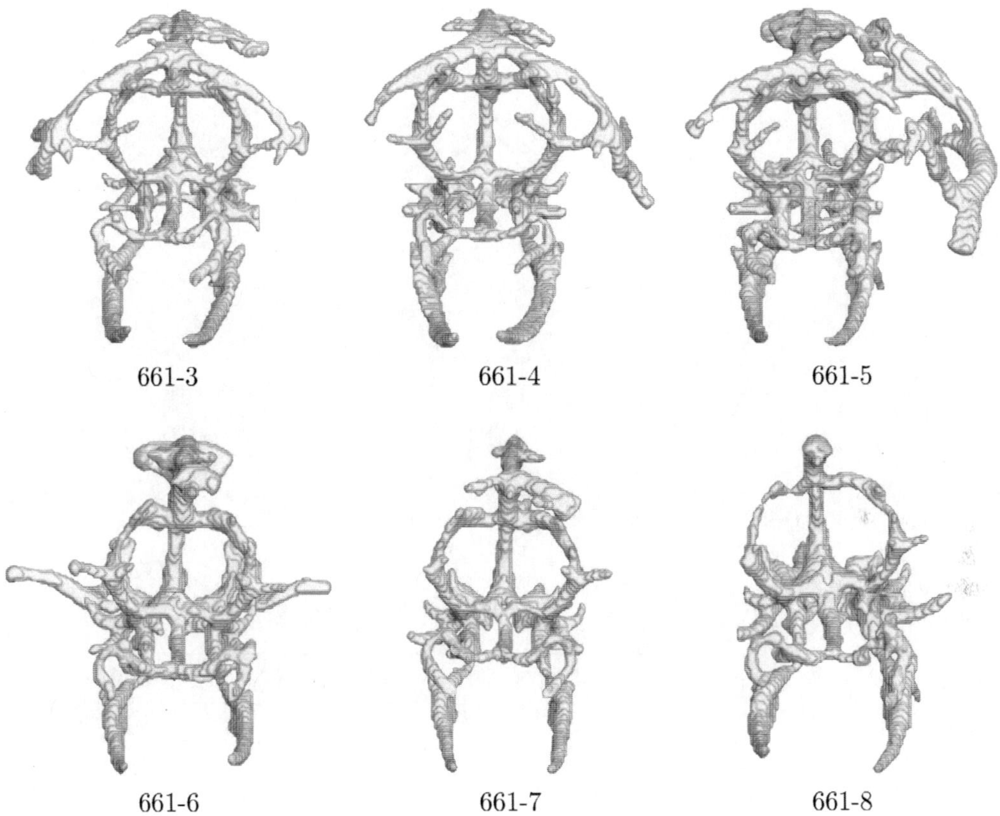

Figure 1. Models of vessel nets for animals of genetic line with TNF knock-out.

Results of measurements of angles between the vessels of Willis' circle (being the part of build three-dimensional models) shown in Table 1. In the table raws correspond to condition numbers of investigated animals, wherein prefix 661 corresponds to animals from genetic line with TNF knock-out, and prefix BL — to animals from genetic line C57Bl/6. The values of angles between blood vessels in places of their confluence adduce in columns. The angles between left and right anterior cerebral arteries (ACA) are adduced in the second column. The angles between left and right ACA and posterior communication arteries (PCoA) are adduced in the third and fourth columns. The angle between left and right posterior cerebral arteries (PCeA) is adduced in the fifth column.

Journal of Physics: Conference Series **722** (2016) 012029 doi:10.1088/1742-6596/722/1/012029

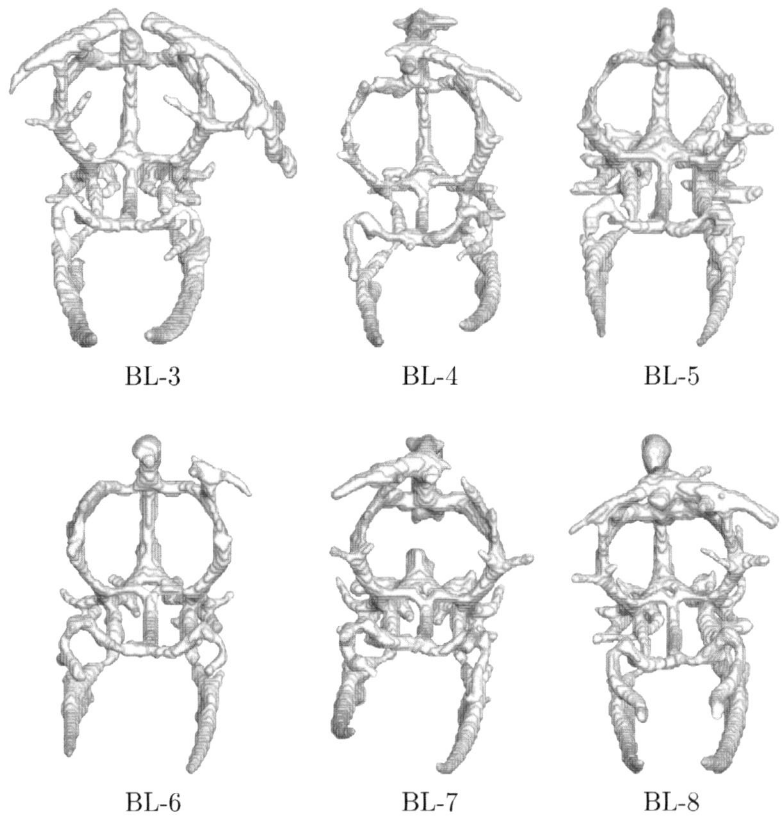

Figure 2. Models of vessel nets for animals of genetic line C57Bl/6.

Table 1. Angles of Willis' circle (in degrees).

Animal	ACA	ACA and PCoA (l.)	ACA and PCoA (r.)	PCeA
661-3	75	145	145	80
661-4	88	153	150	96
661-5	58	160	150	110
661-6	65	145	144	90
661-7	60	150	145	95
661-8	60	150	150	92
BL-3	60	145	150	96
BL-4	70	145	135	90
BL-5	45	155	155	85
BL-6	60	155	158	100
BL-7	50	160	150	90
BL-8	70	150	145	95

Table 2. Boundary conditions for blood flow velocity (in cm^3/s) in numerical simulation.

CCA (l.)	CCA (r.)	VA (l.)	VA (r.)
8.0	8.0	5.	5.

Statistical analysis of the measured angles between the arteries of Willis' circle was performed. No statistically significant distinctions in construction of Willis' circle between the two genetic lines were found. In spite of normal distribution of the mark but in connection with small sample the data analysis was carried out using parametric and non-parametric method. Application of Mann–Whitney–Wilcoxon test gives statistical significance of distinctions between the genetic lines less than 0.4 (despite the fact that values are statistically significant if they are less than 0.05). Results of analysis using Student's t-test considered to be similar and statistical significance is no less than 0.2.

4. Numerical analysis of hemodynamics

Software ANSYS/CFX is used for numerical analysis of hemodynamics. Computation carried out on the base of supercomputer centre of Novosibirsk State University. Model of incompressible viscous liquid (viscosity and density of liquid is given equal to viscosity and density of water) is used for numerical modelling. Walls of blood vessels considered to be rigid and condition of adhesion is given to them. Flow considered to be laminar, so the mathematical model of the blood flow consists of four equations expressing conservation laws of mass and three pulse component.

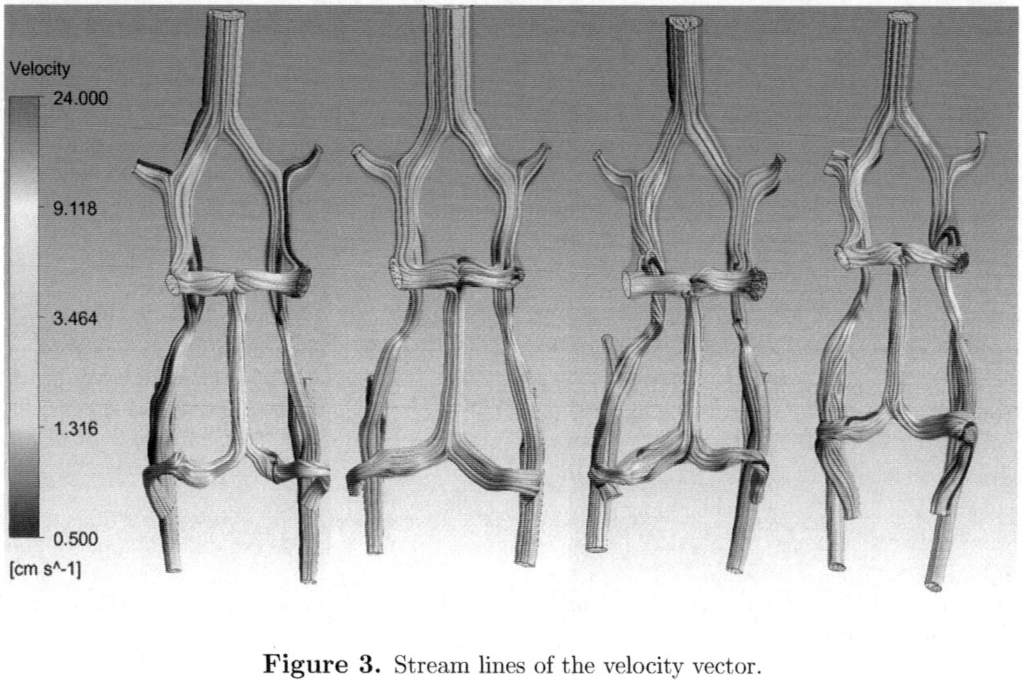

Figure 3. Stream lines of the velocity vector.

At the inlets of vertebral and carotid arteries we set blood flow velocity, at the outlet of others cerebral arteries we set flowrate. Values of the blood flow velocity used in numerical simulations

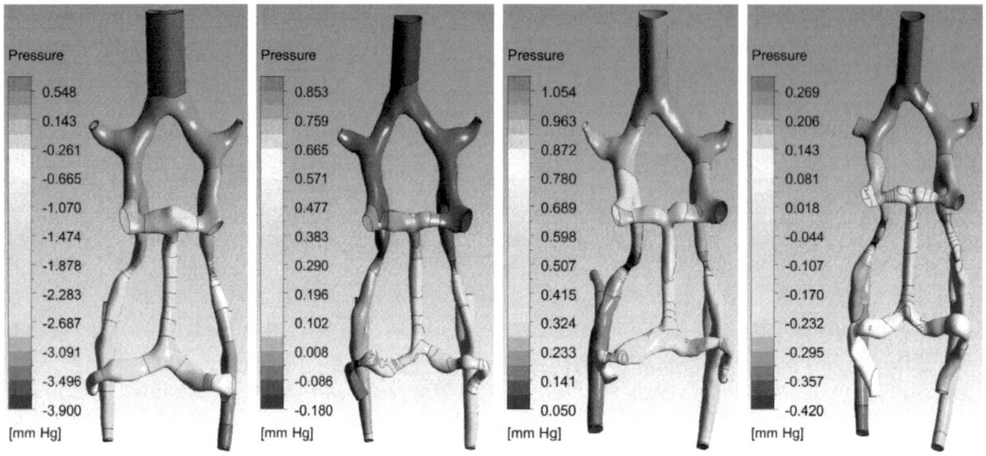

Figure 4. Variations from mean pressure distribution on the vessel wall.

are shown in Table 2. Values of the blood flow velocity in left and right Common cerebral artery (CCA) are adduced in the first and second columns. Values of the blood flow velocity in left and right Vertebral artery (VA) are adduced in the third and forth columns. In outlets we set volumetric rate different from the vessel outlet in the cross-section area.

Results of numerical simulations are shown below. In Figures 3, 4 you can see stream lines and variations in mean pressure distribution of the vessel wall near Willis' circle. In both figures the first and second models correspond to the animals of the genetic line with TNF knock-out, third and forth correspond to the animals of the genetic line C57Bl/6.

5. Conclutions

In the paper the construction of the models of vascular channel for the laboratory animals of two genetic lines realised using the method of variation of slope of scanning plane. The method showed its efficiency and allowed for construction of the models of vascular channel suitable for hemodynamic calculations and statistical analysis of vessel architectonics. Also it was shown that there is no influence of TNF knock-out on morphological and hydrodynamical characteristics of Willis' circle of adult animals.

Acknowledgments

The study was completed thanks to the support of Russian Science Foundation (project No. 14-35-00020, all MRI experimentation studies using Bruker BioSpec 117/16 USR scanner), Russian Foundation for Basic Research (project No.14-01-00036, mathematical modelling) at the site of the Centre for Laboratory Animal Genetic Resources, Institute of Cytology and Genetics, Siberian Branch of the Russian Academy of Sciences (RFMEFI61914X0005 and RFMEFI62114X0010).

References

[1] Maltseva S V, Cherevko A A, Khe A K, Akulov A E, Savelov A A, Tulupov A A, Derevtsov E Yu, Moshkin M P and Chupakhin A P 2016 *Appl. Magn. Reson.* **47** 23–39

[2] Yushkevich P A, Piven J, Hazlett H C, Smith R G, Ho S, Gee J C and Gerig G 2006 *Neuroimage* **31** 1116–28

[3] Vorobtsova N A, Yanchenko A A, Cherevko A A, Chupakhin A P, Krivoshapkin A L, Orlov K Yu, Panarin V A and Baranov V I 2013 *Russ. J. Numer. Anal. M.* **28** 505–16

[4] Yanchenko A A, Cherevko A A, Chupakhin A P, Krivoshapkin A L and Orlov K Yu 2014 *Russ. J. Numer. Anal. M.* **29** 307–17

[5] Angenent S, Pichon E and Tennenbaum A 2006 *B. Am. Math. Soc.* **43** 365–96
[6] Beckmann N 2000 *Magn. Reson. Med.* **44** 252–8
[7] Fursenko D V, Khotskin N V, Kulikov V A and Kulikov A V 2015 *Vavilov Journal of Genetics and Breeding* **19** 394–8

Differential properties of Van der Pol – Duffing mathematical model of cerebrovascular haemodynamics based on clinical measurements

D V Parshin[1], I V Ufimtseva[1,2], A A Cherevko[1,2], A K Khe[1,2], K Yu Orlov[3], A L Krivoshapkin[3] and A P Chupakhin[1,2]

[1] Lavrentyev Institute of Hydrodynamics of the Siberian Branch of the Russian Academy of Sciences, 630090 Lavrentyev av 15, Novosibirsk, Russia
[2] Novosibirsk State University, 630090 Pirogova st 2, Novosibirsk , Russia
[3] E. N. Meshalkin Research Institute of circulation pathology, 630055 Rechkunovskaya st 15, Novosibirsk, Russia

E-mail: *danilo.skiman@gmail.com*, *i_ufimtseva@mail.ru*

Abstract. The present paper discusses the method of identification (diseased/healthy) human cerebral vessels by using of mathematical model. Human cerebral circulation as a single tuned circuit, which consists of blood flow, elastic vessels and elastic brain gel tissue is under consideration. Non linear Van der Pol-Duffing equation is assumed as mathematical model of cerebrovascular circulation. Hypothesis of vascular pathology existence in some position of blood vessel, based on mathematical model properties for this position is formulated. Good reliability of hypothesis is proved statistically for 7 patients with arterial aneurysms.

1. Introduction

Human cerebral circulation is a complex haemodynamics system. It consists of elastic vessels, surrounded by elastic brain tissue and the blood in vessels, which is a liquid with complex reology[1]. Cerebrovascular pathologies are not as frequent as heart or aorta pathologies [2, 3], however, they are extremely dangerous because of it's mortality and disability risks. Arterial aneurysms (AA) are rather common for all nationalities and genders and occur in about 2 cases for 10 000 people. They are extremely increasing risk of morbidity for pregnant women in case of rupture [4].

All aneurysms are dangerous because of rupture risk. There are two different approaches in treatment. When such kind of pathology is located in exterior brain vessels– neurosurgeons usually apply microsurgery methods with using clips [5]. Aneurysm of proximal brain vessels is usually treated by using noninvasive methods, for example like in [6].

Present study is about features of Van der Pol-Duffing oscillator mathematical model in applications to cerebrovascular haemodynamics in presence of arterial aneurysm. The whole research is devised on several stages: clinical measurements, data processing, analytical research. The current study deals with analysis of data processed. Clinical data was obtained at Meshalkin institute of circulation pathology, Novosibirsk. This data was obtained during real-time cerebrovascular measurements for neurosurgery operations. The pressure and velocity were measured with ComboWire sensor and ComboMap station (Volcano Corp.)[7, 8]. The data for

7 patients (totally 99 measurements) was collected and processed. All measurements were made in arteries of human brain. By ethical reasons we will use only first letters of patients last names to identify them: G1,K1,P1,P2,T1,R1,S1. Data set of the clinical measurements for one patient consists of pressure and velocity values obtained on the interval from 5 sec till 2 minutes, for different ComboWire sensor positions in blood vessel (minimum 5 sec per each position). The data obtained is processed as in [8].
So the mathematical model which is under consideration is empirical and described by Van der Pol-Duffing nonlinear ordinary differential equation:

$$\varepsilon q'' + (a_1 + a_2 q + a_3 q^2)q' + (b_1 q + b_2 q^2 + b_3 q^3) = ku, \tag{1}$$

where q–pressure and u–velocity in some position in blood vessel with respect to time. Clinical data obtained describes the pressure and velocity values on the same intervals, on which values are seen during the operation. The mathematical model constructed in [9] allows to expand the region of haemodynamics parameters that can be investigated and to try different modes of external force-right hand side of equation (1).

In equation (1) quantity ε is a small parameter, q, u- are dimensionless pressure and velocity in a flow. Comparison of the numerical solution of the equation (1) to the clinical data shows that this equation has good approximation of pressure and velocity values. So we may assume that generally $a_3 \neq 0$. Also point that the region of interest of variables is $\{q \subset [-1, 1], u \subset [-1, 1]\}$, because equation 1 is dimensionless, and upload is assumed as $u = sin(2\pi\omega t)$.

2. Mathematical model and methods

By using of several substitutions like in [10] or more modern approach [11] the equation (1) leads us to dynamic system:

$$p' = -(b_1 q + b_2 q^2 + b_3 q^3) + ksin(2\pi\theta),$$

$$\varepsilon q' = a_3^2 p - a_1 a_3 q - a_2 a_3 q^2/2 - a^3 q^3/3, \tag{2}$$

$$\theta' = \omega.$$

Here the prime means the differentiation with respect to t and $p = \varepsilon q'/a_3^2 + a_1 q/a_3 + a_2 q^2/a_3^2 + q^3/3$. When value of ε is small, the system (2) is slow-fast [10, 11, 12]. The right hand side of the second equation of this system defines *slow s rface*. It is pure fact [13] that, the solution of equation (1) is in vicinity $\delta(\varepsilon)$ of slow surface S and consists of the combination of the solutions of slow and fast subsystems (view Fig. 1). For system (2) surface S is defined by formula:

$$f(p, q, \theta) \equiv a_3^2 p - a_1 a_3 q - a_2 a_3 q^2/2 - a_3^2 q^3/3 = 0. \tag{3}$$

As seen from Fig. 1 study of slow motions of system (2) is very important. Slow motions for system (2) are motions of integral curves through the surface S. As for points of S the equation (3) is done. System 2 over surface S seems like:

$$q' = -(b_1 q + b_2 q^2 + b_3) + k \sin 2\pi\theta,$$

$$\theta' = (q^2 + q a_2/a_3 + a_1/a_3)\omega \tag{4}$$

Here the prime means the differentiation with respect to new modified "time" $t_1 = t(q^2 + q a_2/a3 + a_1/a_3)$. To simplify calculations let us denote new variable t_1 just t again.

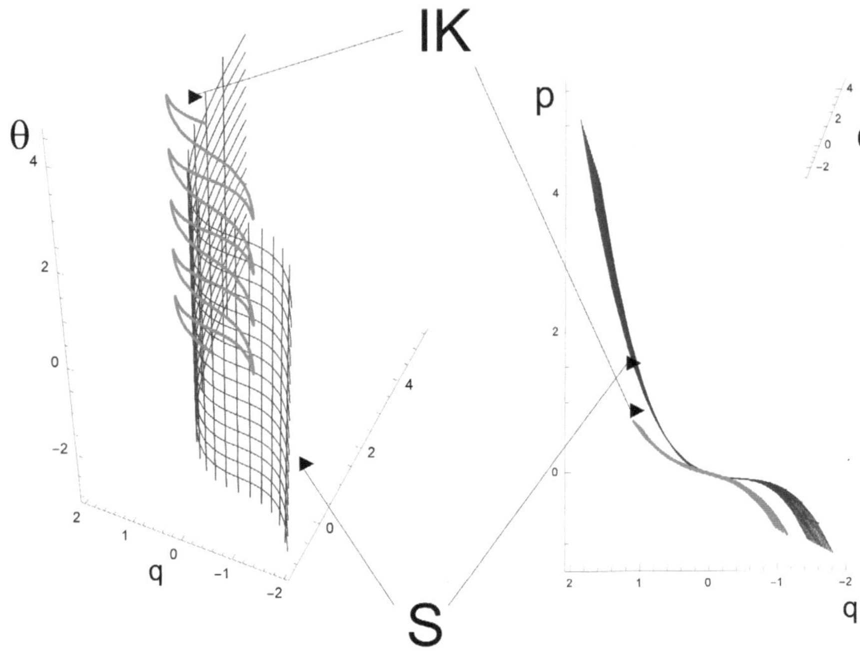

Figure 1. Slow surface S and integral curve IK of system (2) where ($k = 1.4422, a1 = 0.095381, a2 = -0.027, a3 = 0.056639, b1 = 1.1068, b2 = 0.1231, b3 = -0.14913$).

To investigate the properties of the solution of system (4) it is important to find the singularities and stationary points of this system. In order to calculate the stationary points it is necessary to solve the system of equations:

$$\sin 2\pi\theta = (b_1 q + b_2 q^2 + b_3)/k, \qquad q^2 + q a_2/a_3 + a_1/a_3 = 0. \tag{5}$$

To obtain the singularities it is necessary to calculate eigenvalues of Jacoby-matrix for system (4). Here this Jacoby-matrix is presented:

$$J = \begin{pmatrix} -(b_1 + 2b_2 q + 3b_3 q^2) & -2\pi k \cos 2\pi\theta \\ 2q + a_2/a_3 & 0 \end{pmatrix}. \tag{6}$$

Let us consider eigenvalues of J on the solution of system (5):

$$q_0^{\pm} = (-a_2/a_3 \pm \sqrt{a_2^2/a_3^2 - 4a_1 a_3}), \theta_n^{\pm} = \frac{\arcsin\left(b_1 q_0^{\pm} + b_2 q_0^{\pm^2} + b_3 q_0^{\pm^3}\right)/k}{2\pi} - n, \tag{7}$$

where n-whole numbers. It is useful to denote:

$$\sigma = k^2 - (b_1 q_0 + b_2 q_0^2 + b_3 q_0^3)^2, tr_0 = -(b_1 + 2b_2 q_0^{\pm} + 3b_3 \left(q_0^{\pm}\right)^2),$$

$$\Gamma(q, \theta) = s(\theta) 8\pi (2q + a_2/a_3)\sqrt{k^2 - (b_1 q + b_2 q^2 + b_3 q^3)^2}, C_0 = b_1 + 2b_2 q_0^{\pm} + 3b_3 \left(q_0^{\pm}\right)^2, \tag{8}$$

where

$$s(\theta) = 1|\theta \in [\theta_1, \theta_2], s(\theta) = -1|\theta \in [\theta_3, \theta_4], \quad (9)$$

and $\theta_1 = 0+l, \theta_2 = 1/2+l, \theta_3 = 1/2+n, \theta_4 = 1+n$, and l, n-whole numbers. It is of importance that singularities exist only when:

$$\sigma \geq 0. \quad (10)$$

The condition (10) is used for numerical analysis to define the existence of the singularities. Further assumption is very essential:
Ass mption. It is reasonable to consider only hyperbolic singularities for living systems.
Really, this assumption is very essential because for living system some variability of parameters is necessary. This variability allows the system to be stable under small parameters perturbations.
Note. Only hyperbolic singularities of system (4) were found by numerical investigation with varying of ω.

The cases when one or both of eigenvalues of Jacoby-matrix equal to zero or pure imaginary are out of interest for current study for the reasons mentioned above, but anyway they should be considered as critical modes of living system. However, this problem is very complicated and deserves special consideration.
So the classification of all hyperbolic singularities of system (4) was completely obtained (view Table. 1):

Table 1. Classification of hyperbolic singularities of system (4)

Values of C, Γ	Value of tr_0	Phase of θ	Type of singularity at $q_0^{\pm}$
$C^2 > \Gamma$	$tr_0 > 0$	$\theta \in (\theta_1, \theta_2]$	q_0^+-unstable node,q_0^--saddle
$C^2 > \Gamma$	$tr_0 < 0$	$\theta \in (\theta_1, \theta_2]$	q_0^+-stable node,q_0^--saddle
$C^2 > \Gamma$	$tr_0 > 0$	$\theta \in (\theta_3, \theta_4]$	q_0^+-saddle,q_0^--unstable node
$C^2 > \Gamma$	$tr_0 < 0$	$\theta \in (\theta_3, \theta_4]$	q_0^+-saddle,q_0^--stable node
$C \neq 0, C^2 < \Gamma$	$tr_0 < 0$	–	stable focus
$C \neq 0, C^2 < \Gamma$	$tr_0 > 0$	–	unstable focus

3. Protocol, data development and statistics

In present study the clinical data of 7 patients is under consideration. Generally 99 measurements were performed and 73 packs of coefficients of equation (1) were obtained. Only one patient is discussed in details. For other only final results will be presented in the table.
Definition. Let us talk that measurement position has index n if the system (4) has n different singularities.

And now let us present the changing of measurement index for patient $G1$ before treatment, during the operation and after the operation for different positions in blood vessel, containing arterial aneurysm.

Angiogram and measurements scheme are presented (view Fig. 2). The number in box/pentagon/round means the ordinal number of measurement. The boxes correspond to the case, when the index of measurement is equal to zero, pentagons– the case when index is

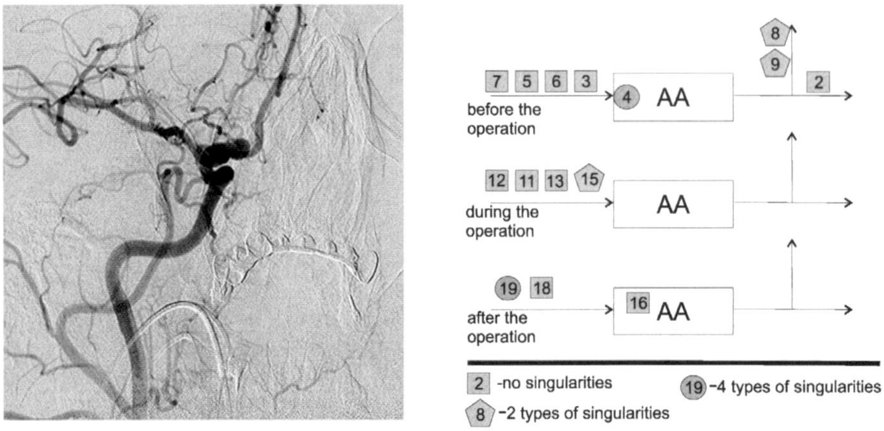

Figure 2. Angiogram and scheme of singularities for patient G1.

equal 2, and the circles, for the case when index is 4. The measurement scheme (view Fig. 2 at the right) consists of three parts. The top part corresponds to the measurements before the operation, the center part corresponds to the operation time measurements and the bottom part corresponds to the measurements after the operation (20-40 minutes).

It is seen from Fig. 2 that before the operation the measurement near arterial aneurysm has 2 or 4 types of singularities in terms discussed above.

4. Results and discussion

Comparison of the current analysis results to the clinical visualization data allows us to formulate hypothesis.

H pot esis(Test). The presence of arterial aneurysm in blood vessel could be detected by index value of the measurement for this position.

In terms of Van der Pol-Duffing mathematical model it means the existence of singularities.

Some statistics obtained is shown here (view Table. 2) and proofs the Test. In the column "Test" + means that test is proved, — test is refuted and ± results obtained do not contradict to the test. It is seen from Fig. 2 that test is proved in 71, 5% of cases, and refuted in 14, 3% of

Table 2. Test statistics

Ptient id	Gender, Age	Aneurysm type	Aneurysm location	Test
$G1$	$M, 42$	normal	ICA	+
$K1$	$F, 65$	giant	ICA	+
$P1$	$M, 68$	normal	MCA	+
$P2$	$F, 55$	giant	ICA	+
$R1$	$F, 47$	giant	bifurcation of BA	±
$S1$	$M, 40$	normal	bifurcation of the basilar apex	+
$T1$	$F, 67$	giant	ICA	-

cases considered. This statistics shows rather high reliability of the test. For patient $R1$ the test could not be proved or refuted, because there were no measurements near aneurysm obtained before the operation. For the most frequent location of aneurysm considered (ICA-interior carotid artery) the test has reliability 75%. Concerning to the size of aneurysm it is seen that test is proved for all (100%) of normal aneurysms and has reliability only 50% for giant aneurysms.

From physiological point of view this test is reasonable. The living system can not live normally under complicated conditions. The existence of such systems completely depends on their ability of energy saving and optimal energy transformation. The presence of singularities (aneurysms) makes such systems less effective and that is bad from the medical point of view, because the blood vessel with aneurysm can not transport blood from aneurysm to distal parts of the brain circulation system optimally. Hence, the existence of aneurysm coincides with the existence of singularities system (4)based on the clinical measurements taken near the aneurysm.

The results obtained during and after the operations do not show any common behavior of singularities system (4). Further research is necessary to answer the question whether there are any addictions to differential properties of system (4). This difficulty is due firstly to the perturbations– occur during and immediately after the treatment. These perturbations occur because regeneration of a normal blood flow function could lead to substantial restructuring of the cerebral circulation.

5. Conclusions

The analysis of the non linear oscillator model, based on clinical data obtained during neurosurgery treatment monitoring was performed. As result of the analysis the test on presence of arterial aneurysm in blood vessel was developed. Good reliability of this test was statistically proven for ICA cerebral aneurysms. The test is proven by the measurements obtained before the operation. The influence of treatment on differential properties of Van der Pol-Duffing model could not be discussed from the data considered.

The test is a challenge for medical and engineering society, which should pay attention to 0-d models of cerebral haemodynamics and use of modern measurement devices for developing of personalized treatment protocols in vascular neurosurgery. Also this test can be used for perspective aneurysm diagnostics (without MRI or CT) and for developing of perspective embolization devices.

Acknowledgments

The study was completed thanks to the support of Russian Foundation for Basic Research (project No 14-01-00036).

References

[1] Sankara D S and Hemalathab K 2006 *J. App. Math. Mod.* **31** 9 1847–64
[2] Criqui M H and Aboyans V 2015 *irc. Res.* **116** 1509-26
[3] Poirier P, Giles T D, Bray G A, Hong Y, Stern J S, Pi-Sunyer F X and Eckel R H 2006 *irc. 113* 898-918
[4] Meyersa P M, Halbacha V V, Maleka A M, Phatourosa C C, Dowda C F, Lawtona M T, Lemperta T E and Higashida R T 2000 *AJNR.* **21** 1306-11
[5] Hanel R A and Spetzler R F 2008 *Neurosur ery* **6** 3 1289–99.
[6] Alderazi Y J, Shastri D, Kass-Hout T, Prestigiacomo C J and Gandhi C D 2014 *Stro e Research and Treatment* (2014)415653
[7] Ferns S P, Schneiders J J, Siebes M, van den Berg R, van Bavel E T and Majoie C B 2010 *AJNR.* **31** 324-26
[8] Chupakhin A P, Cherevko A A, Khe A K, Telegina N J, Krivoshapkin A L, Orlov K J, Panarin V A and Baranov V I 2012 *irculation atholo y and ardiac Sur ery* (Novosibirsk, Meshalkin Research institute of circulation pathology). **16** 44 27-31 (in Russian)
[9] Mikhailova A V, Cherevko A A, Chupakhin A P, Krivoshapkin A L, Orlov K Yu and Panarin V A 2013 Modeling of the brain haemodynamic based on experimental clinical data *roceedin s of nternational*

conference on Mathematical ontrol Theory and Mechanics (Suzdal, Vladimir state university) 166-67 (in Russian)

[10] Mischenko E F, Rozov N H 1975 *Differential equations ith small parameter and relaxation oscillations* (Moscow, Science) (in Russian)

[11] Guckenheimer J, Ilyashenko Yu S 2001 *Mosco Math. J.* **1** 1 27—47

[12] Kozlov A K, Sushchik M M, Molkov Y I, Kuznetsov A S 1998 *Bulletin of University of Nizhny Nov orod. Radiophysics* **1** 89-104 (in Russian)

[13] Arnold V I, Afraymovich V S, Ilyashenko Yu S, Shilnokov L P 1986 *Dynamic systems* (Moscow, Allrussian institue of science and technical information, modern problems of mathematics. Fundamental directions) **5**

Stratification parameters and dispersion of internal solitary waves

E G Perevalova[1] **and N I Makarenko**[1,2]

[1]Lavrentyev Institute of Hydrodynamics SB RAS, 630090 Novosibirsk, Russia
[2]Novosibirsk State University, 630090 Novosibirsk, Russia

E-mail: evperev@gmail.com

Abstract. A theoretical model of internal solitary waves of large amplitude in a weakly stratified fluid is considered. It is assumed that the background density profile depends linearly or exponentially on the fluid depth. It is demonstrated that inverse problem on determining fine-scale structure of the density profile by a known curve of amplitude dispersion is reduced to solving a linear Fredholm integral equation of the first kind having special form of the kernel. The one-to-one correspondence between the density coefficient and the dispersion function is established in the case of analytical stratification.

1. Introduction

We study in this paper the relationship between the density stratification of a non-homogeneous fluid and kinematic characteristics of nonlinear internal waves. The attention is focused on analysis of the possibility to restore the density profile by the known curves of amplitude dispersion for internal solitary waves. This interest is stimulated by the problems of remote sensing internal waves with straight thermohaline measurements. At present, these operations employ a set of experimental oceanographic methods which include the observations from the ships and long-term buoy stations, monitoring satellite data and measurements by deep research devices [1]. The problem on determining the density of vertically stratified environment by the known curves of the phase velocity of linear dispersive waves has already been considered [2, 3]. These studies used the theory of inverse spectral problem based on the method of integral equations. We propose to consider mathematical model of a strongly nonlinear soliton-type waves in order to formulate the inverse problem on the density stratification. Numerous observation data [4] show that the trains of internal solitary waves form regularly by the interaction of tidal currents with continental slope. Spatial amplitudes of these nonlinear disturbances are comparable with total depth of sea water on the shelf [5, 6]. Therefore the asymptotic analysis which involves non-linear equations of dispersive long waves [7] seems to be suitable for the modelling. This method uses asymptotic expansion on powers of the small Boussinesq parameter which characterizes vertical gradient of the fluid density. By those circumstances, it is supposed that the fluid density disturbs slightly the linear- or exponential profile, and there are no hypotheses on small amplitude of the waves. As a result, the inverse problem is reduced to solving the Fredholm integral equation of the first kind which couples the density profile with the function describing amplitude dispersion of solitary wave.

2. Basic equations

We consider two-dimensional flow of an ideal incompressible non-homogeneous fluid in a horizontal layer bounded by a flat bottom $y = 0$ and rigid lid $y = h$. The flow is stationary in the frame of the reference associated with the solitary wave. Therefore, the stationary Euler equations are used here as a basic mathematical model for velocity field $u = (u, v)$, density ρ and pressure p:

$$u_x + v_y = 0, \quad u\rho_x + v\rho_y = 0, \quad uu_x + vu_y + \rho^{-1}p_x = 0, \quad uv_x + vv_y + \rho^{-1}p_y = -g \tag{1}$$

where g is the gravity acceleration. The boundary conditions at the bottom and the lid require the equality $v = 0$ which should be satisfied at $y = 0$ and $y = h$. Furthermore, the velocity vector $u(x, y)$ should tend to the constant vector $u_0 = (u_0, 0)$ as $|x| \to \infty$. Here u_0 is the wave speed with respect to quiescent fluid. Similarly, the fluid density $\rho(x, y)$ should tend to its undisturbed profile $\rho_\infty(y)$. It is well known [8] that the introduction of a stream function ψ by the formulae $u = \psi_y$, $v = -\psi_x$ leads after eliminating the pressure p from the system (1) to single scalar PDE equation of the second order. This is the Dubreil-Jacotin — Long equation

$$\rho(\psi)(\psi_{xx} + \psi_{yy}) + \rho'(\psi)\Big(gy - \frac{y\psi}{u_0} + \frac{1}{2}(\psi_x^2 + \psi_y^2 - u_0^2)\Big) = 0. \tag{2}$$

The coefficient $\rho(\psi)$ depends here on the stream function due to the formula $\rho(\psi) = \rho_\infty(\psi/u_0)$. Boundary conditions on rigid walls and radiation condition at infinity have the form

$$\psi(x, 0) = 0, \quad \psi(x, h) = u_0 h; \quad \psi(x, y) \to u_0 y \quad (x \to -\infty). \tag{3}$$

The direct problem on solitary wave requires a solution ψ of equations (2), (3) which differs from uniform flow $\psi_\infty(y) = u_0 y$. By that, the dependence of density $\rho_\infty(y)$ on y is assumed to be known far upstream. In this context, the problem is formulated as a nonlinear eigenvalue problem with spectral parameter u_0 which determines the speed of solitary wave. Let us note that the equation (2) admits a variational formulation with the Lagrangian

$$L = -\frac{1}{2}\int \rho(\psi)(\psi_x^2 + \psi_y^2 - u_0^2) + g\int_{u_0 y}^{\psi} (\rho(\chi) - \rho(\psi))d(\chi). \tag{4}$$

The undisturbed upstream flow is characterized by two dimensionless parameters, i.e. the Boussinesq parameter σ and the parameter λ which is a square of inverse densimetric Froude number,

$$\sigma = \frac{{N_0}^2 h_0}{\pi g}, \quad \lambda = \frac{\sigma g h_0}{\pi {u_0}^2}.$$

Here the constant N_0 characterizes the buoyancy frequency N defined by the formula $N^2(y) = -g\rho'_\infty(y)/\rho_\infty(y)$. Thus, the parameter σ is an appropriate small parameter appearing in the case of weak stratification. According to the well-known ideas about the thermohaline stratification of sea water [7, 9–11], the fluid density can be modeled at the equilibrium state by the equation

$$\rho_\infty(y) = \rho_\infty(0)\Big(1 - \sigma\rho_*(\pi y/h) - \sigma^2\rho_1(\pi y/h, \sigma)\Big)$$

where the function ρ_1 defines the fine-scale structure of the density profile. The scaling factor π is entered here in order to simplify the form of trigonometric eigenfunctions related to the normal modes of internal waves.

Note that there is a relatively small number of background density profiles ρ_* which can be realized for sea water by natural conditions. They include the linear- and the exponential profiles, and also the layered stratification with piecewise constant density can be observed, as well as all these profiles can be combined. In contrast, the fine-scale density ρ_1 appears to be much more diverse. However, the typical time of its evolution can be clearly longer than the temporal periods of internal waves in a coastal zones. Therefore we can simulate this stratification by stationary profile having the dimensionless form

$$\rho(y,\sigma) = 1 - \sigma y - \sigma^2\rho_0(y) + O(\sigma^3) \qquad (5)$$

with $0 \leq y \leq \pi$. The function $\rho_0(y) = \rho_1(y,0)$ characterizes here the fine-scale stratification at the leading order on σ. The class of the profiles (5) include the linear density $\rho = 1 - \sigma y$ and the exponential stratification $\rho = exp(-\sigma y)$.

3. Direct problem on solitary waves in a weakly stratified fluid

Following [7], we consider here the long-wave approximation having the ratio of the vertical- and horizontal scales of the order $\sqrt{\sigma}$. It can be shown that this asymptotic order follows immediately from the dispersion relation for the modes of long waves corresponding to the density profile (5). Using the dimensionless variables with "slow" horizontal variable $x \to \sqrt{\sigma}x$ in equations (2), (3), we can consider only the terms up to the order $O(\sigma)$ inclusively,

$$\sigma\psi_{xx} + \psi_{yy} + \lambda(\psi - y) = \sigma\Big(\psi\psi_{yy} + (y-\psi)\rho_0'(\psi) + \frac{1}{2}(\psi_y^2 - 1)\Big) + O(\sigma^2), \qquad (6)$$

$$\psi(x,y) = 0, \quad \psi(x,\pi) = \pi; \quad \psi(x,y) = y \quad (x \to \infty).$$

We look for the solution in the form

$$\psi(x,0) = y + v_0(x,y) + \sigma v_1(x,y) + ...,$$

and the parameter λ should be also presented by the power series $\lambda = \lambda_0 + \sigma\lambda_1 + ...$ with small parameter σ. Thus, we obtain from the equation (6) the recursive set of equations for the coefficients v_i $(i = 1,2,...)$

$$v_{iyy} + \lambda_0 v_i = f_i \quad (0 < y < \pi),$$

$$v_i(x,0) = v_i(x,\pi) = 0,$$

where the right-hand terms f_0 and f_1 have the form

$$f_0 = 0, \quad f_1 = -v_{0xx} + (y + v_0)v_{0yy} - \lambda\rho_0'(y + v_0)v_0 + v_{0y} + \frac{1}{2}v_{0y}^2 - \lambda_1 v_0.$$

The equations with $i = 0$ result $\lambda_0 = m^2$ $(m = 1,2,3,...)$. Thus, the 1-mode wave solution has the form $v_0(x,y) = a(x)\sin y$ while the function $a(x)$ should be determined from the compatibility conditions for a system of subsequent equations with $i = 1$. Namely, the orthogonality condition with the eigenfunction of the first mode should be fulfilled for the right-hand side term f_1 as follows:

$$\int_0^\pi f_1(x,y)\sin y dy = 0.$$

As a result, this gives the nonlinear ordinary differential equation

$$\frac{d^2a}{dx^2} + F'_a(a) = 0, \tag{7}$$

where the nonlinearity $F(a) = s(a) + \frac{1}{2}\lambda_1 a^2$ is determined by the function

$$s(a) = \frac{2}{\pi}\int_0^{\pi}\int_y^{y+a\sin y} (\rho_0(y + b\sin y) - \rho_0(\psi))d\psi dy + \frac{\pi}{4}a^2 + \frac{2}{3\pi}a^3. \tag{8}$$

Note that the specific form of this function inherits the similar structure of the Lagrangian (4) for the Dubreil-Jacotin — Long equation. The function $s(a)$ has the double root at $a = 0$, therefore the function $F(a)$ also has a double root here. As a consequence, the solitary wave-solution is obtained when the value λ_1 belongs to the range of the function

$$\lambda_1(b) = -\frac{2s(b)}{b^2}, \tag{9}$$

and the corresponding value of $b \neq 0$ should be a simple root of the function F. By that, the unknown function $a(x)$ is implicitly determined by the quadrature

$$x = \pm\int_a^b \frac{d\alpha}{\alpha\sqrt{\lambda_1(\alpha) - \lambda_1(b)}}.$$

The parameter $b = a(0)$ characterizes the velocity magnitude in the middle part of the solitary wave when the wave crest places at $x = 0$, and the value $\lambda_1 = \lambda_1(b)$ determines the wave speed by the formula

$$u_0^2 = \frac{\sigma g h_0}{\pi(1 + \sigma\lambda_1(b))}$$

with the accuracy $O(\sigma^3)$. According to this formula, the function $\lambda_1(b)$ presents the amplitude dispersion function of solitary wave.

The perturbation procedure with small density slope described above has been developed first in the original paper [7] which treated the case of linear stratification. In this case, the equation (7) has the non-linearity defined by a cubic polynomial $F(a)$, as well as the classical Korteweg – de Vries model has, and similar situation occurs by an exponential stratification. The including higher-order terms to the density profile can change significantly the structure of equation (7). In particular, the perturbation of the fine-scale stratification leads to the solution like a smooth bore ("kink") and plateau-shape solitary waves ("kink-antikink" solutions). In this concern, an approximate model was proposed in the paper [9] to describe solitary wave with attached vortex zone. Another limit forms of internal solitary waves of finite amplitude such as bores and table-top waves were studied in detail in [10, 11] where the bifurcation of internal waves was associated with monotonic behavior of dispersion function $\lambda_1(b)$. In particular, the solitary wave bifurcates to the plateau-shape wave near the amplitude values b being the local minima of the function $\lambda_1(b)$ which correspond to local maxima of the wave propagation speed u_0.

Journal of Physics: Conference Series **722** (2016) 012031 doi:10.1088/1742-6596/722/1/012031

4. Inverse problem on determining the fine-scale stratification

Returning to the equations (8) and (9), we note that these formulae provide an analytical relation between the coefficient $\rho_0(y)$, that specifies the fine-scale stratification, and the dispersion function $\lambda_1(b)$. These linear relations allow to formulate the inverse problem of reconstructing the fluid density by known dispersion characteristics of nonlinear waves. Let us consider the integral term from the formula (8) having the form

$$I(b) = \frac{2}{\pi}\int\limits_0^{\pi}\int\limits_y^{y+b\sin y}(\rho_0(y+b\sin y)-\rho_0(\psi))d\psi dy \tag{10}$$

Partial integration of the variable y reduces this function $I(b)$ to the form that contains only single integrals:

$$I(b) = \frac{2}{\pi}\int\limits_0^{\pi}[b\sin y + y(1+b\cos y)]\rho_0(y+b\sin y)dy - \frac{2}{\pi}\int\limits_0^{\pi} y\rho_0(y)dy.$$

Further we change the first of these integrals by replacing y with a new integration variable

$$\psi = y + b\sin y. \tag{11}$$

In fact, this is the von Mises transformation which changes the roles of the vertical independent variable y and the stream function ψ. This transformation sets $y = y(\psi, b)$ as an implicit function of ψ and b. It is uniquely provided by the condition $\psi_y = 1 + b\cos y > 0$ which has a simple sense that there are no return flows in the middle section of a solitary wave. This requirement imposes a natural limit $|b| < 1$ on amplitude parameter b. Taking into account the definition (9) of the function $\lambda_1(b)$ and the above-formulated transformation of the function $I(b)$, we obtain from the formula (8) the equation

$$\frac{2}{\pi}\int\limits_0^{\pi}\frac{\sin(2y(\psi,b))}{1+b\cos y(\psi,b)}\rho_0(\psi)d\psi = \lambda_1(b) + \frac{\pi}{2} + \frac{4b}{3\pi}. \tag{12}$$

This relation is a linear Fredholm integral equation of the first kind with respect to the unknown function ρ_0 if the function $\lambda_1(b)$ is given. The kernel of integral operator (12) defined implicitly through the formula (11) has no singularities in view of analyticity of the function $y(\psi, b)$ by $0 \le y \le \pi$, $|b| < 1$. The Fredholm equations of this type are typical for inverse problems of mathematical physics which are generally ill-posed. However, there are effective regularizing algorithms for solving these equations. By that, the main topic is the providing sufficiently broad classes of existence and uniqueness of the solutions. Using the representation of the function $s(b)$ with the double integral (10), we can easily check that the dispersion function $\lambda_1(b)$ can be presented by the power series

$$\lambda_1(b) = \sum_{i=0}^{\infty} l_i b^i \tag{13}$$

if the function $\rho_0(y)$ has the form of a power series

$$\rho_0(y) = \sum_{i=1}^{\infty} r_i y^i. \tag{14}$$

The coefficients $\mathbf{r} = (r_1, ..., r_n, ...)^T$ and $\mathbf{l} = (l_0, ..., l_{n-1}, ...)^T$ of these series are coupled by the linear equation $\mathbf{l} = A\mathbf{r} + \mathbf{d}$ with infinite vector $\mathbf{d} = (-\pi/2, -4/3\pi, 0, ..., 0, ...)^T$ and infinite matrix $A = \|A_{ij}\|$. A more detailed analysis suggests that the matrix A is the upper triangular matrix having the coefficients as follows:

$$A_{n-i+1,n} = -\frac{4}{\pi}\frac{n-i+1}{n-i+2}C_n^{i-1}\int\limits_0^{\pi} y^{i-1}\sin^{n-i+2} y dy.$$

In particular, the diagonal elements of the matrix A have the form

$$A_{nn} = -\frac{8}{\pi}\frac{n}{n+1}\frac{n!!}{(n+1)!!}\frac{\pi}{2} \quad \Big(n \ is \ odd\Big),$$

$$A_{nn} = -\frac{8}{\pi}\frac{n}{n+1}\frac{n!!}{(n+1)!!} \quad \Big(n \ is \ even\Big).$$

These diagonal elements A_{nn} do not vanish for all integer $n \geq 1$. Thus, there is the one-to-one correspondence between the coefficients l_i and r_k which is given by the relation

$$\begin{pmatrix} l_0 \\ l_1 \\ l_2 \\ l_3 \\ \vdots \end{pmatrix} = \begin{pmatrix} -1 & -\pi & -\pi^2+3/2 & -\pi^3+3\pi & \cdots \\ 0 & -32/9\pi & -16/3 & -32\pi/3+1280/27\pi & \cdots \\ 0 & 0 & -9/8 & -9\pi/4 & \cdots \\ 0 & 0 & 0 & -256/75\pi & \cdots \\ \vdots & \vdots & \vdots & \vdots & \ddots \end{pmatrix} \begin{pmatrix} r_1 \\ r_2 \\ r_3 \\ r_4 \\ \vdots \end{pmatrix} + \begin{pmatrix} -\pi/2 \\ -4/3\pi \\ 0 \\ 0 \\ \vdots \end{pmatrix}.$$

One can conclude under these assumptions that the fine-scale density profile (14) is uniquely determined by the function (13) of amplitude dispersion of internal solitary waves, at least in the class of formal power series.

Acknowledgments

This work was supported by the grant of the Russian Foundation for Basic Research (Project No 15-01-03942).

References

[1] Stewart R H 2008 *Introduction to physical oceanography* (Department of Oceanography, TAMU)
[2] Cherkesov L V, Potetunko E N and Shubin D S 2003 *Int. J. Fluid Mech. Res* **30** 13
[3] Potetyunko E N, Cherkesov L V and Shubin D S 2003 *Physic Oceanography* **13** 88
[4] Sabinin K D and A N Serebryanyi 2007 *Acoustical Physics* **53** 357
[5] Duda T F, Lynch J F, Irish J D, Beardsley J D, Ramp S R, Chiu C-S, Tang T Y and Yang Y-J 2004 *IEEE J. Ocean Eng.* **29** 1105
[6] Helfrich K R and Melville W K 2006 *Ann. Rev. Fluid Mech.* **38** 395
[7] Benney D J and Ko D R S 1978 *Stud. Appl. Math.* **59** 187
[8] Yih C S 1980 *Stratified flows* (New York: Academic Press)
[9] Derzho O G and Grimshaw R 1997 *Phys. Fluids* **9** 3378
[10] Maltseva J L 2003 *JOMAE Transactions ASME* **125** 76
[11] Makarenko N I, Maltseva J L and Kazakov A Yu 2009 *Nonlin. Proc. Geophys.* **16** 169

Nonlinear Waves: Theory and New Applications (Wave16) IOP Publishing
Journal of Physics: Conference Series **722** (2016) 012032 doi:10.1088/1742-6596/722/1/012032

Control of Stirling engine. Simplified, compressible model

P I Plotnikov[1,2], **J Sokołowski**[3] **and A Żochowski** [4]

[1] Lavrentyev Institute of Hydrodynamics, Novosibirsk, Russia
[2] Novosibirsk State University, Novosibirsk, Russia
[3] Université de Lorraine, Institut Élie Cartan, Nancy, France
[4] Instytut Badan' Systemowych Polskiej Akademii Nauk, Warszawa, Polska

E-mail: plotnikov@hydro.nsc.ru

Abstract. A one-dimensional free boundary problem on a motion of a heavy piston in a tube filled with viscous gas is considered. The system of governing equations and boundary conditions is derived. The obtained system of differential equations can be regarded as a mathematical model of an exterior combustion engine. The existence of a weak solution to this model is proved. The problem of maximization of the total work of the engine is considered.

1. Problem formulation

The paper deals with a free boundary problem which describes the motion of a heavy piston in a tube filled with viscous gas. It is assumed that the viscous gas occupies the interval $(0, 1)$. The piston is modeled by a material point $a \in (0, 1)$. Its motion is described by a function $x = a(t)$, $t \in (0, T)$. We also assume that the dynamics of the piston is governed by the equation

$$m\ddot{a}(t) + G(\dot{a}(t), a(t)) = F, \tag{1}$$

where F is the hydrodynamical force acting on the piston, G is a given function. This means that the piston is connected with some mechanical device which dynamics is described by the function G. The state of the gas is completely described by the density $\varrho(x, t)$, the velocity $u(x, t)$, and the temperature $\vartheta(x, t)$. Among the other thermodynamic quantities there are the internal energy e, the pressure p, and the entropy s. The three quantities are assumed to be functions of state variable ϱ and θ. Typically, the pressure formula admits the representation

$$p = p_e(\varrho) + p_\theta(\varrho)\theta.$$

The internal energy is defined by the relation

$$\frac{\partial e}{\partial \varrho} = \frac{p_e(\varrho)}{\varrho^2},$$

which yields the representation

$$e = P_e(\varrho) + Q(\theta), \quad P_e(\varrho) = \int_1^\varrho z^{-2} p_e(z)\,dz,$$

where Q is an arbitrary function of θ. The entropy $s(\varrho,\theta)$ is defined by the relations

$$s = \int_1^{\theta} \frac{Q'(z)}{z}dz - P_\theta(\varrho), \quad P_\theta(\varrho) = \int_1^{\varrho} \frac{p_\theta(z)}{z^2}dz.$$

In the simple case of barotropic gas the pressure is a function of the density, $p = p(\varrho)$, and the entropy s equals zero. This eliminates the temperature from the equations.

In this framework, the problem can be formulated as follows. Denote by Ω the rectangular $(0,1) \times (0,T)$ and set

$$\Omega^- = \{(x,t) : 0 < x < a(t), \quad 0 < t < T\}, \quad \Omega^+ = \{(x,t) : a(t) < x < 1, \quad 0 < t < T\}$$

The problem is to find the velocity, the density, the temperature, and the trajectory of the piston satisfying the following equations and boundary conditions

$$\frac{\partial \varrho}{\partial t} + \frac{\partial}{\partial x}(\varrho u) = 0 \quad \text{in} \quad \Omega^\pm, \tag{2}$$

$$\frac{\partial(\varrho u)}{\partial t} + \frac{\partial}{\partial x}(\varrho u^2) + \frac{\partial}{\partial x}(p - \nu u_x) = 0 \quad \text{in} \quad \Omega^\pm, \tag{3}$$

$$\frac{\partial}{\partial t}\left(\varrho\left(\frac{u^2}{2} + e\right)\right) + \frac{\partial}{\partial x}\left(u\left(\varrho\frac{u^2}{2} + \varrho e + p\right)\right) \tag{4}$$

$$-\frac{\partial}{\partial x}(\kappa(\theta)\frac{\partial \theta}{\partial x}) - \nu\frac{\partial}{\partial x}(u u_x) = 0 \quad \text{in} \quad \Omega^\pm,$$

$$u(a(t),t) = \dot{a}(t) \quad \text{in} \quad (0,T), \tag{5}$$

$$m\ddot{a}(t) + G(\dot{a}(t), a(t)) = (\mathbb{T}^+(a(t),t) - \mathbb{T}^-(a(t),t)) \quad \text{in} \quad (0,T), \tag{6}$$

$$\left(\kappa\frac{\partial \theta}{\partial x}\right)^+(a(t),t) - \left(\kappa\frac{\partial \theta}{\partial x}\right)^-(a(t),t) = (\mathbb{T}^-(t) - \mathbb{T}^+(t))u(a(t),t) \quad \text{in} \quad (0,T), \tag{7}$$

$$\theta^+(a(t),t) = \theta^-(a(t),t)). \tag{8}$$

Here $\kappa(\theta)$ is a given positive function, mostly a constant. The stress is defined by the equality

$$\mathbb{T} = \nu u_x - p.$$

We also use the notation

$$\mathbb{T}^\pm(t) := \lim_{x \to a(t) \pm 0} \mathbb{T}(x,t), \quad \vartheta^\pm(t) := \lim_{x \to a(t) \pm 0} \vartheta(x,t). \tag{9}$$

The system of equations (2)-(8) should be supplemented with the boundary conditions. Choose functions $U^\pm : (0,T) \to \mathbb{R}$, $\Theta^\pm : (0,T) \to \mathbb{R}^+$, and $R^\pm : (0,T) \to \mathbb{R}^+$. Denote by $\Sigma^\pm_{\text{in}} \subset [0,T]$ the inlets and by $\Sigma^\pm_{\text{out}} \subset [0,T]$ the outlets defined by U,

$$\begin{aligned} \Sigma^-_{\text{in}} &= \{t \in (0,T) \ : \ U(0,t) > 0\}, \\ \Sigma^+_{\text{in}} &= \{t \in (0,T) \ : \ U(1,t) < 0\}, \\ \Sigma^\pm_{\text{out}} &= (0,T) \setminus \Sigma^\pm_{\text{in}} \end{aligned} \tag{10}$$

We set the boundary conditions

$$u(0,t) = U^-(t), \quad u(1,t) = U^+(t) \quad \text{for} \quad t \in (0,T), \tag{11}$$

$$\varrho(0,t) = R^-(t) \quad \text{on} \quad \Sigma^-_{\text{in}}, \tag{12}$$

$$\varrho(1,t) = R^+(t) \quad \text{on} \quad \Sigma^+_{\text{in}}, \tag{13}$$

$$\theta(0,t) = \Theta^-(t), \quad \theta(1,t) = \Theta^+(t) \quad \text{for} \quad t \in (0,T). \tag{14}$$

At the initial moment $t = 0$ the velocity and the density distributions are prescribed

$$u(x,0) = u_0(x), \quad \varrho(x,0) = R_0(x), \quad \theta(0,x) = \Theta_0(x) \quad \text{for} \quad x \in (0,1). \tag{15}$$

The initial position of the piston is given

$$a(0) = a_0. \tag{16}$$

The equations, boundary, and initial conditions (2)–(16) form a closed system of equations for the velocity, the density, the temperature, and the trajectory of the piston. This system can be regarded as a mathematical model of an exterior combustion engine often called the Stirling engine.

An intensive treatment of one-dimensional compressible Navier-Stokes equations starts with pioneering book by Antontsev, Kazhikhov, and Monakhov [1] on the global theory for one-dimensional homogeneous problems. Although the theory of strong solutions is satisfactory in what concerns the specific boundary conditions, many issues of global behavior of solutions for large data in general boundary conditions are far from being understood.

A global theory of weak solutions was developed by P.L. Lions. We refer the reader to the books by Lions [3], Feireisl [2], and Novotný & Straškraba [4] for the state of the art in the domain. The theory of weak solutions to inhomogeneous boundary value problems for compressible Navier-Stokes equations was developed in the book by Plotnikov & Sokolowski [5]. The problem on the motion of the piston was studied in the paper by Shelukhin [6] under the assumptions that $U^\pm = 0$ and $G = 0$.

In this paper we are focused on the case when the thermodynamical process is adiabatic and the dynamics of the piston is conservative. This means that

$$p = c\varrho^\gamma, \quad \gamma > 1, \quad G(\dot{a}, a) = V'(a), \tag{17}$$

where $V : (0,1) \to \mathbb{R}$ is a mechanical potential. In this case equations (2)–(16) read

$$\frac{\partial \varrho}{\partial t} + \frac{\partial}{\partial x}(\varrho u) = 0 \quad \text{in} \quad \Omega^\pm, \tag{18}$$

$$\frac{\partial(\varrho u)}{\partial t} + \frac{\partial}{\partial x}(\varrho u^2) + \frac{\partial}{\partial x}(p - \nu u_x) = 0 \quad \text{in} \quad \Omega^\pm, \tag{19}$$

$$u(a(t),t) = \dot{a}(t) \quad \text{in} \quad (0,T), \tag{20}$$

$$m\ddot{a}(t) + V'(a(t)) = (\mathbb{T}^+(a(t),t) - \mathbb{T}^-(a(t),t)) \quad \text{in} \quad (0,T). \tag{21}$$

$$u(0,t) = U^-(t), \quad u(1,t) = U^+(t) \quad \text{for} \quad t \in (0,T), \tag{22}$$

$$\varrho(0,t) = R^-(t) \quad \text{on} \quad \Sigma^-_{\text{in}}, \tag{23}$$

$$\varrho(1,t) = R^+(t) \quad \text{on} \quad \Sigma^+_{\text{in}}, \tag{24}$$

$$u(x,0) = u_0(x), \quad \varrho(x,0) = R_0(x) > 0, \quad \text{for} \quad x \in (0,1), \quad a(0) = a_0. \tag{25}$$

Furthermore, we assume that the potential V and the given data satisfy the following conditions

H.1 $V \; : \; (0,1) \to \mathbb{R}^+$ is a smooth positive function such that

$$\lim_{a \to 0+} V(a) = \lim_{a \to 1-0} V(a) = \infty. \tag{26}$$

H.2 The functions $U^{\pm}$ belong to the class $C^2[0,T]$. Every set $\Sigma^{\pm}_{\text{in}}$ consists of finite number of open intervals

H.3 The functions $R^{\pm} \in L^{\infty}(0,T)$ are nonnegative. The function $R_0 \in L^{\infty}(0,1)$ is strictly positive. Moreover,

$$u_0 \in L^{\infty}(0,1), \quad 0 < a_0 < 1.$$

Condition (26) prevents the destruction of the mechanical system.

2. Weak solution. Existence theorem.

The triplet

$$(\varrho, u) : \Omega \to \mathbb{R}^+ \times \mathbb{R}, \quad a : (0,T) \to (0,1)$$

is a weak solution to problem (18)–(25) if the following conditions are satisfied:

- The state variables satisfy

$$\varrho, p, \varrho|u|^2 \in L^{\infty}(0,T; L^1(\Omega)), \quad \mathbf{u} \in L^2(0,T; W^{1,2}(0,1)), \quad a \in W^{1,2}(0,1) \tag{27}$$

$$u(0,t) = U^-(t), \quad u(1,t) = U^+(t) \quad \text{on } (0,T). \tag{28}$$

- The integral identity

$$\int\limits_{\Omega} \varrho u \zeta_t + \varrho u^2 \zeta_x + (p - \nu u_x)\zeta_x dx dt + \int\limits_0^1 \varrho_0(x) u_0(x) \zeta(x,0) dx \tag{29}$$

$$+ m u_0(a_0)\zeta(a_0) + \int\limits_0^T \big(m\dot{a}(t)^2 \partial_x \zeta(a(t),t) - V'(a(t))\zeta(a(t),t)\big) dt = 0$$

holds for all functions $\zeta \in C^1(\Omega)$ such that

$$\zeta(x,T) = 0, \quad \zeta(0,t) = \zeta(1,t) = 0.$$

- The integral identity

$$\int_{\Omega} (\varrho\, \partial_t \psi + \varrho u \cdot \nabla \psi)\, dx dt = \int_0^T \psi R^+ U^+ dt - \int_0^T \psi R^- U^- dt - \int_0^1 R_0 \psi(\cdot, 0)\, dx \tag{30}$$

holds for all functions $\psi \in C^1(\Omega)$ satisfying

$$\psi = 0 \quad \text{in a neighborhood of } \Sigma^{\pm}_{\text{out}} \cup \{(0,1) \times \{T\}\},$$

which means that the support of the test function ψ meets the boundary of the rectangular Ω at the inlet $\Sigma^{\pm}_{\text{in}}$ and the bottom $(0,1) \times \{0\}$.

The first main result of the study is the following

Theorem 1 *Assume that Conditions* (**H.1**)–(**H.3**) *are fulfilled. Then problem* (18)–(25) *has a weak solution which satisfies the estimates*

$$\begin{aligned} \|u\|_{L^2(0,T;W^{1,2}(0,1))} + \|\varrho|u|^2\|_{L^{\infty}(0,T;L^1(0,1))} + \|p\|_{L^{\infty}(0,T;L^1(0,1))} &\le c, \\ \sup_{t \in (0,T)} \dot{a}(t)^2 + V(a(t)) &\le c, \end{aligned} \tag{31}$$

where the constant c depends only on T as well as the boundary and initial data.

The proof is based on the methods developed in [5]. Following [5] we use the multistep regularization. The weak solutions is obtained as a limit of a sequence of approximate solutions to regularized equations. The main difficulty is the derivation of estimate (31).

3. The optimization problem.

Recall that the power of the hydrodynamical forces acting on the piston is equal to $(\mathbb{T}^+(a(t),t) - \mathbb{T}^-(a(t),t))\dot{a}(t)$. Assume that the functions $U^\pm$ and u_0, R_0 are fixed and satisfy conditions (**H.1**)–(**H.3**). In this case the total work W of hydrodynamical forces acting on the piston becomes the function of $R^\pm$,

$$W(R^\pm) = \int_0^T (\mathbb{T}^+(a(t),t) - \mathbb{T}^-(a(t),t))\dot{a}(t)dt \tag{32}$$

In view of Theorem 1 the functional W is well defined on the set of bounded nonnegative functions. Fix an arbitrary $N > 0$ and consider the set of admissible data

$$\mathfrak{M} = \{R^\pm \in L^\infty(0,T) : \ \|R^\pm\|_{L^\infty(0,T)} \le N\}.$$

The natural question is to find the boundary data which maximize the total work of the engine. This problem can be formulated as follows: to find an admissible couple $R^{\pm *} \in \mathfrak{M}$ such that

$$W(R^{\pm *}) = \sup_{R^\pm \in \mathfrak{M}} W(R^\pm), \tag{33}$$

where the cost function is given by (32). The following theorem on solvability of variational problem (33) is the second main result of the study.

Theorem 1 *Assume that conditions* (**H.1**)–(**H.3**) *are fulfilled. Then problem* (33) *has at least one solution* $R^{\pm *} \in \mathfrak{M}$.

The proof is based on Theorem 1. The key observation is that the functional W is star weakly continuous on $\mathfrak{M}$.

References

[1] Antontsev S N, Kazhikhov A V and Monakhov V N 1990 *Boundary Value Problems in Mechanics of Nonhomogeneous Fluids* (Amsterdam: North-Holland)
[2] Feireisl E 2004 *Dynamics of Viscous Compressible Fluids*, (Oxford: Oxford Univ. Press)
[3] Lions P L 1998 *Mathematical Topics in Fluid Dynamics; Compressible Models* (Oxford: Clarendon Press).
[4] Novotný A and Straškraba I 2004 *Introduction to the Mathematical Theory of Compressible Flow* (Oxford: Oxford Univ. Press)
[5] Plotnikov P I and Sokolowski J 2012 *Compressible Navier-Stokes equations. Theory and shape optimization* (Basel: Springer)
[6] Shelukhin V V 1982 Dynamica sploshnoi sredy [in Russian] **57** 131–52

Numerical solution of an equilibrium problem for a membrane with a delaminated thin rigid inclusion

Evgeny M. Rudoy

Lavrentyev Institute of Hydrodynamics of the Russian Academy of Sciences, Novosibirsk 630090, Russia
Novosibirsk State University, Novosibirsk 630090, Russia

E-mail: rem@hydro.nsc.ru

Abstract. The paper deals with the numerical solution of an equilibrium problem for an elastic membrane with a thin rigid inclusion. The thin inclusion is supposed to delaminate, therefore a crack between the inclusion and the membrane is considered. The boundary conditions for nonpenetration of the crack faces are fulfilled. We provide the relaxation of the problem and propose an iterative method for the numerical solution of the approximated problem. The method is based on a domain decomposition and the Uzawa algorithm for finding a saddle point of the Lagrangian. Examples of the numerical solution of the initial problem are presented.

1. Introduction

Rigid inclusions (also called stiffeners or anticracks) are used in solid mechanics to describe fibres embedded in a matrix material. Under compression, rigid inclusions may delaminate from the matrix material, thereby introducing cracks. There are different approaches to model cracks in solids. The classical formulation of the problems with cracks implies linear boundary conditions on the crack faces [1]. It is well known that such models have shortcoming because there can be situations when the crack faces penetrate each other. It is natural to impose boundary conditions that exclude mutual penetration of the crack faces. The book [2] and papers [3, 4] contain results for crack models with the nonpenetration conditions for a wide class of constitutive laws.

In the present paper, we propose an iterative method of the numerical solution an equilibrium problem for an elastic membrane with a delaminated thin rigid inclusion. We assume that the nonpenetration conditions are imposed on the crack faces. To construct an effective numerical algorithm, we provide the relaxation of the initial problem. Then for the approximated problem, we apply the domain decomposition method [5] and the Uzawa algorithm for finding a saddle point of the Lagrangian [6]. To this end, the initial domain is divided into two subdomains. Two linear problems are solved in each subdomain at every iteration. One of problems describes the equilibrium of a membrane having a rigid inclusion on the external boundary. The presence of the rigid inclusion imposes constraints on a solution. In order to find the solution of such problem, we introduce three auxiliary Dirichlet problems without constraints and solve its. For "gluing" the solutions of both linear problems and providing the nonpenetration conditions, we use the Lagrange multipliers. Numerical experiments illustrate the performance of our algorithm.

2. Statement of the problem

Let $\Omega \subset R^2$ be a bounded domain with Lipschitz boundary $\partial\Omega$. Assume that Ω is divided into two subdomains Ω_1 and Ω_2 with Lipschitz boundaries $\partial\Omega_1$ and $\partial\Omega_2$, respectively, such that $\text{meas}(\partial\Omega \cap \partial\Omega_i) > 0$, $i = 1, 2$; and denote by Σ the interface of Ω_i. Let γ_c be a curve lying on the interface Σ such that $\overline{\gamma}_c \cap \partial\Omega = \emptyset$, and let $\Omega_c = \Omega \setminus \overline{\gamma}_c$. Denote by Σ the rest part of γ_g, by ν_i the outward unit normals to Ω_i, $i = 1, 2$. In particular, we have $\nu_2 = -\nu_1 = \nu$ on Σ, see Figure 1 for an illustration of the geometry.

In what follows, an elastic membrane occupies the domain Ω_c, and γ_c corresponds to a crack in the membrane. In the domain Ω_c, we intend to consider a mixed boundary value problem for the displacement field u. Namely, denote by u_i the restriction of u to Ω_i, $i = 1, 2$. We suppose that the traces u_1 and u_2 on γ_c are, in general, different and satisfying nonlinear boundary conditions that prevent mutual penetration of the crack faces. Furthermore, due to the presence of a thin rigid inclusion the function u_2 is a constant function on γ_c. The equilibrium problem for the elastic membrane having the crack and the thin rigid inclusion reads as follows. For a given external force $f \in L^2(\Omega)$ acting on the membrane, we have to find a displacement u and a real number A such that

$$-\Delta u = f \quad \text{in} \quad \Omega_c, \tag{1}$$

$$u = 0 \quad \text{on} \quad \partial\Omega, \tag{2}$$

$$[u] \geq 0 \quad \text{on} \quad \gamma_c, \tag{3}$$

$$\frac{\partial u_1}{\partial \nu} \geq 0, \quad \frac{\partial u_1}{\partial \nu}[u] = 0 \quad \text{on} \quad \gamma_c, \tag{4}$$

$$u_2 = A \quad \text{on} \quad \gamma_c, \tag{5}$$

$$\int_{\gamma_c} \left[\frac{\partial u}{\partial \nu}\right] ds = 0, \tag{6}$$

where $[u] = u_2 - u_1$ is a jump of the function u on γ_c.

To give a weak formulation of the problem (1)–(6), we introduce the functional space

$$V = \{\, v \in H^1(\Omega_c) \mid v = 0 \text{ on } \partial\Omega \,\}$$

and the set of admissible displacements

$$K_c = \{\, v \in V \mid [v] \geq 0 \text{ on } \gamma_c,\ v_2 \text{ is a constant on } \gamma_c \,\},$$

where $v_2 = v|_{\Omega_2}$. The potential energy of the system is represented by the functional

$$\Pi(v) = \frac{1}{2}\int_{\Omega_c} |\nabla v|^2 \, dx - \int_{\Omega_c} f v \, dx.$$

Then the weak formulation of the problem (1)–(6) is the following minimization problem:

$$\text{Find} \quad u \in K_c \quad \text{such that} \quad \Pi(u) = \inf_{v \in K_c} \Pi(v). \tag{7}$$

Note that the functional Π is coercive and weakly lower semicontinuous on the space V. Moreover, the set K_c is weakly closed. Hence, the minimization problem (7) has a unique solution $u \in K_c$ (see, e.g., [6, 2]), which satisfies the variational inequality

$$\int_{\Omega_c} \nabla u(\nabla v - \nabla u) \, dx \geq \int_{\Omega_c} f(v - u) \, dx \quad \forall v \in K_c. \tag{8}$$

Journal of Physics: Conference Series **722** (2016) 012033 doi:10.1088/1742-6596/722/1/012033

3. Domain decomposition

Define the functional spaces $V_i = \{ v_i \in H^1(\Omega_i) \mid v_i = 0 \text{ on } \partial\Omega \cap \partial\Omega_i \}$ that we endow with the norms $\|v_i\|_{V_i}^2 = \int\limits_{\Omega_i} |\nabla v_i|^2 \, dx, \quad i = 1, 2$, which, due to the Poincaré inequality are equivalent to the usual H^1-norms. Moreover, we introduce the one-dimensional subspace V_2^g of V_2 given by

$$V_2^g = \{ v_2 \in V_2 \mid v_2 \text{ is a constant on } \gamma_c \}$$

and the set $K_{gc} \subset V^1 \times V_g^2$ given by

$$K_{gc} = \{ (v_1, v_2) \in V_1 \times V_2^g \mid v_2 - v_1 \geq 0 \text{ on } \gamma_c, \ v_2 - v_1 = 0 \text{ on } \gamma_g \}.$$

Let us consider now the constrained minimization problem:

$$\text{Find} \quad (u_1, u_2) \in K_{gc} \quad \text{such that} \quad \Pi_1(u_1) + \Pi_2(u_2) = \inf_{(v_1,v_2)\in K_{gc}} (\Pi_1(v_1) + \Pi_2(v_2)), \tag{9}$$

with the energy functionals Π_i are defined on V_i by the relations

$$\Pi_i(v_i) = \frac{1}{2} \int\limits_{\Omega_i} |\nabla v_i|^2 \, dx - \int\limits_{\Omega_i} f v_i \, dx, \quad i = 1, 2.$$

Theorem 1. Let u be a solution of the minimization problem (7). Then (9) has a unique solution $(u_1, u_2) \in K_{gc}$, which satisfies the conditions

$$u_i = u|_{\Omega_i}.$$

We omit details, the essential ideas behind the proof of Theorem 1 are the same as in [7].

4. Regularized problem

For the efficient numerical solution of (7), we propose a domain decomposition method applied to a properly regularized problem.

For any positive number p, let us introduce the sets

$$U_1^p = \{ v_1 \in V_1 \mid v_1\|_{V_1} \leq p \}, \quad U_2^p = \{ v_2 \in V_2^g \mid \|v_2\|_{V_2} \leq p \},$$

$$\Lambda_c^p = \{ \lambda_c \in L_2(\gamma_c) \mid 0 \leq \lambda_c \leq p \text{ on } \gamma_c \}, \quad \Lambda_g^p = \{ \lambda_g \in L_2(\gamma_g) \mid -p \leq \lambda_g \leq p \text{ on } \gamma_g \},$$

and the Lagrange function L defined on $U_1^p \times U_2^p \times \Lambda_c^p \times \Lambda_g^p$ by

$$L(v_1, v_2, \lambda_c, \lambda_g) = \Pi_1(v_1) + \Pi_2(v_2) + \int\limits_{\gamma_c} \lambda_c(v_1 - v_2) \, ds + \int\limits_{\gamma_g} \lambda_g(v_1 - v_2) \, ds.$$

Consider the family of saddle point problems depending on the parameter p:

$$\text{Find} \quad (u_1^p, u_2^p, \mu_c^p, \mu_g^p) \in U_1^p \times U_2^p \times \Lambda_c^p \times \Lambda_g^p \quad \text{such that}$$

$$L(u_1^p, u_2^p, \lambda_c, \lambda_g) \leq L(u_1^p, u_2^p, \mu_c^p, \mu_g^p) \leq L(v_1, v_2, \mu_c^p, \mu_g^p) \ \forall (v_1, v_2, \lambda_c, \lambda_g) \in U_1^p \times U_2^p \times \Lambda_c^p \times \Lambda_g^p. \tag{10}$$

The existence theory of saddle points implies that, for every $p > 0$, the problem (10) has a solution $(u_1^p, u_2^p, \mu_c^p, \mu_g^p) \in U_1^p \times U_2^p \times \Lambda_c^p \times \Lambda_g^p$.

Theorem 2. There exists a constant $c > 0$ such that, for all $p > c$, the saddle point $(u_1^p, u_2^p, \mu_c^p, \mu_g^p)$ satisfies the system of variational equalities and inequalities

$$\int\limits_{\Omega_1} \nabla u_1^p \nabla v_1\, dx + \int\limits_{\gamma_c} \mu_c^p v_1\, ds + \int\limits_{\gamma_g} \mu_g^p v_1\, ds = \int\limits_{\Omega_1} f v_1\, dx \quad \forall v_1 \in V_1, \tag{11}$$

$$\int\limits_{\Omega_2} \nabla u_2^p \nabla v_2\, dx - \int\limits_{\gamma_c} \mu_c^p v_2\, ds - \int\limits_{\gamma_g} \mu_g^p v_2\, ds = \int\limits_{\Omega_2} f v_2\, dx \quad \forall v_2 \in V_2^g, \tag{12}$$

$$\int\limits_{\gamma_c} \lambda_c(u_1^p - u_2^p)\, ds \le \int\limits_{\gamma_c} \mu_c^p(u_1^p - u_2^p)\, ds \quad \forall \lambda_c \in \Lambda_c^p, \tag{13}$$

$$\int\limits_{\gamma_g} \lambda_g(u_1^p - u_2^p)\, ds \le \int\limits_{\gamma_g} \mu_g^p(u_1^p - u_2^p)\, ds \quad \forall \lambda_g \in \Lambda_g^p. \tag{14}$$

Remark. Theorem 2 means that the saddle point of the Lagrangian L over the set $U_1^p \times U_2^p \times \Lambda_c^p \times \Lambda_g^p$ coincides with the saddle point of L over the set $V_1 \times V_2 \times \Lambda_c^p \times \Lambda_g^p$ for all $p > c$.

Applying the same arguments as in [7], we can prove the following theorem.

Theorem 3. The sequence (u_1^p, u_2^p) is such that

$$(u_1^p, u_2^p) \to (u_1, u_2) \quad \text{strongly in} \quad V_1 \times V_2 \quad \text{as} \quad p \to \infty.$$

5. Linear equilibrium problem for a membrane with a thin rigid inclusion

The function u_2 satisfying the variational equality (13) is a solution of a linear equilibrium problem for an elastic membrane with a rigid inclusion located on the external boundary. Despite the fact that this problem is linear, it is a constraint problem. Now we specify the method of solving of (13).

Let ω be a bounded domain with Lipschitz boundary $\partial\omega$, which is divided on two disjoint parts γ_N and γ_D. Let γ be a part of γ_N such that $\overline{\gamma} \cap \overline{\gamma}_D = \emptyset$. Denote by n the outward normal vector to $\partial\omega$.

In the domain ω, let us consider the boundary value problem with nonlocal boundary conditions. For given functions $f \in L_2(\omega)$ and $g \in H^{-1/2}(\gamma_N)$, we have to find a function u and a real number A such that

$$-\Delta u = f \quad \text{in} \quad \omega, \tag{15}$$

$$u = 0 \quad \text{on} \quad \gamma_D, \tag{16}$$

$$\frac{\partial u}{\partial n} = g \quad \text{on} \quad \gamma_N \setminus \overline{\gamma}, \tag{17}$$

$$u = A \quad \text{on} \quad \gamma, \int\limits_{\gamma} \frac{\partial u}{\partial n}\, ds = \int\limits_{\gamma} g\, ds. \tag{18}$$

The problem (15)–(18) corresponds to the equilibrium state of an elastic membrane occupying the domain ω and containing a thin rigid inclusion γ on the external boundary $\partial\omega$. The volume force f and the boundary traction g act on the membrane. In particular, the traction g acts on the rigid inclusion γ.

Let us give a weak formulation of (15)–(18). For this purpose, we introduce the set of admissible displacements $K = \{v \in H^1_{\gamma_D}(\omega) \mid v \text{ is a constant on } \gamma\}$, where $H^1_{\gamma_D}(\omega) = \{v \in$

$H^1(\omega) \mid \; v = 0 \text{ on } \gamma_D \}$. A function $u \in K$ is a weak solution of (15)–(18), if the variational equality

$$\int\limits_{\omega} \nabla u \nabla v \, dx = \int\limits_{\omega} f v dx + \int\limits_{\gamma_N} g v \, ds \quad \forall v \in K \tag{19}$$

holds.

The space $H^1_{\gamma_D}(\omega)$ is a Hilbert space with the inner product

$$(u, v) = \int\limits_{\omega} \nabla u \nabla v \, dx$$

in virtue of the Poincaré inequality. Let us consider the following problem:

$$\text{Find} \quad u_f \in H^1_{\gamma_D}(\omega) \quad \text{such that} \quad \int\limits_{\omega} \nabla u_f \nabla v \, dx = \int\limits_{\omega} f v \, dx + \int\limits_{\gamma_N} g v \, ds \quad \forall v \in H^1_{\gamma_D}(\omega). \tag{20}$$

There exist a unique solution u_f of the problem (20); therefore we can rewrite the variational equality (19) in the form

$$(u - u_f, v) = 0 \quad \forall v \in K,$$

which means that the function u is an orthogonal projection of u_f onto the set K.

Since K is a subspace of $H^1_{\gamma_D}(\omega)$, the decomposition

$$H^1_{\gamma_D}(\omega) = K \oplus K^{\perp}$$

holds, where $K^{\perp}$ is an orthogonal complement for K in $H^1_{\gamma_D}(\omega)$.

Let $u_f \in H^1_{\gamma_D}(\omega)$ be an arbitrary function. Describe the algorithm that allow us to find the decomposition $u_f = u + w$, where $u \in K$, $w \in K^{\perp}$.

It is possible to show that any function $w \in K^{\perp}$ satisfies the following nonnlocal boundary value problem:

$$-\Delta w = 0 \quad \text{in} \quad \omega, \tag{21}$$

$$w = 0 \quad \text{on} \quad \gamma_D, \tag{22}$$

$$\frac{\partial w}{\partial n} = 0 \quad \text{on} \quad \gamma_N \setminus \overline{\gamma}, \tag{23}$$

$$\int\limits_{\gamma} \frac{\partial w}{\partial n} \, ds = 0. \tag{24}$$

Obviously, the converse is also true, i.e., any function satisfying (21)–(24) belongs to $K^{\perp}$.

Choose a real number $a \in R$ and consider the variational problem:

$$\text{Find } y^a \in H^1_{\gamma_D}(\omega) \text{ such that } y^a = u_f|_{\gamma} - a \text{ on } \gamma \text{ and } \int\limits_{\omega} \nabla y^a \nabla v \, dx = 0 \;\; \forall v \in H^1_{\gamma_D \cup \gamma}(\omega). \tag{25}$$

For any a, the problem (25 has a unique solution. Furthermore, y^a is a weak solution to the boundary value problem

$$-\Delta y^a = 0 \quad \text{in} \quad \omega,$$

$$y^a = 0 \quad \text{on} \quad \gamma_D,$$

$$y^a = u_f|_{\gamma} - a \quad \text{on} \quad \gamma,$$

$$\frac{\partial y^a}{\partial n} = 0 \quad \text{on} \quad \gamma_N \setminus \overline{\gamma}.$$

Denote by

$$h(a) = \int_\gamma \frac{\partial y^a}{\partial n} \, ds,$$

and show that there exists $A \in R$ such that $h(A) = 0$. It means that the function y^A satisfies the nonlocal condition (24). Since the solution y^a of (25) is an affine function of a, we have the decomposition

$$y^a = ay + y^0, \tag{26}$$

where y^0 is the solution of (25) corresponds to $a = 0$ and $y \in H^1_{\gamma_D}(\omega)$ satisfies the variational equality

$$(y, v) = 0 \quad \forall v \in H^1_{\gamma_D \cup \gamma}(\omega), \tag{27}$$

while $y = -1$ on γ. The differential formulation of the problem (27) reads as follows. We have to find a function y such that

$$-\Delta y = 0 \quad \text{in} \quad \omega,$$

$$y = 0 \quad \text{on} \quad \gamma_D,$$

$$y = -1 \quad \text{on} \quad \gamma,$$

$$\frac{\partial y}{\partial n} = 0 \quad \text{on} \quad \gamma_N \setminus \overline{\gamma}.$$

In turn, the function h is an affine function of a, and the representation

$$h(a) = ka + b$$

takes place with

$$k = \int_\gamma \frac{\partial y}{\partial n} \, ds, \quad b = \int_\gamma \frac{\partial y^0}{\partial n} \, ds.$$

Due to the Green formula, we have

$$k = -(y_1, y_1) < 0, \quad b = -(y^0, y_1),$$

therefore the function h vanishes when $a = A := -b/k$.

Finally, we obtain that any function $u_f \in H^1_{\gamma_D}(\omega)$ can be decomposed into a sum of two functions u and w, where

$$w = y^A \in K^\perp, \quad u = u_f - y^A \in K$$

and y^A is the solution of (25) with $a = A$.

Remark. Due to (26), we have $y^A = Ay + y^0$. It means that we do not need to solve the problem (25) for $a = A$ again. It suffices to use the functions y and y^0.

6. Domain decomposition algorithm

With the result of the previous section, we present a domain decomposition method to solve (7) numerically. Let us introduce the space

$$V_{2,\gamma} = \{\, v_2 \in V_2 \mid v_2 = 0 \text{ on } \gamma \,\} = \{\, v_2 \in H^1(\Omega_2) \mid v_2 = 0 \text{ on } \gamma \cup (\partial\Omega \cap \partial\Omega_2) \,\}.$$

Algorithm.

1. Initialization. $\mu_{c,0} \in \Lambda_c^p$ and $\mu_{g,0} \in \Lambda_g^p$ are given.
2. Iteration $k \geq 0$. Compute successively $u_{1,k}$, $u_{2,k}$ as follows:

– find $u_{1,k} \in V_1$ such that

$$\int_{\Omega_1} \nabla u_{1,k} \nabla v_1 \, dx + \int_{\gamma_c} \mu_{c,k} v_1 \, ds + \int_{\gamma_g} \mu_{g,k} v_1 \, ds = \int_{\Omega_1} f v_1 \, dx \quad \forall v_1 \in V_1.$$

– find $w_k \in V_2$ such that

$$\int_{\Omega_2} \nabla w_k \nabla v_2 \, dx = \int_{\Omega_2} f v_2 \, dx + \int_{\gamma_c} \mu_{c,k} v_2 \, ds + \int_{\gamma_g} \mu_{g,k} v_2 \, ds, \quad \forall v_2 \in V_2.$$

– find $y_k^0 \in V_2$ such that $y_k^0 = w_k$ on γ and

$$\int_{\Omega_2} \nabla y_k^0 \nabla v_2 \, dx = 0 \quad \forall v_2 \in V_{2,\gamma}.$$

– find $y_k \in V_2$ such that $y_k = -1$ on γ and

$$\int_{\Omega_2} \nabla y_k \nabla v_2 \, dx = 0 \quad \forall v_2 \in V_{2,\gamma}.$$

– compute the auxiliary real numbers A_k, b_k, c_k as follows:

$$b_k = -\int_{\Omega_2} \nabla y_k \nabla y_k \, dx, \quad c_k = -\int_{\Omega_2} \nabla y_k \nabla y_k^0 \, dx, \quad A_k = -c_k / b_k.$$

– set $y^{A_k} = A_k y_k + y_k^0$ and $u_{2,k} = w_k - y^{A_k}$.
– update the Lagrange multipliers

$$\mu_{c,k+1} = P_{\Lambda_c^p}(\mu_{c,k} + \theta(u_{1,k} - u_{2,k})),$$
$$\mu_{g,k+1} = P_{\Lambda_g^p}(\mu_{g,k} + \theta(u_{1,k} - u_{2,k})).$$

3. Iteration continue until the relative error

$$\varepsilon = \max\left(\frac{\|u_{1,k} - u_{1,k-1}\|_{V_1}^2}{\|u_{1,k}\|_{V_1}^2}, \frac{\|u_{2,k} - u_{2,k-1}\|_{V_2}^2}{\|u_{2,k}\|_{V_2}^2} \right)$$

becomes sufficiently small.

The convergence of Algorithm follows from the general convergence theorem for the Uzawa algorithm (see [6]). Moreover, note that the proof of the convergence is similar to that in the articles [7].

Theorem 4. There exists θ^* such that, for all $\theta \in (0, \theta^*)$, the sequence $(u_{1,k}, u_{2,k})$ generated in Algorithm 1 (and, respectively, in Algorithm 2) is such that

$$(u_{1,k}, u_{2,k}) \to (u_1, u_2) \quad \text{strongly in} \quad V_1 \times V_2$$

as $k \to \infty$.

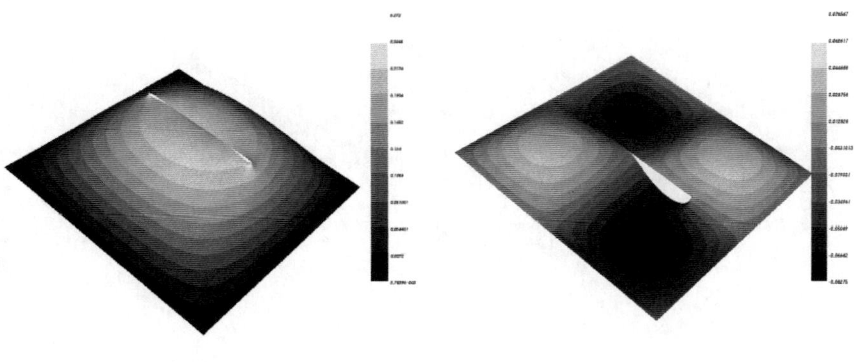

Figure 1. On the left: Example 1; On the right: Example 2.

7. Numerical experiments

Using piecewise linear finite elements ($P1$-Lagrange elements), Algorithm was implemented in FreeFEM++. The test problems with different external forces are used to illustrate the behavior of our algorithm. Dimensionless units are used, and deformed configurations are plotted with amplification factors.

In all numerical experiments, we set $\mu_{c,0} = 0$ and $\mu_{g,0} = 0$ for the initialization of the algorithm, $\varepsilon = 10^{-12}$ for the stopping criterion; the regularized parameter p equals 10^5 and the relaxation parameter θ equals 2.

We consider a special domain configuration. Let the domain Ω consist of two subdomains $\Omega_1 = (-1,1) \times (-1,0)$ and $\Omega_2 = (-1,1) \times (0,1)$ with the interface $\Sigma = (-1,1) \times \{0\}$, while the crack is $\gamma_c = (-0.5, 0.5) \times \{0\}$.

The subdomains Ω_1 and Ω_2 are partitioned into 5700 triangles with 2971 nodes and into 5674 triangles with 2958 nodes, respectively. In particular, 80 nodes are on γ_c and γ_g for both domains. The maximal sizes of the triangles (in a neighborhood of Σ) are 0.077 and 0.078 in Ω_1 and Ω_2, respectively, while the minimal sizes are equal to 0.012 for both domains.

Next, we consider four types of the external force f.

Example 1. Let f is equal to 1 in Ω. The resulting deformations are shown in Figure 1 (left).

Example 2. Assume that f is given by

$$f(x_1, x_2) = \begin{cases} -1, & (x_1, x_2) \in \{(-1,0) \times (0,1)\} \cup \{(0,1) \times (-1,0)\}, \\ 1, & (x_1, x_2) \in \{(-1,0) \times (-1,0)\} \cup \{(0,1) \times (0,1)\}. \end{cases}$$

In this case, the crack faces contact with each other near the crack tip $(-1,0)$. The resulting deformations and plots of displacements and jump of u along the interface Σ are shown in Figure 1 (right).

Example 3. Assume that f is given by

$$f(x_1, x_2) = \begin{cases} 1, & (x_1, x_2) \in (-1,1) \times (0,1), \\ -1, & (x_1, x_2) \in (-1,1) \times (-1,0). \end{cases}$$

Such external force provides an opening mode of the crack γ_c. The resulting deformations are shown in Figure 2 (left).

Example 4. If f is given by

$$f(x_1, x_2) = \begin{cases} 1, & (x_1, x_2) \in (-1,0) \times (-1,1), \\ -1, & (x_1, x_2) \in (0,1) \times (-1,1), \end{cases}$$

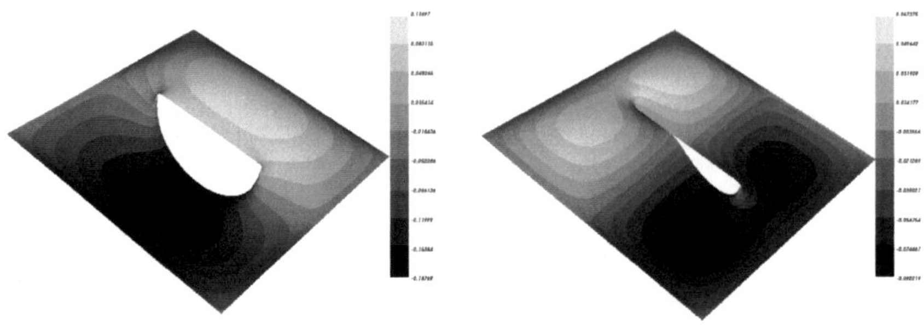

Figure 2. On the left: Example 3; On the right: Example 4.

then the crack faces contact near the crack tip $(-1, 0)$. The resulting deformations are shown in Figure 2 (right).

Acknowledgments

The author is supported by the Grant of the President of the Russian Federation for state support of young Russian researchers (Grant No. MD-3123.2015.1)

References

[1] Cherepanov G P 1979 *Mechanics of Brittle Fracture* (McGraw-Hill)
[2] Khludnev A M and Kovtunenko V A 2000 *Analysis of Cracks in Solids* (WIT-Press)
[3] Lazarev N P and Rudoy E M 2014 Shape sensitivity analysis of Timoshenko's plate with a crack under the nonpenetration condition *Z. Angew. Math. Mech.* **94** 730–739.
[4] Itou H, Khludnev A M, Rudoy E M and Tani A 2012 Asymptotic behaviour at a tip of a rigid line inclusion in linearized elasticity *Z. Angew. Math. Mech.*, **96** 716–730.
[5] Quarteroni A and Valli A 1999 *Domain Decomposition Methods for Partial Differential Equations* (Clarendon Press)
[6] Ekeland I and Temam R 1999 *Convex Analysis and Variational Problems* (SIAM)
[7] Rudoy E M 2015 Domain decomposition method for a model crack problem with a possible contact of crack edges *Comput. Math. Math. Phys.* **55** 305–316

Journal of Physics: Conference Series **722** (2016) 012034 doi:10.1088/1742-6596/722/1/012034

Model isothermal internal erosion of soil

A A Papin and A N Sibin
Altai State University, Barnaul, Russia
E-mail: papin@math.asu.ru, sibin_anton@mail.ru

Abstract. The process of internal erosion in a three-phase saturated soil is studied. The problem is described by the equations of mass conservation, Darcy's law and the equation of capillary pressure. The original system of equations is reduced to a system of two equations for porosity and water saturation. In general, the equation of water saturation is degenerate. The degenerate problem in a one-dimensional domain and one special case of the problem in a two-dimensional domain are solved numerically using a finite-difference method. Existence and uniqueness of a classical solution of a nondegenerate problem is proved.

1. Introduction

Various models of internal erosion are used for describing applied problems of the formation of cavities under dam reservoir, the destruction of the wellbore walls as a result of soil erosion, the suffusion craters formation on surface topography. Evaluation of suffusion removal is also a relevant problem in many environmental works. Advanced models of internal erosion are based on approaches of mechanics of multiphase media [1, 2, 3, 4]. These models are intended to describe more detailed picture of the movement of water and fluidized solid particles mixture in soil. Precisely, the determination of a velocity field, porosity, water saturation and pressure of each phase.

These models include phase transition and use filtration approximation. The basic equations are mass conservation law for each phase and Darcy's law for moving phases. This system of equations is similar in structure to the system of Masket-Leverett equations of two-phase filtration for immiscible fluids; for detailed mathematical theory description see [5, 6]. We did not find any study related to the justification problem of the system of equations with porosity to be determined, except a few special cases [7, 8].

The key points of the problem are the degeneracy of the equations of the obtained solution (relative phase permeability coefficients $\overline{k}_{0i}$ can be equal to 0 if saturation of the ith phase $s_i \leq 0$ in the equation (2)) and the unknown porosity of soil. The problem in this formulation is very difficult to study analyticaly, especially it is necessary to justify the physical principles of maximum for the porosity and the water saturation ($0 \leq \phi \leq 1$, $0 \leq s_i \leq 1$).

This approach is also used in the study of two-phase filtration, dissociation of hydrates in natural layers with iced parts, heat and mass transfer in freezing or melting soil.

2. Formulation of the problem

We consider a mathematical model of internal erosion of soil in a finite domain $Q \subset R^n$ (where Q and n will be specified further). Saturated soil is a three–phase porous medium [1] consisting of water ($i = 1$, the first phase), fluidized solid particles ($i = 2$, the second phase) and solid skeleton of soil (i =3, the third phase). The process is described by the equations of mass conservation for each phase with phase transitions, generalized Darcy's law and the equation of capillary pressure for water and fluidized solid particles [2, 3, 4]:

$$\frac{\partial \rho_i}{\partial t} + \nabla \cdot (\rho_i \vec{u}_i) = \sum_{j=1}^{3} I_{ij}, \quad i = 1,2,3, \quad I_{ji} = -I_{ij}, \quad \sum_{i,j=1}^{3} I_{ij} = 0; \tag{1}$$

$$\vec{v}_i = s_i \phi \vec{u}_i = -K_0(\phi)\frac{\overline{k}_{0i}}{\mu_i}(\nabla p_i + \rho_i^0 \vec{g}),\ i = 1,2; \quad p_2 - p_1 = p_c(s_1) \geq 0, \quad s_1 + s_2 = 1. \tag{2}$$

Here $\vec{u}_i$ is a velocity of the ith phase ($\vec{u}_3 = 0$, the third phase is considered to be stationary), ρ_i is a reduced density related to a true density ρ_i^0 and a volumetric concentration α_i by the formula $\rho_i = \alpha_i \rho_i^0$ (the condition $\sum_{i=1}^{3} \alpha_i = 1$ is a consequence of the definition of ρ_i), I_{ji} is a rate of mass transfer from the jth to the ith phase per unit volume in unit of time, $\vec{v}_i = \phi s_i \vec{u}_i$ is a filtration velocity of the ith phase (i=1,2), ϕ is porosity, s_1 and s_2 are water and fluidized solid particles saturations ($\alpha_1 = \phi s_1$, $\alpha_2 = \phi s_2$, $\alpha_3 = 1 - \phi$), K_0 is the filtration tensor; k_{0i} is a relative phase permeability ($\overline{k}_{0i} = \overline{k}_{0i}(s_i) \geq 0$, $\overline{k}_{0i}|_{s_i=0} = 0$), μ_i is a dynamic viscosity, p_i is a phase pressure (i=1,2); p_c is the capillary pressure, $\vec{g}$ is the acceleration vector due to gravity.

The functions of the system (1), (2) satisfy the assumptions: $\rho_i^0 = const$, $\rho_2^0 = \rho_3^0$, $I_{ij} = -I_{ji}$. Moreover, $I_{12} = I_{13} = 0$, $I_{23} = \rho_3^0 I$, I is a prescribed function of s_1, ϕ, v_1, v_2.

If porosity is a known function, then system (1), (2) will read as the Masket — Leverett equations of two-phase filtration for immiscible fluids. We shall investigate the problem with the porosity as a function to be determined.

We shall prove the unique solvability of the nondegenerate problem (it means $a(s) > 0$ for any s) in a one-dimensional domain, check validity of physical maximum principles for the porosity and the water saturation in the same problem and solve numerically the one-dimensional degenerate problem and one special case of the problem in a two-dimensional domain.

In the one-dimensional domain the system (1), (2) reads [4, 9]

$$\frac{\partial s\phi}{\partial t} = \frac{\partial}{\partial x}(K_0(\phi)a(s)\frac{\partial s}{\partial x} - b(s)v(t) + F(s,\phi)), \tag{3}$$

$$\frac{\partial \phi}{\partial t} = I(s,\phi). \tag{4}$$

Here $s \equiv s_1$, $v(t) = v_1 + v_2$ is a mixture filtration velocity (prescribed function of time) [5]. The rate of mass transfer is given as $I = \lambda\delta(s)R(\phi)\max\{|v(t)| - v_k, 0\}$ [2, 10], where $\delta(s) = 0$ if $s \geq 1$, $\delta(s) = 1 - s$ if $0 < s < 1$, $\delta(s) = 1$ if $s \leq 0$; $R(\phi) = 1$ if $\phi \geq 1$, $R(\phi) = \phi(1-\phi)$ if $0 < \phi < 1$, $R(\phi) = 0$ if $\phi \leq 0$, $\lambda > 0$ is a dimension constant $[1/m]$, v_k is the critical velocity. The critical velocity indicates whether the erosion process has started or has not. If the velocity v reaches this critical speed, the theory suggests the erosion process will begin (the critical speed can be evaluated using experimental methods, see, e.g. [11]).

The coefficients of the system (3), (4) are given in following form

$$a(s) = -\frac{k_{01}k_{02}}{k}\frac{\partial p_c}{\partial s} > 0, \quad k_{0i}(s) = \frac{\overline{k}_{0i}}{\mu_i} \geq 0, \quad k(s) = k_{01} + k_{02} > 0,$$

$$F(s,\phi) = K_0(\phi)\frac{k_{01}k_{02}}{k}(\rho_1^0 - \rho_2^0)g, \quad b(s) = \frac{k_{01}}{k} \geq 0.$$

The functions $s(x,t)$ and $\phi(x,t)$ satisfy the initial and boundary conditions

$$s(0,t) = s_0(t), \quad s(1,t) = s_1(t), \quad s(x,0) = s^0(x), \quad \phi(x,0) = \phi^0(x). \tag{5}$$

In further analysis we use notation of functional domains described in [12].

Definition. *We call a pair of f nctions $s(x,t)$ and $\phi(x,t)$ as a classical sol tion of t e pro lem ( )–(5), w ere $s(x,t), \phi(x,t) \in C^{2+\alpha,(2+\alpha)/2}(Q_T)$, $Q_T = Q \times (0,T)$, $Q = (0,1)$ and satisf t e eq ations ( ), (4) and t e initial and o ndar conditions (5) as contin o s f nctions in Q_T.*

Theorem *We s ppose t at f nctions, coefficients and parameters of t e pro lem ( )–(5) satisf t e following conditions:*

1) t e f nctions $K_0(\phi), a(s), b(s), F(s,\phi), p_c(s)$ and t eir deri ati es p to t e second order are contin o s in $s \in [0,1]$, $\phi \in [0,1]$ and satisf t e conditions

$$0 < m \leq K_0(\phi), a(s) \leq M < \infty, \quad |F(s,\phi)|, b(s) \leq M, \quad p_c \geq 0, \quad \frac{\partial p_c}{\partial s} < 0,$$

w ere $F(s,\phi) = 0$ if $s < 0, s > 1$;

2) t e f nctions $v(t)$, $s_0(t)$, $s_1(t)$, $s^0(x)$, $\phi^0(x)$ satisf following smoot ness conditions

$$v(t), s_0(t), s_1(t) \in C^{2+\alpha}[0,T]; \quad s^0(x), \phi^0(x) \in C^{2+\alpha}(\overline{Q})$$

and t e matc ing conditions

$$s_0(0) = s^0(0), \quad s_1(0) = s^0(1),$$

as well as t e ineq alities

$$|v(t)| \geq v_k, \quad 0 \leq s^0(x) \leq 1, \quad 0 < m_0 \leq \phi^0(x) \leq M_0 < 1,$$

$$0 \leq s_0(t) \leq 1, \quad 0 \leq s_1(t) \leq 1,$$

w ere m_0, M_0, m_1, M_1 are gi en positi e constants.

T en t e pro lem ( )–(5) as onl one classical sol tion for an finite inter al $(0,T]$. oreo er,

$$0 \leq s(x,t) \leq 1, 0 < \phi(x,t) < 1, (x,t) \in Q_T.$$

We shall establish the proof of the theorem with the help of the Schauder fixed-point theorem [13] and standard auxiliary constructions [14].

Lemma 1. *Let t e ass mptions of t e t eorem e satisfied and a pair of f nction (s,ϕ) e t e classical sol tion of t e pro lem ( )–(5). T en ϕ and s satisf ineq alities $\phi_0 \leq \phi \leq 1$ and $0 \leq s \leq 1$.*

Proof. The inequality for ϕ is easily obtained from the fact that $I(s,\phi)$ is nonnegative. In the purpose to obtain the inequality for s we introduce a cut–off function $\overline{s} = max\{s-1, 0\}$. Multiplying the equation (3) by $\overline{s}$ and integrating the result in Q_t, we derive the equality

$$\int_0^t \int_0^1 s\phi_\tau \overline{s} dx d\tau + \int_0^t \int_0^1 (\phi \overline{s})_\tau \overline{s} dx d\tau = \int_0^t \int_0^1 (K_0 a \overline{s}_x + F)\overline{s}_x dx d\tau - \int_0^t \int_0^1 v(\tau) b_s \overline{s}_x \overline{s} dx d\tau. \tag{6}$$

Due to the definition of $\overline{s}$, all integrals in (6) should be taken in the domain $Q_t^* = \{(x,t) \in Q_t, s > 1\}$, where $\overline{s} = s - 1$, $s_t = \overline{s}_t$, $s_x = \overline{s}_x$, $\phi_t > 0$ and $F = 0$. Taking this into account in (6), we arrive at the following estimate

$$||\sqrt{\phi}\overline{s}||_{2,Q}^2 \leq C \int_0^t ||\sqrt{\phi}\overline{s}||_{2,Q}^2 d\tau,$$

where C is a positive constant. Hence, it follows that $\overline{s} \equiv 0$. The inequality $s \geq 0$ is proved in complete analogy using the same method with $\overline{s} = max\{-s, 0\}$. Lemma 1 is proved.

Following [4], we come to the Lemma 2.

Lemma 2. *Let two pairs (s_i, ϕ_i), $i = 1, 2$, e two di erent classical sol tions of t e s stem ( )-(4) wit initial and o ndar conditions eq al to s_i^0, ϕ_i^0 and $s_0^{(i)}, s_1^{(i)}$ respecti el . Let*

$$||s_0^1 - s_0^2|| + ||s_1^1 - s_1^2|| + ||s_1^0 - s_2^0|| + ||\phi_1^0 - \phi_2^0|| = \delta.$$

T en we a e t e estimate

$$\int\limits_0^1 (s^2 + \phi_t^2 + \phi^2)dx \leq C\delta, \qquad \int\limits_0^t \int\limits_0^1 (\phi_x^2 + \phi_{xt}^2 + s_x^2)dxd\tau \leq C\delta,$$

w ere C is a constant depends onl on m_0, m, M, v_k.

Using the substitution $V = \int\limits_0^s a(\tau)d\tau$ in (3) we arrive at the equation

$$V_t - \frac{K_0(\phi)}{\phi} a(s(V))V_{xx} = f,$$

$$f = -a(s(V))s(V)\frac{\phi_t}{\phi} + a\frac{1}{\phi}\frac{\partial K_0}{\partial \phi}\phi_x V_x + \frac{1}{\phi}\frac{\partial F}{\partial V}V_x + a\frac{1}{\phi}\frac{\partial F}{\partial \phi}\phi_x - a\frac{v(t)}{\phi}\frac{\partial b}{\partial V}V_x.$$

Following [5], we come to the Lemma 3.

Lemma 3. *Let ϕ and s e t e classical sol tion of t e pro lem ( )-(4), t en we a e t e ineq alit*

$$\int\limits_0^1 (V_x^2 + V_t^2)dx + \int\limits_0^t \int\limits_0^1 (V_{xx}^2 + V_{xt}^2)dxdt \leq C_1(m_0, m, M, v_k).$$

Proving of the Lemmas 1 and 2 is discussed in [4].

As a consequence of lemma 3 we have the estimate

$$(||s||_{Q_T}^{(\alpha)} + ||\phi||_{Q_T}^{(\alpha)}) \leq C_2(m_0, m, M, v_k).$$

Proof of the theorem. Substituting any continuous function $\widetilde{s}$, which satisfies the inequality $|\widetilde{s}| \leq M_1$ (where the constant $M_1 > 0$ is any nonnegative constant and will be chosen further), into coefficients of the equation (4), we arrive at the problem for $\widetilde{\phi}$

$$\frac{\partial \widetilde{\phi}}{\partial t} = I(\widetilde{\phi}, \widetilde{s}), \quad \widetilde{\phi}|_{t=0} = \phi^0(x). \tag{7}$$

Considering that the problem (7) has the unique classical solution and following the Lemma 1 we come to the estimate $\phi_0 \leq \widetilde{\phi} \leq 1$. Substituting $\widetilde{s}(x,t)$ and $\widetilde{\phi}(x,t)$ into coefficients of the equation (3), we arrive at the linear equation for $s(x,t)$

$$\frac{\partial s\widetilde{\phi}}{\partial t} = \frac{\partial}{\partial x}(K_0(\widetilde{\phi})a(\widetilde{s})\frac{\partial s}{\partial x} - b(\widetilde{s})v(t) + F(\widetilde{s}, \widetilde{\phi})), \tag{8}$$

with the boundary and initial conditions described by the equation (5). The solvability of this problem is evidently established following the well-known results about the solvability of the linear parabolic equations [12]. Moreover, following the proof of the Lemmas 1–3, we obtain, in particular, the estimate of the Holder constant

$$||s||^{(\alpha)}_{Q_T} \leq C_3(m_0, m, M, v_k). \tag{9}$$

Taking $M_1 = C_3$ due to the fact that M_1 can be chosen freely, we shall consider compact mapping $U : \widetilde{s} \to s$. From the inequality (9) it follows that this mapping in the domain $C_{0,0}(Q)$ transforms the solid sphere $\{|\widetilde{s}| \leq M_1\}$ into itself. Besides, the mapping is continuous in the domain $C_{0,0}(Q)$. Moreover, the mapping U is a completely continuous mapping: any continuous function is transformed to a function from the class $C^{\alpha,\alpha/2}(Q)$ satisfying the inequality (9). Consequently, all conditions of the Schauder fixed-point theorem are satisfied for our mapping U. Hence, there exist at least one fixed-point $s(x,t) \equiv \widetilde{s}(x,t)$ of the mapping U and the corresponding classical solution $\phi(x,t) \equiv \widetilde{\phi}(x,t)$ of the problem (7)–(8). It is evident that the pair $(\phi(x,t)s(x,t))$ is the classical solution of the problem (3)–(5). The uniqueness is achieved with the help of the Lemma 2.

3. Method of the numerical solution

The system of the equations (1)–(4) is solved numerically with the given initial and boundary conditions in the form (5) in a finite one-dimensional domain. Numerical study aims to investigate the behavior of the porosity and water saturation as functions of space variables and time depending on the given parameters of the problem.

The numerical solution of the problem (3)–(5) is obtained with the help of finite-difference method. Taking this into account, it is convenient to introduce a finite-difference grid with constant step $x_i = ih$, $i = 0...N$ and $t_n = n\tau$, $n = 0...T$, h is a mesh size, N is the number of spatial steps, τ is a time step, T is the number of time steps. Hence, the system (3), (4) reduced to the system of finite-difference equations

$$\phi_i^n \frac{s_i^{n+1} - s_i^n}{\tau} = \frac{1}{h^2}(K^n_{0i+\frac{1}{2}} a^n_{i+\frac{1}{2}}(s^{n+1}_{i+1} - s^{n+1}_i) - K^n_{0i-\frac{1}{2}} a^n_{i-\frac{1}{2}}(s^{n+1}_i - s^{n+1}_{i-1})) + \tag{10}$$

$$+ \left(\frac{\partial F}{\partial s}(s_i^n, \phi_i^n) - v(t_n)\frac{\partial b}{\partial s}(s_i^n)\right) \frac{s^{n+1}_{i+1} - s^{n+1}_{i-1}}{2h} - s_i^{n+1} I_i^n,$$

$$a^n_{i+\frac{1}{2}} = \frac{a(s_i^n + s^n_{i+1})}{2}, \quad a^n_{i-\frac{1}{2}} = \frac{a(s_i^n + s^n_{i-1})}{2},$$

$$K^n_{0i+\frac{1}{2}} = \frac{K_0(\phi_i^n + \phi^n_{i+1})}{2}, \quad K^n_{0i-\frac{1}{2}} = \frac{K_0(\phi_i^n + \phi^n_{i-1})}{2}, \quad I_i^n = I(s_i^n, \phi_i^n).$$

Here the index i refers to space and index n refers to time.

We shall find ϕ in the form [15]

$$\phi = \frac{\phi^0}{\phi^0 + (1 - \phi^0)e^{-\int_0^t \delta(s) max\{|\vec{v}(\tau)| - v_k, 0\} d\tau}}, \tag{11}$$

which can be obtained from (4) with $R(\phi) = (1 - \phi)\phi$.

The numerical algorithm consists of several steps. First, we substitute the initial condition for the porosity ϕ^0 into the coefficients of the finite-difference scheme (10). Using sweep method in the scheme (10) we find the values of the saturation s. Then substituting found the values of the saturation into the given equation (11) for ϕ, we find the values of the porosity ϕ_i^1 in each grid point. Finally, repeating the algorithm, we find the values of porosity and concentration at next time step and so on.

Numerical results

The porosity and the water saturation are calculated with ($g = 0$, $v_k = 0$, $v(t) = 1$, $p_c(s) = (1/s^2 - 1)$, $K_0(\phi) = \phi^3$, $k_{0i} = s_i^2$ if $0 \leq s \leq 1$, $k_{0i} = 0$ if $s_i \leq 0$, $k_{0i} = 1$ if $s_i \geq 1$). The initial and boundary conditions are $s(x, 0) = (x0.5)/l + 0.5$, $s(0, t) = 0.5$, $s(l, t) = 1$, $\phi^0(x) = 0.3$. At the points $x = l$ there is no fluidized solid particles. The initial condition for s assumed to be a linear function. In our calculations we keep $h = 0.01$, $\tau = 0.01$, $N = 100$, $T = 100$, $l = 1$. The numerical results are shown in figure 1–2. It is shown that if saturation s reaches 1 the erosion process will stop and the porosity will be unchanged.

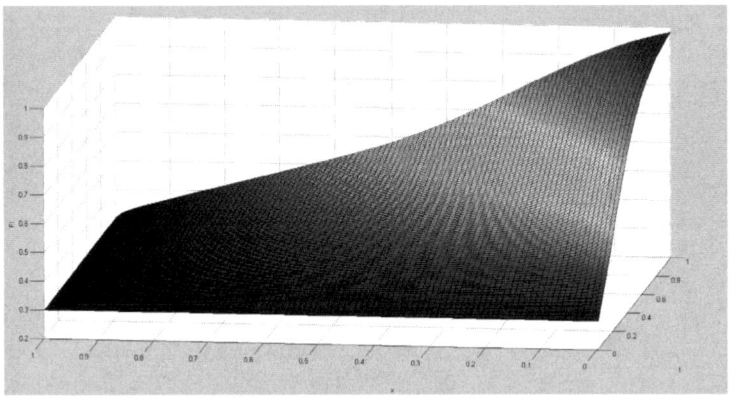

Figure 1. Porosity as a function of time and x.

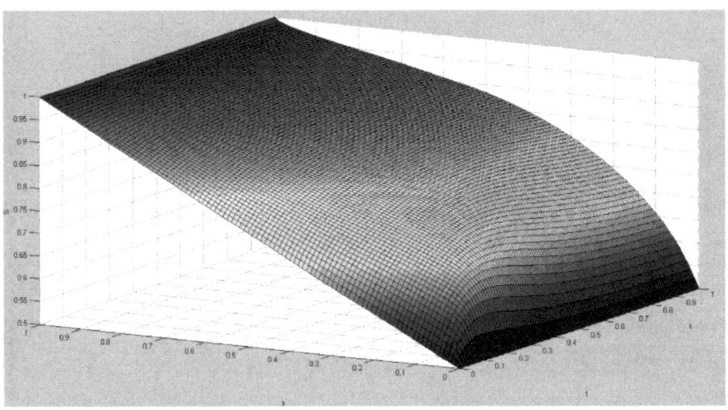

Figure 2. Water saturation as a function of time and x.

4. Two dimensional case of the problem

We shall describe one special case of the profile problem and solve it numerically in the purpose of finding 2D area affected by the erosion process.

We consider a two-dimensional movement of the water and fluidized solid particles with given concentration of the fluidized solid particles $s_2 = c = const \in [0, 1]$ and with the equality of both pressures $p_1 = p_1 = p$. The intensity of the phase transition is taken in the form $I_{23} = \rho_3^0 \lambda \phi c(1-\phi) max(|v_1| - v_k, 0)$.

With such assumptions in the system (1)–(2) we arrive at the following system

$$div(a(c,\phi)(\nabla p + \rho_1^0 \vec{g})) = 0, \quad \frac{\partial \phi}{\partial t} = I, \tag{12}$$

where $a(c,\phi) = (c\lambda + 1 - c)\phi k_f$ and all derivatives are taken with respect to two space variables. The flow region Q is a rectangular cross section with sides $0 \le x \le 25$, $0 \le y \le 1$. The boundary of Q consists of 5 parts: $\Gamma_1 = \{0 \le x \le 25, y = 0\}$, $\Gamma_2 = \{0 \le y \le 0.25, x = 0\} \bigcup \{0.29 \le y \le 1, x = 0\}$, $\Gamma_3 = \{0 \le x \le 25, y = 1\}$, $\Gamma_4 = \{0 \le y \le 1, x = 25\}$, $\Gamma_5 = \{0.25 \le y \le 0.29, x = 0\}$. The velocity is satisfied to the conditions of impermeability on $\Gamma_1, \Gamma_2, \Gamma_3$. Deriving equations for p from these conditions with the help of the Darcy's law and adding conditions for p on Γ_4, Γ_5, we come to the boundary conditions for pressure p

$$\frac{\partial p}{\partial y}\Big|_{\Gamma_1} = -g\rho_1^0; \quad \frac{\partial p}{\partial x}\Big|_{\Gamma_2} = 0; \quad \frac{\partial p}{\partial y}\Big|_{\Gamma_3} = -g\rho_1^0; \quad p\Big|_{\Gamma_4} = -g\rho_1^0 y + p_0; \quad p\Big|_{\Gamma_5} = -g\rho_1^0 y.$$

Here p_0 is a given pressure of the saturated soil in the point of intersection of Γ_1 and Γ_4. The equation (12) is solved numerically using the similar method described in the section 1, but with respect to two space variables, i.e. at first, we solve the finite-difference equation for pressure p in x direction at a half time step $n + 1/2$, then we solve the equation for p in y direction at next time step $n + 1$. The equation (12) is investigated numerically for $\phi^0 = 0,25$, $g = 9.81$ m/c^2, $k_f = 10^{-2}$ m/s. The critical speed v_k is 0.11 m/s. Numerical results are shown in figure 3. The area affected by the process of erosion is shown by the orange color.

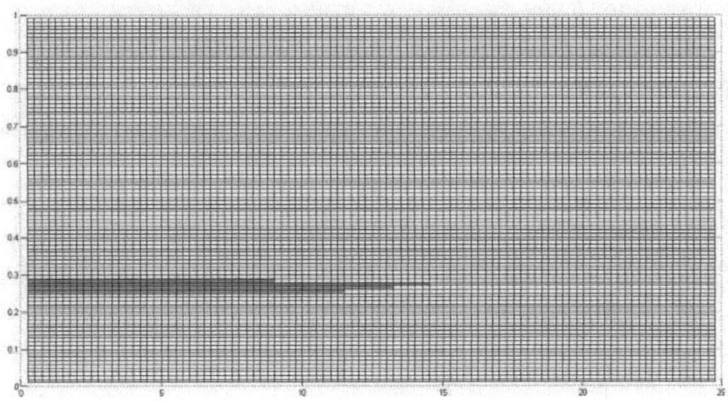

Figure 3. The area affected by the process of erosion (orange).

5. Conclusion

The global solvability of the nondegenerate one dimensional problem is proved. The degenerate one dimensional problem and one special case of the problem in a two dimensional domain with given concentration of the fluidized solid particles are solved numerically by the finite-difference method. The area affected by the process of erosion is founded.

Acknowledgments

In conclusion the author express their profound gratitude to K. Shishmarev for his helpful comments and constructive suggestions.

References

[1] Nigmatulin R I 1987 *Dynamics of multiphase media, arts 1 and 2* (Moscow: Nauka)
[2] Vardoulakis I, Stavropoulou M and Papanastasiou R 2001 *J. Transp. orous Med.* **45** 267–81
[3] Bonelli S 2012 *Erosion of Geomaterials* (London: John Wiley & Sons, Inc.)
[4] Papin A A, Weigant W A and Sibin A N 2015 *zvestiya of ASU* **85** 89–103 (in Russian)
[5] Antontsev S N, Kazhikhov A V and Monakhov V N 1990 *Boundary value pro lems in mechanics of nonhomo eneous fluids* (Amsterdam: North Holland)
[6] Antontsev S N and Papin A A 1979 *Soviet Math. Do l.* **247** 722–6
[7] Papin A A 2008 *J. Appl. Mech. Tech. hys.* **49** 527–36
[8] Tokareva M A 2015 *J. Si . Fed. Univ. Math. hys.* **8**(4) 467–77
[9] Bocharov O B, Rudyak V Ya and Seryakov A V 2014 *J. Min. Sci.* **50** 235–48
[10] Garde R J and Ranga Raju K G 2000 *Mechanics of Sediment Transportation and Alluvial Stream ro lems* (New Delhi: New Age International)
[11] Wang J, Walters D A, Settari A and Wan R G 2006 *Simulation of cold heavy oil production usin an inte rated modular approach ith emphasis on foamy oil flo and sand production effects. World Heavy Oil onf.* (Calgary: AB)
[12] Ladyzhenskaya O A, Solonnikov V A and Uraltseva N N 1967 *Linear and Quasilinear Equations of ara olic Type* (Moscow: Nauka)
[13] Edwards R 1965 *Functional Analysis: Theory and Applications* (New York: Dover Public. Inc.)
[14] Kruzhkov S N and Sukorjanskii S V 1977 *Mat. USSR S .* **33**(1) 62–80
[15] Kuzikov S S, Papin A A and Sibin A N 2014 *zvestiya of ASU* **81** 41–4 (in Russian)

Free-surface potential flow of an ideal fluid due to a singular sink

A A Mestnikova[1,2] **and V N Starovoitov**[1,2]

[1] Lavrentyev Institute of Hydrodynamics, Novosibirsk, Russia
[2] Novosibirsk State University, Novosibirsk, Russia

E-mail: `starovoitov@hydro.nsc.ru`

Abstract. A two-dimensional problem of a potential free-surface flow of an ideal incompressible fluid caused by a singular sink is considered. The sink is placed at the horizontal bottom of the channel. By employing a conformal map, the problem is equivalently rewritten in the unit circle. After that, it is investigated by the Levi — Civita technique with the extraction of the singular part of the flow that corresponds to the sink. We derive a Nekrasov type equation that describes exactly the form of the free boundary. This equation is studied at first numerically and then by an exact mathematical technique. It is shown that for the Froude number greater than some particular value, there exists a unique solution of the problem such that the free surface decreases monotonically when moving from the infinity to the sink. At the point over the sink, the free surface has a cusp.

1. Introduction

In this paper, we study a two-dimensional steady problem of a potential free surface flow of an ideal incompressible fluid over a flat horizontal bottom. The flow is caused by a singular sink of the strength $m > 0$ that is located at the bottom. We assume that at infinity the flow velocity tends to a constant V that is determined by the strength of the sink and by the depth h of the fluid layer. The problem has only one non-dimensional parameter, the Froude number $\mathrm{Fr} = V/\sqrt{gh}$, where g is the gravity acceleration. Since the statement of the problem does not include the value of V, it would be more appropriate to define the Froude number in terms of the sink strength m. Due to the incompressibility of the fluid, $m = 4hV$ and we have

$$\mathrm{Fr}^2 = \frac{m^2}{16gh^3}.$$

In this paper, for brevity, we use the parameter $\alpha = \mathrm{Fr}^2/2$ instead of Fr.

In the non-dimensional statement, the problem reads as follows. Let (x, y) be rectangular Cartesian coordinate system with the origin at the point O, where the sink is located. The flow domain D is then between the free boundary $\Gamma = \{(x,y) \mid y = \eta(x)\}$ and the bottom Γ_0 that coincides with the horizontal axis Ox (see Fig. 1). The problem is to find the function η as well as the velocity vector field $\boldsymbol{v} = (v_x, v_y)$ and the pressure p that satisfy in D the stationary Euler equations of an ideal incompressible fluid. These functions satisfy also the following conditions:

$$\eta(x) \to 1 \quad \text{and} \quad \boldsymbol{v}(x,y) \to (\mp 1, 0) \quad \text{as} \quad x \to \pm\infty,$$

$$\Gamma_0 \setminus \{O\}: \quad v_y = 0,$$

$$\left(\boldsymbol{v}(x,y) + \frac{2}{\pi}\frac{\boldsymbol{r}}{|\boldsymbol{r}|^2}\right) \to 0 \quad \text{as} \quad |\boldsymbol{r}| \to 0, \quad \boldsymbol{r} = (x,y) \in D,$$

$$\Gamma: \quad \boldsymbol{v}\cdot\boldsymbol{n} = 0, \quad p = 0,$$

where $\boldsymbol{n}$ is the normal vector to Γ. We assumed that the pressure is equal to zero at Γ since the pressure is constant on the free boundary and the Euler equations include only the gradient of p. Notice that, as a consequence of the Euler equations and the conditions at the boundary Γ, the functions $\boldsymbol{v}$ and η satisfy the Bernoulli law:

$$\alpha|\boldsymbol{v}(x,\eta(x))|^2 + \eta(x) = \alpha + 1 \quad \text{for} \quad x \in (-\infty,\infty). \tag{1}$$

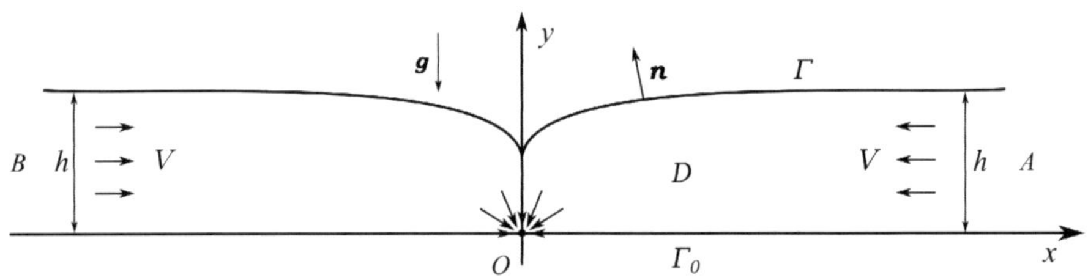

Figure 1. General flow pattern in the physical plane.

Due to the incompressibility of the fluid and the potentiality of the flow we can exclude the pressure from the statement of the problem. Namely, there exist defined in D functions φ and ψ called the potential and the stream function respectively such that

$$v_x = \partial_x\varphi = \partial_y\psi \quad \text{and} \quad v_y = \partial_y\varphi = -\partial_x\psi \quad \text{for} \quad (x,y) \in D. \tag{2}$$

These equations mean that φ and ψ are adjoint harmonic functions in D. Notice also that the potential and the stream function are defined up to an additive constant.

There are quite a large number of works in which this and similar problems are studied numerically and from a mechanical point of view. A lot of remarkable results are obtained and we are not able to mention all of them. We restrict ourselves to papers directly related to our study. It seems to us that there are two key properties of the flow under consideration. The first one consists in the fact that, for sufficiently large values of the Froude number, there are no waves going to infinity. The monotonicity of the free surface has already been obtained numerically in the first papers on the subject (see, e.g., [1–4]). It can be explained by the following reason. The wave velocity is equal to $\sqrt{gh}$ and, for large Froude numbers, this quantity is less than the velocity of the flow which is directed towards the sink. Notice that in the case of the source the waves do exist (see [5]). In general, the waves occur in the flows with small Froude numbers, i.e., in the so called subcritical case. This is not the subject of the present paper and we refer to the monograph [6], where an extensive bibliography on the question and the solution of various free surface flow problems can be found. In particular, the existence of waves for the flow around the point vortex was proved. We have another point singularity but the technique suggested there seems to be applicable also in our case.

The second and not so obvious property of the flow is that the free boundary has a cusp over the sink for the sufficiently large Froude number. This fact was discovered numerically in the above-cited works [1–4]. Here, we should notice that for small Froude numbers the stagnation point can occur over the sink. At this point the velocity of the fluid is equal to zero. The presence of the stagnation point is typical for the problem with the sink at the sloping bottom (see [7] and also [1, 2]).

In this paper, we consider the case of the Froude number greater than some particular value. Thus, the free boundary possesses both these properties. The novelty of our work consists in the following. The papers cited above investigate numerically approximate equations describing the free surface. Usually they use a truncated power series. We derive a Nekrasov type equation that is exactly satisfied by the free boundary and then look for its numerical solutions. Moreover, we are able to proof with the mathematical rigor that this equation has a unique solution. This fact enables us to be sure that our numerical solution is close to the exact solution of the problem. The mathematical part of this work will be published later. In the present paper, we formulate the results only.

2. The statement of the problem in the unit semicircle

Since we consider a two-dimensional problem, it is convenient to make use of the complex variable approach. Together with the complex variable $z = x + iy$ we introduce the complex velocity $v = v_x - iv_y$ and the complex potential $w = \varphi + i\psi$. As it follows from (2), w is a holomorphic function in D and $v = dw/dz$. Let us denote by F the conformal mapping of the flow domain D onto the upper unit semicircle D^*. We denote by G the inverse mapping and by t the complex variable in D^*. The images of the point of the sink $z = 0$ and the points A and B at infinity are the points $t = 0$, $t = 1$ and $t = -1$ respectively (see Fig. 2). Besides that, $F(\Gamma) = \Gamma^* = \{t \mid |t| = 1,\ \operatorname{Im} t > 0\}$. The mapping F as well as G is unknown and should be found. However, we can easily determine the flow in the semicircle. It is not difficult to see that

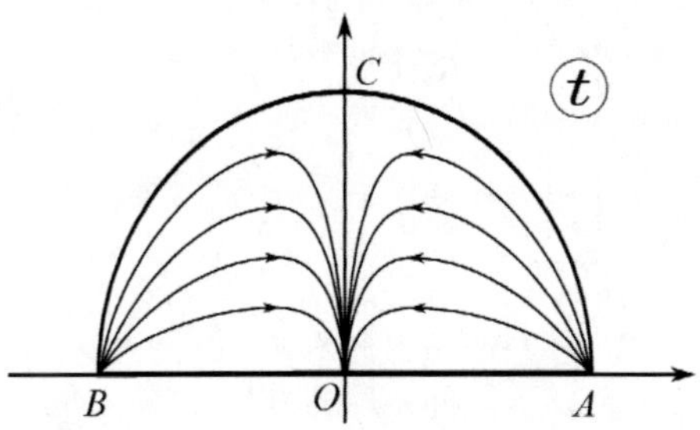

Figure 2. The flow in the unit semicircle D^*.

we have a singular sink at the point $t = 0$ and the same singular sources at $t = \pm 1$. Therefore, the complex potential w^* in D^* is given by the following expression:

$$w^*(t) = \frac{2}{\pi} \log\left(\frac{t^2 - 1}{2t}\right) - i = \frac{2}{\pi}\left(\log(t+1) + \log(t-1) - \log 2t - i\pi/2\right).$$

Notice, that $w(z) = w^*\big(F(z)\big)$ and

$$v\big(G(t)\big) = \frac{dw^*}{dt}(t)\, \frac{dF}{dz}\big(G(t)\big) = \frac{dw^*}{dt}(t) \left(\frac{dG(t)}{dt}\right)^{-1}.$$

This implies that

$$\frac{dG(t)}{dt} = \frac{1}{v\big(G(t)\big)}\, \frac{dw^*}{dt}(t). \tag{3}$$

If we could determine the function $v\big(G(t)\big)$, then the last equation would make it possible to determine G and, as a consequence, to solve the problem. The function $v\big(G(t)\big)$ is holomorphic in D^* and we denote by $u(t)$ its holomorphic extension to the unit circle, which is possible due to the Schwarz reflection principle. Notice that $u(t)$ has a pole at the point $t = 0$. Following the Levi-Civita approach, we introduce a holomorphic in the unit circle $\{t \mid |t| < 1\}$ function $\Omega(t) = \hat{\theta}(t) + i\hat{\tau}(t)$ with real $\hat{\theta}$ and $\hat{\tau}$ such that

$$u(t) = -\frac{2}{\pi}\frac{1}{t}\, e^{-i\Omega(t)}.$$

The function $\hat{\theta}$ is defined up to an additive constant $2\pi k$, where k is an integer number. We make this function single-valued by fixing it at the point $t = 1$: $\hat{\theta}(1) = 0$. In order to find Ω in the unit circle, it is sufficient to know its boundary value at $|t| = 1$, i.e., for $t = e^{i\sigma}$, $\sigma \in [0, 2\pi)$. Denote

$$\tau(\sigma) = \hat{\tau}(e^{i\sigma}), \quad \theta(\sigma) = \hat{\theta}(e^{i\sigma}).$$

These functions satisfy the following boundary conditions:

$$\theta(0) = \theta(\pi/2) = \theta(\pi) = \theta(3\pi/2) = 0 \quad \text{and} \quad \tau(0) = \tau(\pi) = \log\frac{\pi}{2}. \tag{4}$$

Besides that, due to the symmetry of the problem, we have:

$$\tau(\pi/2 + \sigma) = \tau(\pi/2 - \sigma), \quad \theta(\pi/2 + \sigma) = -\theta(\pi/2 - \sigma),$$
$$\tau(\sigma) = \tau(-\sigma), \quad \theta(\sigma) = -\theta(-\sigma)$$

for all $\sigma \in [0, 2\pi]$.

In order to obtain an equation for τ and θ on Γ^*, we employ the Bernoulli equation (1). The substitution $t = e^{i\sigma}$ and the differentiation with respect to σ give:

$$\alpha\, \frac{d|u(e^{i\sigma})|^2}{d\sigma} + \operatorname{Im}\frac{dG(e^{i\sigma})}{d\sigma} = 0 \quad \text{for} \quad \sigma \in (0, \pi).$$

As it follows from (3),

$$\frac{dG(e^{i\sigma})}{d\sigma} = -e^{-\tau(\sigma)}\, e^{i(\sigma+\theta(\sigma))}\cot\sigma. \tag{5}$$

After some calculation we obtain the Nekrasov type equation:

$$\tau'(\sigma) = \frac{\sin(\sigma + \theta(\sigma))\cot\sigma}{\alpha\pi + 3\int_0^\sigma \sin(s + \theta(s))\cot s\, ds} \quad \text{for} \quad \sigma \in (0, \pi). \tag{6}$$

This equation contains two unknown functions. To obtain one more equation, we can use the Hilbert transform. Taking into account the symmetry properties of the functions τ and θ, we derive the following relation:

$$\theta(\sigma) = \frac{1}{\pi}\int_0^{\pi/2} \tau(s)\big(\cot(s - \sigma) - \cot(s + \sigma)\big)\, ds, \quad \sigma \in [0, \pi/2]. \tag{7}$$

Equations (6) and (7) together with the boundary conditions (4) enable us to find the functions τ and θ and to solve the problem. However, it would be more convenient to rewrite (7) in another form. Since $\cot\sigma = \big(\log|\sin\sigma|\big)'$, we have

$$\theta(\sigma) = \frac{1}{\pi}\int_0^{\pi/2} K(s,\sigma)\,\tau'(s)\, ds, \quad \sigma \in [0, \pi/2], \tag{8}$$

where

$$K(s,\sigma) = \log\left|\frac{\sin(s+\sigma)}{\sin(s-\sigma)}\right| = \log\left|\frac{\tan s + \tan\sigma}{\tan s - \tan\sigma}\right| = 2\sum_{k=1}^{\infty}\frac{\sin 2ks\, \sin 2k\sigma}{k}.$$

Notice that $K(s,\sigma) \geq 0$ for all s and σ in $[0, \pi/2]$.

3. Numerical simulations

If we find the functions τ and θ from equations (6) and (8), we can determine the velocity vector field in D and the free boundary Γ. In this section, we present the numerical calculations of Γ. This boundary can be defined in the parametric form:

$$\Gamma = \{(x,y) \mid x = x_\Gamma(\sigma),\ y = y_\Gamma(\sigma),\ \sigma \in (0,\pi)\}.$$

This definition in the complex variables is as follows: $\Gamma = \{z \in \mathbb{C} \mid z = G(e^{i\sigma}),\ \sigma \in (0,\pi)\}$. Due to (5) we have

$$\frac{dx_\Gamma(\sigma)}{d\sigma} = \mathrm{Re}\,\frac{dG(e^{i\sigma})}{d\sigma} = -e^{-\tau(\sigma)}\cos\big(\sigma+\theta(\sigma)\big)\cot\sigma, \quad x_\Gamma(\pi/2)=0,$$
$$\frac{dy_\Gamma(\sigma)}{d\sigma} = \mathrm{Im}\,\frac{dG(e^{i\sigma})}{d\sigma} = -e^{-\tau(\sigma)}\sin\big(\sigma+\theta(\sigma)\big)\cot\sigma, \quad y_\Gamma(0)=1.$$

These equations, in particular, yield that

$$\frac{dy_\Gamma(\sigma)}{d\sigma}\Big(\frac{dx_\Gamma(\sigma)}{d\sigma}\Big)^{-1} = \tan\big(\sigma+\theta(\sigma)\big).$$

It means that $\sigma+\theta(\sigma)$ is the slope angle of the tangent to Γ at the point $\big(x_\Gamma(\sigma), y_\Gamma(\sigma)\big)$.

In order to find the solution of equations (6) and (8), we have used the method of successive approximations. We take $\theta \equiv 0$ as the first approximation and calculate until the difference between two successive approximations is greater than 10^{-6}. In Fig. 3, the free boundary for

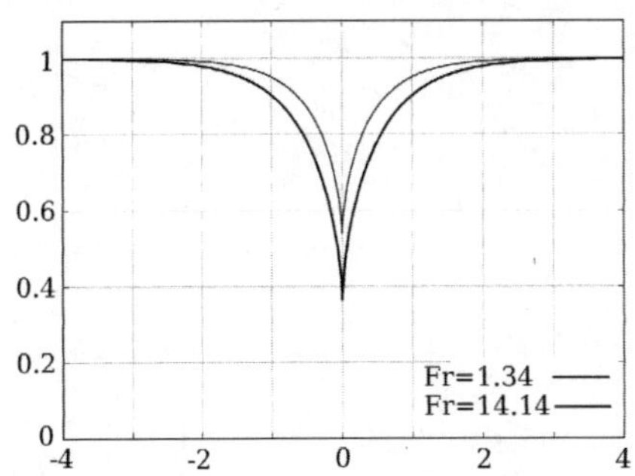

Figure 3. The free boundaries for two values of the Froude number.

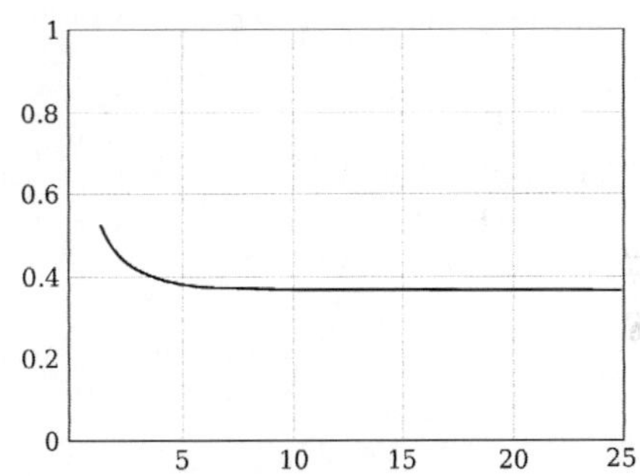

Figure 4. The dependence of the distance between the sink and the cusp point at the free boundary on the Froude number.

$\alpha = 0.9$ and $\alpha = 100$ is shown. It should be noted that the iterative process diverges for sufficiently small α, namely, when α is less than 1/2 approximately. By this reason, the graph in Fig. 4 begins at a non-zero Froude number. This graph shows the dependence of the distance between the sink and the cusp point at the free boundary on the Froude number. As seen, this distance decreases very slowly starting from the value 5 of the Froude number.

4. Mathematical treatment of the problem

In this section, we formulate the mathematical results related to the solvability of the problem. These results will be proved in a separate publication. Notice that we have essentially used some ideas from [8] and especially from [9].

Let us rewrite equations (6) and (8) as an operator equation for the function $\zeta(\sigma) = 3\tau'(\sigma)$:

$$\zeta = \Phi(\zeta), \tag{9}$$

where Φ is a nonlinear operator that acts as follows

$$\Phi(\zeta)(\sigma) = \frac{3}{\alpha\pi}\,\frac{\sin\big(\sigma + \theta(\sigma)\big)\,\cot\sigma}{\exp\int_0^\sigma \zeta\,ds}, \qquad \sigma \in [0, \pi/2],$$

$$\theta(\sigma) = \frac{1}{3}\,(H\zeta)(\sigma), \quad (H\zeta)(\sigma) = \frac{1}{\pi}\int_0^{\pi/2} K(s,\sigma)\,\zeta(s)\,ds.$$

The kernel K is defined after equation (8). We look for a solution of equation (9) in the Banach space $L^2(0, \pi/2) = \{\zeta(\sigma) \mid \|\zeta\| < \infty\}$,

$$\|\zeta\|^2 = \int_0^{\pi/2} |\zeta(\sigma)|^2\,d\sigma.$$

In order to prove the solvability of equation (9), we have used two classical fixed point theorems. The first one is the Schauder theorem. Let us take the following closed convex set in $L^2(0, \pi/2)$:

$$B_R^+ = \{\zeta \in L^2(0, \pi/2) \mid \|\zeta\| \leqslant R,\ \zeta \geq 0\}.$$

It is not difficult to establish that Φ is continuous and compact as an operator in $L^2(0, \pi/2)$. It remains only to prove the existence of $R > 0$ such that $\Phi(B_R^+) \subset B_R^+$. By employing the Fourier analysis, we have found that this condition is satisfied if

$$R \geq \frac{3}{2}\sqrt{\frac{\pi}{2}}\,\frac{1}{\alpha\pi - 2}.$$

The positiveness of Φ follows from the positiveness of the kernel K. Thus, equation (9) is solvable if

$$\alpha > \frac{2}{\pi}. \tag{10}$$

This inequality in terms of the Froude number approximately looks as Fr > 1.128.

The Schauder fixed point theorem says nothing about the uniqueness of the solution. In order to establish this fact, we have to use other methods, for instance, the Banach contraction mapping principle. This principle gives not only the existence but also the uniqueness of the solution. We have proved that

$$\|\Phi(\zeta_1) - \Phi(\zeta_2)\| \leq \frac{8}{\alpha\pi}\,\|\zeta_1 - \zeta_2\|, \quad \text{for all} \quad \zeta_1, \zeta_2 \in B_R^+,$$

where R is the same as above. Thus, Φ is a contraction and the solution of (9) is unique if

$$\alpha > \frac{8}{\pi}. \tag{11}$$

This inequality in terms of the Froude number approximately looks as Fr > 2.257.

We do not claim that (10) and (11) are the best possible conditions for the existence and the uniqueness. Surely, more precise estimates will improve these conditions. Notice that, in the numerical simulations, we were able to calculate the solution for smaller values of α. Of course, it can also be explained by a not sufficient accurate approximation of the problem.

References

[1] Vanden-Broeck J-M and Keller J B 1987 *J. Fluid Mech.* **175** 109–17
[2] Hocking G C 1985 *J. Austr. Math. Soc.* B **26** 470–86
[3] Tuck E O and Vanden-Broeck J-M 1984 *J. Austr. Math. Soc.* B **25** 443–50
[4] Forbes L K and Hocking G C 1990 B **32** 231–49
[5] Lustri C J, McCue S W and Chapman S J 2013 *MA J. Appl. Math.* **78**(4) 697–713
[6] Maklakov D V 1997 *Nonlinear pro lems of hydrodynamics of potential flo s ith un no n oundaries* (Moscow: Yanus-K.)
[7] Dun C R and Hocking G C 1995 *J. En . Math.* **29** 1–10
[8] Plotnikov P I and Toland J F 2004 *Arch. Rat. Mech. Anal.* **171** 349–416
[9] Fraenkel L E 2007 *Arch. Rat. Mech. Anal.* **183** 187–214

Roll waves in two-layer Hele-Shaw flows

I V Stepanova[1,2], A A Chesnokov[1,3] and V Yu Liapidevskii[1,3]

[1] Novosibirsk State University, Novosibirsk, Russia
[2] Institute of Computational Modelling SB RAS, Krasnoyarsk, Russia
[3] Lavrentyev Institute of Hydrodynamics SB RAS, Novosibirsk, Russia

E-mail: stepiv@icm.krasn.ru, chesnokov@hydro.nsc.ru, liapid@hydro.nsc.ru

Abstract. In this paper we study the emergence and development of roll waves in two-layer fluid flow in a Hele-Shaw cell. We propose the mathematical model of such flow and define the conditions of transition from stable state to instability in the form of the roll waves. We find out the physical parameters of flows at which the roll waves exist. A linear stability analysis and the Whitham criterion of roll waves existence are used for solving the problem and arrive to identical conclusions on depths of upper and lower layers at which violation of flow stability occurs. The numerical calculations for the obtained mathematical model at found ratios of densities, viscosities and depths of layers are performed. They confirm development of the roll waves of finite amplitude from small oscillations of the interface.

1. Introduction

A wide class of inhomogeneous fluid flows in long channels allow arising nonlinear regime leading to roll waves. This regime is a quasi periodic channel flow, in which smooth regions of flow are separated by hydraulic jumps. A feature of these flows is a transition from sub-critical flow to supercritical one in a coordinate system moving with the wave. For the first time roll waves were described by R. Dressler [1] in 1949. He found out that a class of travelling waves for shallow water equations consisted of periodic discontinuous solutions. This flow is easy to reproduce experimentally in natural or laboratory conditions. Despite of the long history of roll waves study, there are many open questions. For example it is interested to know how the roll waves arise and develop from small perturbations of interface of two-layer flow under lid. The roll waves change stable stationary flow and can lead to slug regime. That is why their study has a practical value. Different theoretical and experimental aspects of study of roll waves in liquids are considered in monograph [2] and papers [3, 4, 5]. A description and nonlinear analysis of roll waves in the framework of Burgers, Sen-Venane and Kuramoto–Sivashinsky equations can be found in [6, 7].

The mathematical model describing the motion of inhomogeneous immiscible fluid in a Hele-Shaw cell is proposed in recent paper [8]. Moreover, different simple modifications of layered flows are considered in that paper. Now we pay our attention to the study of two-layer flows of fluid and modelling of roll waves. A linear stability analysis is provided for two-layer Hele-Shaw flows. Values of physical parameters are found out for the development of flow instabilities. Velocities of propagation of nonlinear perturbations are defined and hyperbolicity conditions are formulated. Critical parameters and depths, at which the transition to roll wave begins, are found using the Whitham criterion [9]. A numerical modelling of perturbation development on the interface is provided. A comparison of conditions of stability violation with numerical results is carried out.

2. Mathematical model of inhomogeneous Hele-Shaw flow

A Hele-Shaw cell is the area between two parallel plates with a small gap between them. Equations of motion of viscous incompressible liquid in the Hele-Shaw cell have the form

$$\rho(\mathbf{v}_t + (\mathbf{v}\cdot\nabla)v) + \nabla p = \mu\mathbf{v}_{zz} + \rho\mathbf{g}, \quad \nabla\cdot\mathbf{v} = 0. \tag{1}$$

Here $\mathbf{x} = (x, y, z)$ is the coordinate vector, t is the time, $\mathbf{v} = (u, v, w)$ is the velocity vector, p is the pressure, ρ is the density, μ is the viscosity and $\mathbf{g} = (0, -g, 0)$ is the acceleration of gravity vector. The operator ∇ is calculated with respect to the coordinate vector. The characteristic sizes of cell (L, H) in the x and y direction, respectively, are significantly higher than the cell gap a. That is why the summands $\mathbf{v}_{xx}$ and $\mathbf{v}_{yy}$ vanish in the momentum equations. They are negligible compared to $\mathbf{v}_{zz}$.

We consider the velocity field in the form

$$u = \frac{3}{2}\left(1 - \left(\frac{2z}{a}\right)^2\right)u'(t, x, y), \quad v = \frac{3}{2}\left(1 - \left(\frac{2z}{a}\right)^2\right)v'(t, x, y), \quad w = 0.$$

It provides the fulfilment of the no-slip conditions on the cell walls $z = \pm a/2$. We also assume that the function p does not depend on z. Integrating equations (1) from $-a/2$ to $a/2$ leads to the system

$$\begin{aligned} &\rho(u_t + \beta(uu_x + vu_y)) + p_x = -\mu u, \\ &\rho(v_t + \beta(uv_x + vv_y)) + p_y = -\mu v - \rho g, \\ &u_x + v_y = 0. \end{aligned} \tag{2}$$

The primes are omitted. Here and below μ denotes the modified fluid viscosity $12\mu/a^2$. The coefficient β is equal to $6/5$. Listed above simplifications are connected with the inequalities $a << L$, $a << H$ and often used for modelling of Hele-Shaw flows [10, 11]. We assume that the horizontal size L of the cell is much more than its depth H, i.e. $H/L = \varepsilon << 1$. Supposing that

$$t \to \varepsilon^{-1}t, \quad x \to \varepsilon^{-1}x, \quad v \to \varepsilon v, \quad \mu \to \varepsilon\mu$$

and omitting all terms of order ε^2 we have the long-wave approximation model

$$\begin{aligned} &\rho(u_t + \beta(uu_x + vu_y)) + p_x = -\mu u, \quad p_y = -g\rho, \\ &u_x + v_y = 0; \quad v\big|_{y=0} = v\big|_{y=H} = 0. \end{aligned} \tag{3}$$

Here $y = 0$ and $y = H$ are the lower and upper boundaries in the y-direction.

2.1. Equations of two-layer flow

Let us consider the class of layered flows with the depths $h_i(t, x)$, the velocities $u_i(t, x)$, the constant densities ρ_i and the viscosities μ_i in the layers $i = 1, 2$ (index "1" is for the lower layer and "2" is for the upper one). We assume that the cell height is constant, i.e. $h_1 + h_2 = H = \text{const}$. We find the pressure from the second equation of (3) and take into account the kinematic condition on the interface. After some amount of calculations we arrive at the system describing the two-layer flow of viscous liquid in a Hele-Shaw cell [8]

$$\begin{aligned} &\rho_1(u_{1t} + \beta u_1u_{1x} + gh_{1x}) + g\rho_2h_{2x} + p_{0x} = -\mu_1u_1, \\ &\rho_2(u_{2t} + \beta u_2u_{2x}) + p_{0x} = -\mu_2u_2, \\ &h_{1t} + (u_1h_1)_x = 0, \quad h_{2t} + (u_2h_2)_x = 0, \end{aligned} \tag{4}$$

where $p_0(t,x)$ is the pressure on an upper lid.

Further we use the following notations for the unknown functions, the density ratio and the viscosity ratio in the layers

$$u = u_1, \quad w = u_2, \quad h = h_1, \quad \rho = \rho_2/\rho_1, \quad \mu = \mu_1/\mu_2.$$

Taking into consideration that $h_2 = H - h$ we multiply the second equation of (4) by ρ and subtract from the first one. As a result we obtain the system

$$(u - \rho w)_t + \beta(uu_x - \rho w w_x) + bh_x = \frac{\mu_2}{\rho_1}\Big(w - \mu u\Big), \quad h_t + (uh)_x = 0,$$

where $b = g(1-\rho)$. According to the last equations of (4) the flow rate in the channel does not depend on x. We consider flows with known constant flow rate

$$hu + (H - h)w = H\, u_m = \text{const},$$

where u_m is the average flow velocity. We can take $H = 1$, $b = 1$ after using corresponding dimensionless variables. The additional scaling of independent variables helps to exclude the parameter μ_2/ρ_1. Summarizing given assumptions we take the equations for specification of the depths and velocities in the form

$$\begin{aligned} &(u - \rho w)_t + \beta(uu_x - \rho w w_x) + h_x = w - \mu u, \\ &h_t + (uh)_x = 0, \quad hu + (1-h)w = u_m. \end{aligned} \tag{5}$$

Further theoretical analysis and modelling of the roll waves are provided in the framework of equations (5).

3. Linear stability analysis

Let us consider the constant solution of system (5) $u = u_0$, $w = w_0$, $h = h_0$. We should emphasize that u_0, w_0 and h_0 have to satisfy the equations

$$w_0 = \mu u_0, \quad h_0 u_0 + (1 - h_0) w_0 = u_m. \tag{6}$$

The velocity components and depth of perturbed flow deviate from the basis flow weakly. That is why these values take the form

$$h = h_0 + \tilde{h}, \quad u = u_0 + \tilde{u}, \quad w = w_0 + \tilde{w}. \tag{7}$$

Here $\tilde{h}$, $\tilde{u}$, $\tilde{w}$ are the functions of all independent variables.

Substitution of solution (7) into equations (5) and linearization of the system obtained give the equations for small perturbations

$$\begin{aligned} &(u - \rho w)_t + \beta(u_0 u_x - \rho w_0 w_x) + h_x = w - \mu u, \\ &h_t + u_0 h_x + h_0 u_x = 0, \quad h_0 u + (1 - h_0) w + (u_0 - w_0) h = 0. \end{aligned} \tag{8}$$

The tildes are omitted. A solution of these equations is sought in the form

$$(h, u, w) = (\hat{h}, \hat{u}, \hat{w}) \exp(ik(x - ct)), \tag{9}$$

where c is the phase velocity, $k > 0$ is the wave number, i is the imaginary unit.

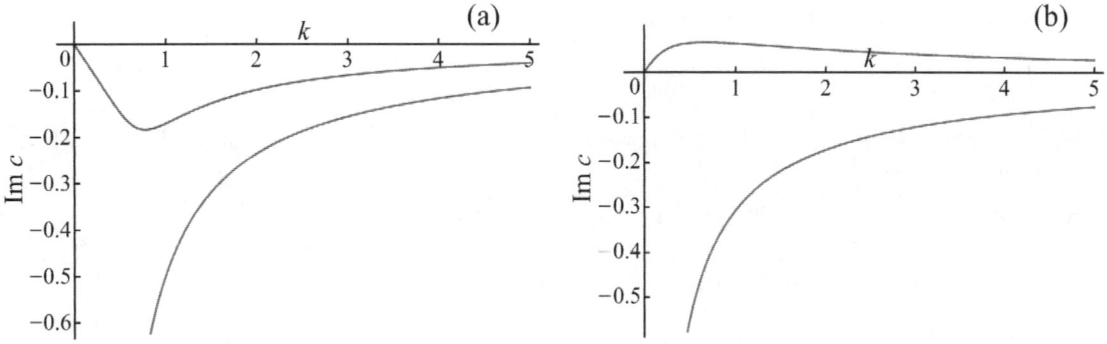

Figure 1. Imaginary parts of solutions of equation (10) at $\beta = 1.2$, $u_m = 0.3$, $\mu = 0.2$ and $\rho = 0.8$. They correspond to stable flow at $h_0 = 0.5$ (a) and unstable flow at $h_0 = 0.05$ (b).

Substituting anzats (9) into equations (8) and eliminating $\hat{u}$ and $\hat{w}$ with the help of $\hat{h}$ by the formulas

$$\hat{u} = -\frac{u_0 - c}{h_0}\hat{h}, \quad \hat{w} = \frac{w_0 - c}{1 - h_0}\hat{h}$$

we obtain the dispersion relation

$$c^2 - \frac{(\beta+1)(1-(1-\rho\mu)h_0)u_0 - iu_m/(ku_0)}{1-(1-\rho)h_0}c + \\ + \frac{(1-(1-\rho\mu^2)h_0)u_0^2\beta - (1-h_0)h_0 - iw_0/k}{1-(1-\rho)h_0} = 0. \tag{10}$$

This relation should be fulfilled for constructing a non-trivial solution of equations (8) in form (9). Here the velocities of the basic stream u_0 and w_0 depend on the depth h_0 only with respect to conditions (6) by the formulas

$$w_0 = \mu u_0, \quad u_0 = \frac{u_m}{\mu + (1-\mu)h_0}.$$

According to classical linear stability theory [12] the flow under study is stable if $\operatorname{Im} c < 0$ ($c = c(k)$ is the root of the dispersion relation). The curves $\operatorname{Im} c(k)$ are shown in Fig. 1. Here $c = c(k)$ are the branches of solution of quadratic equation (10) at $\beta = 1.2$, $u_m = 0.3$, $\mu = 0.2$, $\rho = 0.8$ and different values of depth h_0 of the lower layer in the basic flow ($h_0 = 0.5$ and $h_0 = 0.05$). It can be seen from Fig. 1 (a) that the basic flow is stable when the values of h_0 are close to half of channel width. A development of instability is possible when one of the layers is thin enough at a fixed average velocity, density ratio and viscosity ratio. It is shown in Fig. 1 (b).

In the following sections we demonstrate that proposed linear analysis is in a good agreement with the Whitham conditions of roll waves existence and with numerical experiments.

4. Characteristics of equations (5). The Whitham conditions

Using the last equation of (5) we eliminate the velocity in the lower layer

$$w = \frac{u_m - uh}{1-h}$$

and introduce the equations of motion in the form

$$\mathbf{U}_t + \mathbf{A}\mathbf{U}_x = \mathbf{F},$$

where $\mathbf{U} = (u, h)^{\mathrm{T}}$ is the vector of unknowns, $\mathbf{F} = (F, 0)^{\mathrm{T}}$ is the right part. The matrix of coefficients takes the form

$$\mathbf{A} = \begin{pmatrix} M_1 & M_2 \\ h & u \end{pmatrix}.$$

Values M_1, M_2 and F are the following

$$M_1 = \frac{((1-h)\beta - \rho h)u + (\beta+1)\rho h w}{1-(1-\rho)h}, \quad M_2 = \frac{1 - h - (u-w)(u-\beta w)\rho}{1-(1-\rho)h},$$
$$F = \frac{(w-\mu u)(1-h)}{1-(1-\rho)h}.$$

Eigenvalues $\lambda^{\pm}$ of matrix $\mathbf{A}$ have the form

$$\lambda^{\pm} = \frac{1}{2}\Big(M_1 + u \pm \sqrt{(M_1+u)^2 - 4(uM_1 - hM_2)}\Big)$$

and define the velocities of characteristics propagation ($dx/dt = \lambda^{\pm}$). It is necessary to note that equations (5) generate a mixed type system. When the liquid physical parameters are arbitrary, the characteristics $\lambda^{\pm}$ can be complex. A development of long-wave instability is possible beyond of the interval of system hyperbolicity. The statement of Cauchy problem requires an additional analysis and justification [2].

Together with equations (5) we consider more simple kinematic model, in which the inertial effects are not taken into account. Then the first equation in (5) is replaced by the condition $w = \mu u$. So we suppose that the flow rate is constant and the velocity u can be performed with the help of the lower layer of depth h. Make allowance for all aforesaid, we write the kinematic model as one equation

$$\frac{\partial h}{\partial t} + \frac{\partial f(h)}{\partial x} = 0, \quad f(h) = \frac{u_m h}{\mu + (1-\mu)h}. \tag{11}$$

The velocity of characteristic of model (11) is

$$\hat{\lambda} = f'(h) = \frac{\mu u_m}{(\mu + (1-\mu)h)^2}.$$

With respect to the Whitham criterion [9], the roll waves exist in the intervals, in which one of inequalities is valid

$$\hat{\lambda} \geq \lambda^{+} \quad \text{or} \quad \hat{\lambda} \leq \lambda^{-}. \tag{12}$$

In addition, the solution of equations (5) should not cross over the domain of hyperbolicity. Therefore, a correct statement of Cauchy problem requires not to pass beyond the intersection point of the characteristics λ^{+} and λ^{-}.

Characteristics $\lambda^{\pm}$ and $\hat{\lambda}$ for different average velocity u_m, density ratio ρ and viscosity ratio μ are given in Fig. 2. For all cases $\beta = 1.2$. For constructing the dependencies $\lambda = \lambda^{\pm}(h)$ we use the relation $u = f(h)/h$. Fig. 2 (a) shows that the Whitham criterion (12) is fulfilled for small values of h. Furthermore, the hyperbolicity domain covers the whole interval of depths $h \in (0, 1)$. It is necessary to note that the obtained interval for roll waves existence is in a good agreement with the linear stability analysis discussed in the previous section. We can see from both Fig. 1 (b) and Fig. 2 (a) that the instability arises at small depth of the lower layer.

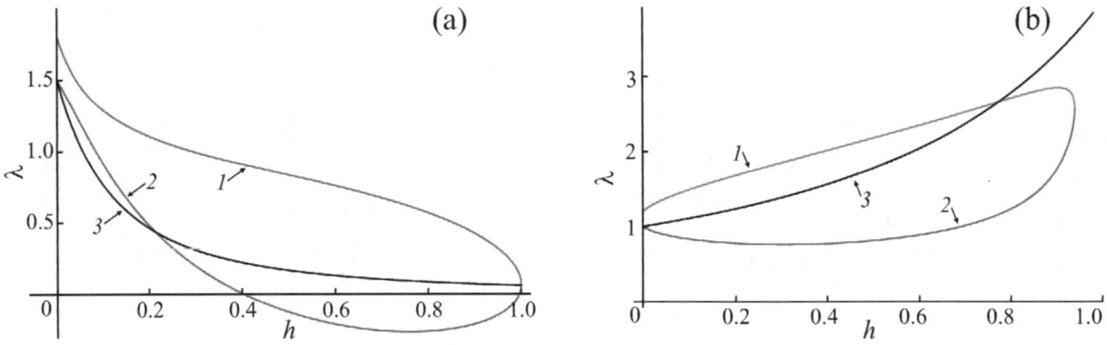

Figure 2. Velocities of characteristics $\lambda^{\pm}$ and $\hat{\lambda}$ (curves *1*, *2* and *3*) depending on h at $u_m = 0.3$, $\rho = 0.8$, $\mu = 0.2$ (a) and at $u_m = 2$, $\rho = 0.02$, $\mu = 2$ (b).

The values for constructing velocities of characteristics in Fig. 2 (b) correspond to gas-liquid media. In this case the Whitham criterion $\hat{\lambda} \geq \lambda^+$ is valid for big enough h. Moreover, the hyperbolicity conditions are kept in not the whole flow region. It requires a more detailed analysis of problem correctness. The study of physical parameters allows to conclude that density ratio affects the critical layer depth, while the viscosity ratio and given flow rate bring to a bare influence on the hyperbolicity conditions of the governing equations.

5. Numerical modelling of the roll waves

Further we carry out a numerical analysis of small non-stationary perturbation development using nonlinear equations of two-layer flow (5). We rewrite equations (5) in conservative form for the calculations. We denote $s = u - \rho w$ and express the velocities in the layers with the help of s and h:

$$u = \frac{1-h}{1-(1-\rho)h}\Big(s + \frac{\rho u_m}{1-h}\Big), \qquad w = \frac{u_m - uh}{1-h}.$$

Then, for the functions s and h we have the conservative form of equations

$$s_t + \big((u^2 - \rho w^2)\beta/2 + h\big)_x = w - \mu u, \quad h_t + (uh)_x = 0. \tag{13}$$

Calculations corresponding to model (13) are carried out using MATLAB and TVD Nessyahy–Tadmor scheme of the second-order approximation.

We suppose that the channel length $L = 200$ and height $H = 1$. At the initial moment the channel is occupied by the two-layer liquid moving with the average velocity $u_m = 0.3$. We assume also that $\rho = 0.8$, $\mu = 0.2$ and $\beta = 1.2$. The depth of the lower layer is $h_0 = 0.05$. The initial velocities in the layers are found from (6) with the help of given h_0. They equal $u_0 = 1.25$ and $w_0 = 0.25$. As mentioned above the emergence of the roll waves is possible in this interval of physical parameters with respect to the Whitham criterion. The soft boundary conditions $\mathbf{u}_N = \mathbf{u}_{N-1}$ are to be valid at the right boundary of the computational domain. Here $\mathbf{u}_j$ is the value of function in the nodal point x_j. Small perturbations $h(t,0) = h_0(1 + 0.05\sin(\omega t))$ and $s(t,0) = u_0 - \rho w_0 = 1.05$ are given at the left boundary. A uniform grid with respect to the spatial variable x is used for the calculations, the number of nodes $N = 4000$. The time step is determined by the Courant condition. The results of calculations at $t = 250$ are shown in Fig. 3. It can be observed that the roll waves are generated in the considered intervals of parameters. The lengths of the waves depend on frequency of oscillations at the left boundary.

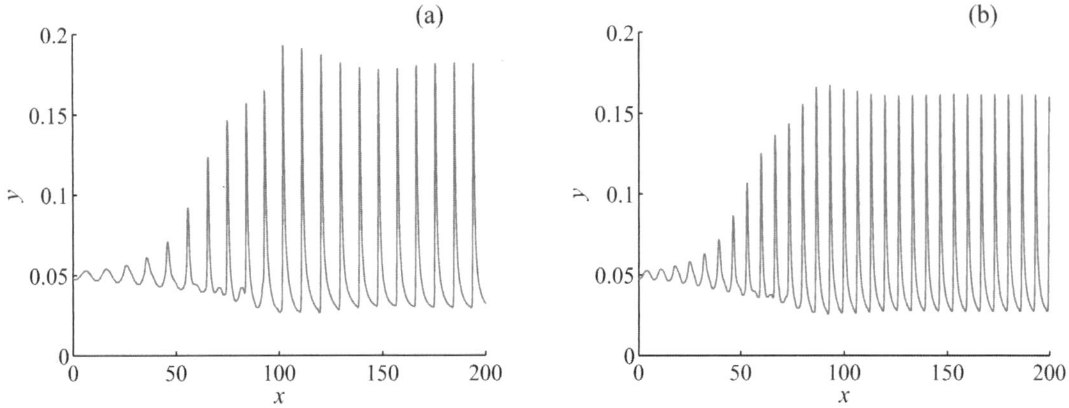

Figure 3. Interface of two-layer flow $y = h$ at $t = 250$ for $u_m = 0.3$, $\rho = 0.8$, $\mu = 0.2$ and $h_0 = 0.05$. There is small periodic interface perturbation with oscillation frequency $\omega = 0.7$ (a) and $\omega = 1$ (b) in the left boundary.

Note that the roll waves move against the flow in the regime under consideration. Thus the calculation using equations (13) demonstrates development of the interface perturbations and the roll waves formation if the Whitham criterion (12) is satisfied.

6. Conclusion

The study is focused on the analysis of two-layer flow equations (5). The solution in the form of normal modes is constructed and dispersion relation (10) is treated for linearized system (8). Stability domains of the flow are identified depending on the depth of the lower layer at a given flow rate and known density ratio and viscosity ratio (Fig. 1). Characteristics for nonlinear system (5) are found. The Whitham conditions for roll waves existence are formulated (Fig. 2). It is shown that the results on linear stability analysis and the Whitham criterion are in a good agreement. A numerical modelling of the roll waves is carried out using TVD scheme of high-order accuracy (Fig. 3). The calculation of interface position demonstrates the emergence of the roll waves for physical parameters found by means of the theoretical analysis.

Acknowledgments

The work is supported by Russian Science Foundation (project 15-11-20013).

References

[1] Dressler R F 1949 *Comm Pure Appl Math* **2** 149–94
[2] Liapidevskii V Yu and Teshukov V M 2000 *Mathematical Models of Propagation of Long Waves in a Non-Homogeneous Fluid* (Novosibirsk: Siberian Branch of Russian Academy of Sciences) [in Russian]
[3] Needham D J and Merkin J H 1981 *Proc Roy Soc A* **394** 259–78
[4] Liapidevskii V Yu 2002 *J Appl Mech Tech Phys*, **43**(2) 204–7
[5] Boudlal A and Liapidevskii V Yu 2015 *J Appl Mech Tech Phys* **56**(4) 541–8
[6] Johnson M A, Zumbrun K and Noble P 2011 *SIAM J Math Analysis* **43**(2) 577–611
[7] Boutounet M, Noble P and Vila J P 2012 *Math Models Methods Appl Sci* **22** 1250006
[8] Chesnokov A A and Stepanova I V 2015 *Appl Math Comput* **265** 320–8
[9] Whitham G B 1999 *Linear and Nonlinear Waves* (John Wiley & Son)
[10] Gondret P and Rabaud M 1997 *Phys Fluids* **9** 3267–74
[11] Dias E O and Miranda J A 2011 *Phys Rev E* **83** 066312
[12] Drazin P G 2002 *Introduction to Hydrodynamic Stability* (Cambridge University Press)
[13] Nessyahy H and Tadmor E 1990 *J Comp Phys* **87** 408–63

Solvability of initial boundary value problem for the equations of filtration in poroelastic media

M A Tokareva

Altai State University, Barnaul, Russia

E-mail: tma25@mail.ru

Abstract. The study is devoted to the mathematical model of fluid filtration in poroelastic media. The laws of conservation of mass for each phase, Darcy's law for fluid phase, the rheological law and the general equation of conservation of momentum for system describe this process. The local solvability of the problem is proved in this paper for the case in which the density of the mass forces is equal to zero and the fluid is compressible.

1. Introduction

The problems of filtration in porous media are of a practical importance for studies related to the forecast of distribution of pollution filtration near river dams, reservoirs and other hydraulic structures, drainage, foundations and basements of buildings, irrigation drainage and agricultural fields, water, oil and gas production, movement of magma in the crust, etc. We consider a mathematical model for viscous compaction. The model is used for description of processes in sedimentary basins [1],[2]. For describing this process we use the equation of general medium momentum conservation (force balance) [3]–[5]. Some exact solutions have been obtained for the model with this ratio in the case of a viscous-elastic media [6]. Also, the localization property of a solution have been obtained for a model with an elastic compressibility [7]. Finite-time stabilization to solution has been established in [8]. The numerical results for this system of equations are given in [4],[9]. For proving the solvability of the problem we use the methods developed in the works [10],[11]. Structurally similar systems of equations were considered in [12]–[15]. The local solvability of the Cauchy problem in Sobolev spaces was established in [13]. Travelling wave solutions have been studied in [14], [15].

2. Statement of the problem and formulation of the main result

We study the quasi-linear system of equations of composite type [4], [9]:

$$\frac{\partial(1-\phi)\rho_s}{\partial t}+\frac{\partial}{\partial x}((1-\phi)\rho_s v_s)=0,\quad \frac{\partial(\rho_f\phi)}{\partial t}+\frac{\partial}{\partial x}(\rho_f\phi v_f)=0, \tag{1}$$

$$\phi(v_f-v_s)=-k(\phi)(\frac{\partial p_f}{\partial x}-\rho_f g),\quad \frac{\partial v_s}{\partial x}=-\frac{1}{\xi(\phi)}p_e,\quad p_e=p_{tot}-p_f, \tag{2}$$

$$\frac{\partial}{\partial x}\left(2\eta(1-\phi)\frac{\partial v_s}{\partial x}\right)=\rho_{tot}g+\frac{\partial p_{tot}}{\partial x},\qquad p_{tot}=\phi p_f+(1-\phi)p_s,\quad \rho_{tot}=\phi\rho_f+(1-\phi)\rho_s, \tag{3}$$

considered in the domain $(x,t) \in Q_T = \Omega \times (0,T)$, under the following boundary and initial conditions

$$v_s\mid_{x=0,x=1}= v_f\mid_{x=0,x=1}= 0, \qquad \phi\mid_{t=0}= \phi^0(x), \quad \rho_f\mid_{t=0}= \rho^0(x). \tag{4}$$

This initial boundary-value problem describes 1D non-stationary isothermal motion of fluid in a poroelastic medium. Here ρ_s, ρ_f, v_s, v_f are, respectively, the real density and velocity of solid and fluid phases, ϕ is the porosity, p_f, p_s are pressures of the fluid and solid phases, respectively; p_e is the effective pressure, p_{tot} is the total pressure, ρ_{tot} is the density of the two-phase medium, g is the density of the mass forces, $k(\phi)$ is the coefficient of filtration, $\xi(\phi)$ is the coefficient of bulk viscosity (specified function), η is the constant of shear viscosity. The problem is written in the Eulerian coordinates x, t. The real density of the solid particles ρ_s is assumed constant. The unknown quantities are ϕ, v_s, v_f, ρ_f, p_f, p_s. The system of equations (1)–(3) is closed by using the equation of state of the fluid phase $p_f = p_f(\rho_f)$. In this paper we use the notation of functional spaces and the relevant norms adopted in [16].

By a solution of problem (1)–(4) we mean the set of functions $v_s \in C^{3+\alpha,1+\alpha/2}(Q_T)$, $(\phi, \rho_f, p_f, p_s) \in C^{2+\alpha,1+\alpha/2}(Q_T)$, $v_f \in C^{1+\alpha,1+\alpha/2}(Q_T)$, such that $0 < \phi < 1$, $\rho_f > 0$, $p_f > 0$. These functions satisfy equations (1)–(3), the boundary and initial conditions (4) as continuous functions in $\overline{Q}_T$. Let us state the main result of the paper.

Theorem. *Let $g = 0$ and the data of problem (1)–(4) satisfy the following conditions:*

1) the functions $k(\phi), \xi(\phi), p_f(\rho_f)$ and their derivatives up to the second order are continuous for $\phi \in (0,1), \rho_f > 0$, and satisfy the conditions

$$k_0^{-1}\phi^{q_1}(1-\phi)^{q_2} \le k(\phi) \le k_0\phi^{q_3}(1-\phi)^{q_4}, \quad 1/\xi(\phi) = a_0(\phi)\phi^{\alpha_1}(1-\phi)^{\alpha_2-1},$$

$$0 < R_1 \le a_0(\phi) \le R_2, k_0^{-1}\rho_f^{q_5} \le p_f(\rho_f) \le k_0\rho_f^{q_6}, \quad k_0^{-1}\rho_f^{q_7} \le \frac{\partial p_f(\rho_f)}{\partial \rho_f} \le k_0\rho_f^{q_8},$$

where $k_0, \alpha_i, R_i, i = 1,2$ are positive constants, $q_1, ..., q_8$ are fixed real parameters,

2) the initial functions ϕ^0, ρ^0 satisfy the following smoothness conditions: $\phi^0 \in C^{2+\alpha}(\overline{\Omega}), \rho^0 \in C^{2+\alpha}(\overline{\Omega})$ and the matching conditions $\frac{dp_f(\rho^0)}{dx}\mid_{x=0,x=1}= 0$, as well as satisfy the inequalities $0 < m_0 \le \phi^0(x) \le M_0 < 1, 0 < m_1 \le \rho^0(x) \le M_1 < \infty, \quad x \in \overline{\Omega}$, where m_0, M_0, m_1, M_1 are given positive constants.

Then problem (1)–(4) has a unique local solution, i.e., there exists a value of t_0 such that $v_s(x,t) \in C^{3+\alpha,1+\alpha/2}(\overline{Q}_{t_0}), (\phi(x,t), p_s(x,t), p_f(x,t), \rho_f(x,t)) \in C^{2+\alpha,1+\alpha/2}(\overline{Q}_{t_0}), v_f(x,t) \in C^{1+\alpha,1+\alpha/2}(\overline{Q}_{t_0})$. Moreover, $0 < \phi(x,t) < 1$, $\rho_f(x,t) > 0$ in $\overline{Q}_{t_0}$.

3. Proof of solvability

Following [10], we rewrite the system (1)–(3) in the dimensionless Lagrange variables $x \in (0,1), t \in (0,T)$:

$$\frac{\partial(1-\phi)}{\partial t} + (1-\phi)^2\frac{\partial v_s}{\partial x} = 0, \quad \frac{\partial}{\partial t}\left(\rho_f\frac{\phi}{1-\phi}\right) + \frac{\partial}{\partial x}(\rho_f\phi(v_f - v_s)) = 0,$$

$$\phi(v_s - v_f) = k(\phi)(1-\phi)\frac{\partial p_f}{\partial x}, \quad (1-\phi)\frac{\partial v_s}{\partial x} = -a_1(\phi)(p_{tot} - p_f),$$

$$\frac{\partial p_{tot}}{\partial x} = \frac{\partial}{\partial x}(2\eta(1-\phi)^2\frac{\partial v_s}{\partial x}).$$

Using the rheological relationship, Darcy's law and the conditions $v_s|_{x=0,1} = 0$, we find that

$$p_{tot} = \frac{2\eta a_1(\phi)(1-\phi)p_f + p^0(t)}{1 + 2\eta a_1(\phi)(1-\phi)},$$

$$p^0(t) = \int_0^1 \frac{a_1(\phi)p_f}{(1-\phi)(1+2\eta a_1(\phi)(1-\phi))}dx \left(\int_0^1 \frac{a_1(\phi)}{(1-\phi)(1+2\eta a_1(\phi)(1-\phi))}dx\right)^{-1}.$$

The system of equations can be reduced to the following form (to determine ρ_f, ϕ):

$$\frac{\partial}{\partial t}(a(\phi)\rho_f) - \frac{\partial}{\partial x}(K(\phi)b(\rho_f)\frac{\partial \rho_f}{\partial x}) = 0, \tag{5}$$

$$\frac{\partial G(\phi)}{\partial t} = p_f(\rho_f) - p^0(t), \tag{6}$$

$$\frac{\partial \rho_f}{\partial x}\mid_{x=0,x=1} = 0, \quad \rho_f\mid_{t=0} = \rho^0(x), \quad \phi\mid_{t=0} = \phi^0(x). \tag{7}$$

Lemma. *Let the data of problem (5)–(7) satisfy the conditions of the theorem. Then problem (5)–(7) has a unique local solution, i.e., there exists a value of* t_0 *such that* $(\phi(x,t), \rho_f(x,t)) \in C^{2+\alpha,1+\alpha/2}(\overline{Q}_{t_0})$. *Furthermore* $0 < \phi(x,t) < 1$, $\rho_f(x,t) > 0$ *in* $\overline{Q}_{t_0}$.

The solvability of problem (5)–(7) is established by using the Tikhonov — Schauder fixed-point theorem [17]. We set $\rho(x,t) = \rho_f(x,t) - \rho^0(x)$ and express equations (5), (6) as follows:

$$\frac{\partial}{\partial t}\left(a(\phi)(\rho+\rho^0)\right) = \frac{\partial}{\partial x}\left(K(\phi)b(\rho+\rho^0)\frac{\partial(\rho+\rho^0)}{\partial x}\right), \tag{8}$$

$$\frac{\partial G(\phi)}{\partial t} = p_f(\rho+\rho^0) - p^0(t). \tag{9}$$

Here

$$\rho\mid_{t=0} = \frac{\partial(\rho+\rho^0)}{\partial x}\mid_{x=0,x=1} = 0, \quad \phi\mid_{t=0} = \phi^0(x). \tag{10}$$

For the Banach space, we choose the space $C^{2+\beta,1+\beta/2}(\overline{Q}_{t_0})$, where β is any number from the interval $(0,\alpha)$, $\alpha \in [0,1)$. Let

$$V = \{\bar\rho(x,t), \bar\phi(x,t) \in C^{2+\alpha,1+\alpha/2}(\overline{Q}_{t_0}) | \quad \bar\rho\mid_{t=0} = \frac{\partial\bar\rho}{\partial x}\mid_{x=0,x=1} = 0,$$

$$0 < \frac{m_1}{2} - \rho^0(x) \le \bar\rho(x,t) \le 2M_1 - \rho^0(x) < \infty, \quad 0 < \frac{m_0}{2} \le \bar\phi(x,t) \le \frac{M_0+1}{2} < 1, \quad (x,t) \in Q_{t_0},$$

$$(|\bar\phi|_{1+\alpha,(1+\alpha)/2,Q_{t_0}}, |\bar\rho|_{1+\alpha,(1+\alpha)/2,Q_{t_0}}) \le K_1, \quad (|\bar\phi|_{2+\alpha,(2+\alpha)/2,Q_{t_0}}, |\bar\rho|_{2+\alpha,(2+\alpha)/2,Q_{t_0}}) \le K_1 + K_2\},$$

where K_1 is an arbitrary positive constant, while the positive constant K_2 will be given later. Let us construct an operator Λ, mapping V in V. Suppose that $\bar\phi, \bar\rho \in V$. Using (9) we define the function ϕ by the equality

$$G(\phi) = G(\phi^0) + \int_0^t \Big\{p_f(\bar\rho(x,\tau)+\rho^0(x)) - \int_0^1 \frac{a_1(\bar\phi)}{(1-\bar\phi)(1+2\eta a_1(\phi)(1-\phi)))}p_f(\bar\rho(x,\tau)+\rho^0(x))dx\cdot$$
$$\cdot\left(\int_0^1 \frac{a_1(\bar\phi)}{(1-\bar\phi)(1+2\eta a_1(\phi)(1-\phi)))}dx\right)^{-1}\Big\}d\tau. \tag{11}$$

From the definition of functions $G(\phi)$ and $a_1(\phi)$, we have for $0 < \phi_1 \le \phi_2 < 1$

$$0 < \Delta G \equiv G(\phi_2) - G(\phi_1) = \int_{\phi_1}^{\phi_2} \frac{ds}{(1-s)a_1(s)} \ge \frac{1}{R_2}(\phi_2 - \phi_1).$$

In this way, we have estimate $|\phi(x,t)-\phi^0(x)|\le\delta(t),\quad \delta(t)\longrightarrow 0\quad \text{as}\quad t\longrightarrow 0$, which implies, that there exists a value $t_1=t_1(m_0,M_0,m_1,M_1)$, such that for all $t_0\le t_1$ the following inequality holds

$$0<\frac{m_0}{2}\le\phi(x,t)\le\frac{M_0+1}{2},\quad (x,t)\in Q_{t_0}. \tag{12}$$

From (11), (12) we also have the inequalities

$$|\phi|_{\alpha,Q_{t_0}}\le C_1(m_0,M_0,m_1,M_1)(1+t_0|\bar\rho|_{\alpha,\alpha/2,Q_{t_0}}),$$

$$|\phi_t|_{\alpha,Q_{t_0}}\le C_2(m_0,M_0,m_1,M_1)(1+|\bar\rho|_{\alpha,Q_{t_0}}+|\rho^0|_{\alpha,\Omega}+|p^0|_{\alpha,[0,T]}),$$

$$|\phi_x|_{\alpha,\alpha/2,Q_{t_0}}\le C_3(m_0,M_0,m_1,M_1)(1+t_0|\bar\rho_x|_{\alpha,\alpha/2,Q_{t_0}}),$$

$$|\phi|_{2+\alpha,1+\alpha/2,Q_{t_0}}\le C_4(m_0,M_0,m_1,M_1,K_1)(1+|\phi_x|_{\alpha,\alpha/2,Q_{t_0}}+t_0|\bar\rho_{xx}|_{\alpha,\alpha/2,Q_{t_0}}).$$

Moreover from [17], $|\phi_x|_{\alpha,\alpha/2,Q_{t_0}}\le C|\phi|^{\kappa}_{2+\alpha,1+\alpha/2,Q_{t_0}}|\phi|^{1-\kappa}_{\alpha,\alpha/2,Q_{t_0}}$, $\kappa=\frac{1+2\alpha}{3+2\alpha}$. Therefore, $|\phi|_{2+\alpha,1+\alpha/2,Q_{t_0}}\le C_5(m_0,M_0,m_1,M_1,K_1)(1+t_0|\bar\rho_{xx}|_{\alpha,\alpha/2,Q_{t_0}})$.

Using (8), (10) and $\bar\phi(x,t)$ we find the function $\rho(x,t)$ as a solution of the problem (here and elsewhere, we assume that the initial and boundary conditions are matched):

$$\tfrac{\partial}{\partial t}(a(\bar\phi)(\rho+\rho^0))=\tfrac{\partial}{\partial x}(K(\bar\phi)b(\bar\rho)\tfrac{\partial(\rho+\rho^0)}{\partial x}),\quad \rho\,|_{t=0}=\tfrac{\partial\rho_f}{\partial x}\,|_{x=0,x=1}=0,\quad \tfrac{\partial\rho^0}{\partial x}\,|_{x=0,x=1}=0. \tag{13}$$

The equation for $\rho(x,t)$ is uniformly parabolic. In view of the properties of $\bar\phi(x,t)$ and $\rho^0(x)$ problem (13) has a classical solution [16]. In addition, we have the following estimate:

$$\left|\frac{1}{a(\bar\phi)}\frac{\partial a(\bar\phi)}{\partial t}\right|\le C_0(m_0,M_0,m_1,M_1,\max_{0\le t\le T}|p^0(t)|).$$

Further, setting $U(x,t)=\rho(x,t)+\rho^0(x)$, we can express problem (13) in the form

$$\frac{\partial}{\partial t}(a(\bar\phi)U)=\frac{\partial}{\partial x}(K(\bar\phi)b(\bar\rho)\frac{\partial U}{\partial x}),\quad \frac{\partial U}{\partial x}\,|_{x=0,x=1}=0,\quad U\,|_{t=0}=\rho^0. \tag{14}$$

First, we show that $U(x,t)\ge 0$, $(x,t)\in Q_{t_0}$. In equation (14), let us make the change $U(x,t)=-z(x,t)$. Then

$$z\frac{\partial a}{\partial t}+a\frac{\partial z}{\partial t}=\frac{\partial}{\partial x}(Kb\frac{\partial z}{\partial x}).$$

Let

$$z^{(0)}(x,t)=\max\{z,0\},\quad z^{(0)}(x,t)\,|_{t=0}=\max\{-\rho^0,0\}=0,$$
$$\sigma_\varepsilon(x,t)=z^{(0)}(x,t)(|z^{(0)}(x,t)|^2+\varepsilon)^{-1/2},\varepsilon>0.$$

Let us multiply the equation for the function z by σ_ε and then integrate over Ω. We obtain the equality

$$\begin{aligned}\tfrac{d}{dt}\int_0^1 a(|z^{(0)}|^2+\varepsilon)^{1/2}dx+\int_0^1\frac{\partial a}{\partial t}(z\sigma_\varepsilon-(|z^{(0)}|^2+\varepsilon)^{1/2})dx+\\ +\varepsilon\int_0^1 Kb\frac{\partial z}{\partial x}\frac{\partial z^{(0)}}{\partial x}(|z^{(0)}|^2+\varepsilon)^{-3/2}dx=0.\end{aligned} \tag{15}$$

Let

$$A^+(t)=\{x\in\Omega|\,z(x,t)>0\},\quad A^-(t)=\{x\in\Omega|\,z(x,t)\le 0\}.$$

Integrating relation (15) with respect to time, we obtain

$$\int\limits_{A^+(t)} a(|z|^2+\varepsilon)^{1/2}dx + \varepsilon^{1/2}\int\limits_{A^-(t)} adx + \varepsilon\int\limits_0^t\int\limits_{A^+(\tau)} Kb|\frac{\partial z}{\partial x}|^2(z^2+\varepsilon)^{-3/2}dxd\tau =$$

$$= \varepsilon\int\limits_0^t\int\limits_{A^+(\tau)} \frac{\partial a}{\partial \tau}(|z|^2+\varepsilon)^{-1/2}dxd\tau + \varepsilon^{1/2}\int\limits_0^t\int\limits_{A^-(\tau)} \frac{\partial a}{\partial \tau}dxd\tau + \varepsilon^{1/2}\int\limits_0^1 a\,|_{t=0}\,dx.$$

Therefore,

$$\min_{0\le x\le 1} a\int\limits_0^1 z^{(0)}dx \le \int\limits_0^1 az^{(0)}dx \le \varepsilon^{1/2}\int\limits_0^t\int\limits_0^1 |\frac{\partial a}{\partial \tau}|dxd\tau + \varepsilon^{1/2}\int\limits_0^1 a\,|_{t=0}\,dx.$$

Passing to the limit as $\varepsilon \to 0$, we find that $z^{(0)} = 0$, i.e.$U \ge 0$. After multiplication by $U^{l-1}(x,t), l > 2$, equation (14) can be expressed as

$$\frac{1}{l}\frac{\partial(aU^l)}{\partial t} + (l-1)KbU^{l-2}(\frac{\partial U}{\partial x})^2 + \frac{l-1}{l}U^l\frac{\partial a}{\partial t} = \frac{\partial}{\partial x}(KbU^{l-1}\frac{\partial U}{\partial x}).$$

Then

$$\frac{1}{l}\frac{d}{dt}\int\limits_0^1 aU^l dx \le \frac{l-1}{l}\max_{0\le x\le 1}|\frac{1}{a}\frac{\partial a}{\partial t}|\int\limits_0^1 aU^l dx.$$

Therefore,

$$y'(t) \le \frac{l-1}{l}\max_{0\le x\le 1}|\frac{1}{a}\frac{\partial a}{\partial t}|y(t), \quad y^l(t) = \int\limits_0^1 (a^{1/l}U)^l dx, \quad y(t) \le y(0)\exp\{\frac{l-1}{l}\int\limits_0^t \max_{0\le x\le 1}|\frac{1}{a}\frac{\partial a}{\partial t}|d\tau\}.$$

After passing to the limit as $l \to \infty$, we obtain

$$\max_{0\le x\le 1} U(x,t) \le \max_{0\le x\le 1}\rho^0(x)\exp\{\int\limits_0^t \max_{0\le x\le 1}|\frac{1}{a}\frac{\partial a}{\partial t}|d\tau\}.$$

Taking into account the inequality $\max\limits_{0\le x\le 1}\rho^0(x) \le M_1$ and choosing t from the condition

$$t \le t_2, \quad \exp\{\int\limits_0^{t_2} \max_{0\le x\le 1}|\frac{1}{a}\frac{\partial a}{\partial t}|d\tau\} \le 2,$$

we obtain upper bound for $\rho+\rho^0$: $\rho(x,t)+\rho^0(x) \le 2M_1$. The lower bound is obtained in the same way with $z(x,t) = 1/U(x,t)$:

$$\max_{0\le x\le 1}\frac{1}{U(x,t)} \le \max_{0\le x\le 1}\frac{1}{\rho^0(x)}\exp\{\int\limits_0^t \max_{0\le x\le 1}|\frac{1}{a}\frac{\partial a}{\partial t}|d\tau\} \le \frac{2}{m_1}.$$

In this way, for $t_0 \leq \min(t_1, t_2)$, $t_2 = ln2/C_0(m_0, M_0, m_1, M_1)$ the solution of problem (13) satisfies the following inequality in Q_{t_0}

$$0 < \frac{m_1}{2} \leq \rho(x,t) + \rho^0(x) \leq 2M_1 < \infty.$$

In view of this inequality and the properties of $\bar{\phi}$, we have the following estimates [16]:

$$|\rho|_{\alpha,\alpha/2,Q_{t_0}} \leq C_6(K_1, m_0, m_1, M_0, M_1), \quad |\rho|_{2+\alpha,1+\alpha/2,Q_{t_0}} \leq C_7(K_1, m_0, m_1, M_0, M_1).$$

Without loss of generality, we assume that $C_5 \geq C_7$. Therefore

$$(|\rho|_{2+\alpha,1+\alpha/2,Q_{t_0}}, |\phi|_{2+\alpha,1+\alpha/2,Q_{t_0}}) \leq C_5(m_0, M_0, m_1, M_1, K_1)(1 + t_0|\bar{\rho}_{xx}|_{\alpha,\alpha/2,Q_{t_0}}).$$

Choose K_2 so that $C_5 \leq (K_1 + K_2)/2$. Then, for $t_0 < \min(t_1, t_2, (K_1 + K_2)^{-1})$ we obtain $|\rho|_{2+\alpha,1+\alpha/2,Q_{t_0}} \leq K_1 + K_2$, $|\phi|_{2+\alpha,1+\alpha/2,Q_{t_0}} \leq K_1 + K_2$. Integrating equation (13) with respect to time, we obtain $|\rho|_{0,Q_{t_0}} \leq C_8 t_0$. From the equation (11), we obtain $|\phi|_{0,Q_{t_0}} \leq C_9 t_0$. Further, using for ρ, ϕ an inequality of the form

$$|u|_{1+\alpha,(1+\alpha)/2,Q_{t_0}} \leq C|u|^c_{2+\alpha,1+\alpha/2,Q_{t_0}}|u|^{1-c}_{0,Q_{t_0}}, \quad c = (1+\alpha)(2+\alpha)^{-1},$$

we find that there exists a sufficiently small value of t_0, depending on K_1 and K_2 such that the required estimates hold: $|\rho|_{1+\alpha,(1+\alpha)/2,Q_{t_0}} \leq K_1$, $|\phi|_{1+\alpha,(1+\alpha)/2,Q_{t_0}} \leq K_1$. Thus, the operator Λ maps the set V into itself for sufficiently small values of t_0. Using the estimates obtained above, we can easily show the continuity of the operator Λ in the norm of the space $C^{2+\beta,1+\beta/2}(\overline{Q}_{t_0})$. By the Tikhonov — Schauder theorem, there exists a fixed point $(\rho, \phi) \in V$ of the operator Λ.

The uniqueness of the solution of the problem (5)–(7) can be proved in the standard way [18]. Lemma is proved.

After finding ϕ and ρ_f, we find $p_{tot} = (2\eta a_1(\phi)(1-\phi)p_f + p^0(t))/(1 + 2\eta a_1(\phi)(1-\phi))$. Then we find $p_s = (p_{tot} - \phi p_f)(1-\phi)^{-1}$. We find v_s from the equation $\partial v_s/\partial x = -a_1(\phi)(1-\phi)^{-1}(p_{tot} - p_f)$, and from the Darcy's law we obtain $v_f = v_s - k(\phi)(1-\phi)\phi^{-1}(\partial p_f/\partial x)$. Since $(\phi, \rho_f) \in C^{2+\alpha,1+\frac{\alpha}{2}}(Q_{t_0})$, we have: $v_s \in C^{3+\alpha,1+\frac{\alpha}{2}}(Q_{t_0})$, $(p_f, p_s) \in C^{2+\alpha,1+\frac{\alpha}{2}}(Q_{t_0})$, $v_f \in C^{1+\alpha,1+\frac{\alpha}{2}}(Q_{t_0})$. Theorem is proved.

4. Conclusion

In summary, we have studied the model of filtration in poroelastic media for viscous compaction. For the case in which the real densities of the solid phase are constant and the density of the mass forces is equal to zero, we establish the local solvability of the problem.

Acknowledgments

In conclusion the author express her profound gratitude to A. A. Papin for his helpful comments and constructive suggestions.

References

[1] Fowler A C and Yang X 1999 *J. Geophys. Res.* **104** 12989–97
[2] Birchwood R A and Turcotte D L 1994 *J. Geophys. Res.* **99** 20051–8
[3] Yang X 2000 *Nonlinear Proc. Geoph.* **7** 1–7
[4] Morency C, Huismans R S, Beaumont C and Fullsack P 2007 *J. Geophys. Res.* **112** B10407
[5] Tokareva M A 2013 *The news of Altai State University* **1-1**(77) 60–2 (in Russian)
[6] Papin A A, Tokareva M A and Shishmarev K A 2015 *Vestnik Altaiskoy Nauki* **1**(23) 161–71 (in Russian)
[7] Tokareva M A 2015 *J. Sib. Fed. Univ. Math. Phys.* **8**(4) 467–77
[8] Tokareva M A 2015 *The news of Altai State University* **1-2**(85) 153–7 (in Russian)

[9] Connolly J A D and Podladchikov Y Y 1998 *Geodin. Acta* **11** 55–84
[10] Papin A A and Akhmerova I G 2010 *Math. Notes* **87**(2) 230–43
[11] Akhmerova I G and Papin A A 2014 *Math. Notes* **96**(2) 166–79
[12] Papin A A 2008 *J. Appl. Mech. Tech. Phys.* **47**(4) 527–36
[13] Simpson M, Spiegelman M and Weinstein M I 2007 *Nonlinearity* **20** 21–49
[14] Abourabia A M, Hassan K M and Morad A M 2009 *Chaos solutions Fract.* **42** 1170–80
[15] Geng Y and Zhang L 2010 *Applied Mathematics and Computation* **217** 1741–8
[16] Ladyzhenskaya O A, Solonnikov V A and Uraltseva N N 1967 *Linear and Quasilinear Eequations of Parabolic Type* (Moscow: Nauka) (in Russian)
[17] Antontsev S N, Kazhikhov A V and Monakhov V N 1983 *Boundary-Value Problems of the Mechanics of Inhomogeneous Fluids* (Novosibirsk: Nauka (Sibirskoe Otdelenie)) (in Russian)
[18] Papin A A 2010 *Math. Notes* **87**(4) 594–8

Localized vortices in a nonlinear shallow water model: examples and numerical experiments

S A Beisel[1] **and A A Tolchennikov**[2]

[1] Institute of Computational Technologies SB RAS, Novosibirsk, Russia
[2] Institute for Problems in Mechanics RAS, Moscow, Russia

E-mail: beisels@gmail.com, tolchennikovaa@gmail.com

Abstract. Exact solutions of the system of nonlinear shallow water equations on paraboloid are constructed by the method of group analysis. These solutions describe fast wave motion of the fluid layer and slow evolution of symmetric localized vortices. Explicit formulae are obtained for asymptotic solution related to the linear shallow water approximation. Numerical methods are used by the modeling the trajectory of the vortex center in the case of asymmetric vortices.

1. Introduction

A system of shallow water equations is used for models of atmospheric and oceanic motion with large length scales [1]. For example, this system can describe propagation of tsunami waves and vortices caused by the uplift of the ocean surface localized in the area of a characteristic size l [2],[3]. It is assumed that l is small in comparison with the size of the basin L (but l is greater than the average depth of the basin). Thus, in dimensionless variables, there is a small parameter $\mu = \frac{l}{L}$ and the solution can be called localized.

In the work [4], the authors concerned the localized vortices (the characteristic size of vortex — μ is a small parameter) moving along the background flow trajectory (the background is the certain solution of the system of shallow water equations). The authors constructed explicit formulas for the asymptotic solution in linear approximation. Also, it was noticed that the vortex has the same form as the initial condition if the Cauchy-Riemann conditions are satisfied along the trajectory of background (i.e., the phase flow is conformal). Nonlinear localized vortices are concerned. For a background, the famous exact solution of the system of shallow water equations on paraboloid from the paper of Thacker [5] was chosen (see also the papers of Ball [6], [7] for common properties of motion of a shallow rotating liquid lying on a paraboloid; and the paper of Rogers [8] for more general exact solutions). In order to construct such a moving vortex, remarkable work [9] is used. Using theorem 2 from [9] (applying one-parametric group of transformations to the stationary localized vortex), exact moving symmetric vortex localized in a vicinity of the center of gravity of liquid is obtained and its properties are described. One applies numerical methods in case the center of vortex does not coincide with the center of gravity and when the vortex is asymmetric.

2. Formulation of problem and results

2.1. Formulation of problem

Let $z = D(x)$ (where $x = (x_1, x_2)$) be the equation of the basin surface underlying the liquid. Let $V(x,t) = (V_1(x,t), V_2(x,t))$ (velocity field) and $H(x,t)$ (free surface elevation) be a solution of the shallow water system:

$$\begin{cases} H_t + ((H + D(x))V_1)_{x_1} + ((H + D(x))V_2)_{x_2} = 0, \\ (V_1)_t + V_1(V_1)_{x_1} + V_2(V_1)_{x_2} + H_{x_1} = 0, \\ (V_2)_t + V_1(V_2)_{x_1} + V_2(V_2)_{x_2} + H_{x_2} = 0. \end{cases} \tag{1}$$

Let $u = (u_1, u_2), \eta$ be a perturbation of (1), i.e. $V + u$, $H + \eta$ also satisfy (1). Then there is a system for u, η:

$$\begin{cases} \eta_t + (\rho u_1 + \eta u_1 + \eta V_1)_{x_1} + (\rho u_2 + \eta u_2 + \eta V_2)_{x_2} = 0, \\ (u_1)_t + (u_1 + V_1)(u_1)_{x_1} + (u_2 + V_2)(u_1)_{x_2} + u_1(V_1)_{x_1} + u_2(V_1)_{x_2} + \eta_{x_1} = 0, \\ (u_2)_t + (u_1 + V_1)(u_2)_{x_1} + (u_2 + V_2)(u_2)_{x_2} + u_1(V_2)_{x_1} + u_2(V_2)_{x_2} + \eta_{x_2} = 0. \end{cases} \tag{2}$$

The linearization of this system:

$$\begin{cases} \eta_t + (\rho u_1 + \eta V_1)_{x_1} + (\rho u_2 + \eta V_2)_{x_2} = 0, \\ (u_1)_t + V_1(u_1)_{x_1} + V_2(u_1)_{x_2} + u_1(V_1)_{x_1} + u_2(V_1)_{x_2} + \eta_{x_1} = 0, \\ (u_2)_t + V_1(u_2)_{x_1} + V_2(u_2)_{x_2} + u_1(V_2)_{x_1} + u_2(V_2)_{x_2} + \eta_{x_2} = 0. \end{cases} \tag{3}$$

$H + D = \rho$ is denoted.

2.2. Background solution

As a background, one takes the exact solution from [8], where depth is the circular paraboloid and velocity field is linear. The Cauchy-Riemann conditions for velocity field are satisfied.

$$D(x_1, x_2) = 1 - \bar{A}(x_1^2 + x_2^2), \;\; \rho(x_1, x_2, t) = (x_1 - c_1(t) \;\;\; x_2 - c_2(t))\, F(t) \begin{pmatrix} x_1 - c_1(t) \\ x_2 - c_2(t) \end{pmatrix} + \frac{\bar{C}}{W(t)},$$

$$V(x_1, x_2, t) = L(t) \begin{pmatrix} x_1 - c_1(t) \\ x_2 - c_2(t) \end{pmatrix} + \begin{pmatrix} c_1'(t) \\ c_2'(t) \end{pmatrix},$$

$$F(t) = \frac{1}{W^2(t)} \begin{pmatrix} \frac{\bar{B}}{2} & 0 \\ 0 & \frac{\bar{B}}{2} \end{pmatrix},$$

$$L(t) = \frac{1}{W(t)} \begin{pmatrix} \frac{W'(t)}{2} & -1 \\ 1 & \frac{W'(t)}{2} \end{pmatrix},$$

$$W(t) = \bar{\lambda} \cos^2(\sqrt{2\bar{A}}\, t) + 2\bar{\mu} \cos(\sqrt{2\bar{A}}\, t) \sin(\sqrt{2\bar{A}}\, t) + \bar{\nu} \sin^2(\sqrt{2\bar{A}}\, t).$$

Constants $\bar{A}, \bar{B}, \bar{C}, \bar{\lambda}, \bar{\mu}, \bar{\nu}$ are as follows

$$\bar{\lambda}\bar{\nu} - \bar{\mu}^2 = \frac{1 - \bar{B} - \bar{\beta}^2}{2\bar{A}}.$$

$(c_1(t), c_2(t))$ (center of gravity) moves around the ellipse

$$c_1(t) = \bar{C}_1 \sin(\sqrt{2\bar{A}}\, t) + \bar{C}_2 \cos(\sqrt{2\bar{A}}\, t), c_2(t) = \bar{C}_3 \sin(\sqrt{2\bar{A}}\, t) + \bar{C}_4 \cos(\sqrt{2\bar{A}}\, t).$$

Two cases will be considered:

example 1: $\bar{B} \neq 0$ (fluid volume is finite, water height is a circular paraboloid).

example 2: $\bar{B} = 0, c_1(t) = 0, c_2(t) = 0$ (fluid volume is infinite, water height depends on time only).

2.3. Numerical experiments

In the following examples, one solves numerically systems (2), (3) with the Cauchy data

$$u|_{t=0} = sgrad\, a \exp\left(-\frac{(x_1 - x_1^0)^2}{\mu_1^2} - \frac{(x_2 - x_2^0)^2}{\mu_2^2}\right), \quad \eta|_{t=0} = 0,$$

where $sgrad = (-\frac{\partial}{\partial x_2}, \frac{\partial}{\partial x_1})$. We use the finite-difference MacCormack scheme of the second order approximation.

2.3.1. Asymmetrical vortex, background from example 2. Let $\mu_1 = 0.025, \mu_2 = 2\mu_1$. In Fig. 1 and Fig. 2 one can see graphics of the free surface elevation η at times $t = 0.16, 0.31, 0.59, 1.06, 2.17$ obtained by linear and nonlinear models (3), (2) respectively. The nonlinearity parameter $a = -0.25$ (i.e., background and vortex twist in the opposite directions).

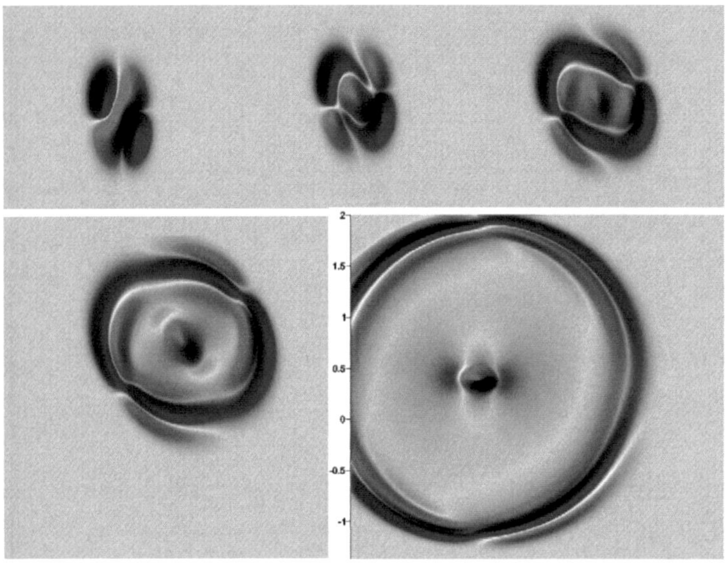

Figure 1. Free surface elevation η (obtained from linear system (3)) at various times. Case $a < 0$

In case of positive a, the free surface elevation graphics for $t = 2.17, a = 0.05, 0.25$ are in Fig. 3.

2.3.2. Symmetrical vortex, background from example 1. Let $\mu_1 = \mu_2 = 0.025$. The main difference from the previous paragraph is in the vortex trajectory. When one solves the linear system (3), then the vortex trajectory coincides with the trajectory of background velocity (in Fig. 4 one can see such a close trajectory). But in the nonlinear model, these two trajectories are not coincide (Fig. 5). The bigger is the nonlinearity parameter, the bigger is the difference between trajectories.

2.4. Model localized vortex

Using method from [9], one can construct a localized vortical solution for backgrounds from examples 1 and 2. Namely:

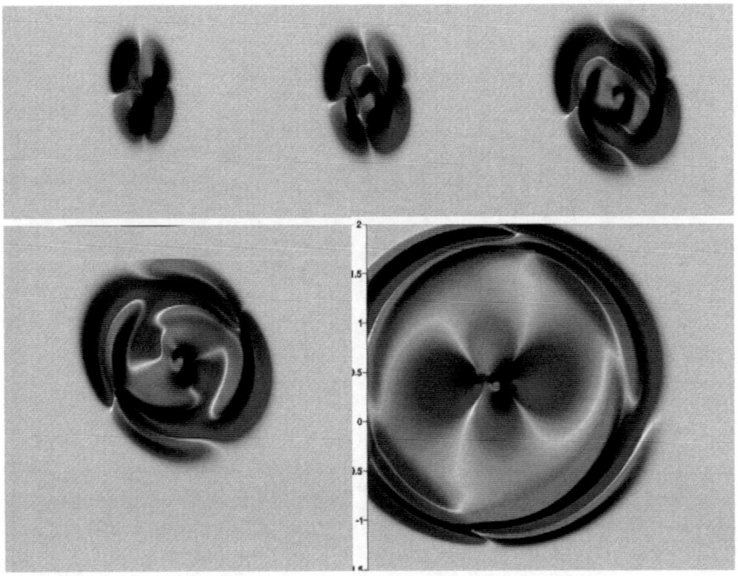

Figure 2. Free surface elevation η (obtained from nonlinear system (2)) at various times. Case $a < 0$

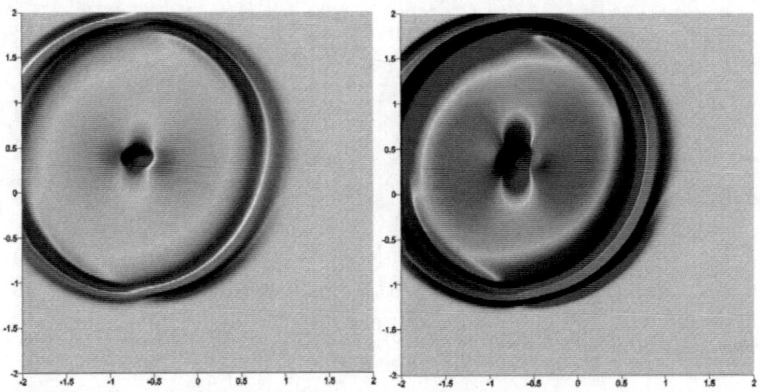

Figure 3. Free surface elevation η (obtained from nonlinear system (2)) for positive $a = 0.05$ (left) and $a = 0.25$ (right)

Let (V, E) be the background from example 1 or 2. $R(t)$ be the trajectory of center of gravity from example 1 or any trajectory from example 2. Let $\mu, a \in \mathbb{R}_+$ (a can be ∞) and functions $f, g \in C^1(\mathbb{R}_+)$ be as follows

$$g(y) = \int_a^y (f^2(x) + 2f(x))dx,$$

then system (2) has the solution

$$u(x,t) = \frac{1}{W(t)} f\left(\frac{|x - R(t)|^2}{2\mu^2 W(t)}\right) \begin{pmatrix} -x_2 + R_2(t) \\ x_1 - R_1(t) \end{pmatrix},$$

$$\eta(x,t) = \frac{\mu^2}{W(t)} g\left(\frac{|x - R(t)|^2}{2\mu^2 W(t)}\right).$$

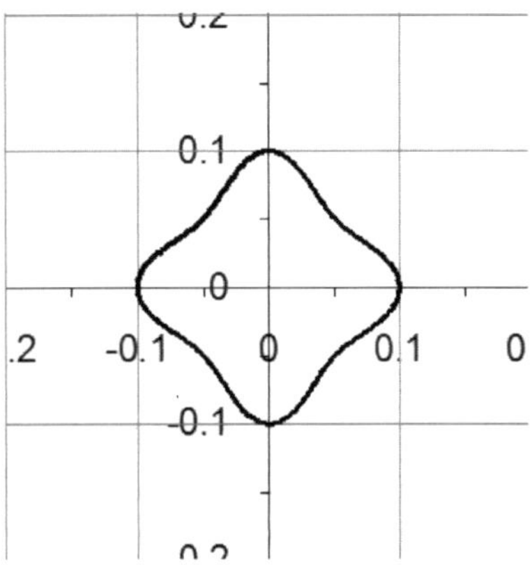

Figure 4. Trajectory of vortex center in linear model

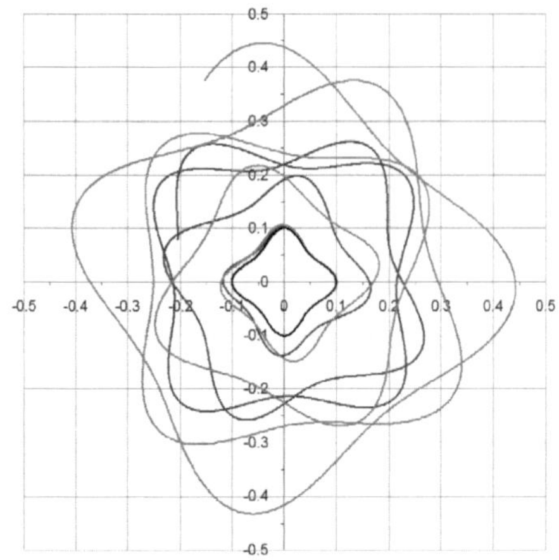

Figure 5. Trajectory of vortex center in nonlinear model. $a = 0.1$ - green line, $a = -0.1$ - blue line

One notes the following:

1. This vortex pulsates, periodically stretches and shrinks.

2. Let $a = \infty$, $\int_0^\infty f^2(x)dx < \infty$, $\int_0^\infty f(x)dx < \infty$, then u, η tend to zero when the distance from the vortex center tends to infinity.

Let $f(x) = \alpha f^0(x), f^0(x) > 0$, α – a nonlinearity parameter.

3. If $\alpha > 0$ (both the vortex and the background twist counterclockwise), then the free surface has a form of cavity. The bigger is α, the deeper is the cavity.

4. If $\alpha < 0$ (the vortex twists clockwise, but the background twists counterclockwise), then for small $|\alpha|$, the free surface has a hump form, and for big $|\alpha|$, the free surface turns to the cavity.

2.4.1. Example. Let $\bar{A} = \frac{1}{4}, \bar{C} = 1, \bar{\lambda} = 2, \bar{\nu} = 1, \bar{\mu} = 0, \mu = 0.1$. Then $W(t) = 1 + \cos^2(\frac{t}{\sqrt{2}})$. Let $a = \infty, f(x) = \alpha e^{-x}$, $g(x) = -\frac{1}{2}\alpha^2 e^{-2y} - 2\alpha e^{-y}$. Then u, η are the fast decaying functions at infinity. The free surface elevation η as function of r (distance to the vortex center) for various nonlinearity parameters is in Fig. 6, 7.

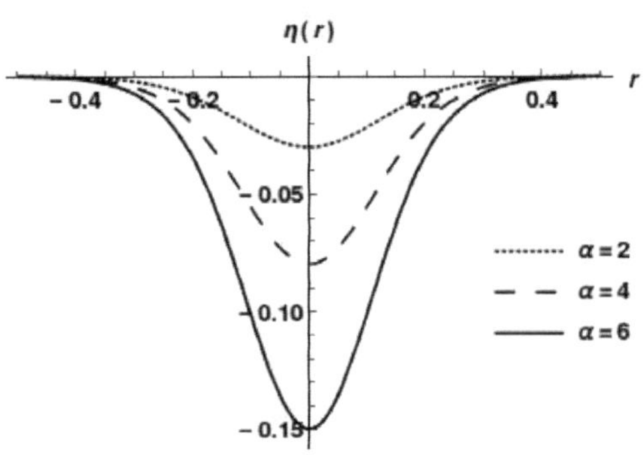

Figure 6. $\eta(r)|_{t=0}, \alpha > 0$

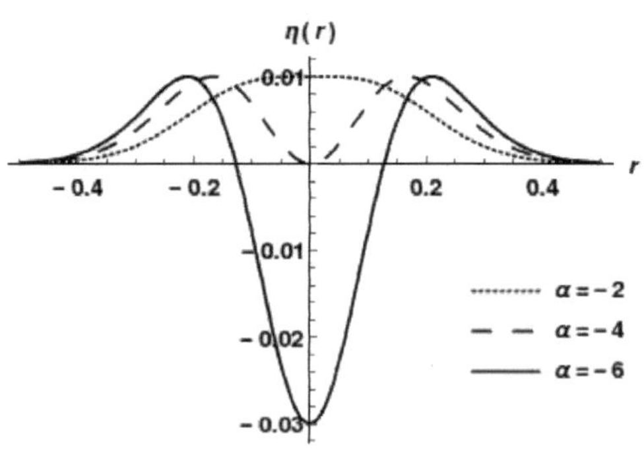

Figure 7. $\eta(r)|_{t=0}, \alpha < 0$

Acknowledgments
The authors are grateful to S.Yu.Dobrokhotov, A.I.Shafarevich, A.A.Chesnokov for useful discussions. This work is supported by the Russian Foundation for Basic Research (Grant No. 16-31-00442).

References
[1] Pedlosky J 1979 *Geophysical Fluid Dynamics* (New York: Springer US)
[2] Pelinovskii E N 1996 *Hydrodynamics of Tsunami Waves*
[3] Berry M V 2005 *New J. Phys.* **7** 129
[4] Dobrokhotov S Y, Shafarevich A I and Tirozzi B 2008 *Russ. J. Math. Phys.* **15** 192–221
[5] Thacker W C 1981 *Journal of Fluid Mechanics.* **107** 499–508
[6] Ball F K 1963 *J. Fluid Mech.* **17** 240–56
[7] Ball F K 1965 *J. Fluid Mech.* **22** 529–45
[8] Rogers C and An H 2010 *Studies in Appl. Math.* **125** 275–99
[9] Chesnokov A 2011 *Appl. Math. and Mech.* **75** 350–6

Traveling circumferential unstable wave of cylindrical flame front

A V Trilis[1], **A A Vasiliev**[1,2] **and S V Sukhinin**[1,2]

[1] Lavrentiev Institute of Hydrodynamics SB RAS, Novosibirsk 630090, Russia
[2] Novosibirsk State University, Novosibirsk 630090, Russia

trilisartie@yandex.ru

Abstract. The researches of stability of cylindrical front of deflagration combustion in an annular combustion chamber were made using phenomenological model. The flame front is described as discontinuity of gasdynamic parameters. It is considered that the combustion products are under chemical equilibrium. The combustible mixture and the combustion products are ideal gases. The velocity of deflagration combustion is determined using the Chapman-Jouget theory. It depends on the temperature of combustible mixture only. It is found that the combustible flame front is unstable for several types of small disturbances in the system. Mechanics of instabilities are examined using both the numeric and analytical methods. The cases of evolution of the unstable waves rotating in circular channel are presented.

1. Introduction

It was discovered in the 20s of XX century that there is a helical sharp trace [1] on sooty inner surface of glass tube, produced by an unknown element of detonation front after the detonation wave (DW) propagation. This effect did not agree with the idealized models of planar detonation front proposed by earlier researchers. Such a wave was called spinning detonation, and the unknown rotating region was called "head of spin". For a long time the nature of spinning DW was unknown and the spinning DW was an exotic phenomenon. Hypothesis of existing unknown element of the spinning DW (transverse wave – TW propagating in induction zone) was suggested in 1950 [2]. Works of Schelkin and Troshin [3] made a valuable contribution to the thorough analysis. It was shown experimentally by Voitsekhovskii [4] that TW does exist.

Discovering of TW made it possible to explain the helical trace produced by spinning configuration that propagates along the tube. It also allowed to find out the inner structure of both the spinning and multifront DW. Thus, contemporary models of the multifront detonation structure with TW in induction zone differ from the idealized structure with planar front.

In order to explain the nature of the spinning detonation in [5] the acoustic theory of the spinning detonation trace was suggested. Later the theory was improved in [6].

The idea of mixture combustion under detonation conditions was suggested many years ago. Voitsekhovskii was the first who suggested to combust a mixture under detonation condition using rotating TW (e.g. transverse wave of spinning configuration). He carried out this idea in practice [7] of annular combustion chamber (circular V-shaped channel), the mixture being delivered radially from the center.

Investigations of rotating DW in annular combustion chamber revealed an important feature. **Namely the propagation velocity of rotating DW never reaches a velocity of the ideal detonation**

D_0, but it belongs to the range between D_0 and sonic speed in detonation products c_d (for gas combustible mixtures $c_d \approx 0.55\, D_0$).

Nowadays there is no explanation for this phenomenon. The transverse detonation wave is called a quasidetonation wave due to this behavior. It is possible to expect that acoustic properties of hot detonation products play important role like it takes place in observing of TW in spinning detonation. The purpose of our research is to investigate the velocity of rotating quasidetonation wave and to corroborate the hypothesis about acoustic nature of this phenomenon. Numeric and analytical methods were used in our research to model several behavior cases of the system that consists of the initial mixture and the products of combustion. Main results are presented.

2. The statement of quasisteady problem

2.1. Modeling of combustion process for calculation

Velocity of gas in annular combustion chamber in quasisteady flow is subsonic and the transverse wave are forced initialized on a certain radius across the flow. The above mentioned facts and the theory of exothermic discontinuity [8] allow us to model the combustion front as a strong discontinuity of the gas-dynamic parameters with a subsonic radial combustion velocity (deflagration wave).

It is necessary to stipulate that the size and the scale of combustion zone is much less than the main geometric scale of the problem in many cases. Therefore it is conventional to replace the finite area of combustion with big gradients of the gas-dynamic parameters with the area of their step-wise changes. Moreover, replacing the line of discontinuity with a combustion zone is fully correct if considered wave length of disturbances are much more than width of combustion area. In other words, this approach is fully correct if processes inside the combustion zone steady much faster than essential changes of gas-dynamic parameters in the flow occur. It is considered that the combustion occurs instantly in the zone of zero width and all essential features of combustion may be reduced to concrete relations for the discontinuity line [9]. It should be emphasized, that the combustion zone is considered as both an area of chemical transformation of mixture to the products and an area of preheating in front of the reaction zone, in which there are gradients of parameters related with the heat conductivity, diffusion and other effects of propagation of the deflagration wave.

2.2. Quasisteady problem

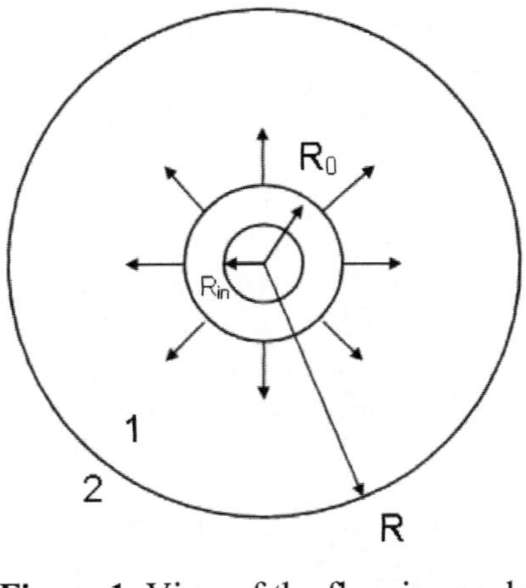

Figure 1. View of the flow in annular combustion chamber

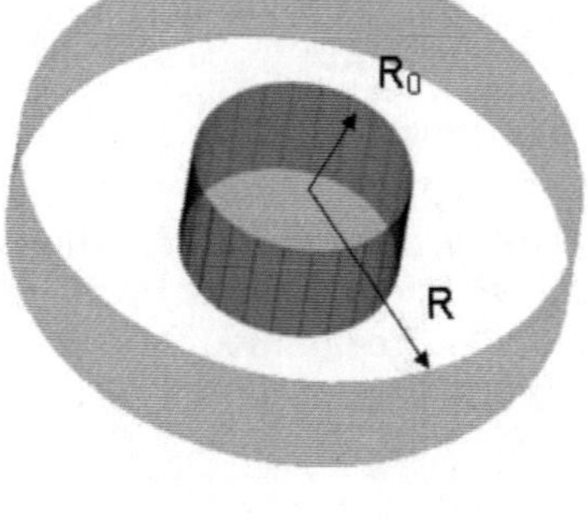

Figure 2. Cylindrical flame front

Let us consider a planar radially diverging subsonic flow of combustible (premixed) mixture from a section with the initial radius R_{in} (Fig. 1). In order to describe the flow let's choose the polar coordinates *(r, φ)*. At initial moment at the output cross-section with the radius R_{in} the ejection velocity is defined by a drop of pressure between the receiver for the initial mixture and one for the products of combustion (the classical problem of break-up of discontinuity). The ejection velocity with the large drop of pressure reaches the critical velocity equal to the local sonic speed. The compression wave in the area $r>R_{in}$ and the rarefaction wave in the area $r<R_{in}$ cause a nonsteady ejection of mixture from the initial receiver. The velocity of ejecting mixture decreases as the coordinate r grows. At some radius R it will reach a value equal to the velocity of combustion of the given combustible mixture. If one ignite the mixture against its movement on this radius, all flow will be divided on two areas: area 1 is the flow of the initial combustible mixture (cold) and area 2 is the area of the flow of high-temperature products of reaction. If the volume of the initial receiver is sufficiently large and the ejection is done in a large-scale volume, the flow will get quasisteady. Then parameters of the flow do not explicitly depend on time (all of derivatives of parameters of the flow with respect to time are equal to zero in equations), and the cylindrical flame front presented in Fig. 2 will rest on (in the laboratory reference system). We shall consider a nonviscous, nonheat-conducting and perfect gas with a constant specific heats (the initial mixture and combustion products). Then, the equations describing the quasisteady flows of combustible mixture and products in polar coordinates for radial ejection will take the form:

$$\begin{aligned} u_{r1,2}\frac{\partial u_{r1,2}}{\partial r} &= -\frac{1}{\rho_{1,2}}\frac{\partial p_{1,2}}{\partial r} \\ \frac{1}{r}\frac{\partial}{\partial r}(r\rho_{1,2}u_{r1,2}) &= 0 \\ p_{1,2} &= A_{1,2}\rho_{1,2}^{\gamma_{1,2}} \end{aligned}, \qquad (2.1)$$

where u is the mass velocity, p is the pressure, ρ is the mass density, γ is the adiabatic ratio, A is the constant depending on thermodynamic parameters at the boundaries R_{in} and R; indices 1 and 2 denote the flow parameters in areas 1 and 2, respectively.

It is well known that the relative changes of density in a flow is linked with the Mach number squared [10]. Thus, if the Mach number equals to 1/3, the relative changes of density of flow are no more than 4%. Thus, they can be neglected. The Mach number in the flow described above steadily decreases as the coordinate r grows, therefore, it reaches 1/3 at some radius R_0 (Fig. 1). Further, the radius R_0 is assumed as the initial one, and the flow in the area 1 ($R_0<r<R$) is assumed to have a constant density ρ_{01}. *The Mach number, thus, is a small parameter of our problem.* Considering all this, equations (2.1) in the area 1 have the following solutions:

$$(u_1(r) = \frac{u_0R_0}{r}; p_1(r) = p_{01} + \frac{\rho_{01}u_0^2}{2} - \frac{\rho_{01}u_0^2R_0^2}{2r^2}; \rho_1 = \rho_{01}; T_1(r) = \frac{\mu \cdot p_1(r)}{\rho_{01}R_g}), \qquad (2.2)$$

where μ is the molar mass, R_g is the universal gas constant, T is the temperature, (u_0, p_{01}, ρ_{01}) are the gasdynamic parameters on the radius R_0 which can be calculated from gasdynamic solution of equations (2.1) [10], specifying the critical parameters on the boundary R_{in} and supposing the Mach number equal 1/3. In turn, the critical parameters depend on the state of combustible mixture in the initial receiver. On the combustion front R the conditions of exothermic discontinuity ($Q>0$) hold. They are conservation laws of mass, momentum, energy and the condition of stationary front in the laboratory reference system:

$$\begin{aligned} \rho_{01}u_1|_R &= \rho_2u_2|_R \\ p_1 + \rho_{01}u_1^2|_R &= p_2 + \rho_2u_2^2|_R \\ h_1 + Q + \frac{u_1^2}{2}|_R &= h_2 + \frac{u_2^2}{2}|_R \\ D_f(p_1, T_1)|_R &= u_1|_R \equiv \frac{u_0R_0}{R} \end{aligned}, \qquad (2.3)$$

where h is the enthalpy and D_f is the combustion velocity. If we know a velocity of combustion, we may calculate the location R of stationary front using the fourth equation in (2.3). If energy release Q is known, the velocity of combustion can be calculated from the well-known Chapman-Jouget condition. This condition means the tangency of Michelson-Rayleigh line to bottom, or deflagration, branch of adiabat of energy release [8] as the flow is subsonic. According to paper [11] the velocity of deflagration combustion linearly depends on temperature of mixture and practically does not depend on pressure of mixture if a chemical equilibrium between the initial stoichiometric mixture and products of combustion is assumed:

$$D_f(T_1) = \alpha \cdot T_1, \tag{2.4}$$

where α is the constant that depends on a concrete combustible mixture. Using equation (2.4) and the Chapman-Jouget condition we may calculate Q if it is necessary. Using the Chapman-Jouget condition we also may calculate all gasdynamic parameters behind the front of combustion located on the radius R and then, using equations (2.1), we may construct a solution in the area 2. However, this solution will not be needed in the sequel as we shall see below.

The condition of equality of the local sonic speed behind the front of combustion to the mass velocity of the flow is one of the equivalent formulations of Chapman-Jouget condition:

$$u_2(R) = c_2(R) \equiv \frac{\gamma_2 p_2(R)}{\rho_2(R)}. \tag{2.5}$$

This relation will be needed in the sequel in order to transform a boundary conditions for nonsteady parameters at the flame front.

3. The statement of nonsteady problem

Let us consider the nonsteady problem. Let add small disturbances harmonically varying according to time to all quasisteady parameters:

$$\begin{gathered}
\tilde{\mathbf{u}}_{1,2}(r,\varphi,t) = u_{1,2}(r)\mathbf{a_r} + (\delta u_{1,2r}(r,\varphi)\mathbf{a_r} + \delta u_{1,2\varphi}(r,\varphi)\mathbf{a}_\varphi)e^{-i\omega t} \\
\tilde{\rho}_{1,2}(r,\varphi,t) = \rho_{1,2} + \delta\rho_{1,2}(r,\varphi)e^{-i\omega t} \\
\delta p_1(r,\varphi)e^{-i\omega t} = c_1^2(r)\delta\rho_1(r,\varphi)e^{-i\omega t} \\
\delta p_2(r,\varphi)e^{-i\omega t} = (c_2^2(r)\delta\rho_2(r,\varphi) + \delta s_2(r,\varphi)\cdot p_2(r)/c_{v2})e^{-i\omega t} \\
\tilde{R}(\varphi,t) = R + A\ (\varphi)e^{-i\omega t}
\end{gathered} \tag{3.1}$$

The third equation in (3.1) between the equations for disturbances of density and pressure means that disturbances propagate isentropically because it is assumed that there is no disturbances of entropy on the boundary R_0. On the front of combustion a heat release occurs, therefore, the disturbances of entropy can appear behind the front and then in the area 2. Therefore, in equations (3.1) the expression for pressure disturbances in the area 2 is presented of the general form. The traveling circumferential transverse wave (along the azimuthal coordinate φ) is simulated by the function $A(\varphi)$ (see the fifth equation in (3.1)).

Substituting equations (3.1) in the well-known conditions of strong discontinuity (the combustion front) [10] marked in the front reference system and neglecting the terms with disturbances of order higher than the first we obtained a conditions on the boundary R which are linear system of equations linking the amplitudes of disturbances in the areas 1 and 2:

$$\frac{1}{M_1}(\bar{u}_{1r} + i\bar{\omega}\cdot\bar{A}(\varphi)) + \bar{\rho}_1\Big|_{\frac{R}{R-R_0}} = \frac{1}{M_2}(\bar{u}_{2r} + \frac{1}{K_2}\cdot i\bar{\omega}\cdot\bar{A}(\varphi)) + \bar{\rho}_2\Big|_{\frac{R}{R-R_0}} \tag{3.2a}$$

$$2\cdot(\bar{u}_{1r}+i\bar{\omega}\cdot\bar{A}(\varphi))+\frac{\bar{\rho}_1}{M_1}(M_1^2+\bar{c}_1^2)|_{\frac{R}{R-R_0}}=2\cdot(K_2\bar{u}_{2r}+i\bar{\omega}\cdot\bar{A}(\varphi))+\frac{\bar{\rho}_2K_2}{M_2}(M_2^2+1)+\frac{\bar{s}_2K_2}{M_2}|_{\frac{R}{R-R_0}} \tag{3.2b}$$

$$\bar{\rho}_1(\bar{c}_1^2+\bar{Q})+M_1(\bar{u}_{1r}+i\bar{\omega}\cdot\bar{A}(\varphi))|_{\frac{R}{R-R_0}}=\bar{\rho}_2K_2^2+M_2K_2^2(\bar{u}_{2r}+\frac{1}{K_2}\cdot i\bar{\omega}\cdot\bar{A}(\varphi))+\frac{\bar{s}_2K_2^2\gamma_2}{\gamma_2-1}|_{\frac{R}{R-R_0}} \tag{3.2c}$$

$$\bar{u}_{1r}-\bar{\rho}_1\frac{\alpha T_{01}\bar{c}_1^2(\gamma_1-1)}{c_{01}}=-i\bar{\omega}\cdot\bar{A}(\varphi)|_{\frac{R}{R-R_0}} \tag{3.2d}$$

$$M_1\frac{R-R_0}{R}\frac{d\bar{A}(\varphi)}{d\varphi}+\bar{u}_{1\varphi}|_{\frac{R}{R-R_0}}=M_2K_2\frac{R-R_0}{R}\frac{d\bar{A}(\varphi)}{d\varphi}+K_2\bar{u}_{2\varphi}|_{\frac{R}{R-R_0}} \tag{3.2e}$$

$$K_2=\frac{c_2(R)}{c_{01}};\bar{Q}=\frac{\rho_{01}}{c_{01}^2}[c_1^2\frac{\partial Q}{\partial p_1}+\frac{\partial Q}{\partial \rho_{01}}], \tag{3.3}$$

where dimensionless quantities were introduced:

$$\bar{\mathbf{u}}_1=\frac{\delta\mathbf{u}_1}{c_{01}};\bar{p}_1=\frac{\delta p_1}{\gamma_1 p_{01}};\bar{\rho}_1=\frac{\delta\rho_1}{\rho_{01}};\bar{c}_1^2=\frac{c_1^2}{c_{01}^2};\bar{T}_1=\frac{\delta T_1}{\gamma_1 T_{01}};\bar{\omega}=\omega\frac{R-R_0}{c_{01}};\bar{A}(\varphi)=\frac{A(\varphi)}{R-R_0}$$
$$\bar{r}=\frac{r}{R-R_0};\bar{t}=t\frac{c_{01}}{R-R_0};\frac{\partial\tilde{\bar{R}}}{\partial\bar{t}}=\frac{1}{c_{01}}\frac{\partial\tilde{R}}{\partial t};M_1(\bar{r})=\frac{u_1((R-R_0)\bar{r})}{c_{01}} \tag{3.4}$$

$$\bar{\mathbf{u}}_2=\frac{\delta\mathbf{u}_2}{c_2(R)};\bar{p}_2=\frac{\delta p_2}{\gamma_2 p_2(R)};\bar{\rho}_2=\frac{\delta\rho_2}{\rho_2(R)};\bar{c}_2^2=\frac{c_2^2}{c_2^2(R)};$$
$$\bar{s}_2=\frac{\delta s_2}{c_{p2}};\bar{T}_2=\frac{\delta T_2}{\gamma_2 T_2(R)};M_2((\bar{r})=\frac{u_2((R-R_0)\bar{r})}{c_2(R)}, \tag{3.5}$$

where $c_{01}=\gamma_1 p_{01}/\rho_{01}$ is the sonic speed on the boundary R_0. In the system of equations (3.2) equations (a), (b) and (c) are the linearized mass, momentum and energy laws, respectively; equation (d) is the linearized kinematic condition, namely the vector sum of the velocity of gas flow and flame velocity relative to the gas flow which equals to the velocity of flame front in laboratory reference system. In equation (e) the disturbances of temperature was expressed through the disturbances of the density using state equation of perfect gas and isentropic condition. If one expresses the disturbances in the area 2 only in terms of the disturbances in the area 1 and uses equation (2.5) in a dimensionless form (Mach number behind combustion front is equal to unit), equations (3.2) will essentially be simplified:

$$\begin{gathered}\bar{\rho}_1|_{\frac{R}{R-R_0}}=0\\ \bar{u}_{1r}|_{\frac{R}{R-R_0}}=-i\bar{\omega}\cdot\bar{A}(\varphi)\\ \bar{s}_2|_{\frac{R}{R-R_0}}=0\\ \bar{u}_{2\varphi}|_{\frac{R}{R-R_0}}=\frac{1}{K_2}\bar{u}_{1\varphi}+\left(\frac{M_1}{K_2}-1\right)\frac{R-R_0}{R}\frac{d\bar{A}(\varphi)}{d\varphi}\end{gathered} \tag{3.6}$$

Thus, we may see that the system of boundary conditions is splitted. Therefore, it is sufficient to solve the unsteady problem in the area 1 in order to find out the behavior of disturbed flame front. On the boundary R_0 we assume no vortical disturbances of velocity of quasisteady flow; therefore, the velocity disturbances can be described by only one function, namely by the potential. The fourth condition in the system of equations (3.6) needs to specify vortical disturbances of velocity behind the flame front and, thus, it will not be needed in the sequel. We may obtain the equations describing the potential disturbances in the area 1 from a general equations of acoustics of moving and inhomogeneous mediums [12]. Thus, we obtain the following equations taking into account that the main quasisteady flow is vortex-free, isentropic and that the Mach number of the main flow is a small parameter:

$$\bar{\omega}^2\bar{f}+\left(\frac{1}{\bar{r}}\frac{\partial}{\partial\bar{r}}(\bar{r}\frac{\partial\bar{f}}{\partial\bar{r}})+\frac{1}{\bar{r}^2}\frac{\partial^2\bar{f}}{\partial\varphi^2}\right)=-2i\bar{\omega}M_1\frac{\partial\bar{f}}{\partial\bar{r}}, \tag{3.7}$$

$$\bar{p}_1=\bar{\rho}_1=i\bar{\omega}\cdot\bar{f}-M_1\frac{\partial\bar{f}}{\partial\bar{r}}. \tag{3.8}$$

where $\bar{f}$ is the dimensionless potential of disturbances of velocity. In the equations (3.7) and (3.8) time has been already separated using a harmonic variation according to time. Thus, in the area 1 we obtained the following problem:

$$\begin{gathered}\bar{\omega}^2\bar{f}+\left(\frac{1}{\bar{r}}\frac{\partial}{\partial\bar{r}}(\bar{r}\frac{\partial\bar{f}}{\partial\bar{r}})+\frac{1}{\bar{r}^2}\frac{\partial^2\bar{f}}{\partial\varphi^2}\right)=-2i\bar{\omega}M_1\frac{\partial\bar{f}}{\partial\bar{r}}\\ \bar{p}_1\equiv i\bar{\omega}\cdot\bar{f}-M_1\frac{\partial\bar{f}}{\partial\bar{r}}|_{\frac{R}{R-R_0}}=0\\ \bar{u}_{1r}\equiv\frac{\partial\bar{f}}{\partial\bar{r}}|_{\frac{R}{R-R_0}}=-i\bar{\omega}\bar{A}(\varphi)\end{gathered} \tag{3.9}$$

In order to determine the problem in area 1 completely, it is necessary to identify a condition on the boundary R_0. A form of this condition depends on combustible mixture supply. We shall consider a three following cases: mass flow is constant, pressure is constant and velocity is constant. Corresponding linearized conditions on *the* boundary R_0 are:

$$i\bar{\omega}\cdot M_1\cdot\bar{f}+\frac{\partial\bar{f}}{\partial\bar{r}}(1-M_1^2)|_{\frac{R_0}{R-R_0}}=0 \tag{3.10}$$

$$\bar{p}_1\equiv i\bar{\omega}\cdot\bar{f}-M_1\frac{\partial\bar{f}}{\partial\bar{r}}|_{\frac{R_0}{R-R_0}}=0 \tag{3.11}$$

$$\bar{u}_{1r}\equiv\frac{\partial\bar{f}}{\partial\bar{r}}|_{\frac{R_0}{R-R_0}}=0 \tag{3.12}$$

4. Results

Let us carry out calculations for the annular combustion chamber in which experiments for burning of stoichiometric acetylene-oxygen mixture were carried out [7]. Problems governed by equations (3.9) and (3.10), (3.9) and (3.11), (3.9) and (3.12) with the uniform boundary conditions has been solved using variable separation method (Fourier method). Consequently, we have obtained dimensionless eigenfrequencies $\bar{\omega}$ of cylindrical flame front oscillations and have carried out a modal stability analysis.

Table 1. Eigenfrequencies of modes of the problem with constant mass flow: l is the radial mode number, k is the azimuthal mode number

		l		
		1	2	3
k	0	1.82-0.144i	4.73-0.161i	7.76-0.164i
	1	2.11-0.122i	4.84-0.158i	7.82-0.163i
	2	2.79-0.094i	5.14-0.149i	8.00-0.160i
	3	3.59-0.080i	5.63-0.135i	8.29-0.155i

Table 2. Eigenfrequencies of modes of the problem with constant pressure: l is the radial mode number, k is the azimuthal mode number

		L		
		1	2	3
k	0	3.03+0.199i	6.13+0.194i	9.21+0.193i
	1	3.15+0.163i	6.20+0.184i	9.26+0.190i
	2	3.51+0.089i	6.40+0.155i	9.40+0.182i
	3	4.02-0.021i	6.73-0.114i	9.63+0.169i

In Table 1 the frequencies for the problem with constant mass flow are presented. It can be seen from Table 1 that all frequencies have the negative imaginary part. We used $e^{-i\bar{\omega}\bar{t}}$; therefore, **the modes with frequencies represented in table 1 are stable to small harmonic disturbances.** In Table 2 the frequencies for the problem with constant pressure are presented. It can be seen from Table 2 that there are frequencies with the positive imaginary part, i.e. **there are modes unstable to small harmonic disturbances.** In the problem with constant velocity the unstable modes also exist.

Disturbed flame front of mode of order *k-l* is a closed curve which has *k* loops rotating together along the circumference of radius *R* with the angular velocity $Re(\bar{\omega}_{kl})$. We also may calculate a linear phase velocity of loops rotation:

$$V_{kl} = \frac{c_{01}\,\mathrm{Re}(\bar{\omega}_{kl})}{R - R_0} \cdot R \qquad (3.13)$$

Table 4. Linear phase velocity of loops for various boundary conditions on R_0

		V_{kl}, m/sec		
		mass flow is constant	velocity is constant	pressure is constant
k	1	1138	1155	1702
	2	1504	1526	1892
	3	1940	1962	2171
	4	2379	2395	2499

In Table 4 the phase velocities of rotation of disturbed flame front are presented for radial number $l=1$ and various boundary conditions on R_0. The sonic speed behind flame front, c_2, according to the Chapman-Jouget parameters equals 1126 m/sec.

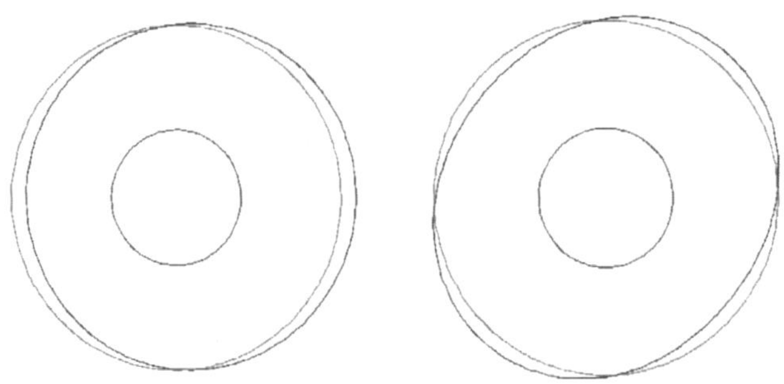

(a) (b)

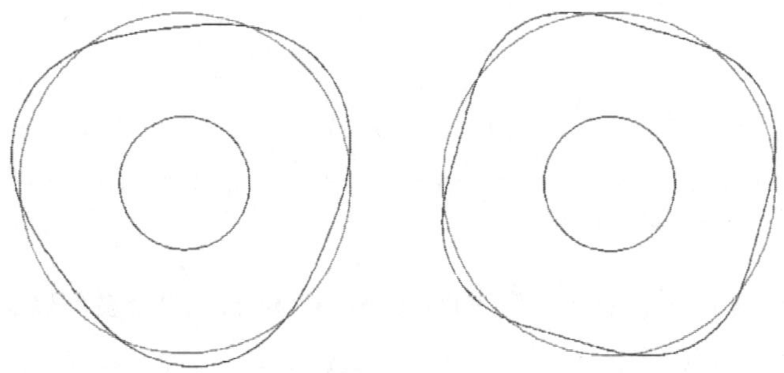

Figure 3. Shapes of disturbed flame front for various modes: (a) is the mode 1-1, (b) is the mode 2-1, (c) is the mode 3-1 and (d) is the mode 4-1. Green line is the boundary R_0, red line is the stationary flame front and blue line is the disturbed one.

In Fig. 3 the shapes of disturbed flame front are visualized.

5. Conclusion

The researches of stability of cylindrical front of deflagration combustion in annular combustion chamber were made using phenomenological model of the combustion and acoustic approximation. The Mach number of main flow was the small parameter of the problem.

It is shown that if the mass flow of combustible mixture supply is strictly constant and does not depend on the acoustic disturbances, the flame front is always stable.

It is found that if the disturbance of pressure or velocity in combustible mixture supply is negligible quantity, then there are discrete sets of frequencies and corresponding modes of oscillations for which the flame front is unstable. Spatial shapes of the oscillations were obtained using both the numeric and analytical technique. Rotating waves with a finite number of local loops were discovered. Linear phase velocities of rotation were calculated.

References

[1] Campbell C and Finch A C 1928 *J. Chem. Soc.* 2094-106
Bone W A Fraser R P and Wheeler W H 1936 *Phil. Transact. Royal Soc*. VII **235** 29-67
[2] Voinov A N 1950 *Rep Ac. Sci USSR* **73** 125-28
[3] Schelkin K I and Troshin Ya K 1965 *Gasdynamics of combustion* (Baltimore: Mono Book Corp.)
[4] Voitsehovskii B V 1957 *Rep Ac. Sci USSR* **114** 717-20
[5] Manson N 1946 *Propagation des detonations et des deflagrations dans les mélanges gazeux* Dr. Sc. Thesis University of Paris
[6] Chu B T 1956 *Proc. Gas Dyn. Aerothermochem* **49** 95-111
[7] Voitsehovskii B V 1959 *Rep Ac. Sci USSR* 6 **129** 1254-56
Voitsehovskii B V 1960 *App. Mech Tech Phys* 3 157-64
[8] Mitrofanov V V 2003 *Detonation of homogeneous and heterogeneous systems* (Novosibirsk: Lavrentiev Institute of Hydrodynamics SB RAS)
[9] Rauschenbakh B V 1961 *Vibration combustions* (Moscow: State Publ. Office Phys. Math. Lit.)
[10] Chyorniy G G 1988 *Gasdynamics* (Moscow: Science)
[11] Vasiliev A A and Trilis A V 2013 *Thermophys. Aeromech* **20**(5) 605-12
[12] Blokhintsev D I 1946 *Acoustics of moving and inhomogenious mediums* (St. Petersburg: State Publ. Office Tech. Theo. Lit.)

Presentation of special series with computed recurrently coefficients of solutions of nonlinear evolution equations

M Filimonov[1,2] **and A Masih**[3]

[1]Ural State University, Ekterinburg, Russia
[2]Institute of Mathematics and Mechanics of UrB RAS, Ekaterinburg, Russia
[3]Quaid-e-Azam University Islamabad, Pakistan

E-mail: fmy@imm.uran.ru

Abstract. One of the analytical methods of presenting solutions of nonlinear partial differential equations is the method of special series in powers of specially selected functions called basis functions. The coefficients of such series are found successively as solutions of linear differential equations. To find recurrence, the coefficient is achieved by the choice of basis functions, which may also contain arbitrary functions. By using such functional arbitrariness, it allows in some cases to prove the global convergence of the corresponding constructed series, as well as the solvability of the boundary value problem.

1. Introduction

Method of special series [1, 2, 3] is a method of representation of solutions of nonlinear partial differential equations in the form of series by the powers of one or several functions chosen in a special way which allows the series' coefficients to be calculated recurrently without applying any truncation procedures. In the sequel, these functions are called as *basic functions* [4]. When the well-known Fourier method is used for linear equations, the coefficients are also calculated recurrently. For nonlinear case, the variables cannot be usually separated and the superposition principle is inapplicable. If the solutions are represented by Fourier series, one or another truncation procedure is usually used to derive the final system of equations, from which the coefficients are to be determined. In the most cases, a characteristic feature of special series is that the construction is independent of the type of nonlinear equation being solved, although the study of convergence of these series is sure to be closely related to the type of the equation. The approaches more closed to the special series method of obtaining solutions of partial differential equations are related to characteristic expansions investigated in the works of R. Courant [5].

In contrast to the power Taylor series, which converge only for the Cauchy-Kovalevskaya equations under the conditions of analyticity of the problem initial data, the constructed series can converge for wider classes of equations and systems.

In some cases, it is possible to exactly satisfy zero boundary conditions by choosing the basis functions (for example, for description of nonlinear vibrating string with fixed end points [1], or membrane with fixed edges [6]). In other cases, it is possible to satisfy a predetermined boundary condition by using the functional arbitrariness contained in the basic functions [7].

In some cases, it is possible to prove global convergence of the constructed series, including in unlimited domains, where the use of numerical methods meets essential difficulties.

Construction of solutions in the form of series with recurrently calculated coefficients allows to investigate properties for meaningful problems. For example, these series were used for study new properties of the audio line for Lin–Reissner–Tsien equation [8]. Note that the various representations of the solutions of this equation were also obtained in the form of convergent series in powers of special basic functions with functional arbitrariness [9].

In some cases, it was possible to construct the basis functions that take into account presence of known exact solutions, as well as the specific character of non-linear equations. For example, these series were built and studied in [10, 11]. Sometimes, constructed series turn into finite sums and then obtain the exact solution [12], which allows their use for testing numerical methods [13, 14], and for numerical algorithms, taking into account the asymptotics of solutions in these problems [15]. In this paper, new results are obtained related to investigation of the convergence of series for a class of nonlinear evolution equations.

2. General scheme of the series with recurrently calculated coefficients

Let us consider one of constructions of special series for solving the Cauchy problem for nonlinear partial differential equations of the form

$$u_t = F\left(t, u, \frac{\partial u}{\partial x}, \dots, \frac{\partial^m u}{\partial x^m}\right), \quad u(0,x) = u^0(x), \tag{1}$$

where F is a polynomial of the unknown function $u(t,x)$ and its derivatives with respect to the space variable. The solution is represented by the series

$$u(t,x) = \sum_{n=0}^{\infty} u_n(t) P^n(t,x) \tag{2}$$

by the powers of basic function $P(t,x)$ satisfying the overdetermined system

$$P_x = A(t,P), \quad P_t = B(t,P) \tag{3}$$

with functions $A(t,P)$ and $B(t,P)$ being analytic respect to P and such that $A(t,0)\equiv 0$ and $B(t,0)\equiv 0$. It was shown that if the initial conditions are written in the form

$$u^0(x) = \sum_{n=0}^{\infty} u_n^0 P^n(0,x), \tag{4}$$

then substituting series (2) into equation (1), collecting similar terms, and taking into account relations (3), we obtain the sequence of first-order ordinary differential equations for the coefficients $u_n(t)$

$$u_n' = F_n(t, u_n, \dots, u_0), \qquad u_n(0) = u_n^0, \quad n = 1, 2, \dots \tag{5}$$

where the right-hand sides F_n include only u_j with $j \le n$ and the coefficients u_n may be linearly contained only.

3. Investigation of convergence of special series to the solution of nonlinear evolution equations

The use of special series with basic functions with functional arbitrariness to construction of solutions of various types of nonlinear partial differential equations is considered. Proof of convergence of the constructed series depends on the type of equations to be solved.

3.1. Generalized Korteweg–de Vries equations

As an example, we mention the construction of the series (2) for Korteweg-de Vries (KdV) equation with initial condition $u(0,x){=}u^0(x)$ in the form of (4). Basic function

$$P(x,t){=}\frac{1}{\exp bx{+}f(t)}, \quad f(t) \in C^1[0,\infty), \quad b>0 \tag{6}$$

is used, which satisfies system (3) for

$$A(t,P){=}{-}bP{+}bf(t)P^2, \quad B(t,P){=}{-}f(t)'P^2$$

with an arbitrary function $f(t)$. Convergence of series (2), (6) is proved for all $x{\geq}0$, $t{\geq}0$, if $0{\leq}f(t){\leq}(\sqrt[3]{2}{-}1)/(2{-}\sqrt[3]{2})$ [2]. At $x{=}0$, the arbitrary function $f(t)$ gives rise to a new boundary condition $u(0,t){=}h(t)$. We can try to find the function $f(t)$ from the function $h(t)$ [7].

Consider an initial-boundary value problem for the nonlinear evolutionary equation

$$\frac{\partial u}{\partial t}+\gamma\frac{\partial^{2r+1}u}{\partial x^{2r+1}}+F\left(t,u,\ldots,\frac{\partial^{2r}u}{\partial x^{2r}}\right)=0, \quad \gamma=\text{const} \tag{7}$$

with initial conditions (4) and boundary condition at $x=0$

$$u(0,t)=h(t), \quad h(t)\in C^1[0,\infty), \quad h(t)-u_0\neq 0, \quad t\geq 0. \tag{8}$$

The well-posedness of the initial-boundary value problem for the generalized KdV equation ($r{=}1$, $F{=}(g(u))_x$) was considered for the class of generalized functions in [16, 17]. One of the conditions of solvability of the mixed problem is the condition

$$|g(u)|\leq(|u|^{10/3}+|u|), \quad \forall u\in R. \tag{9}$$

We find a solution of equation (7) in the form of series (2) with basic function (6) and coefficient $u_0(t)=u_0=\text{const}$. Then the solution of equations (5) has the form

$$u_n(t){=}\exp(b_nt)\left\{u_n^0{+}\int_0^t\exp(-b_n\tau)[(n{-}1)f'(\tau)u_{n-1}{+}R_n(\tau,u_\nu)]d\tau\right\},$$

where $b_n=\text{const}$ and the expressions $R_n(\tau,u_\nu)$ include only functions u_ν with $\nu<n$ [7].

For $x=0$ and $t=0$ the compatibility condition

$$h(0)=h_0=u_0+\sum_{n\geq1}\frac{u_n^0}{(1+f_0)^n}, \quad f(0)=f_0$$

must be fulfilled. To satisfy boundary condition (8) for $t\geq0$, it is necessary to show that there exists a function $f(t)$ such that the equality

$$h(t)=u_0+\sum_{n\geq1}\frac{u_n(t,f(t))}{(1+f(t))^n}$$

is valid. Taking into account the estimates obtained for the proof of the convergence of the constructed series in powers of the basic functions with a functional arbitrariness to construct the solution of a generalized KdF equation [18], we can prove the following theorem.

Theorem 1. *Let the following conditions be fulfilled:*

1. $|u_n^0| \le Mn^{-m}, \quad 0 < M \le M_0, \quad n \ge 1, \quad m = 2r+4, \quad u_0 > 0;$
2. $0 < f(t) \le M_1, \quad t \ge 0.$

Then there exists a unique function $f(t) \in C^1[0, T_0]$, $M_0, M_1, T_0 = \text{const}$ *such that problem* (7), (4) *and* (8) *has a solution for* $-\infty < x \le 0$ *and* $0 \le t \le T_0$.

Remark 1. Thus, it is shown that using special series with functional arbitrariness one may solve initial-boundary value problems for an important class of evolutionary equations (7), for which restriction (9) on the nonlinear function increasing may be not fulfilled.

3.2. Nonlinear wave equations

Let consider the Cauchy problem for nonlinear wave equation

$$u_{tt} = u_{xx} + G(t, u), \tag{10}$$

$$u_t(x, 0) = u_1(x), \quad u(x, 0) = u_0(x). \tag{11}$$

Here, $G(t, u)$ is a polynomial by the unknown function $u(x, t)$ with the coefficients which are continuous functions of t. Let initial conditions (11) be represented in the form of convergent series by the powers of function (6)

$$u_\nu(x) = \sum_{n=0}^{\infty} u_{n,\nu} P^n(0, x), \quad u_{n,\nu} = \text{const}, \quad \nu = 1, 2. \tag{12}$$

Then a solution of problem (10), (10), (12) also may be constructed in the form of the series (2), (6). The coefficients $u_n(t)$ after substitution of series (2), (6) into (10) and equating the expressions of the same powers $P(x, t)$, will be found from a sequence of linear second order differential equations

$$u_n'' - b^2 n^2 u_n = L_n(t, f) + R_n(t), \quad n \ge 0, \tag{13}$$

where $L_n(t, f)$ is the following expression

$$L_n(t, f) = f'' u_{n-1} + 2(n-1) f' u_{n-1}' - b^2 f(n-1) n u_{n-1} - b^2 (n-1)^2 u_{n-1}$$
$$+ [b^2 f^2 - (f')^2](n-2)(n-1) u_{n-2},$$

related to the linear terms in equation (10), and $R_n(t)$ is an expression, related to nonlinear function $G(t, u)$. By this, the coefficients $u_k(t)$ of series (2), (6) are included into expressions $L_n(t, f)$ and $G(t, u)$ with $k \le n-1$, that allows to find the coefficients of the series successively.

Initial conditions for the equation (13) are determined by constant $u_{n,\nu}$, $\nu = 1, 2$ in initial conditions (11)

$$u_n(0) = u_{n,0}, \quad u_n'(0) = u_{n,1} + (n-1) f'(0), \quad n \ge 1. \tag{14}$$

The following theorem is valid.

Theorem 2. *Let the following conditions be fulfilled:*

1. $|u_{n,0}| \le \frac{M}{6n^4}, \quad |u_{n,1}| \le \frac{Mb}{6n^3}, \quad M > 0, \quad n \ge 1;$
2. $f(t) \in C^2[0, \infty)$, $0 < q_1 \le f(t) \le q$, $|f'(t)| \le q$, $|f''(t)| \le q$, $\quad q_1, q > 0, \quad t \ge 0.$

Then there are exist constants $M_0 > 0$ *and* $q_0 > 0$, *that for* $M \le M_0$ *and* $q \le q_0$ *series* (2), (6) *converges to a solution of the Cauchy problem* (10), (11), (12) *for all* $x \ge 0$ *and* $0 \le t \le T$, *where* $T = \frac{\ln(1+q_0)}{2b}$.

Proof. Solutions of the sequence of linear second order differential equations (13), (14) have the form

$$u_n(t) = A_n \exp(bht) + B_n \exp(-bnt) + \frac{1}{bn} \int_0^t \operatorname{sh}(bn(\tau - t))(L_n(\tau, f) + R_n(\tau)) d\tau, \quad n \ge 1, \tag{15}$$

where constants A_n and B_n are determined from initial conditions (14)

$$A_n + B_n = u_{n,0}, \quad A_n bn - B_n bn = u_{n,1}.$$

To investigate convergence of series (2), (6) we estimate the coefficients of the series $u_n(t)$. In contrast to estimations of the series which is used to solve equation (7), for nonlinear wave equation (10) it is necessary to estimate and the derivatives of these coefficients up to the second order. At the same time in the assessments, we cannot use the same inductive assumption that for equation (7). It is related to the difference of types of ordinary differential equations for the series coefficients $u_n(t)$.

In the considered case of nonlinear wave equation the following estimations are valid:

$$|u_n(t)| \le \frac{M\exp(2bnt)}{n^4}, \quad n \ge 1, \tag{16}$$

$$|u_n'(t)| \le \frac{Mb\exp(2bnt)}{n^3}, \quad n \ge 1, \tag{17}$$

$$|u_n''(t)| \le \frac{4Mb^2\exp(2bnt)}{n^2}, \quad n \ge 1, \tag{18}$$

which are proved by induction.

We carry out the proof on the base of example of inequality (16). One notes that for $n = 0$ $u_0(t) = A_0 + B_0 t$. Obviously, an estimation like (16) is valid for $u_0(t)$. We can assume that inequality $|u_0(t)| \le M\exp(2bt)$ is valid.

For $n = 1$, taking into account (15), there is

$$u_1(t) = A_1\exp(bt) + B_1\exp(-bt) + \frac{e^{-bt}}{2b}\int_0^t f''(\tau)(A_0\tau + B_0)e^{b\tau}d\tau - \frac{e^{bt}}{2b}\int_0^t f''(\tau)(A_0\tau + B_0)e^{-b\tau}d\tau.$$

Let estimate the series coefficient $u_1(t)$.

$$|u_1(t)| \le M\exp(2bt)\Big(\frac{|A_1|e^{-bt}}{M} + \frac{|B_1|e^{-3bt}}{M} + \frac{q}{6b^2} + \frac{q}{2b^2}\Big) \le M\exp(2bt)\Big(\frac{1}{6} + \frac{1}{6} + \frac{2q}{3b^2}\Big). \tag{19}$$

For $q_0 \le b^2/2$ from inequality (19), it follows that inequalities (16) are valid for $n = 1$. Thus, induction assumption (16) is proved for $n = 1$. Assuming that the inequality (16) is valid for $n = N$, it is easy to prove this inequality and $n = N + 1$. Let estimate expression $L_{N+1}(t)$, which contains the series coefficients $u_k(t)$ for $k \le N$ corresponding the linear part of (10).

$$|L_{N+1}(t)| \le Mqe^{2bNt}\Big(\frac{2b}{N^3} + \frac{b^2qN}{(N-1)^3} + \frac{1}{N^3} + \frac{2b}{N^2} + \frac{qN}{(N-1)^3}\Big). \tag{20}$$

Using inequality (20), we obtain the following estimation for the integral:

$$\frac{1}{b(N+1)}\int_0^t \mathrm{sh}(b(N+1)(\tau - t))|L_{N+1}(\tau, f)|d\tau \le \frac{M\exp(2b(N+1)t)}{3(N+1)^4}. \tag{21}$$

Note, that for $b \ge 1$ value of q_0 it is possible to estimate. In this case for $q_0 = 1/27$, estimation (21) is valid. T.i. the second condition of Theorem 2 is used only for estimating terms of $L_{N+1}(t, f)$, which correspond to the linear part of the equation (10). These restrictions on arbitrary function $f(t)$ are similar to those of an arbitrary function in Theorem 1.

Estimating expression $R_{N+1}(t)$, we use the condition $M \leq M_0$, where M_0 is determined by the form of nonlinear function $G(t,u)$. Constant M_0 we may choose so that the following inequality will be fulfilled

$$\frac{1}{b(N+1)}\int\limits_0^t \mathrm{sh}(b(N+1)(\tau - t))|R_{N+1}(\tau, f)|d\tau \leq \frac{M\exp(2b(N+1)t)}{3(N+1)^4}. \qquad (22)$$

Thus, inequality (21), (22) prove inequality (16). Similarly, inequality (17), (18) can be proved.

Inequalities (16)–(18) allow to prove convergence of series (2), (6). Indeed,

$$|u(x,t)| \leq \sum_{n=0}^{\infty} |u_n(t)|P^n(t,x) \leq |u_0| + \sum_{n=1}^{\infty} \frac{M\exp(2bnt)}{n^4} P^n(t,x) \leq |u_0|$$

$$+M\sum_{n=1}^{\infty} \frac{\exp(2bnt)}{n^4(1+q_0)^n} \leq |u_0| + M\sum_{n=1}^{\infty} \frac{e^{2bTn}}{n^4(1+q_0)^n}.$$

Similarly one can prove the convergence of the series corresponding derivatives u_t, u_{tt}, u_t, u_{xx}.

Consequently, series (2), (6) converge to a solution of the Cauchy problem (10), (11), (12) for all $x \geq 0$ and $0 \leq t \leq T$, where $T = \frac{\ln(1+q_0)}{2b}$. Theorem 2 is proved.

Remark 2. Proven estimates for the coefficients and convergence of (2), (6) to a solution of the Cauchy problem in a semi-infinite region in x allow to hope to prove existence of solutions of the initial-boundary value problem (10), (11), (12) with the given boundary condition at $x = 0$ by choosing an arbitrary function $f(t)$ and the second boundary condition, given in the form $u(+\infty, 0) = u_0$ due to the specificity of the base function (6) tends to zero when $x \to +\infty$. The choice of such a function $f(t)$ is currently still open.

Acknowledgments

The work was supported by Act 211 Government of the Russian Federation, contract No. 02.A03.21.0006 and project 15–16–1–10.

References

[1] Filimonov M Y 1991 *Diff. Equations* **27** 1158–63
[2] Filimonov M Y, Korzunin L G and Sidorov A F 1993 *Russ. J. Numer. Anal. Math. Model.* **8** 101–25
[3] Filimonov M 2014 *AIP Conf. Proc.* **1631** 218–24
[4] Filimonov M Y 2007 *Trudy Inst. Mat. i Mekh. UrO RAN* **13** 138–53
[5] Courant R and Hilbert D 1962 *Methods of Mathematical Physics* vol II (New York: Wiley-Interscience)
[6] Filimonov M Y 2003 *Diff. Equations* **39** 1159–66
[7] Filimonov M Y 2004 *Proc. of the Steklov Inst. of Math.* **27**(suppl. 1) 1625–32
[8] Kurmaeva V V and Titov S S 2015 *Math. Simulation in Natural Sci.* **1** 240–3
[9] Filimonov M Y 2008 *Proc. of the Steklov Inst. of Math.* **261**(suppl. 1) 55–76
[10] Filimonov M Y 2003 *Diff. Equations* **39** 844–52
[11] Vaganova N A 2008 *Proc. of the Steklov Inst. of Math.* **261**(suppl. 1) 260–71
[12] Bashurov V V, Kropotov A I, Vaganova N A, Filimonov M Y, Pchelintsev M V and Skorkin N A 2012 *J. of Appl. Mech. and Techn. Phys.* **53**(1) 43–8
[13] Filimonov M Y and Vaganova N A 2012 *Proc. of ASME 45158* **4** 133–41
[14] Vaganova N A and Filimonov M Y 2015 *AIP Conf. Proc.* **1690**
[15] Vaganova N A and Filimonov M Y 2015 *Lect. Notes in Comp. Sci.* **9045** 393–9
[16] Bona J L and Luo L 1999 *Contemp. Math.* **221** 59–125
[17] Faminskii A V 2001 *Math. Model.* **13** 115–25
[18] Filimonov M Y 2015 *AIP Conf. Proc.* **1690** 1625–32

Endovascular blood flow measurement system

A K Khe[1,2], A A Cherevko[1,2], A P Chupakhin[1,2], A L Krivoshapkin[3] and K Yu Orlov[3]

[1] Lavrentyev Institute of Hydrodynamics, Novosibirsk, Russia
[2] Novosibirsk State University, Novosibirsk, Russia
[3] Meshalkin Novosibirsk Research Institute of Circulation Pathology, Novosibirsk, Russia

E-mail: alekhe@hydro.nsc.ru

Abstract. In this paper an endovascular measurement system used for intraoperative cerebral blood flow monitoring is described. The system is based on a Volcano ComboMap Pressure and Flow System extended with analogue-to-digital converter and PC laptop. A series of measurements performed in patients with cerebrovascular pathologies allows us to introduce "velocity–pressure" and "flow rate–energy flow rate" diagrams as important characteristics of the blood flow. The measurement system presented here can be used as an additional instrument in neurosurgery for assessment and monitoring of the operation procedure. Clinical data obtained with the system are used for construction of mathematical models and patient-specific simulations. The monitoring of the blood flow parameters during endovascular interventions was approved by the Ethics Committee at the Meshalkin Novosibirsk Research Institute of Circulation Pathology and included in certain surgical protocols for pre-, intra- and post-operative examinations.

1. Introduction

Real-world data play a crucial role in mathematical modelling since it is the main ingredient for verification and justification of the theory being developed. Designing a proper experimental set-up and a measurement system is a challenging task in engineering sciences. And it is even more difficult in natural and medical sciences. In this work we describe the measurement system we use in the context of mathematical modelling of cerebral haemodynamics. Clinical data are collected during endovascular neurosurgical treatments of cerebral blood vessel diseases (arteriovenous malformations and aneurysms).

A number of lesions of the brain lead to a change in the blood flow régime. Cerebrovascular diseases include atherosclerosis, embolism, aneurysms, vascular malformations, and others [1]. All these pathologies can be a cause of a stroke with high disability and mortality rates. According to the World Health Organization, cardiovascular diseases are "the top major killers" during the past decade [2]. The second leading cause of death in the world is the stroke following the ischaemic heart disease. Despite the great technological advances in diagnostic and operating medical equipment, treatment planning and a choice of a particular surgery heavily depend on the surgeon's experience and common medical practice.

For these reasons, mathematical modelling of the circulatory system is one of the urgent and topical questions in biomechanics. Patient-specific simulations will allow the clinicians to perform pre-operative modelling and to assess the risks of post-operative complications. At the

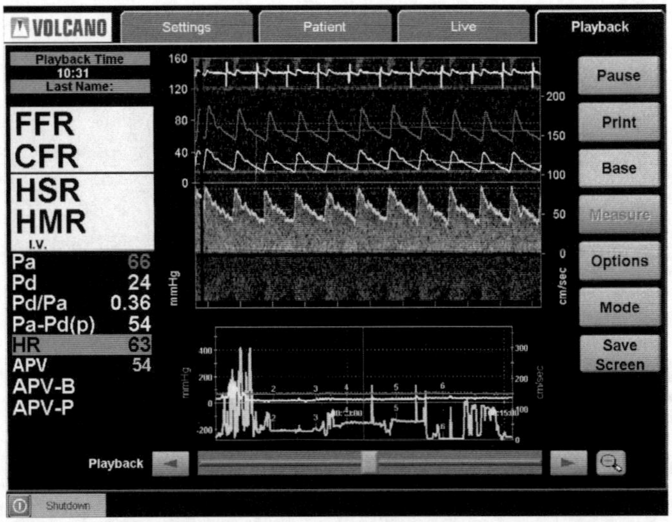

Figure 1. ComboMap screen showing pressure (*yellow*) and velocity (*cyan*); external data: systemic pressure (*red*), ECG (*white*). *Top*: current readings; *bottom*: current measurement session

same time, in order to construct a proper mathematical model one has to base on a reliable set of clinical or laboratory data.

A widely used method to measure the cerebral blood flow is the transcranial Doppler ultrasonography [3]. The method is applicable for relatively large vessels—carotid and vertebral arteries and their first level branches: anterior, middle, and posterior cerebral arteries. Another non-invasive technology is the magnetic resonance (MR) velocimetry [4]. The latter provides also with 3D geometry of the vasculature which can be reconstructed from the MR angiography. The MR imaging depends on the magnetic field strength and the finer resolution is needed the longer scanning time it takes.

The main shortcoming of any non-invasive technique is its inability to measure the blood pressure (at least in the vessels not near the body surface), which is one of the essential characteristics of the fluid flow. In our work we use endovascular flow and pressure measurement system Volcano ComboMap (Volcano Corp., USA). The measurements are carried out during minimally invasive endovascular neurosurgical operations in patients with cerebral arteriovenous malformations and arterial aneurysms.

2. Endovascular measurement system

The flow monitoring system is based on a ComboMap Pressure and Flow System, which is a PC-based portable machine with 15 inch LCD touch screen and remote control. The information about current readings along with the averaged values are shown on the screen (figure 1). The measurements are performed using a ComboWire guide wire (Volcano Corp., USA). The guide wire is 0.36 mm in diameter and 185 cm in length (figure 2). Doppler sensor at the tip of the wire measures flow velocity approximately 5 mm from the tip. The pulsed Doppler beam angle is 45 degrees and insonifies the maximum diameter of approximately 4 mm [5].

The ComboMap system has additional input and output plugs allowing for connection of other medical equipment and measuring instruments. In our flow monitoring procedures we use two additional input signals—systemic arterial pressure and electrocardiogram. The output plugs of the ComboMap allow one to read the measured pressure and flow velocity, and the systemic arterial pressure (which duplicates the external input signal). We use E14-140M portable USB

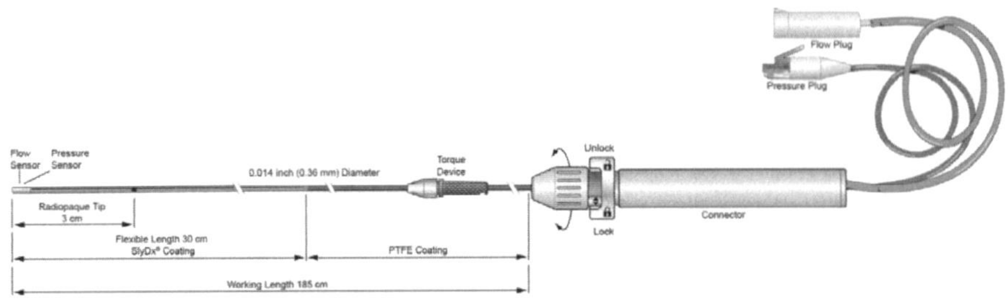

Figure 2. Volcano ComboWire guide wire

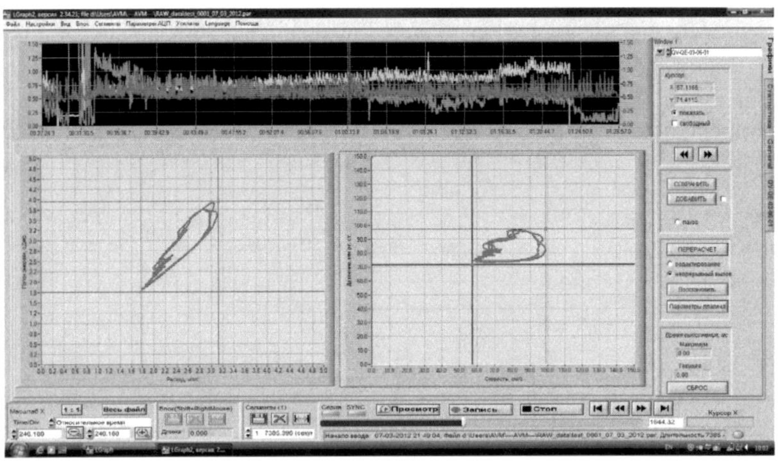

Figure 3. LGraph2 with plug-ins showing real-time data during the measurement session. *Top*: the entire session, *bottom left*: "flow rate–energy flow rate" diagram, *bottom right*: "velocity–pressure" diagram

analogue-to-digital converter (L-Card, Russia) connected to a PC laptop. The E14-140M is a 14-bit ADC with maximum sampling rate of 200 kHz, 16/32 commutation lines, input voltage ranges of ± 10, ± 2.5, ± 0.6, ± 0.15 V; the data acquisition can be carried out in synchronous or asynchronous modes [6]. During endovascular blood flow monitoring we use 198.216 Hz sampling rate.

Data acquisition is performed on a PC laptop with LGraph2 software (L-Card, Russia). The software is designed to collect, store and export the data acquired via LCard ADC. The primary functionality of the LGraph2 can be extended with additional plug-ins. To visualise qualitative properties of hydrodynamic parameters of the flow and the relation between the velocity and pressure, we developed and implemented a plug-in showing "velocity–pressure" and "flow rate–energy flow rate" diagrams for current measurement data (figure 3). The information is shown on the laptop screen in real time in addition to the data on the ComboMap display.

The data acquired during the pre-, intra- and post-operative blood flow monitoring is a single tabulated function representing the time series of the flow parameters. In order to mark the site of a measurement, we minute down the time of the measurement, its location, average pressure and velocity along with the series number of the digital subtraction angiography with the position of the ComboWire (figure 4).

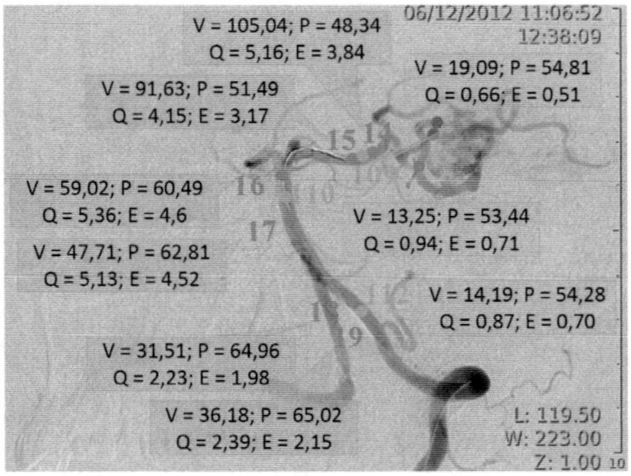

Figure 4. Digital subtraction angiography with measurement locations and average values of the velocity, pressure, flow rate and energy flow rate. *Magenta*: before the operation, *green*: after the operation

3. Post-processing

The post-processing of the blood flow monitoring data includes several steps. The resulting patient record contains all the information on the cerebral flow which is used in the future analysis and mathematical modelling.

The output of the data acquisition procedure is a one block of data containing the measurements and irrelevant noise due to wire repositioning or neurosurgical activities. The initial step of the post-processing consists in the extraction of the time intervals and data ranges corresponding to the actual measurements. Using the minutes of the intraoperative blood flow monitoring and digital subtraction angiography, we identify and map the data ranges to the measurement sites in the cerebral vessels. A typical duration of a measurement at one site in pre- and post-operative examinations is 30 to 60 seconds. During the surgery the ComboWire is usually placed at some location proximal to lesion, so we can monitor the flow changes during the intervention.

Once the data are divided into appropriate intervals, each of these individual data ranges is processed in a home-made software written in Wolfram Mathematica.

We use methods of wavelet analysis for noise filtering and frequency analysis. In particular, we compute the continuous wavelet transform of the measured data and zero out the coefficients corresponding to high frequency waves. The transform is computed using the complex non-orthogonal Gabor wavelet of frequency 6: $\psi(x) = \pi^{-1/4}\exp(iwx)\exp(-x^2/2)$ with $w = 6$. The wavelet scalograms along with the denoised time series are placed into the patient's measurement record.

The qualitative characteristics of the cerebral haemodynamics can be described with the help of "velocity–pressure" and "flow rate–energy flow rate" diagrams. The former is constructed as a trajectory of a point with coordinates $(v(t), p(t))$ on the (v, p) plane, where $v(t)$ and $p(t)$ are time series of the measured velocity and pressure in the blood flow.

The flow rate $Q(t)$ and energy flow rate $E(t)$ are computed by formulae: $Q(t) = \pi d^2 v(t)/4$ and $E(t) = Q(t)(p(t) + \alpha\rho v^2/2)$. The diameter of the vessel d is measured using the angiography. Coefficient α is the Coriolis coefficient describing the non-uniformity of the velocity cross-sectional profile: $\alpha = 1$ for a flat profile, $\alpha = 2$ for a parabolic velocity profile. In our computations we take $\alpha = 1.5$. The exact value of the Coriolis coefficient does not significantly change the total pressure and is often considered to be equal to 1.

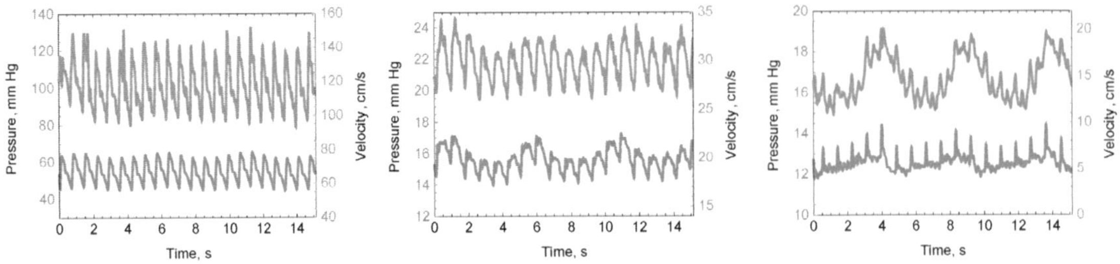

Figure 5. Velocity and pressure in blood flow in different cerebral vessels. *Left*: artery, *middle*: vein, *right*: sinus

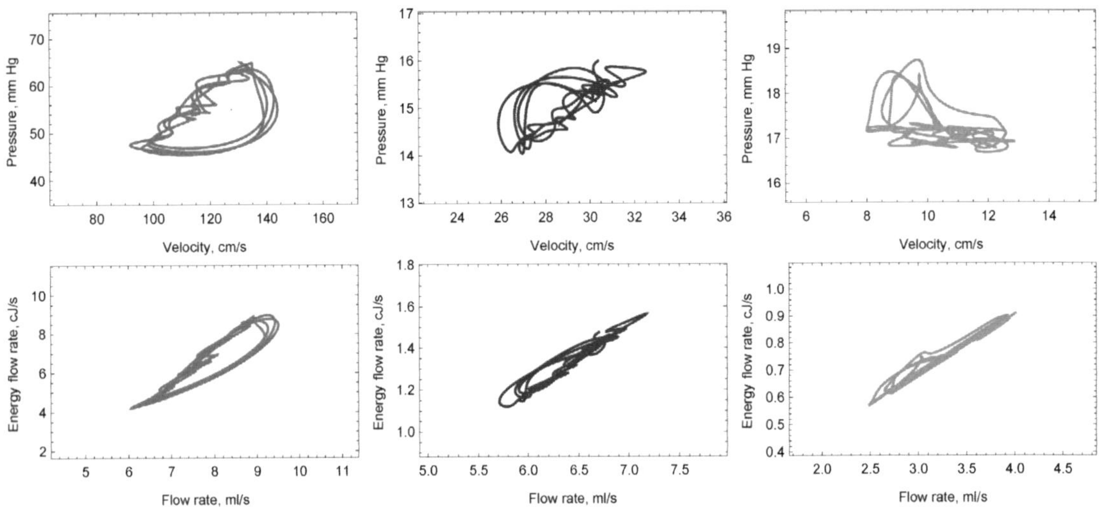

Figure 6. Velocity–pressure and flow rate–energy flow rate diagrams in different blood vessels. *Top*: *vp*-diagrams, *bottom*: QE-diagrams; *left*: artery, *middle*: vein, *right*: sinus

4. Results

In this section we present several characteristic examples of the endovascular blood flow monitoring in patients with arteriovenous malformations and arterial aneurysms.

Figure 5 shows time series of the measured hydrodynamic parameters—velocity and pressure—in different blood vessels: arteries, veins and sinuses. In distal cerebral arteries, the pressure wave, in general, still carries the information about the systemic pressure and their wave forms are more or less qualitatively similar. However, in veins and sinuses they are significantly different. The amplitude of the pulse wave decreases and the respiratory wave becomes more apparent.

The velocity–pressure and flow rate–energy flow rate diagrams for the flows in different blood vessels are shown in figure 6. There are several distinct features characteristic to different types of the vessels. The *form* of the *vp*-diagrams varies depending on the compliance of the vessel: it has a half-circle shape for the arteries and veins and a triangle shape for the sinuses. The *movement direction* of a point along the diagram is counterclockwise for the arteries and clockwise for the veins and sinuses.

A series of measurements in patients with different cerebral vascular pathologies reveals that the shape of the diagram can characterise the type of the lesion [7]. In patients with arteriovenous malformations the diagrams are more semi-circular, while in the presence of the

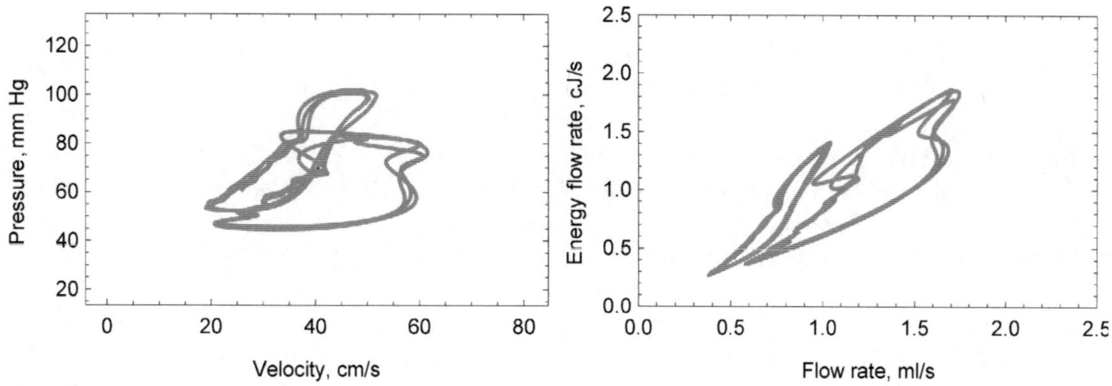

Figure 7. Velocity–pressure (*left*) and flow rate–energy flow rate (*right*) diagrams in the cerebral arteries with aneurysm. *Magenta*: before the operation, *green*: after the operation

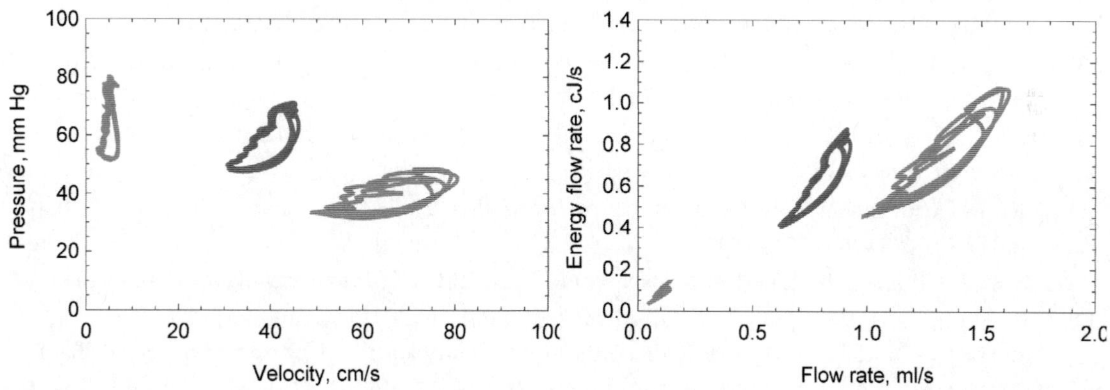

Figure 8. Velocity–pressure (*left*) and flow rate–energy flow rate (*right*) diagrams during the embolisation. *Magenta*: before operation, *brown*: 90% partial embolisation, *green*: total embolisation

arterial aneurysms they have more rectangular shape with S-like sides (figure 7).

In case of arteriovenous malformations, the vp-diagrams constructed for the flow in the afferent vessel (incoming vessel supplying the AVM nidus with the blood) can be used to describe the degree of the embolisation during the treatment procedure. The goal of an embolisation is to fill the AVM nidus with the embolisation agent (adhesive or non-adhesive material) in order to block the blood flow through it. During this procedure the velocity and flow rate decrease and the pressure increases leading to a shift of the vp-diagram in the upper-left direction (figure 8).

5. Conclusion

The data retrieved during the operation can be used for patient-specific numerical modelling and evaluation of the operation. In particular, the common medical recommendation for the AVM treatment is to embolise the fistula compartment, if present, at first, and to limit the volume of the embolisation to 60%. Using the clinical data, these recommendations are confirmed in [8, 9]. Applications to patient-specific numerical simulations of the blood flow in cerebral arterial aneurysms are presented in [10, 11].

The measurement system presented in this work can be used as an additional instrument in neurosurgery for assessment and monitoring of the operation procedure. The monitoring of the blood flow parameters during endovascular interventions was approved by the Ethics Committee at the Meshalkin Novosibirsk Research Institute of Circulation Pathology and included in certain surgical protocols for pre-, intra- and post-operative examinations [12].

Acknowledgments
The endovascular measurements of the blood flow were financially supported by the Russian Foundation for Basic Research (Project No. 14-01-00036). The theoretical research of this paper was financially supported by the Russian Science Foundation (Project No. 15-11-20013).

References

[1] Sacco R L, Kasner S E, Broderick J P, Caplan L R, Connors J B, Culebras A, Elkind M S, George M G, Hamdan A D, Higashida R T, Hoh B L, Janis L S, Kase C S, Kleindorfer D O, Lee J M, Moseley M E, Peterson E D, Turan T N, Valderrama A L and Vinters H V 2013 *Stroke* **44** 2064–2089

[2] World Health Organization 2014 The top 10 causes of death Fact sheet No. 310 updated May 2014 URL `http://www.who.int/mediacentre/factsheets/fs310/en/`

[3] Gaidar B V, Semenyutin V B, Parfenov V E and Svistov D V 2008 *Transcranial Doppler ultrasonography in neurosurgery* (ELBI-SPb)

[4] Ku D N, Biancheri C L, Pettigrew R I, Peifer J W, Markou C P and Engels H 1990 *Journal of Biomechanical Engineering* **112** 464–472

[5] Volcano Corporation 2012 ComboWire pressure/flow guide wire: Instructions for use URL `http://www.volcanocorp.com/`

[6] L-Card 2010 E14-140/E14-140-M user guide URL `http://lcard.ru/ldocs/E14-140M`

[7] Chupakhin A P, Cherevko A A, Khe A K, Telegina N Y, Krivoshapkin A L, Orlov K Y, Panarin V A and Baranov V I 2012 *Circulation Pathology and Cardiac Surgery* **16** 27–31

[8] Panarin V A, Orlov K Y, Krivoshapkin A L, Chupakhin A P, Cherevko A A, Khe A K, Telegina N Y and Baranov V I 2012 *Circulation Pathology and Cardiac Surgery* **16** 39–43

[9] Krivoshapkin A L, Panarin V A, Orlov K Y, Berestov V V, Shayakhmetov T S, Gorbatykh A V, Kislitsin D S, Chupakhin A P, Cherevko A A, Khe A K, Sergeev G S and Chebykin D V 2013 *Bulletin of SB RAMS* **33** 65–73

[10] Vorobtsova N A, Yanchenko A A, Cherevko A A, Chupakhin A P, Krivoshapkin A L, Orlov K Y, Panarin V A and Baranov V I 2013 *Russian Journal of Numerical Analysis and Mathematical Modelling* **28** 505–516

[11] Yanchenko A A, Cherevko A A, Chupakhin A P, Krivoshapkin A L and Orlov K Y 2014 *Russian Journal of Numerical Analysis and Mathematical Modelling* **29** 307–317

[12] Orlov K Y, Panarin V A, Berestov V V, Krivoshapkin A L, Kislitsin D S, Chupakhin A P, Baranov V I, Cherevko A A, Khe A K and Telegina N J 2014 Method for intraoperative doppler control of radicality of arteriovenous malformation embolisation Russian Federation Patent No. 2511235, Int. Cl. A61B 8/06; Appl. No. 2012123062/14, 04.06.2012; public. date: 10.04.2014, Bull. No. 10

Haemodynamics of giant cerebral aneurysm: A comparison between the rigid-wall, one-way and two-way FSI models

A K Khe[1,2], A A Cherevko[1,2], A P Chupakhin[1,2], M S Bobkova[2], A L Krivoshapkin[3] and K Yu Orlov[3]

[1] Lavrentyev Institute of Hydrodynamics, Novosibirsk, Russia

[2] Novosibirsk State University, Novosibirsk, Russia

[3] Meshalkin Novosibirsk Research Institute of Circulation Pathology, Novosibirsk, Russia

E-mail: alekhe@hydro.nsc.ru

Abstract. In this paper a computer simulation of a blood flow in cerebral vessels with a giant saccular aneurysm at the bifurcation of the basilar artery is performed. The modelling is based on patient-specific clinical data (both flow domain geometry and boundary conditions for the inlets and outlets). The hydrodynamic and mechanical parameters are calculated in the frameworks of three models: rigid-wall assumption, one-way FSI approach, and full (two-way) hydroelastic model. A comparison of the numerical solutions shows that mutual fluid–solid interaction can result in qualitative changes in the structure of the fluid flow. Other characteristics of the flow (pressure, stress, strain and displacement) qualitatively agree with each other in different approaches. However, the quantitative comparison shows that accounting for the flow–vessel interaction, in general, decreases the absolute values of these parameters. Solving of the hydroelasticity problem gives a more detailed solution at a cost of highly increased computational time.

1. Introduction

A description of the blood flow in a circulatory system is one of the challenging and topical problems in fluid dynamics. Mathematical modelling based on real clinical data allows one to obtain qualitative and quantitative characteristics of the blood flow in healthy and pathological circulation. This brings possibilities to solve such important medical problems as determination of the laws of circulatory system functioning, prognosis for consequences of surgical interventions and natural history of various diseases. One of such dangerous vascular diseases is a cerebral arterial aneurysm (CA), which is an abnormal dilation of a blood vessel wall [1].

Surgical treatment of the CA is carried out mainly with two kinds of interventions: an open surgery and an endovascular operation. In the open surgery, clipping is the most used procedure, in which the aneurysmal sac is removed from the flow domain with a clamp placed over the aneurysmal neck. The endovascular treatment is a minimally invasive procedure, which can be applied to patients with heavy concomitant somatic pathologies, or older patients for which the traditional neurosurgical intervention is contraindicated. Moreover, endovascular operations are used in cases of CA with difficult access or no possibility for the clipping [1]. Medical instruments in the endovascular surgery are delivered through a small cut in the femoral artery or femoral vein. Among the tools used for the operation are catheters, guide wires, stents, coils, balloons and other.

When choosing the right technology and technique for the treatment, a neurosurgeon does not have all the information needed for the exact prognosis of the operation outcome. That includes the characteristics of all elements of the system, their properties, the range of admissible external influences on the vessels. The neurosurgeon can follow only his/her previous experience and rely on his/her intuition. The lack of the necessary informational support has led to the absence of a common agreement on the technology and strategy for carrying out the surgical intervention [2]. Thus, it is very important to perform patient-specific preoperative modelling for better treatment strategy planning.

Modelling of the blood flow in vessel is a difficult mathematical problem of fluid–structure interaction (FSI). There are still a number of open questions both in analytical methods and in problems of numerical scheme construction [3]. In the present work based on clinical data for the real patient, a numerical simulation of haemodynamics in cerebral vessels with a giant aneurysm at the bifurcation of the basilar artery is performed. Three different approaches are used to simulate the flow: rigid-wall assumption, one-way FSI approximation and two-way FSI. A comparison between the different solutions is carried out.

2. Mathematical models

One of the most widely-used approach to description of the circulation is a mathematical model in which the blood flow is governed by the Navier–Stokes equations and the vessel walls are considered to be rigid. In such a case, only the hydrodynamic problem for the blood flow is solved. In relatively large cerebral arteries one can consider blood as a viscous incompressible Newtonian fluid [4]. Then the system of equations governing a non-stationary 3D flow reads

$$\nabla \cdot \mathbf{v} = 0, \quad \rho\big(\mathbf{v}_t + (\mathbf{v} \cdot \nabla)\mathbf{v}\big) + \nabla p = \mu \Delta \mathbf{v}, \quad \mathbf{x} \in \Omega, \tag{1}$$

where $\mathbf{v}$ is the velocity, p is the pressure, μ is the dynamic viscosity, ρ is the constant density, and Ω is the flow domain. The viscosity of blood is taken to be four times larger as that of water: $\mu = 0.004$ Pa s [5].

The computational domain, Ω, is shown in figure 1. The boundary $\partial\Omega$ of the domain consists of the vessel wall Γ_{wall}, the inlets Γ^i_{in}, and the outlets Γ^j_{out}: $\partial\Omega = \Gamma_{\text{wall}} \cup \Gamma^i_{\text{in}} \cup \Gamma^j_{\text{out}}$, $i = 1, 2$, $j = 1, \ldots, 6$.

For the hydrodynamic problem, a time-dependent normal velocity $v_{\text{clin}}(t)$ is prescribed at the inlets Γ^i_{in} and time-dependent pressure $p_{\text{clin}}(t)$ is prescribed at the outlets Γ^j_{out}:

$$\mathbf{v} = v_{\text{clin}}(t)\mathbf{n} \text{ on } \Gamma^i_{\text{in}}, \quad p = p_{\text{clin}}(t) \text{ on } \Gamma^j_{\text{out}}.$$

In our computations we use clinical data obtained during endovascular surgery with Volcano ComboMap Pressure and Flow System (Volcano Corp., USA) for $v_{\text{clin}}(t)$ and $p_{\text{clin}}(t)$ [6].

On the vessel wall we impose no-slip boundary conditions:

$$\mathbf{v} = 0 \text{ on } \Gamma_{\text{wall}}.$$

When performing analysis of the blood flow, it is often very useful to consider the shear stress on the vessel wall determined by the formula [7]:

$$\boldsymbol{\tau} = \mu(\nabla \mathbf{v} \cdot \mathbf{n}_{\text{wall}}),$$

where $\mathbf{n}_{\text{wall}}$ is the normal to the vessel wall.

There are two approaches for determination of the deformations and stresses in the vessel wall: the so called one-way fluid–structure interaction (FSI) and two-way FSI. In the former

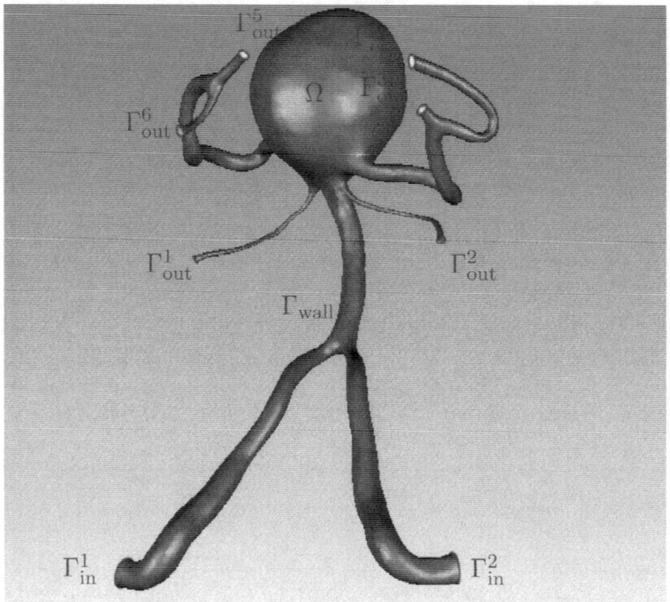

Figure 1. Computational domain

formulation the counter-reaction of the vessel wall onto the blood flow is neglected, while in the latter one mutual interaction between the wall and the flow is taken into account.

In the one-way FSI approximation the solutions of the hydrodynamic and elasticity problems are split into two consecutive steps. Having found the blood flow in the rigid-wall assumption, the deformation of the vessel wall is sought as a reaction to the flow. In this work the vessel wall is modelled by an isotropic linearly elastic material [8]. The system of equations—the equilibrium equations and the Beltrami–Michell equations—reads as follows:

$$\sum_{i=1}^{3} \frac{\sigma_{ij}}{\partial x_i} = 0, \quad \Delta\sigma_{ij} + \frac{1}{1+\nu}\frac{\partial^2(\sigma_{11}+\sigma_{22}+\sigma_{33})}{\partial x_i \partial x_j} = 0, \tag{2}$$

where $\boldsymbol{\sigma} = (\sigma_{ij})$ is the stress tensor, ν is Poisson's ratio. The stress tensor $\boldsymbol{\sigma}$ and the strain tensor $\boldsymbol{\varepsilon}$ are related through Hooke's law [9]:

$$\boldsymbol{\sigma} = \lambda \operatorname{tr} \boldsymbol{\varepsilon} \mathbf{I} + 2G\boldsymbol{\varepsilon}, \quad \lambda = \frac{E\nu}{(1+\nu)(1-2\nu)}, \quad G = \frac{E}{2(1+\nu)},$$

where λ and G are Lamé parameters, E is Young's modulus, tr is the trace operator.

For the boundary conditions for the system (2), the solution for the pressure on the flow domain boundary Γ_{wall} is used for the inner surface of the vessel wall. On the exterior boundary of the wall, an elastic support, which models the surrounding brain tissue, with coefficient of 10^7 N/m^3 is used. The inlet and outlet sides of the vessel are fixed.

In the full (two-way) statement of the FSI problem one needs to find the fluid flow and the wall's deformations coupled through the boundary conditions on the fluid–solid interface.

To describe the interaction of the blood flow with the vessel wall, the Navier–Stokes equations (1) and the elasticity equations (2) are solved simultaneously. In the latter system,

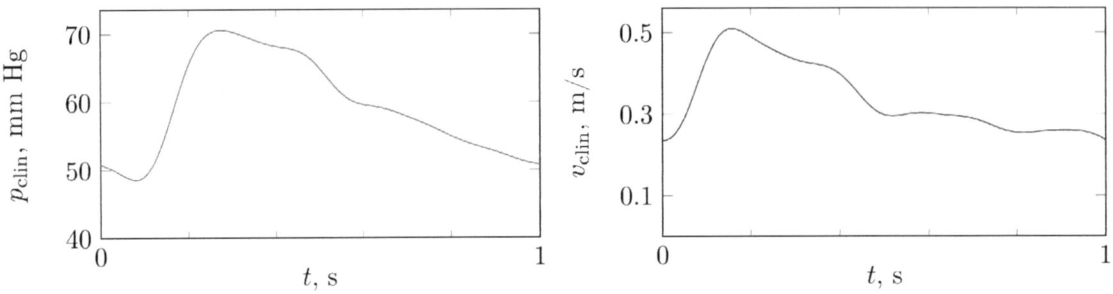

Figure 2. Pressure and velocity profiles used for boundary conditions

the equations of motion instead of the equilibrium equations are used:

$$\rho_{\mathrm{w}}\frac{\partial^2 w_i}{\partial t^2} = \sum_{i=1}^{3}\frac{\sigma_{ij}}{\partial x_i}, \qquad \Delta\sigma_{ij} + \frac{1}{1+\nu}\frac{\partial^2(\sigma_{11}+\sigma_{22}+\sigma_{33})}{\partial x_i \partial x_j} = 0,$$

where $\mathbf{w} = (w_i)$ is the displacement vector, ρ_{w} is the density of the vessel wall.

On the fluid–solid interface the kinematic and dynamic boundary conditions

$$\mathbf{v} = \dot{\mathbf{w}}, \qquad \boldsymbol{\sigma}\cdot\mathbf{n}_{\mathrm{w}} + \boldsymbol{\tau}\cdot\mathbf{n}_{\mathrm{f}} = 0,$$

with stress tensor $\boldsymbol{\sigma}$, normal vector $\mathbf{n}_{\mathrm{w}}$ for the wall and $\boldsymbol{\tau}$, $\mathbf{n}_{\mathrm{f}}$ for the fluid, are prescribed [3].

In our computations we consider the vessel wall having the following parameters: the density $\rho_{\mathrm{w}} = 997$ kg/m^3, Young's modulus $E = 10$ MPa, Poisson's ratio $\nu = 0.49$, and the wall thickness is 0.4 mm [4].

3. Numerical computations

The geometry of the cerebral vessels with the giant aneurysm is reconstructed from the magnetic resonance scans of a real patient. The segmentation is performed with the specialised ITK-SNAP software [10]. The intraoperative endovascular measurements of the velocity and pressure are carried out in this patient during the surgery [6]. The measured clinical data are denoised and rescaled to be a 1 s periodic function of time (figure 2). The calculations are performed on a 3-second interval with a 0.01 s time step. The periodic solution is formed during the first two seconds of the numerical simulation and the solution within the last period is taken to be the sought periodic solution.

The calculations are performed using the finite volume method with the ANSYS CFX solver. The computational mesh has the following parameters: number of nodes is 437 458, number of elements is 2 263 675, maximum size of the element is 0.2 mm, minimum edge length is 0.13 mm. The mesh is unstructured, tetrahedral, and adaptive (the number of elements is increased in the high curvature regions).

The deformation of the vessel wall is found using the finite element analysis with the ANSYS Mechanical solver. The computational mesh consists of 104 295 nodes and 104 878 elements with maximum size of 0.2 mm and minimum edge length of 0.13 mm.

The two-way FSI problem is solved with the CFX and ANSYS solvers with the CFX coordinating the interaction between the solvers. We note that the full hydroelasticity problem is very computationally expensive. Characteristic durations of the simulations on a 6-core station are

- hydrodynamics: 1 hour 51 min 30 s,
- 1-way FSI: 3 hours 36 min 52 s,
- 2-way FSI: 6 days 19 hours 2 min 50 s.

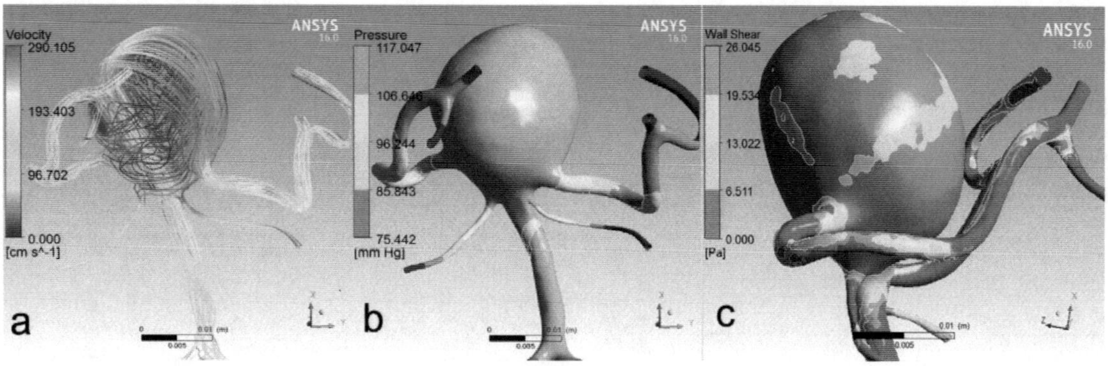

Figure 3. Hydrodynamic parameters at systole ($t = 2.22$ s): (a) streamlines, (b) pressure, (c) wall shear stress

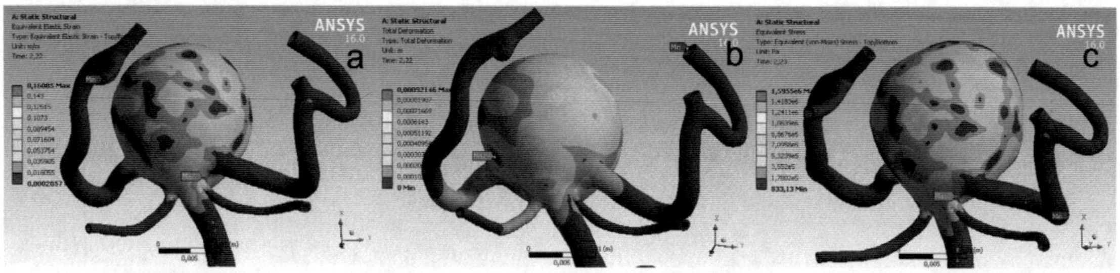

Figure 4. Solution of the one-way FSI at systole ($t = 2.22$ s): (a) strain, (b) displacement, (c) von Mises stress

4. Results

The following hydrodynamic parameters are computed for the hydrodynamic problem (1): velocity and streamlines, pressure, and vessel wall shear stress (figure 3).

The flow in the aneurysmal sac is vortical with a high-speed jet flow with the velocity greater than 1 m/s along the anterior wall of the aneurysm. In the centre of the aneurysmal sac, the flow is slower with the velocity less than 0.3 m/s. The pressure reaches 117 mm Hg on the aneurysm wall.

The region of a high wall shear stress (19.53 Pa) is located along the jet flow in the aneurysm. The wall shear stress is also large (13.02 Pa) at the aneurysmal neck and the upper wall of the aneurysm. It is known that high shear stresses on the vessel wall can result in an emaciation and degradation of the vessel wall, which can contribute to the further aneurysm's growth and rupture [1].

The solution of the one-way FSI problem is depicted in figure 4. The results show that the strains in the aneurysmal wall are larger by 0.11 than those in the vessel wall. Maximum strain of 0.16 is located at the aneurysmal neck. Maximal displacement of 0.92 mm is reached on the aneurysmal wall. In structural analysis, von Mises stress is often used for characterisation of the stress state of the material. The von Mises stress is computed from the stress tensor by formula [11]:

$$\sigma_{\mathrm{vM}} = \sqrt{\frac{(\sigma_{11} - \sigma_{22})^2 + (\sigma_{22} - \sigma_{33})^2 + (\sigma_{33} - \sigma_{11})^2}{2}}.$$

The distribution of the von Mises stress is similar to the strain. Maximum stress of 1.6 MPa is located at the aneurysmal neck.

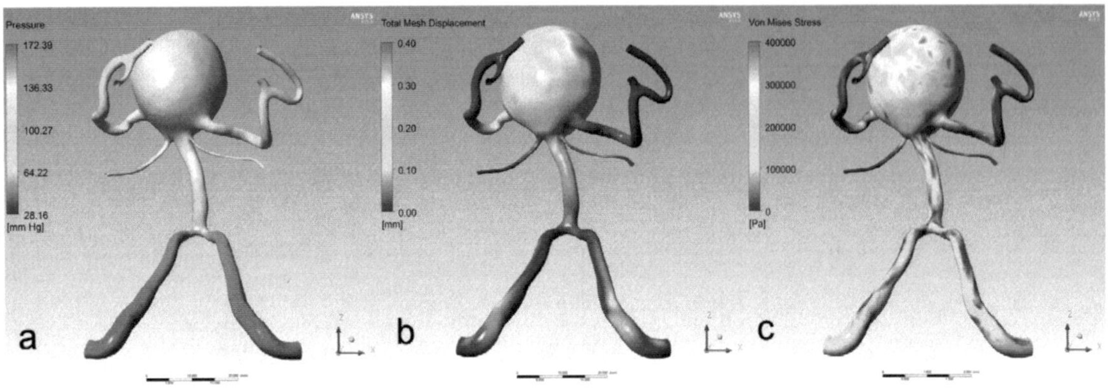

Figure 5. Solution of the two-way FSI at systole (t = 2.22 s): (a) pressure, (b) vessel wall displacement, (c) stress in vessel wall

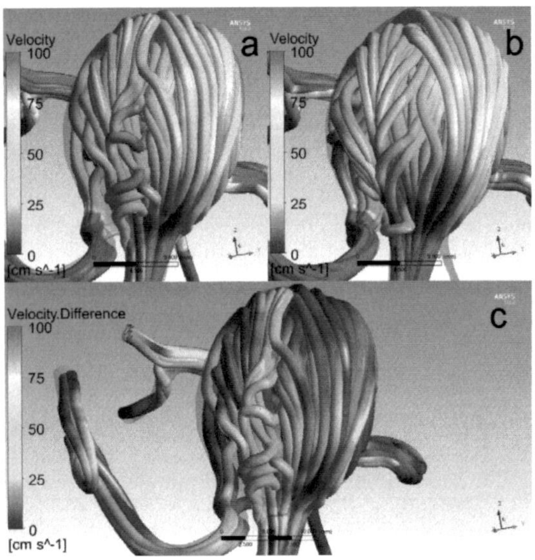

Figure 6. Comparison of the velocity (t = 2.22 s): (a) rigid walls, (b) two-way FSI, (c) the difference of the solutions

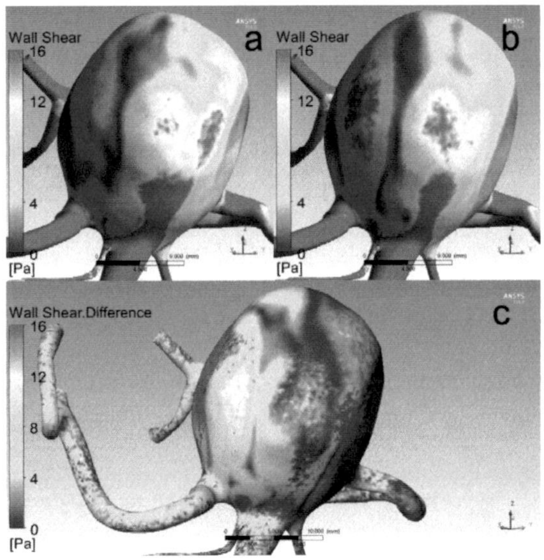

Figure 7. Comparison of the wall shear stress (t = 2.22 s): (a) rigid walls, (b) two-way FSI, (c) difference of the solutions

In figure 5, the solution to the full hydroelasticity problem is shown. The pressure on the aneurysmal wall is distributed uniformly with maximal value of 136 mm Hg. Maximum values of the stress are reached in the aneurysmal sac and neck (0.4 MPa). The displacements reach maxima on the aneurysmal sac and at the junctions with left superior cerebellar artery and left posterior cerebral artery. The solution qualitatively agrees with the two previous ones.

5. Comparison of the solutions

Rigid wall vs two-way FSI In figure 6 the quantitative comparison of the fluid velocity is shown. A difference between the velocity fields clearly shows that the second high speed jet flow in the aneurysmal sac exists when the fluid–solid interaction is taken into account. The difference in velocity magnitude is approximately 0.75 m/s near the aneurysm neck. The existence of the

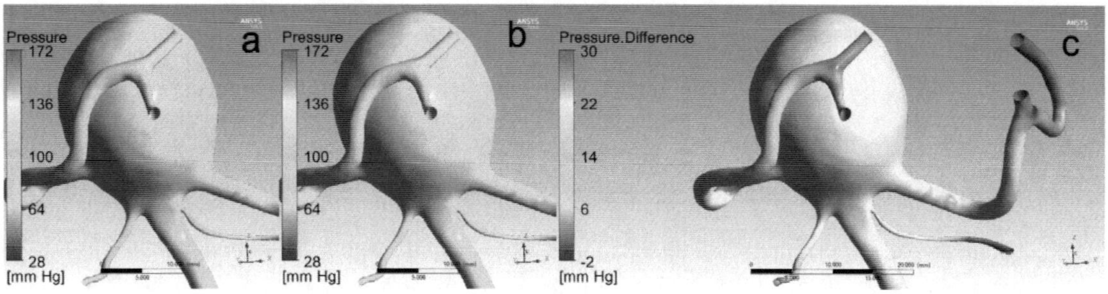

Figure 8. Comparison of the pressure (t = 2.22 s): (a) rigid walls, (b) two-way FSI, (c) difference of the solutions

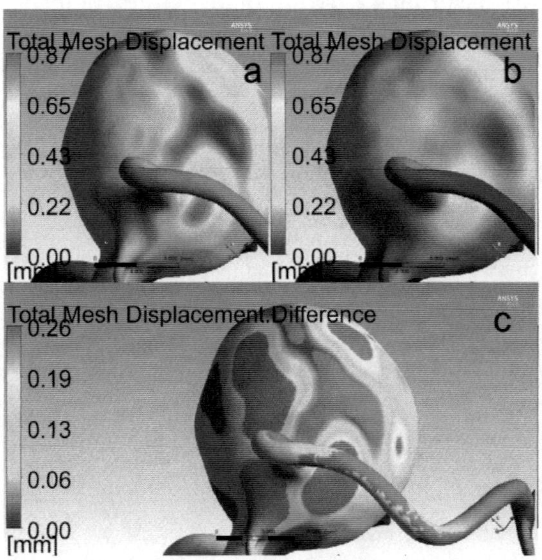

Figure 9. Comparison of the displacement (t = 2.22 s): (a) one-way FSI, (b) two-way FSI, (c) difference of the solutions

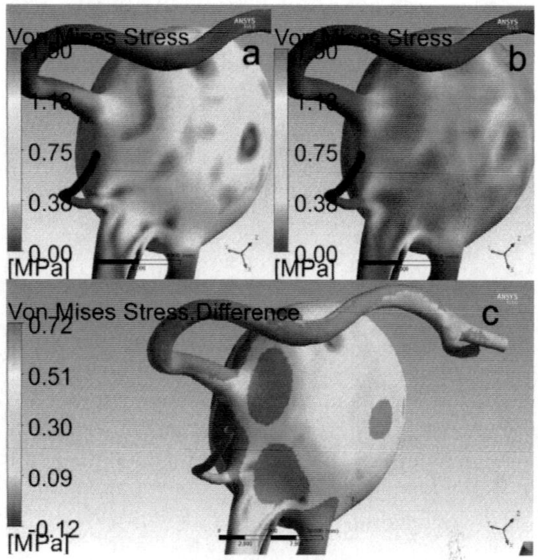

Figure 10. Comparison of the stress (t = 2.22 s): (a) one-way FSI, (b) two-way FSI, (c) difference of the solutions

second jet flow can be also seen in the difference of the wall shear stress shown in figure 7. The difference is approximately 10 Pa.

The pressure in the rigid-wall assumption is higher than that in the two-way hydroelasticity (figure 8). The difference is about 30 mm Hg.

One-way FSI vs two-way FSI The comparison between the displacements in one-way FSI and two-way FSI solutions is shown in figure 9. In case of the one-way interaction, the displacements are higher by approximately 30%. The difference in stress reaches 0.72 MPa (figure 10) with higher values in the one-way FSI approximation.

Thus, the one-way FSI approximation results in higher values of stress, strain and displacement. This is due to the usage of the static equations (2) in which the inertia of the vessel walls is neglected.

6. Conclusion

In this paper a computer simulation of a blood flow in cerebral vessels with a giant saccular aneurysm at the bifurcation of the basilar artery is performed. The modelling is based on patient-specific clinical data (both flow domain geometry and boundary conditions for the inlets and outlets). The hydrodynamic and mechanical parameters are calculated in the frameworks of three models: rigid-wall assumption, one-way FSI approach, and full (two-way) hydroelastic model.

A comparison of the numerical solutions shows that mutual fluid–solid interaction can result in qualitative changes in the structure of the fluid flow. Other characteristics of the flow (pressure, stress, strain and displacement, except the wall shear stress, which depends of the flow pattern) qualitatively agree with each other in different approaches. However, the quantitative comparison shows that accounting for the flow–vessel interaction, in general, decreases the absolute values of these parameters. Thus, solving of the hydroelasticity problem gives a more detailed solution at a cost of highly increased computational time.

Acknowledgments

The theoretical research of this paper was supported by the Russian Foundation for Basic Research (Project No. 16-31-00223). The endovascular measurements of the blood flow were supported by the Russian Foundation for Basic Research (Project No. 14-01-00036). The blood vessel segmentation, 3D reconstruction from MR scans, and numerical simulations were supported by the Russian Science Foundation (Project No. 14-35-00020).

References

[1] Krylov V V (ed) 2011 *Surgery of the cerebral aneurysms* vol 1 (Moscow: T. A. Alexeeva)

[2] Kiselev V S 2010 *Open and endovascular operations in patients with cerebral aneurysms* Ph.D. thesis Polenov Russian Scientific Research Institute of Neurosurgery Saint Petersburg

[3] Bazilevs Y, Takizawa K and Tezduyar T E 2013 *Computational fluid-structure interaction: methods and applications* (John Wiley & Sons Ltd)

[4] Pedley T J 1980 *The Fluid Mechanics of Large Blood Vessels* (Cambridge University Press)

[5] Schmidt R F and Thews G (eds) 1989 *Human Physiology* 2nd ed (Springer-Verlag)

[6] Chupakhin A P, Cherevko A A, Khe A K, Telegina N Y, Krivoshapkin A L, Orlov K Y, Panarin V A and Baranov V I 2012 *Circulation Pathology and Cardiac Surgery* **16** 27–31

[7] van Ooij P, Potters W V, Guédon A, Schneiders J J, Marquering H A, Majoie C B, vanBavel E and Nederveen A J 2013 *Journal of magnetic resonance imaging* **38** 876–884

[8] Caro C G, Pedley T J, Schroter R C and Seed W A 2012 *The mechanics of the circulation* 2nd ed (Cambridge University Press)

[9] Slaughter W S 2002 *The Linearized Theory of Elasticity* (Basel: Birkhäuser)

[10] Yushkevich P A, Piven J, Hazlett H C, Smith R G, Ho S, Gee J C and Gerig G 2006 *NeuroImage* **31** 1116–1128

[11] Love A E H 2013 *A Treatise on the Mathematical Theory of Elasticity* 4th ed (Cambridge University Press)

Journal of Physics: Conference Series **722** (2016) 012043 doi:10.1088/1742-6596/722/1/012043

Junction problems for thin inclusions in elastic bodies

A M Khludnev[1,2]

[1] Lavrentyev Institute of Hydrodynamics, 630090 Novosibirsk, Russia
[2] Novosibirsk State University, 630090 Novosibirsk, Russia

E-mail: `khlud@hydro.nsc.ru`

Abstract. Equilibrium problems for a 2D elastic bodies with thin Euler-Bernoulli and Timoshenko elastic inclusions are considered. It is assumed that inclusions have a joint point, and a junction problem for these inclusions is analyzed. Existence of solutions is proved, and different equivalent formulations of problems are discussed. In particular, junction conditions at the joint point are found. A delamination of the elastic inclusions is also assumed. In this case, inequality type boundary conditions are imposed at the crack faces to prevent a mutual penetration between crack faces. A convergence to infinity of a rigidity parameter of the elastic inclusions is investigated. Limit problems are analyzed.

1. Setting of the problem

Let $\Omega \subset \mathbb{R}^2$ be a bounded domain with Lipschitz boundary Γ. Denote $\gamma_b = (0,1) \times \{0\}$, $\gamma_t = (1,2) \times \{0\}$, $\gamma = \gamma_b \cup \gamma_t \cup \{(1,0)\}$ assuming $\bar{\gamma} \subset \Omega$, see Figure 1 Denote by $\nu = (0,1)$ a unit normal vector to γ; $\tau = (0,1)$, $\Omega_\gamma = \Omega \setminus \bar{\gamma}$.

The domain Ω_γ represents a region with an elastic material, γ_b and γ_t are thin elastic Euler-Bernoulli and Timoshenko inclusions, respectively, incorporated in the elastic material. This means that a behavior of the inclusions is described by the Euler- Bernoulli and Timoshenko equations.

Let $B = \{b_{ijkl}\}, i,j,k,l = 1,2,$ be a given elasticity tensor with the usual properties of symmetry and positive definiteness,

$$b_{ijkl} = b_{jikl} = b_{klij}, \quad i,j,k,l = 1,2, \ b_{ijkl} \in L^\infty(\Omega),$$
$$b_{ijkl}\xi_{ij}\xi_{kl} \geq c_0|\xi|^2 \quad \forall \xi_{ji} = \xi_{ij}, \ c_0 = const > 0 \ .$$

Summation convention over repeated indices is used; all functions with two lower indices are assumed to be symmetric in these indices. Let $f = (f_1, f_2) \in L^2(\Omega)^2$ be a given function.

We first provide a variational formulation of the equilibrium problem for the elastic body with thin inclusions γ_b, γ_t. A space is introduced

$$W = \{(u,v,w,\varphi) \,|\, u \in H^1_0(\Omega)^2, (v,w) \in H^1(\gamma)^2, v \in H^2(\gamma_b),$$
$$\varphi \in H^1(\gamma_t); \ v = u_\nu, \ w = u_\tau \text{ on } \gamma; \ v_x(1-) + \varphi(1+) = 0\}$$

with the norm

$$\|(u,v,w,\varphi)\|^2_W = \|u\|^2_{H^1_0(\Omega)^2} + \|(v,w)\|^2_{H^1(\gamma)^2} + \|v\|^2_{H^2(\gamma_b)} + \|\varphi\|^2_{H^1(\gamma_t)}.$$

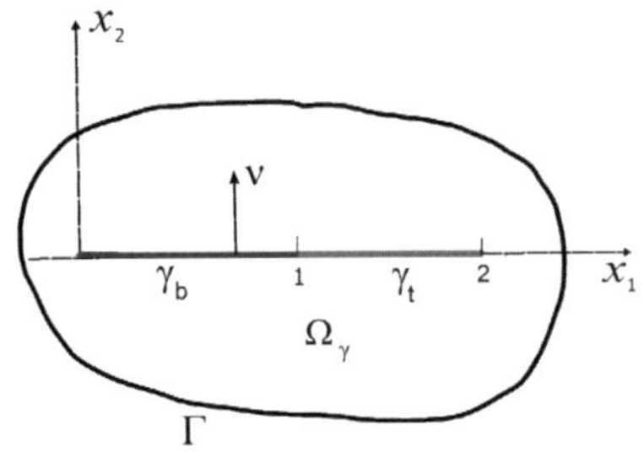

Figure 1. Elastic body with thin inclusions

Here, $u = (u_1, u_2)$, $u_\nu = u\nu$, $u_\tau = u\tau$. The standard notations $H_0^1(\Omega)$, $H^1(\gamma)$, etc., are used for Sobolev spaces. We identify functions defined on γ with functions of the variable x; $x = x_1$, $h_x = \frac{dh}{dx}$, $(x_1, x_2) \in \Omega$. For a simplicity, we write $\sigma(u)\varepsilon(u) = \sigma_{ij}(u)\varepsilon_{ij}(u)$.

The identity is considered

$$(u, v, w, \varphi) \in W, \tag{1}$$

$$\int\limits_{\Omega_\gamma} \sigma(u)\varepsilon(\bar{u}) - \int\limits_{\Omega_\gamma} f\bar{u} + \int\limits_{\gamma_b} v_{xx}\bar{v}_{xx} + \int\limits_{\gamma} w_x\bar{w}_x + \tag{2}$$

$$+ \int\limits_{\gamma_t} \{(v_x + \varphi)(\bar{v}_x + \bar{\varphi}) + \varphi_x\bar{\varphi}_x\} = 0 \quad \forall\,(\bar{u}, \bar{v}, \bar{w}, \bar{\varphi}) \in W.$$

Theorem 1. The problem (1)-(2) has a unique solution.

We are able to provide a differential formulation of the problem (1)-(2). It is necessary to find functions $u = (u_1, u_2), v, w, \varphi$, $\sigma = \{\sigma_{ij}\}, i, j = 1, 2$, as follows

$$-\mathrm{div}\ \sigma = f,\ \sigma - B\varepsilon(u) = 0 \quad \text{in} \quad \Omega_\gamma, \tag{3}$$

$$v_{xxxx} = [\sigma_\nu] \text{ on } \gamma_b,\ -w_{xx} = [\sigma_\tau] \text{ on } \gamma, \tag{4}$$

$$-v_{xx} - \varphi_x = [\sigma_\nu],\ -\varphi_{xx} + v_x + \varphi = 0 \quad \text{on} \quad \gamma_t, \tag{5}$$

$$(u, v, w, \varphi) \in W;\ v_{xxx} = v_{xx} = w_x = 0 \text{ for } x = 0, \tag{6}$$

$$w_x(1-) = w_x(1+);\ v_x + \varphi = \varphi_x = w_x = 0 \text{ for } x = 2, \tag{7}$$

$$-v_{xxx}(1-) = (v_x + \varphi)(1+);\ v_{xx}(1-) = -\varphi_x(1+). \tag{8}$$

Here, $\varepsilon_{ij}(u) = \frac{1}{2}(u_{i,j} + u_{j,i})$, $\sigma_\nu = \sigma_{ij}\nu_j\nu_i$, $\sigma_\tau = \sigma_{ij}\nu_j\tau_i$, and $[h] = h^+ - h^-$ is a jump of a function h on γ, where $h^\pm$ are the traces of h on the faces $\gamma^\pm$. The signs $\pm$ correspond to positive and negative directions of ν.

The function $u = (u_1, u_2)$ describes a displacement field of the elastic body; functions w, v fit to displacements of the inclusions γ_b, γ_t along the axis x_1 and axis x_2 respectively; the function φ describes a rotation angle of the inclusion γ_t. Observe that a part of boundary conditions for functions u, v, w, φ is included in the condition $(u, v, w, \varphi) \in W$.

Relations (3) are the equilibrium equations for the elastic body and Hooke's law; (4)-(5) are the Euler-Bernoulli and Timoshenko equilibrium equations for the inclusions γ_b and γ_t. According to the condition $(u, v, w, \varphi) \in W$, the vertical (along the axis x_2) and tangential (along the axis x_1) displacements of the elastic body coincide with the inclusion displacements at γ. The right-hand sides $[\sigma_\nu], [\sigma_\tau]$ in (4), (5) describe forces acting on γ from the surrounding elastic media.

Theorem 2. *Problem formulations* (3)-(8) *and* (1)-(2) *are equivalent provided that the solutions are smooth.*

By (8) and the first relations of (6), (7), we can write a complete system of junction conditions at the joint point $(1, 0)$:

$$w(1-) = w(1+),\ v(1-) = v(1+),\ v_x(1-) = -\varphi(1+),$$
$$w_x(1-) = w_x(1+),\ -v_{xxx}(1-) = (v_x + \varphi)(1+),\ v_{xx}(1-) = -\varphi_x(1+).$$

2. Delaminated elastic inclusion

Assume that the Euler-Bernoulli part γ_b of the inclusion γ is delaminated, please refer to Figure 1. This means that a crack is located between γ_b and the elastic matrix. To fix a situation, the delamination is assumed to be at the positive side of γ_b. In this case, displacements v, w of the inclusion γ_b should coincide with displacements of the elastic body at γ_b^-. In our model, inequality type boundary conditions are considered at the crack faces to prevent a mutual penetration between the faces.

Denote $\Omega_b = \Omega \setminus \bar{\gamma}_b$ and introduce a set of admissible functions

$$K = \{(u, v, w, \varphi) \mid u \in H^1_\Gamma(\Omega_b)^2, (v, w) \in H^1(\gamma)^2, v \in H^2(\gamma_b),$$
$$\varphi \in H^1(\gamma_t),\ u|_{\gamma^-} = (w, v),\ [u_\nu] \geq 0 \text{ on } \gamma_b;\ v_x(1-) + \varphi(1+) = 0\},$$

where $H^1_\Gamma(\Omega_b) = \{\phi \in H^1(\Omega_b) \mid \phi = 0 \text{ on } \Gamma\}$. Notice that the inequality $[u_\nu] \geq 0$ included in the definition of K provides a mutual nonpenetration between the crack faces $\gamma_b^\pm$. An equilibrium problem for the elastic body with the delaminated Euler-Bernoulli inclusion γ_b and the Timoshenko inclusion γ_t can be formulated as follows. We have to find functions $u = (u_1, u_2), v, w, \varphi, \sigma = \{\sigma_{ij}\}, i, j = 1, 2$, as follows

$$-\text{div}\, \sigma = f,\ \sigma - B\varepsilon(u) = 0 \quad \text{in} \quad \Omega_\gamma, \tag{9}$$
$$v_{xxxx} = [\sigma_\nu] \text{ on } \gamma_b,\ -w_{xx} = [\sigma_\tau] \text{ on } \gamma, \tag{10}$$
$$-v_{xx} - \varphi_x = [\sigma_\nu],\ -\varphi_{xx} + v_x + \varphi = 0 \quad \text{on} \quad \gamma_t, \tag{11}$$
$$(u, v, w, \varphi) \in K;\ v_{xxx} = v_{xx} = w_x = 0 \text{ for } x = 0, \tag{12}$$
$$w_x(1-) = w_x(1+);\ v_x + \varphi = \varphi_x = w_x = 0 \text{ for } x = 2, \tag{13}$$
$$\sigma_\nu^+ \leq 0,\ \sigma_\tau^+ = 0,\ \sigma_\nu^+[u_\nu] = 0 \text{ on } \gamma_b, \tag{14}$$
$$-v_{xxx}(1-) = (v_x + \varphi)(1+);\ v_{xx}(1-) = -\varphi_x(1+). \tag{15}$$

As before, a part of boundary conditions is included in the relation $(u, v, w, \varphi) \in K$.

Remark that the problem (9)-(15) admits an equivalent variational formulation. Indeed, a solution (u, v, w, φ) satisfies a variational inequality

$$\int_{\Omega_b} \sigma(u)\varepsilon(\bar{u} - u) - \int_{\Omega_b} f(\bar{u} - u) + \int_{\gamma_b} v_{xx}(\bar{v}_{xx} - v_{xx}) + \tag{16}$$
$$+ \int_{\gamma_t} \{\varphi_x(\bar{\varphi}_x - \varphi_x) + (v_x + \varphi)(\bar{v}_x + \bar{\varphi} - v_x - \varphi)\} +$$
$$+ \int_\gamma w_x(\bar{w}_x - w_x) \geq 0 \quad \forall (\bar{u}, \bar{v}, \bar{w}, \bar{\varphi}) \in K;\ (u, v, w, \varphi) \in K.$$

3. Rigidity of Euler-Bernoulli beam goes to infinity

In practice, a solution of the problem (9)-(15) should depend on the rigidity parameters of the elastic inclusions. In the model (9)-(15), these parameters were taken to be equal to 1. In this section we introduce a rigidity parameter $\delta > 0$ in the Euler-Bernoulli equations of the problem (9)-(15) and analyze its passage to infinity. For a fixed parameter δ, we have to solve the following problem: to find $u^\delta, v^\delta, w^\delta, \varphi^\delta$, $\sigma^\delta = \{\sigma^\delta_{ij}\}, i,j = 1,2$, such that

$$-\text{div}\,\sigma^\delta = f,\ \sigma^\delta - B\varepsilon(u^\delta) = 0 \quad \text{in} \quad \Omega_\gamma, \tag{17}$$

$$\delta v^\delta_{xxxx} = [\sigma^\delta_\nu] \text{ on } \gamma_b, \ - \ \text{div}(a^\delta w^\delta_x) = [\sigma^\delta_\tau] \text{ on } \gamma, \tag{18}$$

$$-v^\delta_{xx} - \varphi^\delta_x = [\sigma^\delta_\nu],\ -\varphi^\delta_{xx} + v^\delta_x + \varphi^\delta = 0 \quad \text{on} \quad \gamma_t, \tag{19}$$

$$(u^\delta, v^\delta, w^\delta, \varphi^\delta) \in K;\ v^\delta_{xxx} = v^\delta_{xx} = w^\delta_x = 0 \text{ for } x = 0, \tag{20}$$

$$\delta w^\delta_x(1-) = w^\delta_x(1+);\ v^\delta_x + \varphi^\delta = \varphi^\delta_x = w^\delta_x = 0 \text{ for } x = 2, \tag{21}$$

$$\sigma^{\delta+}_\nu \le 0,\ \sigma^{\delta+}_\tau = 0,\ \sigma^{\delta+}_\nu [u^\delta_\nu] = 0 \text{ on } \gamma_b, \tag{22}$$

$$-\delta v^\delta_{xxx}(1-) = (v^\delta_x + \varphi^\delta)(1+);\ \delta v^\delta_{xx}(1-) = -\varphi^\delta_x(1+). \tag{23}$$

Here, $a^\delta(x) = 1$ on γ_t, and $a^\delta(x) = \delta$ on γ_b.

The problem (17)-(23) admits an equivalent variational formulation. A unique solution of the variational inequality (with $\sigma(u^\delta) = \sigma^\delta$) exists

$$(u^\delta, v^\delta, w^\delta, \varphi^\delta) \in K, \tag{24}$$

$$\int\limits_{\Omega_b} \sigma(u^\delta)\varepsilon(\bar u - u^\delta) - \int\limits_{\Omega_b} f(\bar u - u^\delta) + \delta \int\limits_{\gamma_b} v^\delta_{xx}(\bar v_{xx} - v^\delta_{xx}) + \tag{25}$$

$$+ \int\limits_{\gamma_t} \{\varphi^\delta_x(\bar\varphi_x - \varphi^\delta_x) + (v^\delta_x + \varphi^\delta)(\bar v_x + \bar\varphi - v^\delta_x - \varphi^\delta)\} +$$

$$+ \int\limits_{\gamma} a^\delta w^\delta_x(\bar w_x - w^\delta_x) \ge 0 \quad \forall\, (\bar u, \bar v, \bar w, \bar\varphi) \in K.$$

Our aim is to justify a passage to the limit as $\delta \to \infty$ in the problem (24)-(25).

Introduce a set of infinitesimal rigid displacements

$$R(\gamma_b) = \{\rho = (\rho_1, \rho_2) \mid \rho(x_1, x_2) = d(-x_2, x_1) + (d^1, d^2),$$
$$(x_1, x_2) \in \gamma_b\},\ d, d^1, d^2 \in \mathbb{R},$$

and a set of admissible functions

$$K_r = \{(u, v, w, \varphi) \mid u \in H^1_\Gamma(\Omega_b)^2, (v, w, \varphi) \in H^1(\gamma_t)^3,\ [u_\nu] \ge 0 \text{ on } \gamma_b;$$
$$u|_{\gamma_t} = (w, v),\ u|_{\gamma_b^-} = (\rho_1, \rho_2) \in R(\gamma_b),\ \rho_{2x}(1) + \varphi(1) = 0\}.$$

It can be proved that $\delta \to +\infty$,

$$u^\delta \to u \text{ weakly in } H^1_\Gamma(\Omega_b)^2, \varphi^\delta \to \varphi \text{ weakly in } H^1(\gamma_t), \tag{26}$$

$$v^\delta \to v \text{ weakly in } H^1(\gamma), \text{ weakly in } H^2(\gamma_b),\ v_{xx} = 0 \text{ in } \gamma_b, \tag{27}$$

$$w^\delta \to w \quad \text{weakly in } H^1(\gamma),\ w_x = 0 \text{ in } \gamma_b. \tag{28}$$

Moreover,

$$(u,v,w,\varphi) \in K_r, \int\limits_{\Omega_b} \sigma(u)\varepsilon(\bar{u}-u) - \int\limits_{\Omega_b} f(\bar{u}-u)+ \tag{29}$$
$$+\int\limits_{\gamma_t} \{\varphi_x(\bar{\varphi}_x-\varphi_x)+(v_x+\varphi)(\bar{v}_x+\bar{\varphi}-v_x-\varphi)+$$
$$+w_x(\bar{w}_x-w_x)\} \geq 0 \quad \forall (\bar{u},\bar{v},\bar{w},\bar{\varphi}) \in K_r.$$

The problem (29) admits an equivalent differential formulation: find displacements $u = (u_1, u_2)$, v, w, a rotation angle φ, a stress tensor $\sigma = \{\sigma_{ij}\}, i,j = 1,2$, and $\rho^0 \in R(\gamma_b)$ as follows

$$-\text{div}\,\sigma = f,\ \sigma - B\varepsilon(u) = 0 \quad \text{in} \quad \Omega_\gamma, \tag{30}$$
$$-v_{xx} - \varphi_x = [\sigma_\nu],\ -\varphi_{xx} + v_x + \varphi = 0,\ -w_{xx} = [\sigma_\tau] \text{ on } \gamma_t, \tag{31}$$
$$u^- = \rho^0,\ \sigma_\nu^+ \leq 0,\ \sigma_\tau^+ = 0,\ \sigma_\nu^+[u_\nu] = 0 \text{ on } \gamma_b, \tag{32}$$
$$(u,v,w,\varphi) \in K_r;\ v_x + \varphi = \varphi_x = w_x = 0 \text{ for } x = 2, \tag{33}$$
$$\int\limits_{\gamma_b} [\sigma\nu]\rho + (w_x\rho_1)(1) - (\varphi_x\rho_{2x})(1)+ \tag{34}$$
$$+((v_x+\varphi)\rho_2)(1) = 0 \ \forall \rho = (\rho_1,\rho_2) \in R(\gamma_b).$$

Thus, the following statement can be proved.

Theorem 3. *The solutions of the problem* (24)-(25) *converge to the solution of* (29) *in the sense* (26)-(28) *as* $\delta \to \infty$.

The model (30)-(34), or (29), describes an equilibrium state for the elastic body with the rigid inclusion γ_b and elastic Timoshenko inclusion γ_t. The identity (34) provides equilibrium conditions for the rigid inclusion γ_b, i.e. a principal vector of forces and a principal vector of moments acting on γ_b are equal to zero. Indeed, denoting $(\sigma\nu)^\pm$ by $(\sigma^1,\sigma^2)^\pm$ on $\gamma_b^\pm$, the condition (34) can be rewritten in the following form:

$$\int\limits_{\gamma_b} [\sigma^1] = -w_x(1),\ \int\limits_{\gamma_b} [\sigma^2] = -(v_x+\varphi)(1), \tag{35}$$
$$\int\limits_{\gamma_b} ([\sigma^2]x_1 - [\sigma^1]x_2) = \varphi_x(1). \tag{36}$$

We can also write junction conditions included in the definition of K_r:

$$w(1) = \rho_1^0(1),\ v(1) = \rho_2^0(1),\ \varphi(1) + \rho_{2x}^0(1) = 0 \tag{37}$$

and consider (35)-(37) as a complete system of junction conditions at the joint point $(1,0)$. Consequently, nonlocal condition (34) can be seen as a part of junction conditions at the joint point $(1,0)$.

References

[1] Khludnev A M and Negri M 2012 *Z. Angew. Math. Mech.* **92** 341–54
[2] Khludnev A M and Leugering G R 2014 *Math. Mech. Complex Systems* **2** 1–21
[3] Khludnev A M and Leugering G R 2015 *Math. Mech. Solids* **20** 495–511
[4] Itou H and Khludnev A M *Math. Meth. Appl. Sciences* doi 10.1002/mma.3279

Shock wave processes in collisional gas particle mixtures

T A Khmel and A V Fedorov

Khristianovich Institute of Theoretical and Applied Mechanics SB RAS,
630090, Novosibirsk, Russia
E-mail: khmel@itam.nsc.ru, fedorov@itam.nsc.ru

Abstract. Structures and propagation of shock waves in high density particle suspensions in gas are investigated theoretically and numerically. A physical and mathematical model which takes into account integral collisions between the particles on the basis of molecular-kinetic approaches of theory of granular materials is applied. The possibility of different types of shock waves, including double front structures is revealed. The role of particle collisions in the dynamics of particle dense layer expansion under an influence of divergent shock wave and in processes of shock wave diffraction past a backward-facing step is analyzed.

1. Introduction

Problems of interaction of shock waves (further - 'SW') with dust layers arise in connection with detonation suppression or prevention in reactive mixtures. Shock wave interactions with layers of reacting particles can form clouds of explosive concentrations [1]. On the other hand, the explosive dispersion of inert particle layers under the influence of a central charge is one of the ways of detonation damping and combustion extinction.

Describing high density gas particle mixtures, it is necessary to take into account the volume occupied by the particles, as well as the possibility of contact interaction of particles (particle-to-particle collisions).

A physical and mathematical model of two-phase medium to describe the shock-wave dynamics of dense gas mixtures with the averaged description of particle-to-particle collisions on the basis of molecular-kinetic theory has been developed in [2], [3]. The model has been applied in [4], [5] for analysis of shock and detonation wave propagation in collisional particle suspensions in gas and shock wave dispersion of dust layers. A possibility of existence of two governing types of shock waves was revealed in [4]: with leading shock in gas or with leading shock in the particle phase. Only single-front structures were obtained in [4] although existence of double-front structures in mixtures with two pressures was established in [6] in the frame of simple isothermal model.

In this paper, one-dimensional and two-dimensional flows formed in the interaction of shock waves with particle clouds and layers of high-density and in the diffraction of shock waves past a backward-facing step are investigated on the base of the model [2], [3]. The purpose of the study is to analyze an influence of the collisional particle dynamics in the flow patterns and the shock-wave structure formation.

2. Physical and mathematical model

The Euler equations for two-dimensional flow of two-phase mixture of gas and two fractions of particles follow from the conservation laws for mass, momentum, and energy (the subscripts 1, 2, and 3 indicate gas, inert particles, and reactive particles, respectively):

$$\frac{\partial \boldsymbol{W}}{\partial t}+\frac{\partial \boldsymbol{F}}{\partial x}+\frac{\partial \boldsymbol{G}}{\partial y}=\boldsymbol{\Gamma} \tag{1}$$

$$\mathbf{W}=\begin{pmatrix}\boldsymbol{W}_1\\ \boldsymbol{W}_2\end{pmatrix},\ \boldsymbol{F}=\begin{pmatrix}\boldsymbol{F}_1\\ \boldsymbol{F}_2\end{pmatrix},\ \boldsymbol{G}=\begin{pmatrix}\boldsymbol{G}_1\\ \boldsymbol{G}_2\end{pmatrix},\ \boldsymbol{\Gamma}=\begin{pmatrix}-\tilde{\Gamma}_2\\ \boldsymbol{\Gamma}_2\end{pmatrix} \tag{2}$$

$$\mathbf{W}_1=\begin{pmatrix}\rho_1\\ \rho_1u_1\\ \rho_1v_1\\ \rho_1E_1\end{pmatrix},\ \boldsymbol{F}_1=\begin{pmatrix}\rho_1u_1\\ m_1p_1+\rho_1u_1^2\\ \rho_1u_1v_1\\ \rho_1u_1E_1+m_1p_1u_1\end{pmatrix},\ \boldsymbol{G}_1=\begin{pmatrix}\rho_1v_1\\ \rho_1u_1v_1\\ m_1p_1+\rho_1v_1^2\\ \rho_1v_1E_1+m_1p_1v_1\end{pmatrix}. \tag{3}$$

$$\mathbf{W}_2=\begin{pmatrix}\rho_2\\ \rho_2u_2\\ \rho_2v_2\\ \rho_2E_2\\ \rho_2E_c\end{pmatrix},\ \boldsymbol{F}_2=\begin{pmatrix}\rho_2u_2\\ \rho_2u_2^2+m_2p_2\\ \rho_2u_2v_2\\ \rho_2u_2E_2+m_2u_2p_2\\ \rho_2u_2E_c+\eta m_2u_2p_2\end{pmatrix},\ \boldsymbol{G}_2=\begin{pmatrix}\rho_2v_2\\ \rho_2u_2v_2\\ \rho_2v_2^2+m_2p_2\\ \rho_2v_2E_2+m_2v_2p_2\\ \rho_2v_2E_c+\eta m_2v_2p_2\end{pmatrix} \tag{4}$$

$$\Gamma_2=\begin{pmatrix}\tilde{\Gamma}_2\\ \eta p_1(\frac{\partial u_2m_2}{\partial x}+\frac{\partial v_2m_2}{\partial y})-I_0+\eta(f_{2x}u_2+f_{2y}v_2)\end{pmatrix}, \tag{5}$$

$$\tilde{\Gamma}_2=\begin{pmatrix}0\\ p_1\frac{\partial m_2}{\partial x}+f_{2x}\\ p_1\frac{\partial m_2}{\partial y}+f_{2y}\\ q_2+f_{2x}u_2+f_{2y}v_2\end{pmatrix}. \tag{6}$$

Here ρ, p, m are the mean density, pressure, and the particle volume concentration, respectively, $\rho_i=\rho_{ii}m_i$, ρ_{ii} is the true density of phases; u,v are the velocity components; f is the force of interaction between phases. The particles are assumed to be incompressible ($\rho_{22}=\text{const}$). The total energy of the inert particle phase E_2 includes also the energy of chaotic particle motion determined as the average kinetic energy of the particle velocity pulsations:

$$e_c=0.5\upsilon^2,\ \upsilon^2=\frac{1}{N}\sum_{i=1}^{N}[(u_i')^2+(v_i')^2]\ . \tag{7}$$

Here u_i', v_i' are the i-th particle velocity pulsation components and N is the number of particles in the unit volume. The chaotic energy is associated with the discrete phase pressure p_c (intergranular pressure) generated by particle-to-particle collisions. The collisional total energy E_c is determined in the form presented in [3]. The equations of state have the form

$$p_1=\rho_{11}RT_1,\ m_2p_2=m_2p_1+p_c,\ p_c=0.5\alpha_t[1+2(1+e)m_2g(m_2)]\rho_2e_c,$$
$$g(m_2)=[1-(m_2/m_*)^{4m_*/3}]^{-1},\ E_1=c_{v1}T_1+u_1^2/2,\ E_2=e_c+c_2T_2+0.5(u_2^2+v_2^2), \tag{8}$$
$$E_c=e_c+0.5\eta u_2^2.$$

Here, m_* is the limiting maximal value of particle volume concentration, α_t defines a relationship between the energy of the chaotic motion of particles and the collisional pressure in the discrete phase. The dissipative term is assumed to be

$$I_0 = \frac{6}{\pi d_2} C_0 \rho_2 m_2 g(m_2)(e_c^{3/2} - e_{c0}^{3/2}), \tag{9}$$

where C_0 is the dissipation coefficient, e_{c0} is the initial value of the energy of the chaotic motion, which is assumed to be the minimal value of e_c in the following process, and d is the particle diameter. We take into account that the heating of the particles occurs due to convective heat transfer and due to the conversion of the chaotic energy into heat at non-ideal collisions of rough and inelastic particles. The coefficients C_0, η, and the parameter α_t depend on the restitution coefficient ε, the shape coefficient k ($k = 0.4$ for spheres), and the roughness parameter β. Ideal collisions without losses are characterized by $\varepsilon = 1$, $\beta = -1$, $\eta = 1$. The expressions for C_0, η, and α_t are presented in [3]. The function $g(m_2)$ in (8) describes behavior of granular media in a wide range of particle volume concentrations from small values (when it is similar to a monatomic gas with $\gamma = 5/3$) to values of dense packing. The collisional (intergranular) pressure p_c tends to infinity with particle volume concentration approaches its maximal value m_*.

The numerical technique is based on the conservative flux-splitting schemes: the TVD scheme by Harten for gas and the Gentry-Martin-Daly scheme for particles. The numerical method has been tested earlier and applied for 2-D numerical simulations of the shock wave and detonation flows in the frame of a standard collisionless model in [7], [8], [9]. Here, we take similar to Gentry-Martin-Daly scheme approximations for the additional terms related to the gas pressure and the pressure of random motion of particles. The calculations were performed using IBM PC Intel Core2Quad.

3. Types of shock waves in collisional media

The types of possible shock wave structures in dense gas particle mixtures were investigated on the problem of the SW in gas entering into a particle cloud. Two basic types of the SW in collisional mixture are the following [4]: with leading shock in gas and a relaxation zone in the discrete phase (type I), and with collisional shock in the particle phase with following relaxation zone in gaseous phase (type II). Distributions of the parameters in the relaxation zones were obtained in [4] to be continuous for both types analyzed. Varying the amplitude of the incident SW and the mixture parameters (density, collisional constants), additional types of shock wave structures are revealed. Some examples of double-front structures in mixtures of 10-μm particles with leading shock in gas (type I-D) and double-front structures with leading shock of collisional type in particles (type II-D) are plotted in Figures 1-2.

Note that the double shock wave structures shown in Figures 1-2 are different from those obtained under isothermal model for compressible media with two pressures in [6]. The present structures are characterized also by the temperature relaxation zones, as well as non-uniform distribution of the parameters describing the random motion of particles (p_c,e_c). For example, in the structure of type I_D (Figure 1), the presence of the second jump is seen on the particle velocity profile, and in this area the chaotic motion energy generation occurs (see the peak in e_c distribution). For the II_D type structure, the leading shock in particle density and particle velocity is not accompanied by a temperature jump (Figure 2), and the gas temperature is also continuous. The gas temperature shock together with gas density and velocity shocks take place inside the relaxation zone on some distance from the leading collisional shock.

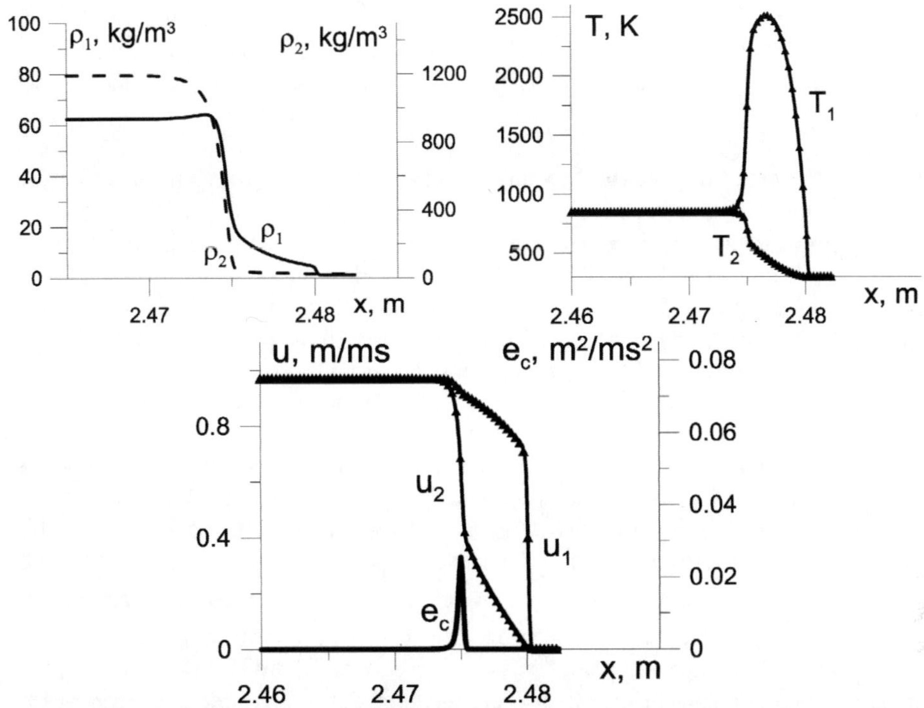

Figure 1. A double-front structure of type I-D ($m_{20} = 0.009$, $u_{SW} = 0.99$ m/ms, $e_{c0} = 0$).

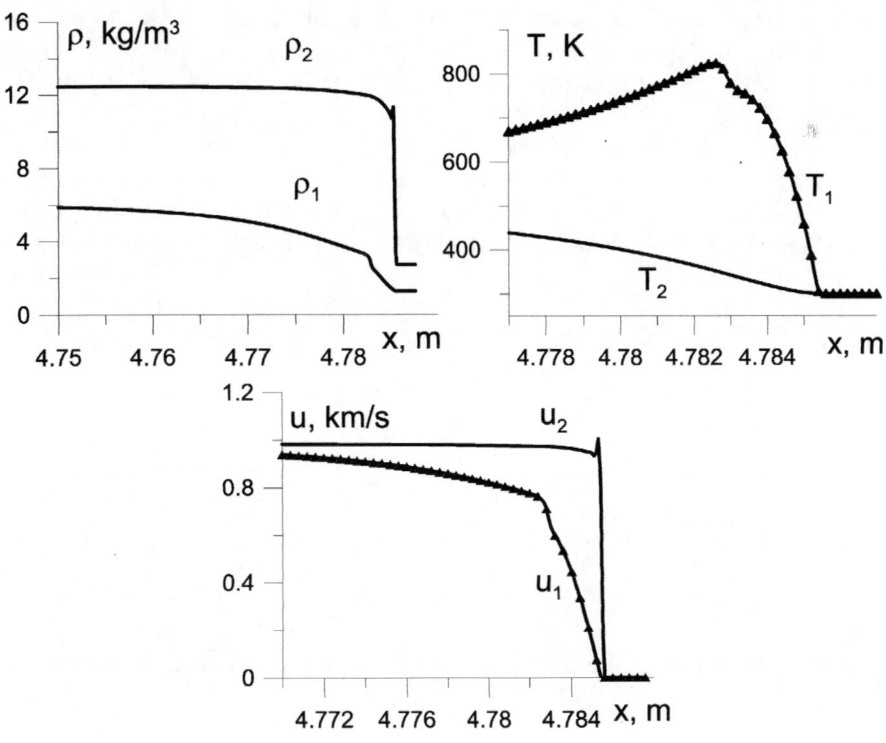

Figure 2. A double-front structure of type II-D
($m_{20} = 0.001$, $u_{SW} = 1.26$ m/ms, $e_{c0} = 0.005$ m^2/ms^2).

Realization of the first or the second type of the shock wave structures depends on the initial parameters of the mixture (particle density, initial level of the chaotic energy), the shock wave propagation velocity, and particle material characteristics (diameter, roughness and restitution coefficients).

4. Collisional effects in interaction of explosive shock waves with particle dense layers

The problem of the interaction of the expanding shock wave with a particles dense layer in one-dimensional formulation (in the approximation of spherical symmetry) taking into account particle-to-particle collisions is considered. The values of gas and particle parameters correspond to experimental conditions in [10] (31 g Comp B booster, 550 g glass spheres with mean particle diameter 10 μm).

Figure 3 shows an initial stage of the SW interaction with the particle layer with the time step 1 μs, numerals indicate the corresponding time moments. Solid and dashed lines indicate gas and particle parameters, respectively. Note that when the SW enters the dense layer it slows and transforms to a dispersive wave in gaseous phase. At the same time, a collisional shock wave in the particle phase of type II forms and propagates ahead the incident wave in gas (compare the solid and dashed lines for equal time moments). This type of the collisional SW with leading shock in particle phase and a relaxation zone in gas has been analyzed in [4]. Here, the particle volume concentration in the relaxation zone behind the front reaches the values of 0.594 which is close to the value of dense packing (m_*=0.6). As the collisional wave front reaches the external boundary of the layer, the leading edge of the particle cloud begins to move, and it happens earlier than the wave front in gas reaches the cloud edge (compare dashed and solid curves 5 in Figure3a). The velocity profiles (Figure 3b) confirm the type II of the shock wave in the mixture (dashed lines are located just to the right of solid lines). A small thickness of the velocity relaxation zones behind the collisional shock front is explained by high differences between gas and particle mean densities here. At the same time, a secondary ordinary shock wave propagates inside the cloud which appears as a description between solid and dashed lines in Figure 3b. Here, gas density is much higher than in the collisional SW, particle density decreases near the internal cloud surface, and the velocity relaxation zone is more extended.

Thus, the collisional pressure arising due to the development of particle chaotic motion plays a decisive role in the scenario of the initial stage of particle motion and formation of shock-wave structures in the cloud at a SW interaction with dense particle layers. Consequently, the collisional particle dynamics is one of the determining factors in processes of explosive dispersion of dusts.

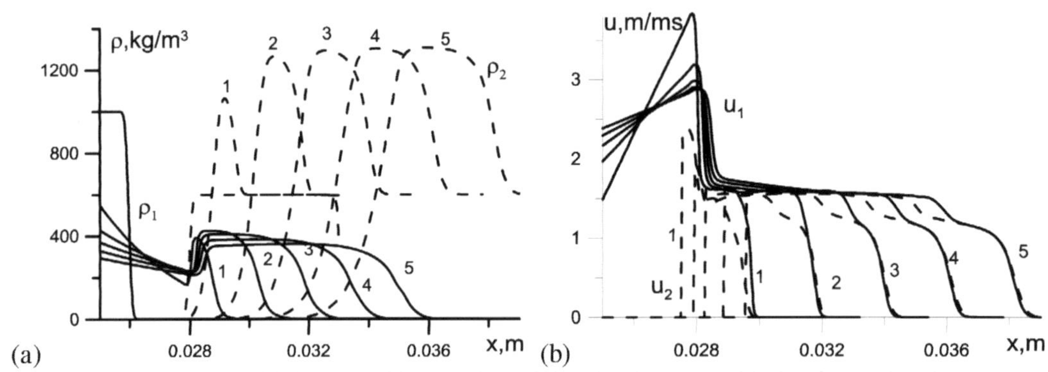

Figure 3. An initial stage of interaction of the shock wave with the dense dust layer.

5. Diffraction of a shock wave on a backward-facing step in collisional gas particle mixture

The problem of shock-wave transition past a backward-facing step in a gas suspension was investigated in [11]. It was shown that the greatest difference between the flow pattern in a two-phase mixture and the corresponding flow in a pure gas is observed in the range of times when the characteristic sizes of the structures being formed are commensurable with the scale of the relaxation zones. It was also established that there was formation of a region with a reduced fraction of particles near the corner that is caused by particle inertia in the vortex flow and accumulation of particles in the layer adjacent to the contact surface in the structure. Collisional dispersion of dense inert particle layers formed in detonation waves was revealed in [4]. Here, one analyzes the problem of shock wave diffraction in gas particle mixture in the frame of the collisional model to estimate the collisional effects in dispersion of the particle layers obtained in [11].

To compare the results for collisional and collisionless models, diffraction of a shock wave of the type I is considered. Figures 4-7 show the flow patterns at the shock wave diffraction in the mixture of oxygen and 10-μm alumina particles with $m_{20} = 0.0006$ and the SW propagation velocity $u_{SW} = 0.82$ km/s. All density fields are scaled in kg/m^3, all collisional (granular) pressure fields are presented in MPa.

Figure 4 presents the gas density and particle density fields obtained in the frame of ordinary collisionless model. The region of low particle concentration (almost free of particles), together with the layer of accumulated particles are obvious in Figure 4b as white and green zones on the right of the corner. Note that maximal particle density takes place in the region of final equilibrium state of the two-phase shock wave (red areas in the picture).

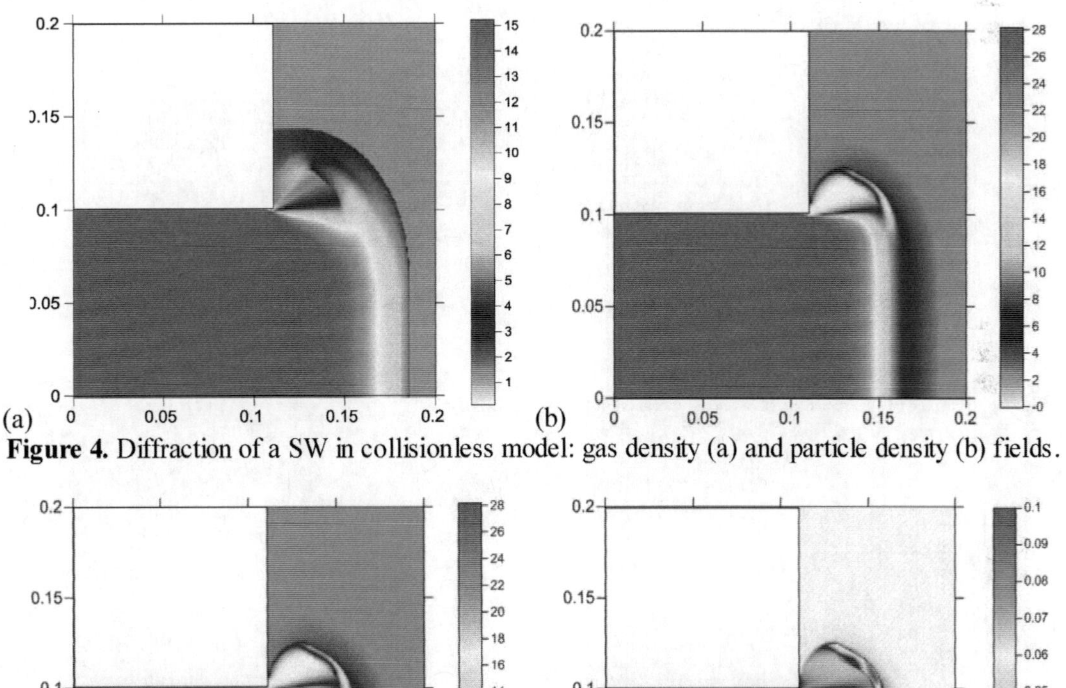

Figure 4. Diffraction of a SW in collisionless model: gas density (a) and particle density (b) fields.

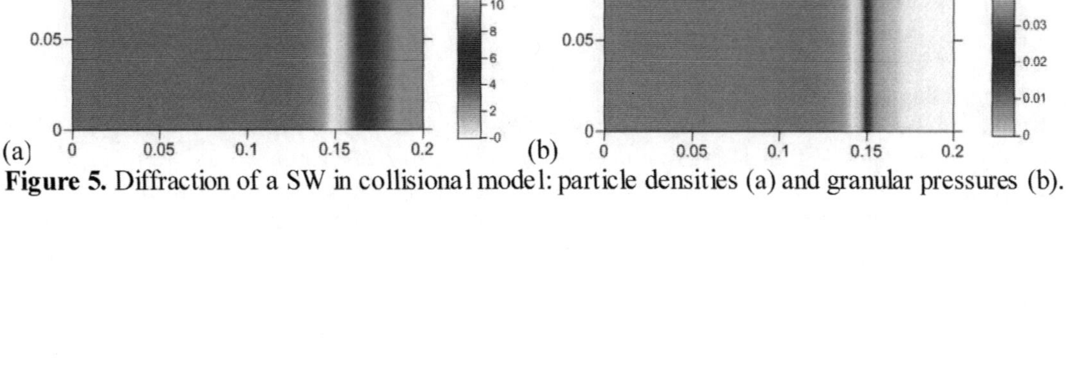

Figure 5. Diffraction of a SW in collisional model: particle densities (a) and granular pressures (b).

The results of calculations obtained in the frame of collisional model ($e_{c0} = 0.0001$ m^2/ms^2) are plotted in Figure 5 for the same time moment as in Figure 4. No difference is obvious between particle density fields in Figure 4b and Figure 5a although the collisional pressure is not negligible (Figure 5b). The maximal values of the granular pressure are comparable with initial gas pressure (0.1 MPa).

Figures 6-7 show the corresponding particle density and collisional pressure fields in a later time moment for two different values of initial level of energy of particle chaotic motion: $e_{c0} =$ 0.0001 m^2/ms^2 (Figure 6) and 0.01 m^2/ms^2 (Figure 7). For convenience, the pictures are presented in the same color scale. One can see that the particle density fields practically coincide that means negligible effect of collisional pressure in dispersion of the particle layer formed behind the backward-facing step. Note that despite two order difference in initial values of e_{c0} the collisional pressure values in the flow behind the SW do not differ so strong. It means that randomization of particle motion generated in non-uniform flow behind two-phase shock wave of the type I is determined mainly by the flow structure, and initial randomization level plays here unimportant role unlike obtained in [4] for the detonation processes.

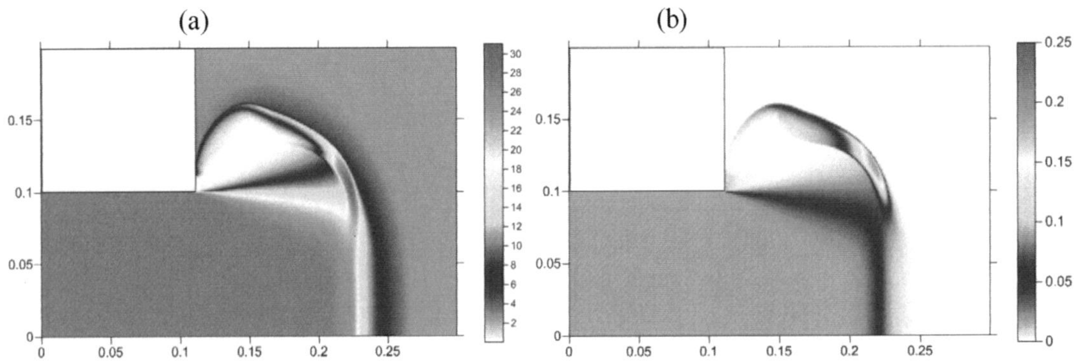

Figure 6. Particle densities (a) and granular pressures (b) at $e_{c0} = 0.0001$ m^2/ms^2.

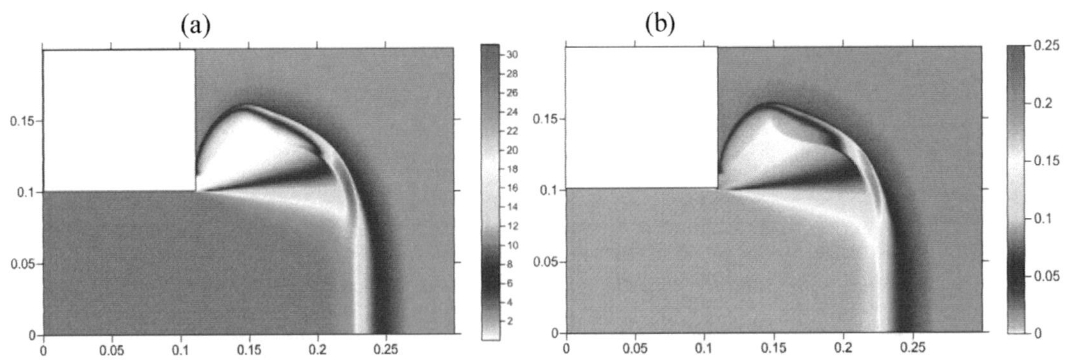

Figure 7. Particle densities (a) and granular pressures (b) at $e_{c0} = 0.01$ m^2/ms^2.

6. Concluding remarks

The comparison of the results of calculations in the frame of ordinary (collisionless) and collisional models for description of shock wave processes of particle suspensions in gas shows that generated in the particle phase collisional pressure may play a significant role. The variety of different types of shock waves in gas particle mixtures including double-front structures is obtained. The shock waves of type II with leading collisional shock in particle phase are responsible for initial stage of dense layer

Journal of Physics: Conference Series **722** (2016) 012044 doi:10.1088/1742-6596/722/1/012044

dispersion under an action of explosive shock of central charge. At the same time, the generated collisional pressure is not a significant factor in shock wave structures and layer dispersion in shock wave diffraction past a backward-facing step.

Acknowledgements
The work was supported by the Russian Scientific Foundation (project № 16-19-00010).

References

[1] Fedorov A V 2004 *Combustion, Explosion, and Shock Waves* **40** 17
[2] Fedorov A V and Khmel T A 2012 *Heat Transfer Research* **43** 95
[3] Khmel T A and Fedorov A V 2014 *Combustion, Explosion, and Shock Waves* **50** 196
[4] Khmel T A and Fedorov A V 2014 *Combustion, Explosion, and Shock Waves* **50** 547
[5] Khmel T A and Fedorov A V 2015 *Journal of Loss Prevention in the Process Industries* **36** 223
[6] Bedarev I A and Fedorov A V 2015 *Computational technologies* **20**(2) 3
[7] Kratova Yu V, Fedorov A V and Khmel T A 2011 *Combustion Explosion and Shock Waves* **47** 572
[8] Kratova Yu V, Fedorov A V and Khmel T A 2009 *Combustion Explosion and Shock Waves* **45** 591
[9] Fedorov A V and Khmel T A 2002 *Combustion Explosion and Shock Waves* **38** 101
[10] Sturtzer S A, Sturtzer M O, Veyssiere B and Khasainov B A 2013 *Proc. 24-th ICDERS* (Taipai)
[11] Fedorov A V, Kratova Yu V and Khmel T A 2008 *Combustion Explosion and Shock Waves* **44** 76

Relaxation oscillation model of hemodynamic parameters in the cerebral vessels

A A Cherevko[1,2], A V Mikhaylova[1], A P Chupakhin[1,2], I V Ufimtseva[1], A L Krivoshapkin[3] and K Yu Orlov[3]

[1] Novosibirsk State University, Novosibirsk, Russia
[2] Lavrentyev Institute of Hydrodynamics, Novosibirsk, Russia
[3] Meshalkin Novosibirsk Research Institute of Circulation Pathology, Novosibirsk, Russia

E-mail: `chupakhin@hydro.nsc.ru`, `ufimtsevaiv@gmail.com`

Abstract. Simulation of a blood flow under normality as well as under pathology is extremely complex problem of great current interest both from the point of view of fundamental hydrodynamics, and for medical applications. This paper proposes a model of Van der Pol - Duffing nonlinear oscillator equation describing relaxation oscillations of a blood flow in the cerebral vessels. The model is based on the patient-specific clinical experimental data flow obtained during the neurosurgical operations in Meshalkin Novosibirsk Research Institute of Circulation Pathology. The stability of the model is demonstrated through the variations of initial data and coefficients. It is universal and describes pressure and velocity fluctuations in different cerebral vessels (arteries, veins, sinuses), as well as in a laboratory model of carotid bifurcation. Derived equation describes the rheology of the "blood stream – elastic vessel wall – gelatinous brain environment" composite system and represents the state equation of this complex environment.

1. Introduction

Modelling of a blood flow in a system such as cerebral vascular network is very complicated and can be realized in different ways. One of them involves building a model based on general equations of composite continuum describing fluid flow (possibly non - Newtonian), in an elastic tube of complex geometry, placed in the viscoelastic gel brain environment. Such models should consider physiology laws and use the parameters obtained during the experiments [1]. Obtaining information about the brain rheology is a complex problem.

Another way is to construct a relatively simple empirical model based on the values obtained directly during the clinical measurements and laboratory experiments. This is a classic approach for inverse incorrect problems of differential equations. It proved to be effective in different applications, and this approach is described and used in this paper [2].

The main preconditions, underlying the construction of relaxation oscillations model of hemodynamic in the cerebral vessels, are the following:

- Experimental clinical data due to measuring system Combo Map device specifics describes the blood rate and pressure connection, implemented in a multicomponent system "the blood stream – the elastic wall of the vessel – the brain surrounding environment". The model should bind experimentally measured values with the parameters (consumption, energy flow) calculated based on their basis [3].

- The experimentally observed velocity and pressure oscillations have a relaxation character. The simplest model for the description of these processes is the equation of a nonlinear oscillator of Van der Pol - Duffing (VdPD), where one of the values – pressure $p = p(t)$ or velocity $v = v(t)$ – acts as the control function, i.e, a right-hand side of the equation. It is important that equation (VdPD) has a stable periodic solution. At the same time, the equation of this form has a large and complex set of solutions, the structure of which is determined by the values of its coefficients. Cartwright and Littlewood obtained a remarkable result on the exact structure of the set of solutions of the equation (VdPD) for a specific selection of the coefficients. If they are changed, the systems of sub-harmonics and solutions that are not periodic appear in the equation [4]. There are a number of works studying the solutions of type (VdPD) equations [5, 6].

2. Intraoperative endovascular monitoring of blood flow

The studies are carried out using ComboWire intravascular guidewire with the diameter of 0.34 mm. Intravascular Doppler sonography is performed by using Combo Map unit (frequency 12 MHz) during the operation. The velocity and pressure in vessels readings are recorded from the Combo Map unit via analog-digital converter at frequency 200 Hz and processed on the computer to produce hydrodynamic quantities. The procedures for monitoring, clinical data processing, data clearing from the noise, and presenting data in the form of "operation map" are developed [3].

Many of various physiological processes occurring in living organisms are periodic. Such living system can be simulated by a system of interconnected oscillators, which produce modeling of heartbeats, breathing cycle, Mayer waves, etc. [7]

These oscillatory processes are observed on the wavelet diagrams, created using Gabor wavelet of frequency 6, obtained from processed monitoring data. Since we are interested in relaxation oscillations associated with the blood pulsing, such wavelet decomposition of the entire spectrum of vibrations is used to filter out the high-frequency noise, the vibrations corresponding to the respiratory rhythm (0.2 Hz) and the low-frequency oscillations [8]. The comparison and analysis of obtained skeylogramms with the original ones show that relaxation oscillation information is completely preserved.

Currently, this procedure of intravascular blood flow monitoring was conducted for 50 neurosurgical operations to cure anomalies in the vascular system of the brain such as arteriovenous malformations and cerebral aneurysms. The operations were performed in the Lavrentyev Institute of Hydrodynamics, and data processing was conducted in the Meshalkin Novosibirsk Research Institute of Circulation Pathology. The measurements were performed in all areas of the vascular network of the brain: in the arteries, veins and sinuses. The monitoring of this type can be considered unique as it was performed for the first time in the world practice. It represents unique experimental data for construction of mathematical models of hemodynamics in brain vessels.

3. Construction of the oscillator model

Van der Pol - Duffing's equation of nonlinear oscillator

$$\varepsilon q'' + P_2(q)q' + Q_3(q) = ku(t), \tag{1}$$

where $q = q(t)$ and $u = u(t)$ - dimensionless normalized values of blood flow, $|q| \leq 1$, $|u| \leq 1$; polynomials $P_2(q) = a_0 + a_1 q + a_2 q^2$, $Q_3(q) = b_1 q + b_2 q^2 + b_3 q^3$ characterize friction and elasticity of the system, real coefficients $\{a_i, b_j, k\}$ $(i = 0, 1, 2; j = 1, 2, 3)$ correspond to the individual patients data, ε - a small relaxation parameter. Control function $u(t)$ is the speed of blood flow on the right hand side of the equation (1) for brain arteries, function $q(t)$ is the normalized pressure. These roles are changing for veins and sinuses, which lie beyond the resistive zone created by capillaries: control function $u(t)$ is the pressure and function $q(t)$ is the normalized velocity. This can be explained by the fact, that the phase shift between pressure and speed in arteries as it was shown in the experiments [3] is positive and it is negative in veins and sinuses. In other words, the speed outpaces the pressure and controls it in arteries, and conversely in veins and sinuses. Accordingly, for the arteries the "pressure-velocity" diagrams traversed anticlockwise with increasing time, and vice versa for the veins and sinuses [3].

Normalized average values (q, u) of blood flow can be expressed through physical ones (p, v), measured during observed time operation in interval I by the formulae

$$q = \alpha_0 + \alpha_1 p, \quad u = \beta_0 + \beta_1 v, \tag{2}$$

where

$$\alpha_0 = -\xi M_p^{-1}, \quad \alpha_1 = M_p^{-1}, \quad \beta_0 = -\eta M_v^{-1}, \quad \beta_1 = M_v^{-1}, \tag{3}$$

$$M_p = \max_{i \in I} |p_I - \xi|, \quad M_v = \max_{i \in I} |v_I - \eta|. \tag{4}$$

The values of ξ and η in (4) are the average values of p and v on the time interval I. During data processing, either integral averages or arithmetic mean values for v and p on I are used, whatever produces the best result. Formulae (2) - (4) are written to normalize the data in the arteries, and they change for veins and sinuses as described above.

The linear replacing (2) transforms (1) into the equation of the same type

$$\varepsilon q'' + \hat{P}_2(p)p' + \hat{Q}_3(p) = \lambda v(t) + \mu, \tag{5}$$

describing the relationship between the physical flow values (p, v). The transformed coefficients $\{\hat{a}_i, \hat{b}_i\}$ in $\hat{P}_2$ and $\hat{P}_3$ polynomials, λ and μ are calculated according to the formulae, the specific form of which we do not present here.

Discretization of equation (1) is performed by transition to normalized clinical data at the time of measurements:

$$q_i = q(t_i), \quad u_i = u(t_i), \quad t_i = t_0 + (i-1)\Delta t \tag{6}$$

$$q' = (\Delta t)^{-1}(q_i - q_{i-1}), \quad q'' = (\Delta t)^{-2}(q_{i+1} - 2q_i + q_{i-1}), \tag{7}$$

where $i = \overline{1, N}$, the N number determines the measurement gap I_N, Δt is the time measuring step.

After substituting (6), (7) in (1), we get the nonlinear ARX model [9] – a system of linear algebraic equations

$$q_i = Ax, \quad (i = \overline{3, N}). \tag{8}$$

Matrix A in (8) has 7 columns and $N-2$ lines:

$$A = (q_{i-1},\ q_{i-2},\ (q_{i-1} - q_{i-2})q_{i-1},\ q_{i-1}^2,\ (q_{i-1} - q_{i-2})q_{i-1}^2,\ q_{i-1}^3,\ u_{i-1}), \tag{9}$$

with seven-dimensional unknown vector $x = (c_1, \ldots, c_7)^T$, the sign T means transposition, so that x is a column vector. And values c_i have relations with unknown $\{a_i, b_j, k\}$, $(i = 0, 1, 2;\ j = 1, 2, 3)$, by the next formulae:

$$\begin{aligned} a_0 &= \varepsilon(1 + c_2)/\Delta t \\ a_1 &= -\varepsilon c_3/\Delta t \\ a_2 &= -\varepsilon c_5/\Delta t \\ b_1 &= \varepsilon(1 - c_1 - c_2)/\Delta t^2 \\ b_2 &= -\varepsilon c_4/\Delta t^2 \\ b_3 &= -\varepsilon c_6/\Delta t^2 \\ k &= \varepsilon c_7/\Delta t^2. \end{aligned} \tag{10}$$

In our case the value of ε is equal to 10^{-3}. So many of unknowns have values of order 1.

The system (8) is strongly overdetermined in the experimental data processing, as rule, $N \gg 7$. Such system has not exact solution. Instead of exact solution, with respect to inverse problems theory [2, 10], we give pseudo-solutions, which minimize the residual norm for system (8). For minimization of residual uses the combination of Gauss-Newton and Levenburg-Marquardt line search, and Trust-region reflective Newton approach.

Applying the algorithm described to (8) - (10) gives the solution as a vector of the coefficients x. The equation (1) is constructed on the basis of this vector. The calculation is made in the Matlab program using the System Identification Toolbox Package.

4. Experimental validation of the model

The algorithm for constructing the equation VdPD (1) described in section 3 has been applied to a variety of experimental data. First of all, it is flow data measured in arteries as well as veins and sinuses of the cerebral vessels during neurosurgery in Meshalkin Novosibirsk Research Institute of Circulation Pathology [3]. Secondly, it is experimental data of pressure and velocity obtained during laboratory experiments on elastic model of carotid artery bifurcation using the Combo Flow pump and Combo Wire sensor with Combo Map apparatus [11]. The adequacy of the model (1) describing the relaxation oscillations in a composite system "blood stream – elastic vessel wall – gelatinous brain environment" is justified by the following factors.

4.1. Approximation of clinical data

After finding the $\{a_i, b_j, k\}(i = 0, 1, 2; j = 1, 2, 3)$ coefficients on I_N interval we write down the equation of relaxation oscillations (1) for a particular patient in a specific location of the vessel. Adequacy check for the solution of this equation to the experimental data is carried out as follows: we solve the equation (1) with the well-known right-hand side $ku(t)$ and obtained coefficients $\{a_i, b_j, k\}$. The obtained data shows a very good quality of approximation. Typical clinical plots (black) and calculated (red) ones for the pressure are shown in Figure 1. It is

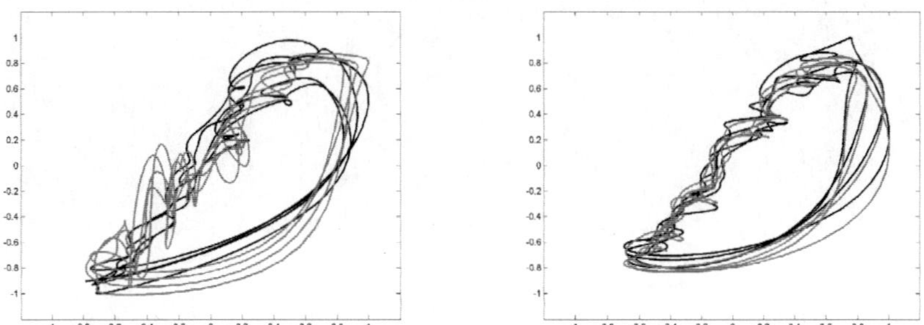

Figure 1. Typical plots of blood pressure in the cerebral artery: the clinical data (black), the calculated data (red).

shown that even the greater approximation property is satisfied, namely the restoration of the coefficients of equation (1) in the long period of time I_{N_1} from the coefficients obtained in the measurement on a smaller interval of I_{N_0}, where $N_1 > N_0$. It is also shown that it is possible to construct the equation (5), which describes the process of embolization. Parameters of blood flow are changing during embolization and, accordingly, the coefficients of equation (5), which describe the physical values of the blood flow, are changing. The equation (1) represents an invariant as the normalized blood flow values do not change. Figure 2 shows the sequence of "pressure – velocity" (p, v) diagrams during the operation of AVM embolization: the clinical data (black), the calculated data (red) [12]. Figure 2 illustrates the success of the operation: diagram drift occurs in the direction of the speed decrease and the pressure increase, and there is a good conformity of clinical and calculated data.

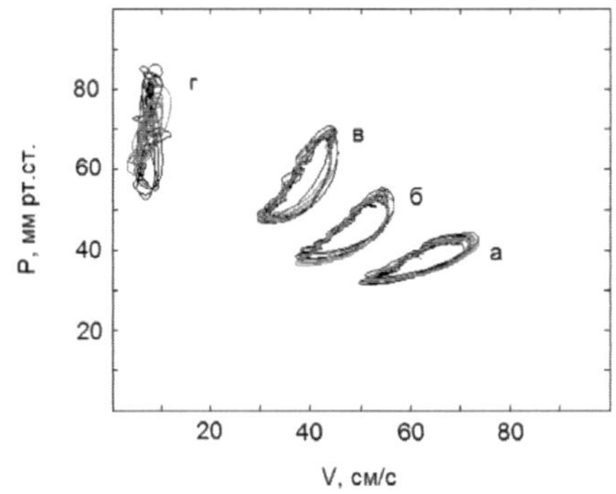

Figure 2. Diagram (p, v) drift during the AVM embolization.

In this case, the coefficients of (5) have been restored to the fifteen minute interval I_{900} according to the five-second interval I_5.

Figure 3 shows plots of velocity and pressure and their average values at 15-minute interval.

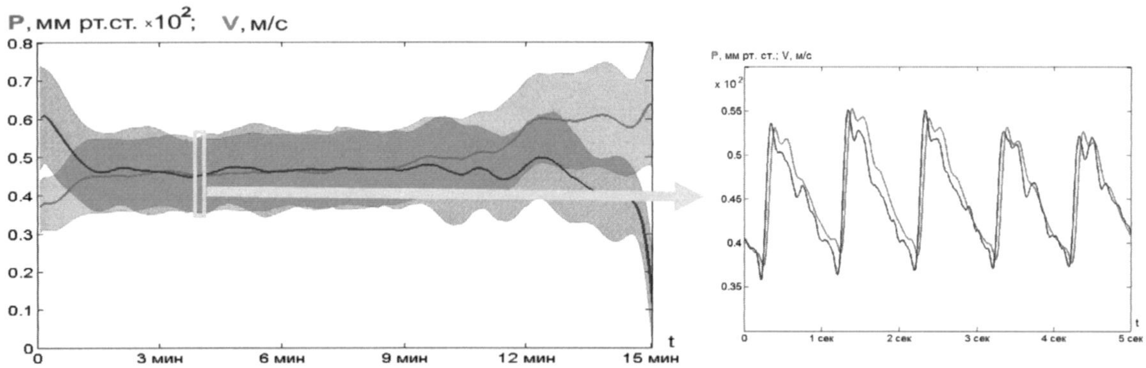

Figure 3. Graphs of velocity and pressure recovered on a large period of time including embolization of AVM.

4.2. The stability of the model relative to the initial data variations

A series of calculations were performed, where the initial data changed for the equation (1): large deviations from the clinical data were set. Calculations have shown that there is a vast area of limit cycle attraction – (q, u) - diagram – corresponding to the diagram (q, q') on the phase plane. This result allows claiming that there is an experimental proof of the asymptotic stability of the solution. A typical calculation of decisions output on the limit attracting cycle with a large deviation from the clinical data is illustrated on Figure 4. Even small perturbations of the initial data are "forgotten" for the time close to the time of the cardiac cycle. A characteristic feature of living systems is the presence of periodic modes of changes in physiological variables, such as rate and blood pressure. This corresponds to the attracting limit cycle presence in the plane (p, v).

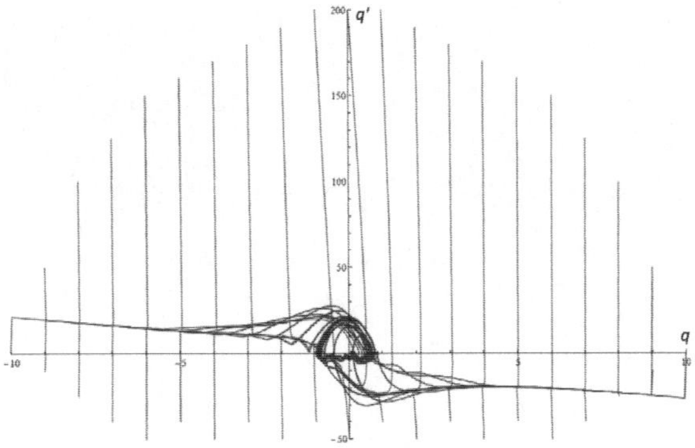

Figure 4. The (q, q') diagram is attractor.

We can assume that a treatable anomaly (such ones recorded during all operations with measurements) is characterized by the presence of attractive cycle and variation of parameters that do not come out of the domain of attraction of the attractor. It appears plausible that the incurable pathology, leading to the disruption of the living system functioning, is responsible for the destruction of periodic solutions.

4.3. The stability of the model relative to the variation of coefficients

Series of calculations was performed for the equation (1), where the deviation of coefficients ai, bj was defined by up to 2% of their values. Calculations showed stability of the solution of equation (1) with respect to such perturbations (Figure 5).

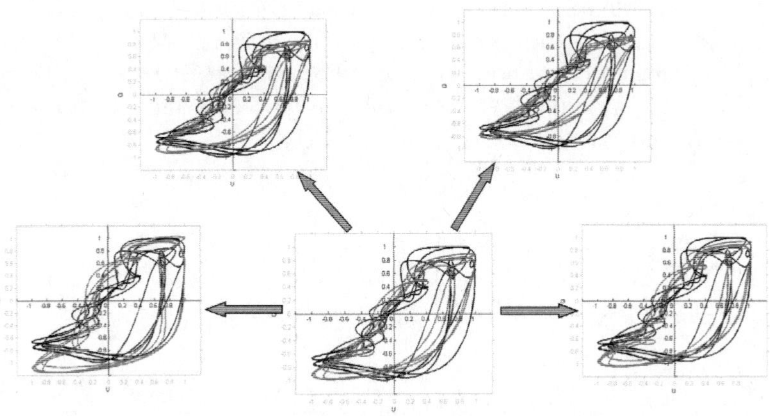

Figure 5. Stability (q, u) of diagrams obtained under the coefficients variation of 2% in the equation (1).

4.4. Model universality for different vessels

Model of relaxation oscillations (1) was tested on the arrays of 137 measurements in the arteries, 9 – in the sinuses, 17 – in the cerebral veins. It accurately describes the relaxation oscillations in the elastic model of the tee with different unsteady flow conditions (40 measurements) [3, 11].

Figure 6 shows (q, u) - diagrams constructed from the experimental (black) and calculated (red) data for various types of composite systems.

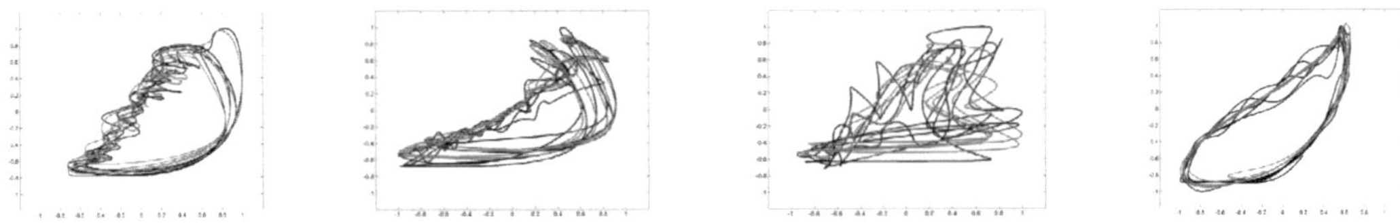

a b c d

Figure 6. (q, u)- diagrams, constructed for arteries (**a**), venous (**b**), sinuses (**c**) of the brain and for the carotid bifurcation elastic model (**d**).

It should be noted that the coincidence of the experimental and calculated data has the highest quality for the arteries and lower quality for the veins and sinuses.

5. Conclusion

In this paper, the mathematical model, which describes the relaxation oscillations of pressure and velocity in the healthy vessels of the brain as well as in the vessels in the presence of abnormalities such as arteriovenous malformations and cerebral aneurysms, was constructed. The model is described by Van der Pol - Duffing type nonlinear oscillator equation (1). It is built implementing the theory of inverse problems for differential equations on the basis of experimental clinical data on pressure and velocity obtained during the endovascular intraoperative monitoring in the Meshalkin clinic. A set of coefficients of the equation (1) describes the individual patient. It is shown that the model is stable with respect to perturbations of the initial data and the coefficients of the equation. The equation for the normalized dimensionless quantities represents the flow invariant. It characterizes the rheology of the composite system "the blood stream – the elastic wall of the vessel – the brain surrounding environment ". The model was tested on data sets of 50 neurosurgical operations and laboratory experiments with elastic tee. Her unconditional advantages are simplicity and versatility. This model allows to describe the characteristics of blood flow in the presence of anomalies.

Acknowledgments

The theoretical research of this paper was supported by the Russian Foundation for Basic Research (Project No. 14-01-00036). The experiments with using Combo Flow were supported by the Russian Science Foundation (Project No. 14-35-00020).

References

[1] Pedley T J 1980 *The fluid Mechanics of Large Blood Vessels*
[2] Tikhonov A N and Arsenin V Ya 1979 *Methods of solving ill-posed problems*
[3] Chupahin A P, Cherevko A A, Khe A K, Telegina N Y, Krivoshapkin A L, Orlov K Y, Panarin V A and Baranov V I 2012 *Circulation Pathology and Cardiac Surgery* **16** 27–31
[4] Cartwright M L and Littlewood J E 1951 *Annals of Mathematics* **54** 1–37
[5] Mishchenko E F and Rozov N Kh 1975 *Differential equations with small parameters and relaxation oscillations*
[6] Pliss V A 1964 *Nonlocal Problems of Vibration Theory*
[7] Schmidt R F and Thews G (eds) 1989 *Human Physiology* **2**
[8] Malla S 2005 *Wavelets in signal processing*
[9] Nelles O 2000 *Nonlinear System Identification*
[10] Redko S F, Ushalov V F and Yakovlev V P 1985 *Identification of mechanic systems*

[11] Boiko A et al. 2015 *Suppl.1* (*Book of Abstracts ESMRMB*) **28** 384
[12] Panarin V A, Orlov K Y, Krivoshapkin A L, Chupahin A P, Cherevko A A, Khe A K, Telegina N Y, and Baranov V I 2012 *Circulation Pathology and Cardiac Surgery* **16** 39–43
[13] Perdikaris P, Grinberg L and Karniadakis G E 2016 *Phys. Fluids* **28** 021304
[14] Krylov V V (ed) 2011 *Surgery of the cerebral aneurysms* **1**

Evolution of localized asymptotic solutions for linearized Navier — Stokes and MHD equations

A I Allilueva[1,2,3] **and A I Shafarevich**[1,2,3,4]

[1]Institute For Problems in Mechanics of RAS, 119526 Moscow, Russia
[2]Moscow Institute of Physics and Technology (State University), 141700 Dolgoprudny, Russia
[3]National Research Centre "Kurchatov Institute", 123182 Moscow, Russia
[4]Moscow State University, 119992 Moscow, Russia

E-mail: esina_anna@list.ru, shafarev@yahoo.com

Abstract. In the paper, the asymptotic behavior of solutions of the Cauchy problem is described for the linearized Navier–Stokes and MHD equations with the initial condition localized in a neighborhood of a two-dimensional surface in three-dimensional space. In particular, conditions for the growth of the perturbation in plane-parallel, two-dimensional, and helical external flows are obtained.

1. Introduction

The evolution of small localized perturbations of the flow of an incompressible fluid was studied in many papers devoted to the theory of hydrodynamic stability. From the mathematical point of view, we speak of properties of solutions of the linearized Euler equations or of the Navier–Stokes equations. In particular, in the works [1] - [5], the asymptotic behavior of solutions of the Cauchy problem was described for the initial condition localized in a neighborhood of a single point (solitary vortex), and the possibility of growth of such a solution in the course of time in diverse external flows was discussed. It turned out that, in plane-parallel flows, only a power-law growth of perturbations is possible, while the growth can be exponential in flows with closed trajectories (a similar result for the spectral problem was obtained earlier by Bailey [6]; the growth conditions generalize the well-known Rayleigh instability conditions [7]).

In the present paper, the asymptotic behavior of a solution of the Cauchy problem for the linearized Navier–Stokes equations was described for the initial condition localized in a neighborhood of a two-dimensional surface in three-dimensional space. The asymptotic solution is also localized near a surface; it is obtained from the initial object by a shift along the trajectories of the external flow. The evolution of the profile of the perturbation is described by a system of linear ordinary differential equations along the trajectories; for diverse classes of external flows, we study in detail the behavior of the leading term of the asymptotic behavior for large times. In particular, conditions for the growth of the perturbation in plane-parallel and two-dimensional external flows are obtained.

We study also analogous localized asymptotic solutions for the linearized MHD equations. The main effect which differs the situation from the previous one, is the effect of interaction of Alfven modes. From the mathematical point of view, we deal here with characteristics of variable multiplicity; for WKB-type solutions this situation was studied in [8]. The solution is

Journal of Physics: Conference Series **722** (2016) 012046 doi:10.1088/1742-6596/722/1/012046

now localized near a number of different surfaces — they correspond to different modes and to the mode interaction. We discuss the structure of the localized solutions as well as their asymptotics as $t \to \infty$. Here we only formulate the results — the proofs are postponed to further publications.

2. Localized solutions for the linearized Navier — Stokes Equations

A small perturbation u of the velocity field of an incompressible viscous fluid (a time-dependent vector field in $\mathbf{R}^3$) satisfies the linearized Navier–Stokes equations

$$\frac{\partial u}{\partial t} + (V, \nabla)u + (u, \nabla)V + \nabla P = \nu\varepsilon^2 \triangle u, \quad (\nabla, u) = 0. \tag{1}$$

Here $V(x)$ is a given smooth divergence-free vector field bounded together with all its derivatives (stationary external flow), P is a perturbation of the pressure, and $\varepsilon^2\nu > 0$ is the viscosity coefficient (below we describe the asymptotic behavior of solutions of this system as $\varepsilon \to 0$ in different situations; it is convenient to single out the parameter ν to clarify the influence of viscosity on the leading part of the asymptotics). We will suppose that V satisfies Euler equations

$$(V, \nabla)V + \nabla P_0 = 0, \quad (\nabla, V) = 0.$$

Let us pose the Cauchy problem for system (1) with a localized initial condition

$$u|_{t=0} = \hat{u}_0 \left(\Phi^0(x)/\varepsilon, x\right) \tag{2}$$

where $\hat{u}_0$ is a smooth compactly supported divergence-free vector field and $\Phi^0(x)$ is a smooth scalar function bounded together with all its derivatives, and the function $\hat{u}_0(y, x)$ decreases as $|y| \to \infty$ more rapidly than any power of $|y|$ (here y stands for the "fast" variables $y = \Phi^0(x)/\varepsilon$). We suppose that the equation $\Phi^0 = 0$ defines a regular smooth compact two-dimensional surface M_0 with $\nabla\Phi^0 \neq 0$; the initial condition as $\varepsilon \to 0$ is localized near this surface.

We study the asymptotic behavior of the Cauchy problem (1) — (2) as $\varepsilon \to 0$.

Remark 1. *The case of initial condition localized near a single point or near a curve was discussed in [1, 4, 5, 9].*

3. Leading term of the asymptotic solution

Introduce the following notation. Let $X(\xi, t)$ be the trajectory of the field V issuing from a point ξ, i.e., the solution of the Cauchy problem

$$\dot{X} = V(X, t), \qquad X|_{t=0} = \xi,$$

and let $x_0(x, t)$ be the initial point of the trajectory which came at time t to the point x, i.e., the solution of the equations $X(x_0, t) = x$ (under the conditions formulated above, $X(\xi, t)$ and $x_0(x, t)$ are smooth functions on $\mathbb{R}^3 \times [0, T]$). The shift of the initial surface M_0 along the trajectories is defined by the equation $\Phi(x, t) = 0$, where $\Phi(x, t) = \Phi^0(x_0(x, t))$. Denote by $k(\xi, t)$ the cotangent vector moving along a trajectory X and annihilating the tangent plane to M_t,

$$k(\xi, t) = \left((\partial X/\partial\xi)^*\right)^{-1} k^0, \qquad k^0 = \nabla\Phi^0(\xi).$$

Finally, denote by $Q(\xi, t)$ the fundamental matrix of the system of equations

$$\dot{A} + Q(k^0, \xi, t)A = 0, \quad Q = \left(E - 2\frac{k \otimes k}{k^2} V_x(\xi, t)\right).$$

Theorem 1. *Let*

$$U(x,t,y) = A(\xi,t)\int_{-\infty}^{\infty} e^{iyp}e^{-\nu p^2\int_0^t k^2dt}\tilde{u}_0(p)dp|_{\xi=x_0(x,t)}$$

where $\tilde{u}_0$ is the Fourier transform of the function $\hat{u}_0$ with respect to the variable y.

The solution $u(x,t,\varepsilon)$ of the Cauchy problem (2.1), (2.2) on every chosen time interval $t \in [0,T]$ can be represented in the form

$$u(x,t,\varepsilon) = U(x,t,\Phi(x,t)/\varepsilon) + w, \quad \text{where} \quad |w| \to 0 \quad \text{as} \quad \varepsilon \to 0.$$

4. Time evolution of localized perturbations

From the point of view of the theory of hydrodynamic stability, the behavior of a localized solution for large times t is of interest, namely, if the perturbation grows, then the external flow V is unstable with respect to these localized perturbations. Below we consider the asymptotic behavior as $t \to \infty$ of the leading term of the asymptotics, i.e., of the vector field U. Recall that an asymptotic solution approximates the exact one on a finite time interval $t \in [0,T]$. Thus, we first proceed to the limit of the exact solution as $\varepsilon \to 0$, and then as $t \to \infty$; from the physical point of view, the problem under consideration concerns the behavior of a solution at times which are large but much less than ε^{-r}, $r > 0$.

The time evolution of a localized perturbation depends substantially on the properties of the external flow $V(x)$; below we consider classes of fields which are well known in hydrodynamics, namely, the plane-parallel flows and two-dimensional flows with closed trajectories.

4.1. Flows with constant direction

Let $V(x) = (0,0,V_3(x_1,x_2))$. The trajectories of this vector field are straight lines parallel to the x_3 axis, and the velocity of the motion of liquid particles varies from line to line.

Assertion 1. *Let V be a flow with constant direction and $k_3^0 = \partial\Phi^0/\partial x_3 = 0$ at some point ξ in a neighborhood of the initial surface M. Then the function U_3 has the following asymptotic behavior as $t \to \infty$ at the point $x = X(\xi,t)$:*

$$U_3 = -t\Big(\hat{u}_0(\Phi^0/\varepsilon,\xi), \nabla V_0(\xi)\Big) + O(1) \quad \text{for} \quad \nu = 0,$$

$$U_3 = -\sqrt{\frac{t}{4\pi\nu}}\frac{e^{-\frac{(\Phi^0)^2(\xi)}{4t\varepsilon^2\nu_0(\nabla\Phi^0)^2}}}{|\nabla\Phi^0(\xi)|}\int_{-\infty}^{\infty}(\hat{u}_0(y,\xi),\nabla V_0(\xi))dy + O(1) \quad \text{for} \quad \nu > 0.$$

The functions U_1, U_2 remain bounded.

Thus, we obtain the following picture of the evolution of a localized perturbation. The leading term of the asymptotic behavior of the localized solution remains bounded on the trajectories of the field V issuing from the points ξ at which $k_3^0 = \partial\Phi^0/\partial x_3 \neq 0$. The vertical component of the asymptotic solution increases on the trajectories of points at which $k_3^0 = \partial\Phi^0/\partial x_3 = 0$ as $\sqrt{t}$ if the viscosity is high enough (if the fluid is perfect, i.e., $\nu = 0$, then the growth is linear). The horizontal components remain bounded.

It is clear that, for $\xi \in M_t$, the growth condition $\partial\Phi^0/\partial x_3 = 0$ means that the tangent plane to M at the given point contains the vertical direction, i.e., the velocity vector of the external flow is tangent to M. For example, if the initial perturbation is concentrated on the standardly embedded sphere, then the growth will be observed along the trajectories issued from the points on the equator (and also from the points placed at a distance of the order of ε from the equator).

4.2. Plane-parallel flows with a shift of the direction

Let the external flow be of the form $V = (V_1(x_3), V_2(x_3), 0)$. Its trajectories are straight lines lying in the horizontal planes $x_3 = \text{const}$; not only the velocity, but also the direction of the flow changes from plane to plane (a shear).

Assertion 2. *Let V be a plane-parallel flow with a shift of the direction, and let $(V', k^0) = (V', \nabla\Phi^0) = 0$ at some point ξ. Then the function U has the following asymptotic behavior:*

$$U = -tV'(\xi)\hat{u}_{03}(\Phi^0(\xi)/c, \xi) + O(1) \quad \textit{for} \quad \nu = 0;$$

$$U = -V'(\xi)\sqrt{\frac{t}{4\pi\nu}}\frac{1}{|\nabla\Phi^0(\xi)|}\,\mathrm{e}^{-(\Phi^0)^2/(4t\varepsilon^2(\nabla\Phi^0)^2\nu)}\int_{-\infty}^{\infty}\hat{u}_{03}(y, \xi)dy + O(1) \quad \textit{for} \quad \nu > 0.$$

At the points at which $(V', \nabla\Phi^0) \neq 0$, the function U remains bounded, while, for $\nu_0 \geq \text{const} > 0$, U decreases as $O(t^{-3/2})$. The condition ensuring the growth of the perturbation also determines a curve (in general) on the initial surface; in particular, if M is a sphere centered at the origin, the growth will occur along the trajectories issuing from the line of intersection of the sphere with the helical surface

$$x_1V_1' + x_2V_2' = 0.$$

The rate of growth on this curve is still $O(\sqrt{t})$ if ν is bounded away from zero and $O(t)$ if $\nu = 0$.

4.3. Two-dimensional flows with closed stream lines

Consider now the third class of stationary two-dimensional Euler flows, namely, two-dimensional vector fields $V = (V_1(x_1, x_2), V_2(x_1, x_2), 0)$. The condition that such a field on a plane is divergence-free means that the field is Hamiltonian,

$$V = (-\frac{\partial\Psi}{\partial x_2}, \frac{\partial\Psi}{\partial x_1});$$

in hydrodynamics, the Hamilton function Ψ is referred to as the *stream function.* As is well known, the curl $\Delta\Psi$ of the two-dimensional Euler flow is constant on its trajectories, and the Euler equations are reduced to the equation $\{\Psi, \Delta\Psi\} = 0$, where $\{,\}$ stands for the Poisson bracket.

Consider a nonsingular closed trajectory $X(\xi, t)$ of this field,

$$X(\xi, t + T) = X(\xi, t),$$

where $T(\xi)$ stands for the period of the trajectory. Assume that, as t increases, the trajectory is going around counterclockwise. Since V is divergence-free, it follows that the nearby trajectories of this field are also closed, and thus, the curve $X(\xi, t)$ belongs to the domain fibered into periodic trajectories.

In this case, the matrix A can grow exponentially; the set on which the growth takes place is defined by the following conditions (see [9]).

(*i*) $X(\xi, t)$ is a convex curve,

(*ii*) the circulation of the field V along the trajectory $X(\xi, t)$ decreases outwards,

$$\Delta\Psi(\xi) < 0;$$

(*iii*)

$$(\nabla\Phi^0, V(\xi)) = 0, \qquad \partial\Phi^0/\partial x_3(\xi) \neq 0.$$

Let us suppose that these conditions are fulfilled. The equations for A now have periodic coefficients and the corresponding monodromy operator has two distinct real eigenvalues of the form $\mathrm{e}^{\pm\sigma(\xi)T}$, $\sigma > 0$, where T is the period of the trajectory. We denote by $e_\pm(\xi, t)\,\mathrm{e}^{\pm\sigma(\xi)t}$, $e_\pm(\xi, t + T) = e(\xi, t)$ the basis of the Floquet solutions.

Assertion 3. *Let V be a two-dimensional Eulerian field with closed current lines and let its trajectory $X(\xi, t)$ issuing from a point ξ of a neighborhood of the surface M satisfy the requirements $(i)-(iii)$. Then the function U has the following asymptotic behavior as $t \to \infty$ at the point $X(\xi, t)$ of the neighborhood of M_t:*

$$U(\xi,t) = \mathrm{e}^{\sigma t}\left(e_+(\xi,t)e^+\left[\hat{u}_0\left(\frac{\Phi^0(\xi)}{\varepsilon},\xi\right)\right] + o(1)\right) \quad for \quad \nu = 0,$$

$$U(\xi,t) = \frac{\mathrm{e}^{\sigma t}}{\sqrt{t}}\Big(\frac{\mathrm{e}^{-(\Phi^0)^2/(4\varepsilon^2\nu_0<k^2>)}}{\sqrt{4\pi\nu<k^2>}}e_+(\xi,t)\int_{-\infty}^{\infty} e^+[\hat{u}_0(z,\xi)]dz + O(1/\sqrt{t})\Big) \quad for \quad \nu > 0,$$

where $\mathrm{e}_0^{\pm}(\xi)$ are the elements of the dual basis to the basis $\mathrm{e}_{\pm}^0(\xi)$, and

$$\langle k^2\rangle(\xi) = \frac{1}{T}\int_0^T k^2(\xi,t)dt.$$

Remark 2. *Thus, in the present case, we observe a strong (exponential) growth of a localized perturbation, and the viscous decay cannot hinder this growth. The sufficient conditions for growth, as in the case of flows with constant direction, define a set of points of the initial surface for which the tangent plane contains a vector of the external flow V (generally speaking, this is a curve). For example, consider a radially symmetrical field as an external flow: $\Psi = \Psi(r)$, $V = \omega(r)\partial/\partial\varphi$, where r, φ are the polar coordinates on the plane x_1, x_2. Let the initial surface M be the sphere $|x - x^0|^2 = a^2$, be contained entirely in an area in which the trajectories satisfy condition (ii), which, in this case, coincides with the Rayleigh instability condition $\Phi(r) < 0$, where Φ stands for the Rayleigh function*

$$\Phi(r) = \frac{1}{r^3}\frac{d}{dr}(r^4\omega^2(r)).$$

The field V is tangent to the sphere M at the points of the meridian obtained as the intersection of the sphere with the vertical plane containing the radius vector of the center of the sphere; on the trajectories issuing from this meridian, the initial perturbation grows exponentially.

5. The MHD equations

Now we discuss briefly the linearized MHD equations

$$\begin{cases} \dfrac{\partial v}{\partial t} + (V,\nabla)v + (v,\nabla)V + \nabla P = (B,\nabla)b + (b,\nabla)B + \varepsilon^2\nu\Delta u, \\ \dfrac{\partial b}{\partial t} + (V,\nabla)b + (v,\nabla)B = (B,\nabla)v + (b,\nabla)V + \varepsilon^2\mu\Delta b, \\ (\nabla, u) = (\nabla, b) = 0. \end{cases} \tag{3}$$

We suppose that V, B have the same analytic properties as the vector field V from the previous sections; moreover, we assume that these vector fields satisfy stationary ideal MHD equations

$$(V,\nabla)V - (B,\nabla)B + \nabla P_0 = 0, \quad (V,\nabla)B - (B,\nabla)V = 0, \quad (\nabla, V) = (\nabla, B) = 0.$$

Let us pose the Cauchy problem with a localized initial condition

$$v|_{t=0} = v_0(\frac{\Phi^0(x)}{\varepsilon}, x), \quad b|_{t=0} = b_0(\frac{\Phi^0(x)}{\varepsilon}, x), \tag{4}$$

where the properties of the initial functions are analogous to those from the previous sections. During the evolution of the perturbation, the initial surface is divided into two surfaces M_1^t,

M_2^t defined by the equations $\Phi_1(x,t)=0$ and $\Phi_2(x,t)=0$. The surface M_1 moves along the trajectories of the field $V-B$, and the surface M_2 — along the trajectories of the field $V+B$. Namely, let us denote by $x_{1,2}^0(x,t)$ the initial point of the trajectory of $V\pm B$ which comes to the point x during the time t; the functions $\Phi_{1,2}$ have the form $\Phi_{1,2}(s,t)=\Phi^0(x_{1,2}^0(x,t))$. The interaction of these two modes is described by two functions $\Phi_{12}(x,t,\tau)$ and $\Phi_{21}(x,t,\tau)$ which have the form

$$\Phi_{12}(x,t,\tau)=\Phi^0(x_1^0(x_2^0(x,\tau),t-\tau), \quad \Phi_{12}(x,t,\tau)-\Phi^0(x_2^0(x_1^0(x,\tau),t-\tau).$$

Note that for each $\tau\in[0,t]$ the equation $\Phi_{12}=0$ defines the surface which is obtained from M_0 by the shift during the time τ along the trajectories of $V+B$ and then during the time $t-\tau$ — along the trajectories of $V-B$. The structure of the asymptotic solution has the following form.

Theorem 2. *Solution of the Cauchy problem* (3) — (4) *has the form* $v=v^0+w, b=b^0+q$, $w,q=O(\varepsilon)$, *where*

$$v^0=v_1(\frac{\Phi_1(x,t)}{\varepsilon},x,t)+v_2(\frac{\Phi_2(x,t)}{\varepsilon},x,t)+$$
$$+\int_0^t v_{12}(\frac{\Phi_{12}(x,t,\tau)}{\varepsilon},x,t,\tau)d\tau+\int_0^t v_{21}(\frac{\Phi_{21}(x,t,\tau)}{\varepsilon},x,t,\tau)d\tau,$$
$$b^0=b_1(\frac{\Phi_1(x,t)}{\varepsilon},x,t)+b_2(\frac{\Phi_2(x,t)}{\varepsilon},x,t)+$$
$$+\int_0^t b_{12}(\frac{\Phi_{12}(x,t,\tau)}{\varepsilon},x,t,\tau)d\tau+\int_0^t b_{21}(\frac{\Phi_{21}(x,t,\tau)}{\varepsilon},x,t,\tau)d\tau.$$

The functions v_1,b_1 *and* v_2,b_2 *satisfy the Cauchy problems for ordinary differential equations along the trajectories of* $V-B$ *and* $V+B$, *while the functions* v_{12},b_{12} *and* v_{21},b_{21} *satisfy the Goursat problems for hyperbolic systems of equations with times* τ *and* $\eta=t-\tau$. *The lengths of the vectors* v_1,v_2,b_1,b_2 *are constant along the corresponding trajectories (i.e. the growth of the solutions can be produced only by* $v_{12},b_{12},v_{21},b_{21}$*).*

Remark 3. *The vector fields* v_1,b_1 *and* v_2,b_2 *describe the Alfven modes while the vector fields* v_{12},b_{12} *and* v_{21},b_{21} *describe the interaction of the modes.*

6. Localization and growth of the solutions

Let us consider three different situations.

(i) $(B,\nabla\Phi_0)|_{M_0}\neq 0$ on the support of the initial vector fields. In this case the interaction is small: $\int_0^t v_{12}=O(\varepsilon)$, $\int_0^t v_{21}=O(\varepsilon)$, $\int_0^t b_{12}=O(\varepsilon)$, $\int_0^t b_{21}=O(\varepsilon)$. The leading term of asymptotics is localized near two surfaces M_1^t and M_2^t and there is no growth of the perturbations (the lengths of the corresponding vectors are conserved).

(ii) $(B_0,\nabla\Phi_0)|_{M_0}\equiv 0$. The interaction is very strong — the leading term of asymptotics has the form $\hat{v}(\frac{\Phi(x,t)}{\varepsilon},x,t)$ and is localized near the surface M^t, obtained from M^0 via shift along the trajectories of V. For certain fileds B,V the function $\hat{v}$ can grow in time (exponentially or as some power of t).

(iii) $(B_0,\nabla\Phi_0)|_{M_0}=0$ on a curve $\gamma_0\subset M_0$. In this case $\int_0^t v_{12},\int_0^t v_{21},\int_0^t b_{12},\int_0^t b_{21}=O(\sqrt{\varepsilon})$. These fields are localized near surfaces

$$M_{12}=\cup_{\tau\in[0,t]}g^{t-\tau}h^{\tau}\gamma_0$$

and

$$M_{21} = \cup_{\tau \in [0,t]} g^{t-\tau} h^{\tau} \gamma_0.$$

Here g^s and h^s denote shifts along the trajectories of $V-B$ and $V+B$ during the time s. The functions $\int_0^t v_{12}, \int_0^t v_{21}, \int_0^t b_{12}, \int_0^t b_{21}$ can grow in time for certain fields V, B (exponentially or as some power of t).

Acknowledgments

The work was supported by the Russian Science Foundation (grant 16-11-10282).

References

[1] Dobrokhotov S Yu and Shafarevich A I 1992 *Mathematical Notes* **51** 47–54
[2] Vishik M M and Friedlander S 1991 *Phys. Rev. Lett.* **66**(17) 2204–6
[3] Jain S C, Willander M, Narayan J and Van Overstraeten R 2000 *J. Appl. Phys.* **87** 965
[4] Dobrokhotov S Yu and Shafarevich A I 1993 *Mathematical Notes* **53** 19–26
[5] Shafarevich A I 1994 *Mathematical Notes* **55** 632–47
[6] Bayly B J 1988 *Phys. Fluids* **31** 56–64
[7] Rayleigh J W S 1916 *Proc. R. Soc. Lond. Ser. A.* B **93** 148–54
[8] Kucherenko V V and Kryvko A 2013 *Russian Journal of Math. Phys.* **20**(1) 56–67
[9] Allilueva A I and Shafarevich A I 2015 *Russian Journal of Math. Phys.* **22**(4) 431–6

Singular path-independent energy integrals for elastic bodies with thin elastic inclusions

V V Shcherbakov[1,2]

[1] Lavrentyev Institute of Hydrodynamics, 630090 Novosibirsk, Russia
[2] Novosibirsk State University, 630090 Novosibirsk, Russia

E-mail: sherbakov87@gmail.com

Abstract. An equilibrium problem for a two-dimensional homogeneous linear elastic body containing a thin elastic inclusion and an interfacial crack is considered. The thin inclusion is modeled within the framework of Euler–Bernoulli beam theory. An explicit formula for the first derivative of the energy functional with respect to the crack perturbation along the interface is presented. It is shown that the formulas for the derivative associated with translation and self-similar expansion of the crack are represented as path-independent integrals along smooth contour surrounding one or both crack tips. These path-independent integrals consist of regular and singular terms and are analogs of the well-known Eshelby–Cherepanov–Rice J-integral and Knowles–Sternberg M-integral.

1. Formulation of the problem

Let Ω be a bounded domain in $\mathbb{R}^2$ with smooth boundary $\partial\Omega$, and let γ be the straight-line segment $(0,2)\times\{0\}$, $\overline{\gamma}\subset\Omega$. Denote by γ_0 the set $(0,1)\times\{0\}$, and consider the domain $\Omega_\gamma=\Omega\backslash\overline{\gamma}$. The geometrical setup is shown in Figure 1.

The domain Ω_γ is occupied by an elastic material that obeys the linear Hooke law, and γ corresponds to a thin elastic inclusion. We model a thin inclusion as a beam governed within the framework of Euler–Bernoulli beam theory. The inclusion is delaminated on γ_0^+ so that there is a crack on the interface between two media, while the part $\gamma\setminus\gamma_0$ is adhered perfectly to the elastic body.

The equilibrium problem reads as follows [5]. We want to find a displacement field $u=(u_1,u_2)$, a stress tensor $\sigma=\{\sigma_{ij}\}$, $i,j=1,2$, defined in Ω_γ, and thin inclusion displacements v, w defined on γ such that

$$-\operatorname{div}\sigma=f,\quad \sigma-A\varepsilon(u)=0\quad\text{in}\quad\Omega_\gamma, \tag{1}$$

$$v_{,1111}=[\sigma_{22}]\quad\text{on}\quad\gamma, \tag{2}$$

$$-w_{,11}=[\sigma_{12}]\quad\text{on}\quad\gamma, \tag{3}$$

$$u=0\quad\text{on}\quad\partial\Omega, \tag{4}$$

$$v_{,11}=v_{,111}=w_{,1}=0\quad\text{for}\quad x_1=0,2, \tag{5}$$

$$[u]=0\quad\text{on}\quad\gamma\setminus\gamma_0, \tag{6}$$

$$v=u_2^-,\ w=u_1^-\quad\text{on}\quad\gamma, \tag{7}$$

$$\sigma_{22}^+=0,\ \sigma_{12}^+=0\quad\text{on}\quad\gamma_0. \tag{8}$$

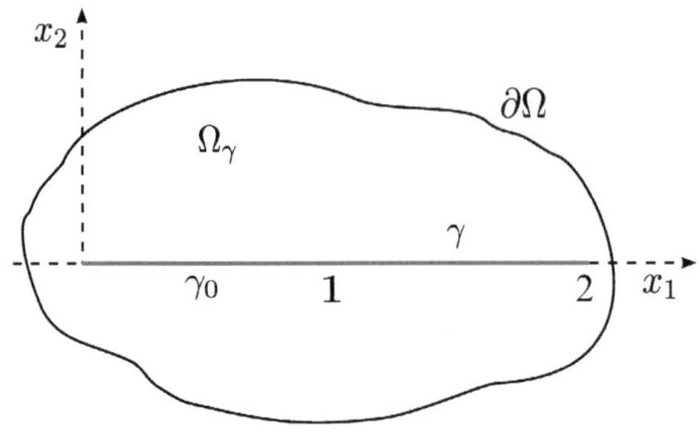

Figure 1. Geometrical setup of the problem.

Here, $\varepsilon = \{\varepsilon_{ij}(u)\}$ is the infinitesimal strain tensor, $\varepsilon_{ij}(u) = 1/2(u_{i,j} + u_{j,i})$. The elasticity tensor $A = \{a_{ijkl}\}$ is assumed to be symmetric and positive definite, i.e.

$$a_{ijkl} = a_{jikl} = a_{klij},$$

$$a_{ijkl}\xi_{ij}\xi_{kl} \geq c_A\xi_{ij}\xi_{ij}, \quad \xi_{ij} = \xi_{ji}, \quad c_A = \text{const} > 0.$$

For simplicity, we assume that a_{ijkl} are constants, and summation over repeated indices is employed. We identify functions defined on γ with the functions of variable x_1. In accordance with the positive and negative directions of the x_2 axis, there are positive face γ^+ and negative γ^-. Then the quantity $[p] = p^+ - p^-$ denotes the jump of p across γ, and $p^\pm$ are the traces of p on $\gamma^\pm$. Finally, $f = (f_1, f_2) \in C^1(\overline{\Omega})^2$ is the density of the applied body forces. The first relations in (1) are the equilibrium equations for the elastic body Ω_γ, while the second relations are the linear Hooke law. Equations (2) and (3) are the Euler–Bernoulli governing equations for the inclusion γ. The right-hand sides of equations (2) and (3) represent the forces acting on γ from the surrounding elastic media. The inclusion displacements coincide with the vertical and the tangential displacements of the elastic body on γ^- in conformity with (7).

The problem (1)–(8) admits a variational formulation. Indeed, let us introduce the space

$$H^1_{\partial\Omega}(\Omega_{\gamma_0}) = \{\, u \in H^1(\Omega_{\gamma_0}) \mid u = 0 \text{ on } \partial\Omega \,\},$$

and the set of admissible displacements

$$K^0 = \left\{\, (u, v, w) \in H^1_{\partial\Omega}(\Omega_{\gamma_0})^2 \times H^2(\gamma) \times H^1(\gamma) \mid v = u_2^-,\ w = u_1^- \text{ on } \gamma \,\right\}.$$

Then problem (1)–(8) is equivalent to minimization of the energy functional

$$\Pi(u, v, w) = \frac{1}{2}\int\limits_{\Omega_{\gamma_0}} \sigma_{ij}(u)\varepsilon_{ij}(u)\, dx - \int\limits_{\Omega_{\gamma_0}} f_i u_i\, dx + \frac{1}{2}\int\limits_{\gamma} v_{,11}^2\, dx_1 + \frac{1}{2}\int\limits_{\gamma} w_{,1}^2\, dx_1$$

on the set K^0 and can be written in the form of the Euler equation

$$(u, v, w) \in K^0, \quad \int\limits_{\Omega_{\gamma_0}} \sigma_{ij}(u)\varepsilon_{ij}(\bar{u})\, dx - \int\limits_{\Omega_{\gamma_0}} f_i\bar{u}_i\, dx$$
$$+ \int\limits_{\gamma} v_{,11}\bar{v}_{,11}\, dx_1 + \int\limits_{\gamma} w_{,1}\bar{w}_{,1}\, dx_1 = 0 \quad \forall(\bar{u}, \bar{v}, \bar{w}) \in K^0. \quad (9)$$

Since the functional Π is coercive and weakly lower semicontinuous on the set K^0, equation (9) has a (unique) solution.

2. Regularity of the solution

We discuss local higher regularity of the solution (u, v, w) to the Euler equation (9). First it should be noted that the H^2-regularity of u inside Ω_γ and near $\partial\Omega$ follows from standard regularity theory for uniformly elliptic differential operators (see, e.g. [4]). We prove the following statement.

Theorem 1. Let $x^0 \in \gamma \setminus \{(1,0)\}$ be any fixed point. Then there exists a neighborhood $S(x^0)$ of the point x^0 such that the solution of (9) satisfies

$$u \in H^2(S(x^0) \cap \Omega_\gamma)^2, \quad v \in H^4(S(x^0) \cap \gamma), \quad \text{and} \quad w \in H^2(S(x^0) \cap \gamma).$$

The proof of Theorem 1 relies upon difference quotient approximations for weak derivatives.

3. Derivative of the energy functional

In this section, we present a formula for the first derivative of the energy functional with respect to the perturbation of the crack γ_0. The first goal of further considerations is to introduce a small perturbation parameter into the problem (9), i.e. to consider a family of perturbed problems depending on the parameter δ. Let the transformation of independent variables

$$y = T_\delta(x), \quad x \in \Omega_{\gamma_0}, \quad y \in \Omega_{\gamma_\delta}, \tag{10}$$

describe the perturbation of the domain Ω_{γ_0}, where $T_\delta(x) = x + \delta V(x)$, $V(x) = (V_1(x), V_2(x))$, and $V \in W^{1,\infty}(\Omega)^2 \cap W^{2,\infty}(\gamma)^2$. The perturbed domain $\Omega_{\gamma_\delta} = T_\delta(\Omega_{\gamma_0})$ containing the perturbed crack $\gamma_\delta = T_\delta(\gamma_0)$ for each fixed δ. The vector field V is supposed to be compactly supported in Ω and such that $\gamma = T_\delta(\gamma)$ and $\gamma_\delta \subset \gamma$. Consequently, we have $\Omega_{\gamma_\delta} = \Omega \setminus \bar{\gamma}_\delta$.

We define the set of admissible displacements associated with the perturbed problem as follows:

$$K^\delta = \{\, (u, v, w) \in H^1_{\partial\Omega}(\Omega_{\gamma_\delta})^2 \times H^2(\gamma) \times H^1(\gamma) \mid v = u_2^-,\ w = u_1^- \text{ on } \gamma \,\}.$$

As before, a unique solution of the Euler equation exists

$$(u^\delta, v^\delta, w^\delta) \in K^\delta, \quad \int\limits_{\Omega_{\gamma_\delta}} \sigma_{ij}(u^\delta)\varepsilon_{ij}(\bar{u})\, dy - \int\limits_{\Omega_{\gamma_\delta}} f_i \bar{u}_i\, dy + \int\limits_{\gamma} v^\delta_{,11}\bar{v}_{,11}\, dy_1 + \int\limits_{\gamma} w^\delta_{,1}\bar{w}_{,1}\, dy_1 = 0 \quad \forall (\bar{u}, \bar{v}, \bar{w}) \in K^\delta. \tag{11}$$

For the solution $(u^\delta, v^\delta, w^\delta)$ to equation (11), we can find the potential deformation energy

$$\Pi^\delta(u^\delta, v^\delta, w^\delta) = \frac{1}{2}\int\limits_{\Omega_{\gamma_\delta}} \sigma_{ij}(u^\delta)\varepsilon_{ij}(u^\delta)\, dy - \int\limits_{\Omega_{\gamma_\delta}} f_i u_i^\delta\, dx + \frac{1}{2}\int\limits_{\gamma} \left(v^\delta_{,11}\right)^2 dy_1 + \frac{1}{2}\int\limits_{\gamma} \left(w^\delta_{,1}\right)^2 dy_1.$$

The main question in this section is as follows: does the first derivative of the energy functional with respect to the crack perturbation exist? More precisely, we refer to the existence of the limit

$$\mathcal{D}_V \Pi(u, v, w) = \lim_{\delta \to +0} \frac{\Pi^\delta(u^\delta, v^\delta, w^\delta) - \Pi(u, v, w)}{\delta}.$$

and an explicit representation for it. The following theorem gives a positive answer to this question.

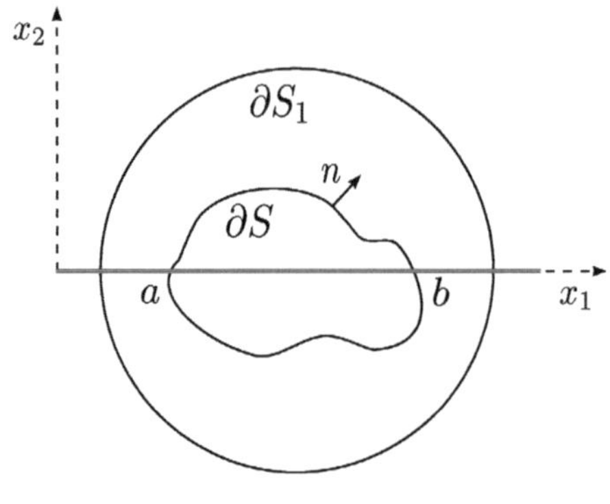

Figure 2. The integration path ∂S surrounding the crack tip.

Theorem 2. The first derivative of the energy functional with respect to the crack perturbation exists and equals

$$\mathcal{D}_V \Pi(u,v,w) = \int\limits_{\Omega_{\gamma_0}} \left(\frac{1}{2} \operatorname{div} V \sigma_{ij}(u) \varepsilon_{ij}(u) - \sigma_{ij}(u) E_{ij}\left(\frac{\partial V}{\partial x}; u\right)\right) dx - \int\limits_{\Omega_{\gamma_0}} \operatorname{div}(V f_i) u_i \, dx$$
$$- \int\limits_{\gamma} \left(\frac{3}{2} V_{1,1} v_{,11}^2 + V_{1,11} v_{,11} v_{,1} + \frac{1}{2} V_{1,1} w_{,1}^2 \right) dx_1, \quad (12)$$

where the components of the transformed strain tensor E_{ij} have the form

$$E_{ij}\left(\frac{\partial V}{\partial x}; u\right) = \frac{1}{2}\left(u_{i,k} V_{k,j} + u_{j,k} V_{k,i}\right), \quad i,j = 1,2.$$

4. Singular path-independent integrals

Let us now consider some particular cases of choosing the vector field V which will yield singular path-independent integrals by means of transformations of formula (12). In all cases, we will have to choose neighborhoods S and S_1 with smooth (Lipschitz) boundaries ∂S and ∂S_1. In what follows, we assume that the boundaries of the domains $(S_1 \setminus S) \cap \Omega_\gamma$ also satisfy the Lipschitz condition and $f \equiv 0$ in $S \cap \Omega_\gamma$. The path-independent integrals are meaningful due to the regularity result from Section 2.

Let the support of a smooth function θ lie in a small neighborhood S_1 of the point $(1,0)$ and $\theta = 1$ in a neighborhood S of the point $(1,0)$, $S \subset S_1$. The path ∂S intersects the axis x_1 at the points with the abscissae a and b as shown in Figure 2. Denote by $n = (n_1, n_2)$ the outward normal vector to ∂S. We choose the coordinate transformation (10) in the form

$$y_1 = x_1 + \delta\theta(x_1, x_2), \quad y_2 = x_2, \quad (13)$$

where $(x_1, x_2) \in \Omega_{\gamma_0}$ and $(y_1, y_2) \in \Omega_{\gamma_\delta}$. The vector field V is determined by the formula $V(x) = (\theta(x), 0)$, and formula (12) can be rewritten as

$$\mathcal{D}_{(\theta,0)} \Pi(u,v,w) = \int\limits_{\Omega_{\gamma_0}} \left(\frac{1}{2} \theta_{,1} \sigma_{ij}(u) \varepsilon_{ij}(u) - \sigma_{ij}(u) u_{i,1} \theta_{,j} \right) dx - \int\limits_{\Omega_{\gamma_0}} (\theta f_i)_{,1} u_i \, dx$$
$$- \frac{1}{2} \int\limits_{\gamma} \left(3\theta_{,1} v_{,11}^2 + 2\theta_{,11} v_{,11} v_{,1} + \theta_{,1} w_{,1}^2 \right) dx_1. \quad (14)$$

The formula (14) is an analog of the well-known Griffith formula for the energy release rate associated with crack extension.

Let the map $J_{\rm B} : (0,2) \to \mathbb{R}$ be given by

$$J_{\rm B}(x_1) = \frac{1}{2}v_{,11}^2(x_1) - v_{,111}v_{,11}(x_1) + \frac{1}{2}w_{,1}^2(x_1).$$

Integrating by parts in (14) and taking into account (1)–(8), we obtain the singular J-integral

$$\mathcal{J}_{\rm tip} = \int\limits_{\partial S} \left(-\frac{1}{2}\sigma_{ij}(u)\varepsilon_{ij}(u)n_1 + \sigma_{ij}(u)u_{i,1}n_j \right) ds + J_{\rm B}(b) - J_{\rm B}(a). \tag{15}$$

It is immediate from the construction that $\mathcal{J}_{\rm tip}$ does not depend on the integration path ∂S (and hence on the points a and b). In other words, the singular J-integral (15) has the same value for all closed curve surrounding the crack tip. The J-integral obtained consists of two parts. The first is a regular part that coincide with the classical Eshelby–Cherepanov–Rice J-integral, and the second is a singular part that depends on the displacement field (v, w) of the inclusion at the points where the path ∂S intersects the segment γ.

Let the coordinate transformation (10) have the form

$$y_1 = x_1(1 + \delta\theta(x_1, x_2)), \quad y_2 = x_2(1 + \delta\theta(x_1, x_2)), \tag{16}$$

where $(x_1, x_2) \in \Omega_{\gamma_0}$ and $(y_1, y_2) \in \Omega_{\gamma_\delta}$. In this case, the vector field V is determined by the formula $V(x) = (x_1\theta(x), x_2\theta(x))$, and formula (12) can be rewritten as

$$\mathcal{D}_{x\theta}\Pi(u,v,w) = \int\limits_{\Omega_{\gamma_0}} \left(\frac{1}{2}(x_l\theta)_{,l}\sigma_{ij}(u)\varepsilon_{ij}(u) - \sigma_{ij}(u)u_{i,l}(x_l\theta)_{,j} \right) dx - \int\limits_{\Omega_{\gamma_0}} (x_l\theta f_i)_{,l}u_i \, dx$$
$$- \frac{1}{2}\int\limits_{\gamma} \left(3(x_1\theta)_{,1}v_{,11}^2 + 2(x_1\theta)_{,11}v_{,11}v_{,1} + (x_1\theta)_{,1}w_{,1}^2 \right) dx_1. \tag{17}$$

Let the map $M_{\rm B} : (0,2) \to \mathbb{R}$ be given by

$$M_{\rm B}(x_1) = \frac{3}{2}(x_1 - 1)v_{,11}^2(x_1) + v_{,1}(v_{,11} - v_{,111})(x_1) + \frac{1}{2}x_1 w_{,1}^2(x_1).$$

Integrating by parts in (17), we obtain the singular path-independent M-integral in the equilibrium problem (1)–(8):

$$\mathcal{M}_{\rm tip} = \int\limits_{\partial S} \left(-\frac{1}{2}(n_l x_l)\sigma_{ij}(u)\varepsilon_{ij}(u) + \sigma_{ij}(u)u_{i,l}x_l n_j \right) ds + M_{\rm B}(b) - M_{\rm B}(a). \tag{18}$$

The M-integral (18) consists of two parts. The first is a regular part coinciding with the classical Knowles–Sternberg M-integral, please refer to [7], and the second is a singular part depending on the displacement field (v, w) of the inclusion γ.

In spite of the fact that transformations (13) and (16) are different, the perturbed domains Ω_{γ_δ} coincide for each fixed δ. Since the solution $(u^\delta, v^\delta, w^\delta) \in K^\delta$ is unique, it follows that the values of the potential deformation energy for the perturbed problems $\Pi^\delta(u^\delta, v^\delta, w^\delta)$ are both equal to each other. Consequently, the corresponding derivatives $\mathcal{D}_{(\theta,0)}\Pi(u,v,w)$ and $\mathcal{D}_{x\theta}\Pi(u,v,w)$ and therefore two singular path-independent integrals $\mathcal{J}_{\rm tip}$ and $\mathcal{M}_{\rm tip}$ are related through the relationships

$$D_{(\theta,0)}\Pi(u,v,w) = \mathcal{D}_{x\theta}\Pi(u,v,w), \quad \mathcal{J}_{\rm tip} = \mathcal{M}_{\rm tip}.$$

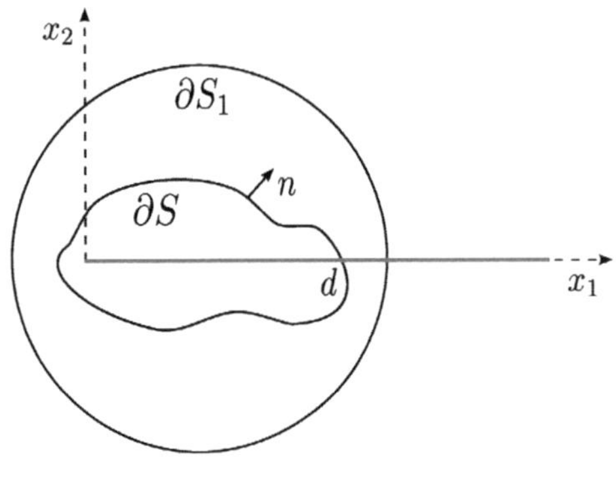

Figure 3. The integration path ∂S surrounding the whole crack.

The next step is to discuss a situation when the integration path surrounds the whole crack. Indeed, let a η be a smooth function with support located in a small neighborhood S_1 of the segment γ_0. Moreover, $\eta = 1$ in a neighborhood S of the segment γ_0, $S \subset S_1$. We assume that the path ∂S intersects the straight-line segment $\gamma \setminus \gamma_0$ at the point with the abscissa d as depicted in Figure 3. As previously, if $f \equiv 0$ in $S \cap \Omega_\gamma$, then we can obtain the path-independent M-integral in the form

$$\mathcal{M}_{\text{crack}} = \mathcal{D}_{x\eta}\Pi(u, v, w) = \int\limits_{\partial S} \left(-\frac{1}{2}(n_l x_l)\sigma_{ij}(u)\varepsilon_{ij}(u) + \sigma_{ij}(u)u_{i,l}x_l n_j \right) ds + M_{\text{B}}(d). \tag{19}$$

In this case, the path ∂S is taken around the whole crack, and n is the outward normal to ∂S. The following interesting observation can be made in relation to (15) and (19):

$$\mathcal{J}_{\text{tip}} = \mathcal{M}_{\text{crack}}.$$

In other words, one quantity $\mathcal{M}_{\text{crack}}$ connected with the whole crack is related to another one, namely $\mathcal{J}_{\text{tip}}$, associated with the crack tip only. A similar relation between the classical M- and J-integrals was firstly found in certain special cases by Freund in [3].

Acknowledgments
The work was supported by the Russian Science Foundation (grant number 15-11-10000).

References
[1] Budiansky B and Rice J 1973 *Trans. ASME J. Appl. Mech.* **40** 201–3
[2] Cherepanov G P 1967 *J. Appl. Math. Mech.* **31** 503–12
[3] Freund L B 1978 *Int. J. Solids Structures.* **14** 241–50
[4] Grisvard P 1991 *Singularities in Boundary Value Problems* (Berlin: Springer-Verlag)
[5] Khludnev A M and Leugering G R 2014 *Math. Mech. Compl. Sys.* **2** 1–21
[6] Rice J R 1968 *Trans. ASME J. Appl. Mech.* **35** 379–86
[7] Sternberg J K and Knowles E 1972 *Arch. Ration. Mech. Anal.* **44** 187–211

Development of instability and turbulence in overturning lee waves: the map of different scenarios in Re–Pr space

S N Yakovenko

Khristianovich Institute of Theoretical and Applied Mechanics, SB RAS, Novosibirsk, Russia

Email: yakovenk@itam.nsc.ru

Abstract. Scenarios of the transition to turbulence in overturning lee waves generated by the two-dimensional obstacle in a stably stratified flow have been explored by visualization of velocity and scalar (density) fields, with analysis of spanwise spectra. The results are obtained by numerical solution of the continuity, Navier–Stokes and scalar equations for stratified fluid with the Boussinesq approximation, for varied Reynolds and Prandtl numbers relating to tank experiments, situations in atmosphere and oceans. Based on the computed data, the dependence of the most unstable perturbation wavelength on Reynolds and Prandtl numbers is derived.

1. Introduction

The results of scanning in the space of physical parameters (Reynolds number Re = UH/ν, Prandtl number Pr = ν/κ) are presented for instability and turbulence development scenarios in overturning internal lee waves generated by the two-dimensional cosine obstacle of height H in a stably stratified flow with the constant values of inflow density gradient and velocity U. Such a phenomenon is studied by visualization of velocity and density fields, with analysis of spectra obtained from DNS/LES data at $10^2 \leq \mathrm{Re} \leq 10^4$, $1 \leq \mathrm{Pr} \leq 700$, relating to tank experiments [1-3], atmospheric and oceanic situations.

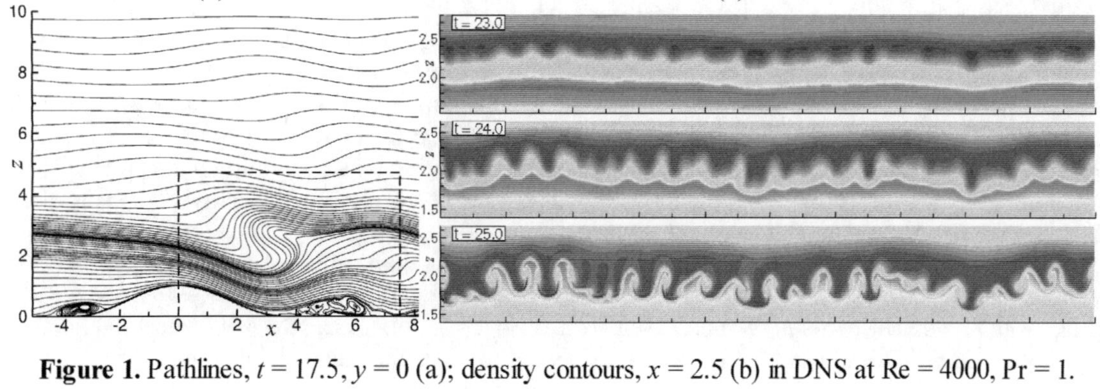

Figure 1. Pathlines, $t = 17.5$, $y = 0$ (a); density contours, $x = 2.5$ (b) in DNS at Re = 4000, Pr = 1.

The single case (Re = 4000, Pr = 1) was recently examined by DNS in [4,5] (figure 1). The grid resolution was sufficient to capture the fine-scale transition processes studied in [4] and the subsequent developed turbulence discussed in [5]. For this set of parameters, secondary instability of the density field arising after wave overturning reveals a range of spanwise modes. The smallest mode ($\lambda_y \sim 0.5H$) represents oscillations of Rayleigh–Taylor instability (RTI), growing and resulting in convective mushroom-like structures with associated Kelvin–Helmholtz instability rolls. At late transition times, the smaller-scale vortices transform into the larger structures, and another noticeable mode ($\lambda_y \sim 2.5H$) is dominant when the large-scale toroidal vortex structures become evident [4]. This large-scale mode can be associated with the most unstable perturbation of the initial two-dimensional vortex pair in the wave-overturning place [4,5] and has also been detected in [2,3,6]. For higher Pr, the earlier and faster RTI growth with smaller-scale structures is observed in LES with Pr = 700 [7].

The more comprehensive study can be performed, with variation of physical/numerical parameters and conditions of computations: in particular, by means of multiple runs with varied Reynolds and Prandtl/Schmidt numbers. Reduction of Re (e.g. runs at Re = 2000, 1000, 500, 200, etc.) will allow us to even further refine the resolution due to larger relative values of Kolmogorov microscale η/H at smaller Reynolds numbers, thus to obtain better statistics in the turbulent patch. Another aim is to see the Re effect on transition details and quasi-steady turbulence. Runs at varied Re may give the light on the ‘critical’ Re value (Re_1) for onset of instability. This means that at Re < Re_1 all perturbations introduced in the region of internal wave overturning (e.g. white-noise seeding) will decay with time, and no structure will appear and transit to turbulence. On the other hand, at Re > Re_2 and depending on the grid resolution $\Delta x/H$, the DNS or implicit LES approach may become inadequate (as shown, e.g. by high-wavenumber parts of spectra), and for parameterization of smaller eddies of sub-grid scales (SGS) we will need then to introduce the SGS model. At Pr = 1, it is evident that $Re_2 > 4000$ [4,5], so extra runs at larger Reynolds numbers, e.g. at Re = 10^4 as in [1,3,8], would be of interest.

The aim of the Pr variation between 1 and 700 is to find the ‘critical’ values too, in particular, for DNS limitation, $Pr_2(\Delta x)$ (for instance, at Re = 4000 we have $1 < Pr_2 < 700$ for the resolution used in [4,5]), and approach to real situations: (1) non-isothermal atmosphere (Pr ~ 1), (2) non-isothermal water reservoir with Pr ≈ 7 at normal temperature, (3) salted water in ocean estuaries (Pr ≈ 700). It could be also interesting to address the dependence on Froude number and geometry of the obstacle, however, such an issue has already been explored in various experimental and numerical studies (e.g. in [1,8]). Moreover, from the point of view of flow physics and relevance for applications, dependence on values of Re and Pr seems to be more crucial.

2. Governing equations

The continuity, the Navier–Stokes and density equations for stratified fluid with the Boussinesq approximation are written in [4,5], as well as details of their realization.

3. Results and discussion

DNS results with fixed Pr = 1 at lower Re value (than in [4,5]) show that, for instance, for Re = 1000 (figures 2–5) multiple peaks in spanwise spectra ($0.6H < \lambda_y \leq 2.0H$) occur during perturbation growth at $25 < t < 40$. The most unstable perturbation wavelength corresponding to the RTI mode during the exponential growth stage ($30 < t < 35$) is $\lambda_y \approx 1.1H$ according to figures 4–5. The large-structure mode $\lambda_y \approx 3.3H$ arises during the turbulent patch stage at $t > 48$. The larger RTI structures (in comparison with the Re = 4000 case in [4,5]) evidently occur due to stronger viscosity/diffusion effects. These effects also cause delay of the instability and turbulence development, and smaller density drop between heavier and lighter fluid tongues (figure 3).

For Re = 500, one can find (figure 6) almost persistent dominance of the spectra peak with wavelength $\lambda_y \approx 2.0H$ (corresponding to the late RTI growth at $45 \leq t \leq 50$), as well as two weaker peaks with $\lambda_y \approx 0.9H$ at $45 \leq t \leq 60$ and $\lambda_y \approx 3.3H$ at $t > 48$. For both cases, Re = 500 and 1000 (as well as for Re = 4000 studied in [1,2]) one can see the large-structure enlargement with time growth.

For Re = 200, the spanwise instability magnitude is quite low (not visible in density contours) and does not lead to turbulence development, i.e. the critical value is within $200 < Re_1 < 500$ at Pr = 1.

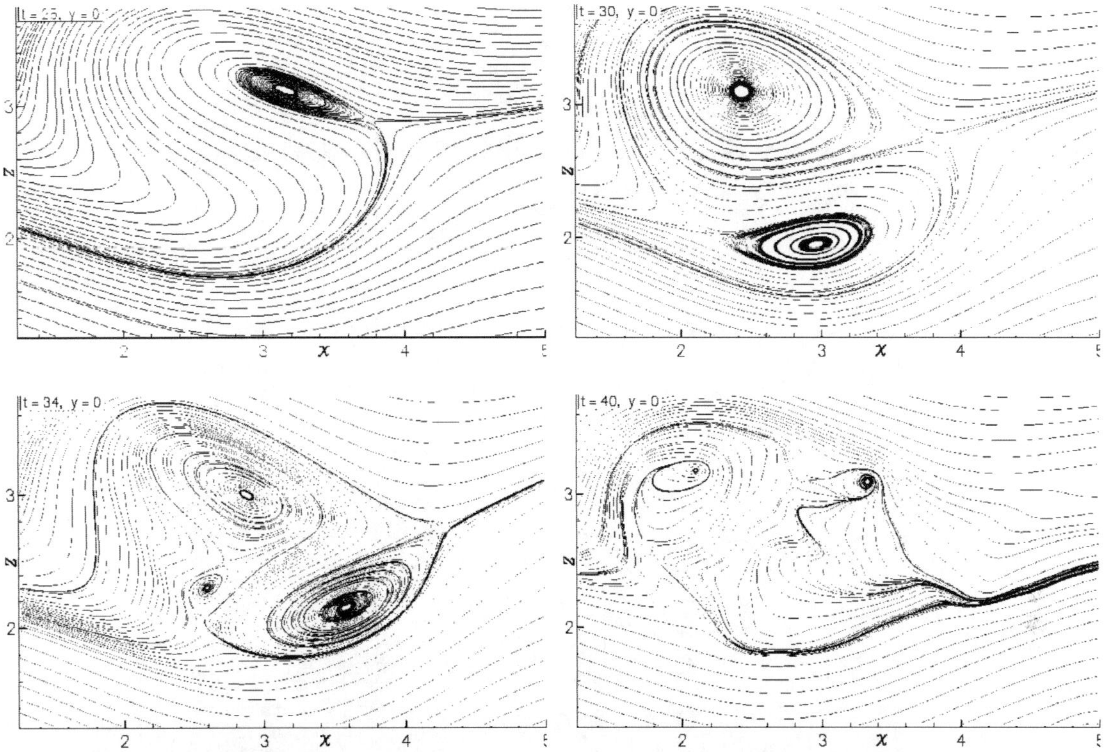

Figure 2. Instantaneous pathlines at Re = 1000, Pr = 1 for $t = 25, 30, 34, 40$ and $y = 0$.

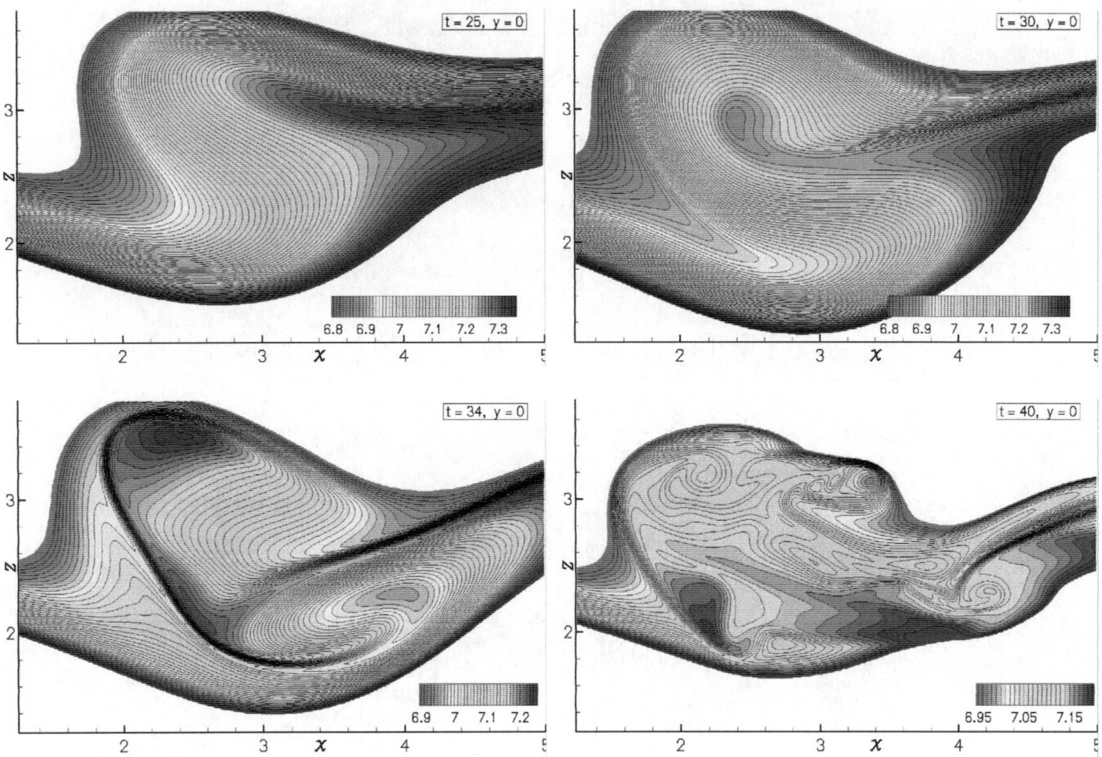

Figure 3. Instantaneous density contours at Re = 1000, Pr = 1 for $t = 25, 30, 34, 40$ and $y = 0$.

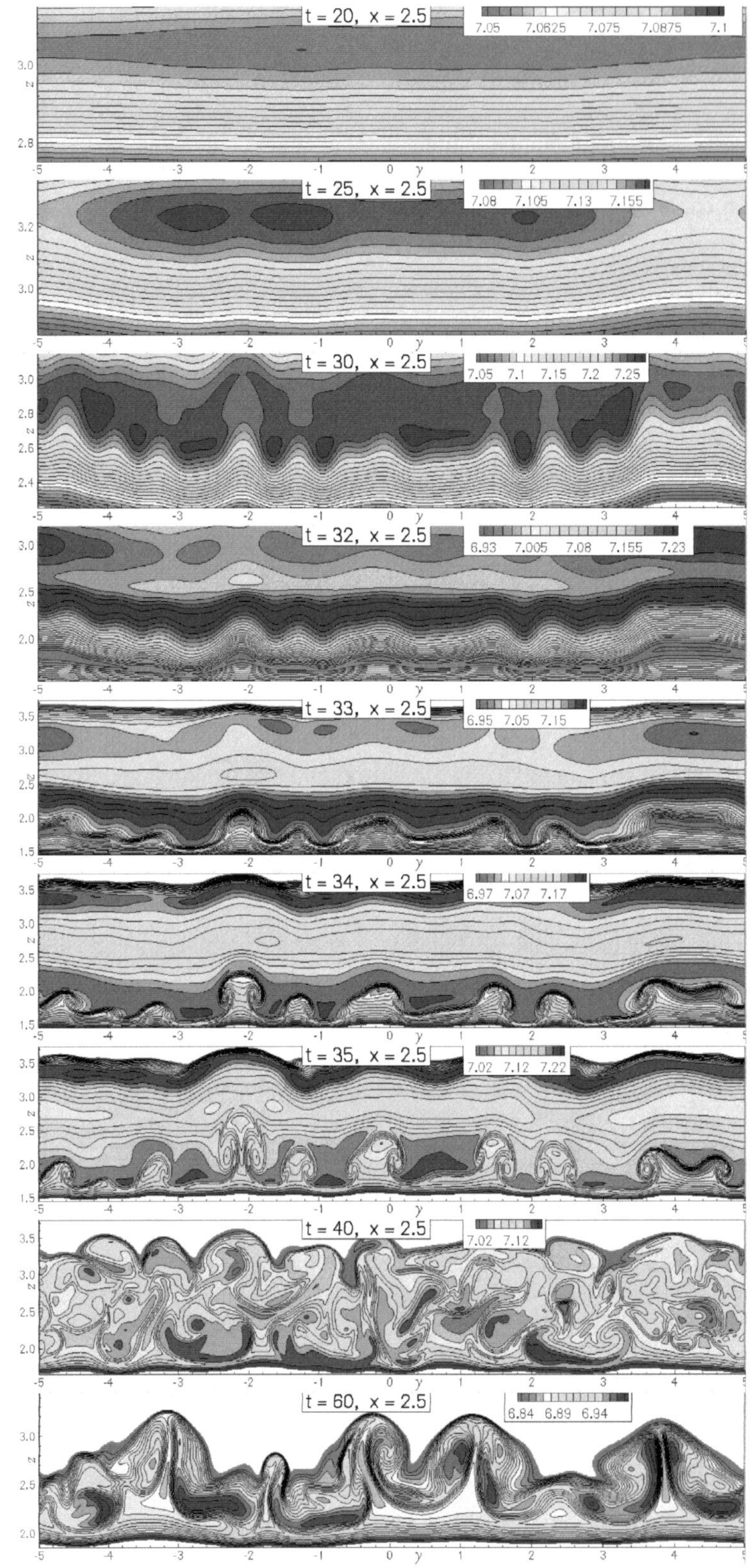

Figure 4. Instant density contours at Re = 1000, Pr = 1 (t = 20, 25, 30, 32, 33, 34, 35, 40, 60, x = 2.5).

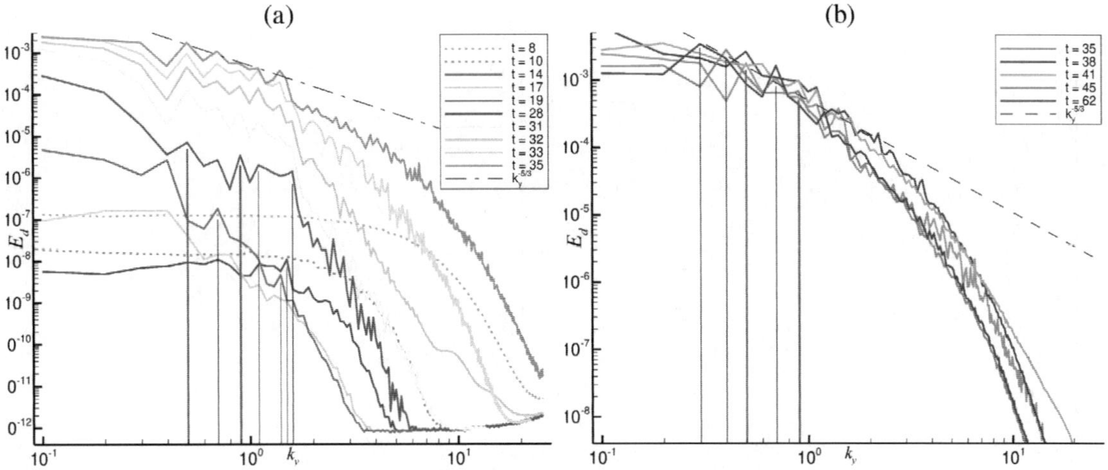

Figure 5. Spanwise spectra of scalar variance at Re = 1000, Pr = 1 (averaged at $1.25 \leq x \leq 5.00$ and $1.25 \leq z \leq 3.75$) during: (a) white-noise perturbation decay (dashed lines) and RTI growth (solid lines), (b) transition to quasi-steady turbulence (with rough relaxation to a single line).

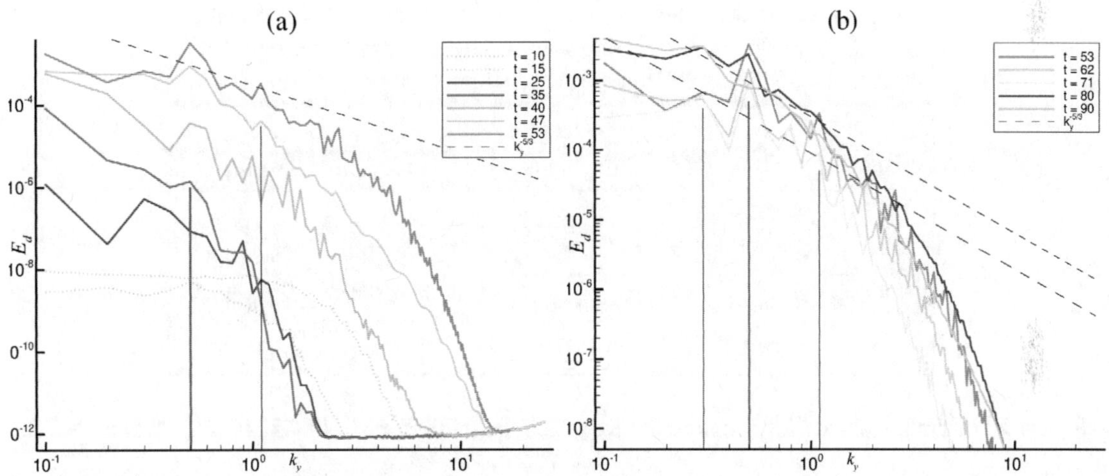

Figure 6. The same as in figure 5 for Re = 500, Pr = 1.

For Re = 10^4, the numerical noise becomes evident at large-wavenumber parts of spectra at late times corresponding to turbulent stages where the inertial range ($k^{-5/3}$) should be followed by larger slopes at higher k corresponding to dissipative ranges for both velocity and scalar spectra. In contrast, after the inertial range at $0.5 < k_y < 2.5$, we observe lower spectra slopes ($\sim k^{-1}$) at higher wave numbers $2.5 < k_y < 10$ (figure 7) which indicates an excessive spurious numerical noise appearing because of poor resolution (or absence of SGS models to suppress such a noise). Therefore, the second critical value is within 4 000 < Re_2 < 10 000 at Pr = 1, i.e. someone has to apply finer grids or LES techniques at Re > 4 000 to predict the small-scale flow features adequately.

In contrast to the Pr = 1 case, runs with the higher value Pr = 700 at Re = 200 do show the RTI growth with the most unstable perturbation wavelength $\lambda_y \approx 2.0H$ (figures 8 and 9), where the scalar spectra at the developed turbulent stage ($t > 60$) reveal not only the inertial range ($k^{-5/3}$), but also the viscous-convective range (k^{-1}) at higher wave numbers. The latter confirms the Batchelor's theory for Pr >> 1, whereas measurements could not capture this fine effect [9]. Note that we should be cautious here since this may be also affected by the density field under-resolution which seems to yield the same effect of smaller spectra slopes for high-wavenumber parts of spectra (figure 7).

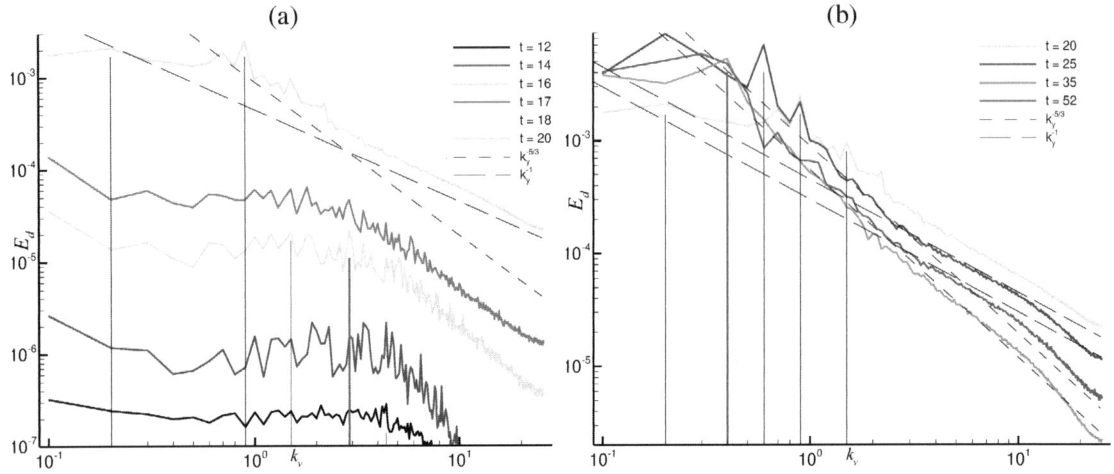

Figure 7. The same as in figure 5 for Re = 10^4, Pr = 1.

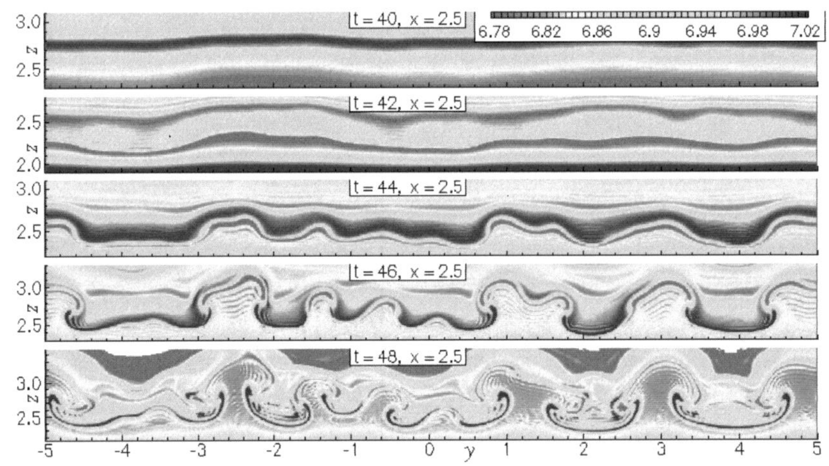

Figure 8. Instantaneous density contours at Re = 200, Pr = 700 for t = 40, 42, 44, 46, 48 and x=2.5.

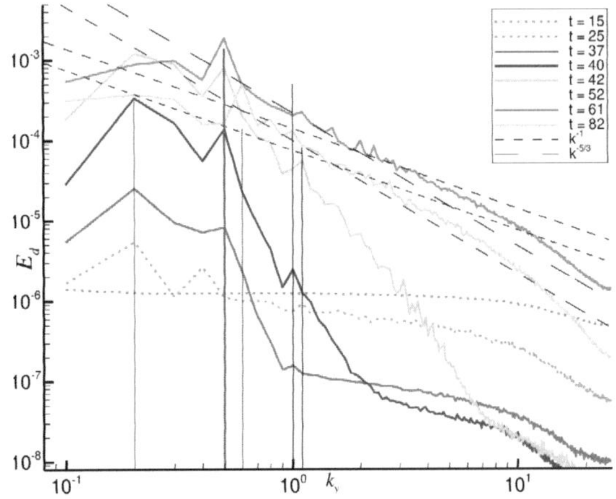

Figure 9. Spanwise spectra of scalar fluctuation variance for Re = 200 and Pr = 700 (averaged at $1.25 \le x \le 5.00$, $1.25 \le z \le 3.75$).

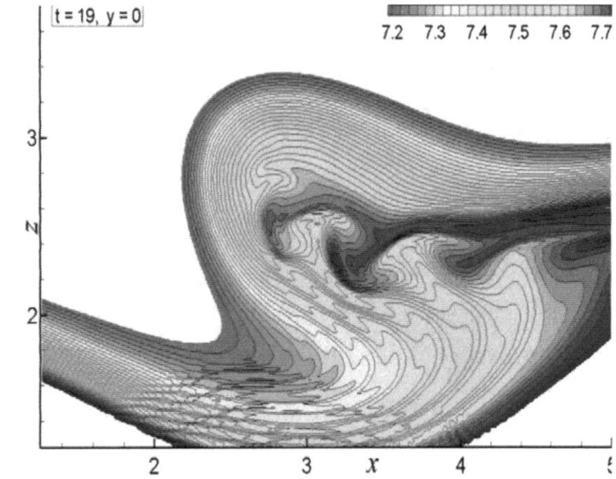

Figure 10. Instantaneous contours of density for Re = 4000, Pr = 7 (t = 19, y = 0), to be compared with figure 3 for Re = 1000, Pr = 1 (t = 25, 30).

For intermediate value Pr = 7, we still observe the transition to turbulence at low Reynolds number Re = 200 due to the RTI growth (which occurs quite late), in contrast to the case of Pr = 1. Note that at Pr = 7 the first critical value is within $100 < Re_1 < 200$, whereas the second critical value is within $2000 < Re_2 < 4000$, since the numerical density-field noise affects the high-wavenumber part of spanwise scalar spectra starting from Re ~ 4000 (and for Re = 2000 such an effect is not visible).

It is surprising that for Re = 4000 and Pr = 7 the RTI periodicity is seen not only along the span, but also in the streamwise direction as shown by visualization of density contours at $y = 0$ (figure 10). The similar effects are observed for Re = 10^4 and Pr = 1. Although, in both cases we have the under-resolved density field resulting in excessive numerical noise, this evidence may still indicate another route of instability growth than the one discussed in [1] with quasi-two-dimensional structures.

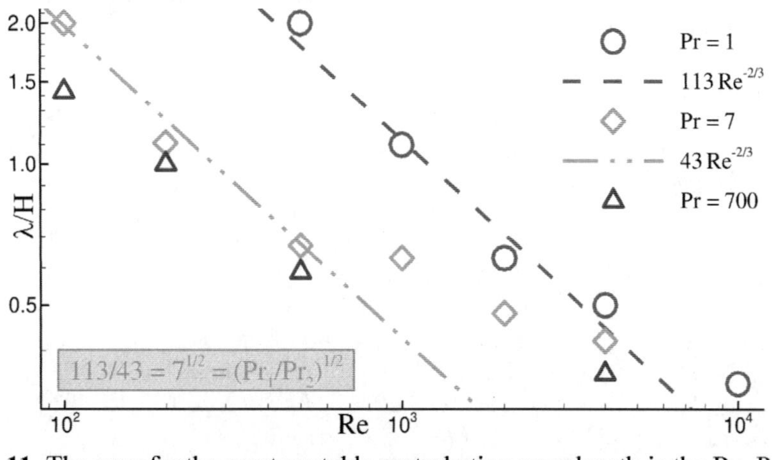

Figure 11. The map for the most unstable perturbation wavelength in the Re–Pr space.

The results can be summarized in figure 11 where the most unstable perturbation wavelength deduced from density snapshot visualizations and spectra peak analysis is given by symbols (each relates to one computation at different Re or/and Pr). One can see that the analytical expression [10]

$$\lambda = \lambda' = 4\pi H[(2\Delta\rho_H/\Delta\rho)(F_H)^2 Re^{-2}]^{1/3}$$

(red dashed line in figure 11) written in dimensionless form [1] at Pr = 1 with the typical density drop, $\Delta\rho \approx 10^{-3}\Delta\rho_H$, between the layers of heavier and lighter fluids related to earlier RTI stages is satisfied. For arbitrary Pr values, the updated relation, $\lambda = f(\mathrm{Pr})\lambda'$, is next suggested according to figure 11 where $f(\mathrm{Pr}) \rightarrow$ const at $\mathrm{Pr} \rightarrow \infty$ and $f(\mathrm{Pr}_1) > f(\mathrm{Pr}_2)$ at $\mathrm{Pr}_1 < \mathrm{Pr}_2$. In particular, the power-law function, $f(\mathrm{Pr}) \sim \mathrm{Pr}^{-1/2}$, may be valid for, at least, relatively low values between Pr ~ 1 and Pr ~ 10.

Note that the results at Pr = 700 are heavily affected by under-resolution of the density field, even at lower value of Re = 200. As the result, the numerical wiggles can be seen in growing mushroom-like structures (figures 8 and 10), and the numerical dissipation effect is a possible reason of obvious over-prediction of the most unstable perturbation wavelength (figure 11), as well as the SGS diffusion effect in the case of LES computation at Re = 4000 and Pr = 700.

4. Concluding remarks

The results of scanning in the space of Reynolds and Prandtl numbers for scenarios of the transition to turbulence in overturning internal lee waves generated by the two-dimensional obstacle in a stably stratified flow have been presented. The instability and turbulence development has been studied by visualization of velocity and density fields, with analysis of spanwise spectra obtained from numerical simulation data at $10^2 \leq \mathrm{Re} \leq 10^4$, $1 \leq \mathrm{Pr} \leq 700$.

The range of Prandtl numbers corresponds to heat stratification in the atmosphere (Pr ~ 1), reservoirs (Pr ~ 7) and to salted water diffusion (Pr ~ 700), whereas the range of Reynolds numbers corresponds to the one in towing-tank measurements [1–3] and concurrent simulations [6,8]. The real atmospheric and oceanic flows have actually much higher levels of Re ~ $10^7 \div 10^{10}$. Nevertheless, as soon as we have quite high Re values with the fully developed turbulence at late stages (as a result of internal lee wave breaking), the behavior of the resulting turbulent patch is essentially the same regardless of transition routes and Re values, due to the well-known Reynolds number independence principle. So both laboratory and numerical experiments [1-8], as well as the present computations at Re ~ $10^2 \div 10^4$ replicate well the much higher Re situations in real geophysical conditions, except for the instability development stage. The latter may indeed have various scenarios with different structures from the zoo of instabilities at different Re. Nevertheless, the map of instability growth scenarios proposed here gives a hint to real flows and allows us to extrapolate results to geophysical situations with Re ~ $10^7 \div 10^{10}$, in particular, for the wavelength of the most unstable perturbation of RTI type which is proportional to $Re^{-2/3}$ and falls with Pr values as $Pr^{-1/2}$.

Acknowledgements
The author thanks Prof I.P. Castro and Dr T.G. Thomas for help in studies of breaking internal lee waves. The work was performed with use of the Iridis3 and Iridis4 computing resources at the University of Southampton.

References
[1] Castro I P and Snyder W H 1993 *J. Fluid Mech.* **255** 195–211
[2] Eiff O F and Bonneton P 2000 *Phys. Fluids* **12** 1073–86
[3] Eiff O, Huteau F and Tolu J 2005 *Dyn. Atmos. Oceans* **40** 71–89
[4] Yakovenko S N, Thomas T G and Castro I P 2014 *J. Fluid Mech.* **760** 466–93
[5] Yakovenko S N, Thomas T G and Castro I P 2011 *J. Fluid Mech.* **677** 103–33
[6] Gheusi F, Stein J and Eiff O F 2000 *J. Fluid Mech.* **410** 67–99
[7] Yakovenko S N, Thomas T G and Castro I P 2014 *Progress in Turbulence V: Proc. of the iTi Conf. in Turbulence, 2012* ed A Talamelli et al. Springer Proceedings in Physics vol 149 (Springer International Publishing) pp 233–36
[8] Paisley M F and Castro I P 1996 *Dyn. Atmos. Oceans* **23** 309–19
[9] Warhaft Z 2000 *Annu. Rev. Fluid Mech.* **32** 203–40
[10] Youngs D L 1984 *Physica* D **12** 32–44